Fluid Mechanics: Key Concepts and Applications

Editor: Donna Braverman

NY RESEARCH PRESS
New York

Published by NY Research Press
118-35 Queens Blvd., Suite 400,
Forest Hills, NY 11375, USA
www.nyresearchpress.com

Fluid Mechanics: Key Concepts and Applications
Edited by Donna Braverman

International Standard Book Number: 978-1-63238-568-0 (Hardback)

Cataloging-in-Publication Data

Fluid mechanics : key concepts and applications / edited by Donna Braverman.
p. cm.
Includes bibliographical references and index.
ISBN 978-1-63238-568-0
1. Fluid mechanics. 2. Fluid dynamics. I. Braverman, Donna.
TA357 .F58 2018
620.106--dc23

Printed in China.

Contents

Preface

Over the recent decade, advancements and applications have progressed exponentially. This has led to the increased interest in this field and projects are being conducted to enhance knowledge. The main objective of this book is to present some of the critical challenges and provide insights into possible solutions. This book will answer the varied questions that arise in the field and also provide an increased scope for furthering studies.

This book unravels the recent studies in the field of fluid mechanics. The branch of physics which studies the behaviour and flow of fluids is known as fluid mechanics. As a subject, fluid mechanics is mainly divided into two branches, fluid statics and fluid dynamics. Fluid statics studies fluids when they are at rest and fluid dynamics studies them in motion. Fluid mechanics is applied in a number of fields like mechanical engineering, chemical engineering, biology and astrophysics. This book studies, analyses and upholds the pillars of fluid mechanics and its utmost significance in modern times. It is an essential guide for both academicians and those who wish to pursue this discipline further.

I hope that this book, with its visionary approach, will be a valuable addition and will promote interest among readers. Each of the authors has provided their extraordinary competence in their specific fields by providing different perspectives as they come from diverse nations and regions. I thank them for their contributions.

Editor

1

A numerical investigation of explicit pressure-correction projection methods for incompressible flows

Ungku Abdul Hafiz[a], Asif Hoda[b] and Waqar Asrar[a]*

[a]*Department of Mechanical Engineering, Kulliyah of Engineering, International Islamic University, P. O. Box 10, 50728, Kuala Lumpur, Malaysia;* [b]*Department of Mechanical Engineering, Jubail University College, Jubail, Kingdom of Saudi Arabia*

A numerical investigation is performed on an explicit pressure-correction projection method. The schemes are fully explicit in time in the framework of the finite difference method. They are tested on benchmark cases of a lid-driven cavity flow, flow past a cylinder and flow over a backward facing step. Comparisons of the numerical simulations have been made with benchmark experimental and DNS data. Based on the results obtained, several numerical issues are discussed; namely, the handling of the pressure term, time discretization and spatial discretization of convective and diffusive terms. The fully explicit projection method is also compared with the fully implicit SIMPLE algorithm. It is observed that the SIMPLE algorithm performs better (faster and produces more accurate results) for laminar flows while the projection method works better for unsteady turbulent flows. Although there have been much research performed using the higher-order pressure incremental projection method, this research work is novel because the schemes employed here are fully explicit, developed in the framework of a finite difference method, and applied to turbulent flows using k-ϵ model. The major difficulty and challenges of this research work is to identify the sources of instability for the higher-order pressure incremental projection method scheme.

Keywords: projection method; explicit schemes; incompressible flows; Navier-Stokes equations; finite difference method

1. Introduction

As technology advances, the field of Computational Fluid Dynamics also went through a major transformation. Solving incompressible Navier-Stokes equations is no longer limited to simple flow problems but now involves complicated and complex flow problems from real-world applications. Even with a simple geometry such as the lid-driven cavity flow, the focus now had shifted to other types of fluids such as nanofluids. For example, Fereidoon, Saedodin, Hemmet, and Noroozi (2013) conducted a study on incompressible flow of nanofluids in a square double lid-driven cavity in the framework of finite volume method using the SIMPLE algorithm. Other than that, the importance of solving incompressible Navier-Stokes equations is extended to complex geometry flow problems. Baniasadi, Aydin, Dincer, and Naterer (2013) conducted a simulation of incompressible turbulent flow to investigate the effects of upstream and downstream blockage on aerodynamic performance of automotive cooling fans. The 3-D simulation was conducted using ANSYS FLUENT (SIMPLE algorithm) in the framework of finite volume method with a realizable k-ϵ turbulence model.

Hence, it should be highlighted that solving incompressible Navier-Stokes equations is crucial for many real-world CFD applications. The main difficulty in dealing with the incompressible Navier-Stokes equations is the handling of the pressure term. This is because there are no explicit equations that describe or relate pressure to the velocity field. For the past few decades, there have been various numerical treatments proposed by researchers to handle this issue. One of the ways is to decouple the pressure and velocity by solving the velocity and pressure in different multiple steps which consist of multiple equations and this approach is referred in the published literature as the projection method.

The Projection method was introduced in the late 1960s by Chorin (1968) and Temam (1969) with a first-order accuracy in time. Since its inception, many efforts have been made to increase the order of accuracy and develop higher-order projection methods. A comprehensive review on the projection method was performed by Guermond, Minev, and Shen (2006). They have divided the projection method into three types: the pressure-correction method, the velocity-correction method and the consistent splitting method.

The pressure-correction method can further be divided into nonincremental, incremental and rotational forms. The first work developed by Chorin was the nonincremental form where the pressure at the previous time step was not considered. Later, Van Kan (1986) and Goda (1979) introduced the incremental-pressure schemes to take into account the old or previous value of pressure. Then, the

*Corresponding author. Email: waqar@iium.edu.my

rotational form of the pressure-correction scheme was introduced by Timmermans, Minev, and Van De Vosse (1996) to overcome the difficulty of handling the Neumann pressure boundary condition. There have been many developments in formulating higher-order pressure-correction schemes, for example Guermond and Shen (2003), Sun, He, and Feng (2011) and a more recent one by Ren, Jiang, Liu, and Zhang (2005) with third-order accuracy.

Also, the vast majority of the investigations on projection methods reported in published literature have been mostly mathematical derivations aiming to develop higher accuracy or higher-order forms of the projection method using techniques such as normal mode analysis. Additionally, the numerical tests reported in these studies have been limited to simple 1-dimensional and 2-dimensional flows that have analytical solutions. The proposed higher-order schemes have rarely been tested on complex flow scenarios as confirmed and elucidated by Guermond et al. (2006). Although some recent studies (Abbasi, Ashrafizadeh, & Shadaram, 2013; Liu, Liu, & Pego, 2010) have reported the application of the pressure-correction schemes to benchmark numerical tests like the 2-D lid-driven cavity and flow past a cylinder, detailed results on of the stability and accuracy of the various forms of the pressure-correction projection method cannot be found in any publication. Additionally, most of the investigations employed implicit or semi-implicit discretization methods for treatment of the convective and diffusive terms and the fully explicit method has rarely been studied in details. The fully explicit scheme is likely to offer substantial advantages in terms of simplicity of implementation and speeding up of calculations which are an important factor as far as unsteady transient flows are concerned. Consequently, the fully explicit approach is likely to offer substantial benefits when dealing with highly transient turbulent flows as well as for carrying out the Reynolds-Averaged Navier Stokes (RANS) simulation. This approach can be compared with the fully implicit SIMPLE scheme to test its viability as an alternate approach. Furthermore, most of the studies reported in the published literature have implemented the pressure-correction projection methods on Cartesian geometries and their use in curvilinear generalized coordinates in the finite difference framework has not been found by the authors.

Based on the literature survey, this work aims to carry out an exhaustive numerical investigation of the pressure-correction projection method. Most of the projection method schemes developed were tested in the context of the finite element method (Guermond et al., 2006). The use of the finite difference and the finite volume framework are few (Abbasi et al., 2013; Ren et al., 2005). Therefore, this work intends to carry out an in-depth numerical investigation of the projection method using the finite difference methodology of discretization. Also, this work aims to compare the various forms of the pressure-correction projection method and to compare the projection method with the commonly employed SIMPLE scheme in the generalized curvilinear coordinates framework.

2. Numerical issues

The continuity and Navier-Stokes equations are widely regarded as the governing equations for Newtonian fluid flows. The incompressible forms of these equations are written in tensor notation as:

Continuity:

$$\frac{\partial (u_i)}{\partial x_i} = 0 \tag{1}$$

Momentum:

$$\frac{\partial (u_i)}{\partial t} + \frac{\partial (u_i u_j)}{\partial x_j} = -\frac{\partial p}{\partial x_i} + \frac{\partial \sigma_{ij}}{\partial x_j} \tag{2a}$$

Substituting the relation for σ_{ij} for Newtonian fluid and using Equation (1), we may alternatively write the momentum equation as;

$$\frac{\partial (u_i)}{\partial t} + \frac{\partial (u_i u_j)}{\partial x_j} = -\frac{\partial p}{\partial x_i} + \nu \frac{\partial^2 u_i}{\partial x_j \partial x_j} \tag{2b}$$

Where p is the static pressure divided by the density ρ. Nondimensionalizing the above equations by $\frac{u}{u_c}$ and $\frac{x}{L_c}$ where u_c and L_c are the characteristics velocity and length, we will get the kinematic viscosity as $= \frac{1}{Re}$.

Equations (1) and (2) can be written in the generalized coordinates as:

$$\frac{\partial}{\partial \xi_j}\left(J \frac{\partial \xi_j}{\partial x_i} u_i\right) = 0 \tag{3}$$

and

$$\frac{\partial (Ju)}{\partial t} + \frac{\partial (u^c{}_j\, u_i)}{\partial \xi_j} = -\frac{\partial P}{\partial \xi_i} + \frac{\partial \sigma^c{}_{ij}}{\partial \xi_j} \tag{4}$$

where

$$u^c{}_j = J \frac{\partial \xi_j}{\partial x_i} u_i \tag{5}$$

and

$$\sigma^c{}_{ij} = J \frac{\partial \xi_j}{\partial x_k} \sigma_{ki} \tag{6}$$

are the contra variant velocities and stresses. The generalized coordinates form of the equations will be used in the test cases considered in this paper.

A number of numerical issues relating to the solution of the above equations will be considered in this paper and are elaborated in the following subsections: time discretization and pressure handling, discretization of the convective and diffusive terms and stability analysis.

2.1. *Time discretization and pressure handling*

There are different methods of handling the pressure term in the incompressible Navier-Stokes equations such as the projection methods, the SIMPLE algorithm, the artificial compressibility approach and the pressure gradient method, to name a few. The projection method and the SIMPLE algorithm will be the focus of this paper.

The projection method and the SIMPLE algorithm are similar in the sense that both schemes do not use a modified continuity equation such as the one used in the artificial compressibility method. However, one fundamental difference between the projection method and the SIMPLE algorithm is the nature of the time-stepping procedure. The projection method is fully explicit in time whereas the SIMPLE algorithm is implicit in time. The projection method utilizes a mathematical technique known as the Helmholtz-Hodge decomposition to split each of the momentum equations into different sub-steps while the SIMPLE algorithm uses an approximation or correction terms to calculate the field variables and uses iterations at each time step until the solution converges.

2.1.1. *Projection methods: standard nonincremental and pressure incremental schemes*

Invoking Equation (2b), the finite difference approximation using first-order accurate Euler backward approximation for time is given as:

$$\frac{u_i^{t+1} - u_i^t}{\Delta t} = -\frac{\partial p}{\partial x_i} - \frac{\partial (u_i u_j)}{\partial x_j} + \frac{\partial \sigma_{ij}}{\partial x_j} \tag{7}$$

Defining an intermediate auxiliary velocity field, u_i*, equation (7) can be split into two different equations:

$$\frac{u_i*^{t+1} - u_i^t}{\Delta t} = -\frac{\partial (u_i u_j)}{\partial x_j} + \frac{\partial \sigma_{ij}}{\partial x_j} \tag{8}$$

and

$$\frac{u_i^{t+1} - u_i*^{t+1}}{\Delta t} = -\left(\frac{\partial p}{\partial x_i}\right)^{t+1} \tag{9}$$

It can be seen in a straightforward way, that adding Equations (8) and (9) will give us Equation (7). This splitting allows us to remove the pressure term from Equation (8) which can be solved explicitly to obtain the intermediate velocity field u_i*. The velocity field at the next time step can be calculated from Equation (9) provided the pressure field is known. The pressure field can be calculated by deriving an equation for the pressure by taking the divergence of Equation (9)

$$\frac{\partial}{\partial x_i}\left(\frac{u_i^{t+1} - u_i*^{t+1}}{\Delta t}\right) = -\frac{\partial}{\partial x_i}\left(\frac{\partial p}{\partial x_i}\right)^{t+1} \tag{10}$$

Using the divergence-free condition (Equation (1)), we obtain

$$\left(\frac{\partial^2 p}{\partial x_i \partial x_i}\right)^{t+1} = \frac{1}{\Delta t}\left(\frac{\partial u_i*^{t+1}}{\partial x_i}\right) \tag{11}$$

The velocities at the next time step can be calculated using Equation (9)

$$u_i*^{t+1} = u_i^{t+1} + \Delta t\left(\frac{\partial p}{\partial x_i}\right)^{n+1} \tag{12}$$

The above formulation is also known as the nonincremental pressure-correction method (Chorin, 1968; Temam, 1969) in which the star velocity field is calculated first (Equation (4)) followed by determination of the pressure field (Equation (11)) and finally the velocity field at the next time step is calculated (Equation (12)).

The simplistic nonincremental pressure-correction method has been developed further to obtain a higher-order time accurate scheme as well as to include the influence of pressure from the previous time step and is referred to as the pressure-incremental projection method. The general formulation of the pressure-incremental projection methods scheme can be written as follows:

$$\frac{\partial (u_i)}{\partial t} + \frac{\partial (u_i u_j{}^t)}{\partial x_j} = -\frac{\partial p^{t+1}}{\partial x_i} + \frac{\partial \sigma_{ij}{}^t}{\partial x_j} \tag{13a}$$

$$\frac{\partial (u_i)}{\partial t} + \frac{\partial (u_i u_j{}^t)}{\partial x_j} = -\frac{\partial p^{t+1}}{\partial x_i} + \frac{\partial \sigma_{ij}{}^{*,t}}{\partial x_j} \tag{13b}$$

The first step of this scheme involves neglecting the pressure term or using the previous pressure values. For generality we will replace the p^{t+1} term in Equation (13) with the $p*^{t+1}$ term in the following manner:

$$p*^{t+1} = \sum_{j=0}^{r-1} \gamma_j p^{t-j} \tag{14}$$

where γ_j represents the coefficients of each pressure term for the different time (t, t-1 and so forth) and the expansion of the above equation will reveal that:

$$p*^{t+1} = \begin{cases} o & if\ r = 0 \\ p^t & if\ r = 1 \\ 2p^t - p^{t-1} & if\ r = 2 \end{cases}$$

Note that if r = 0, then we will obtain the standard nonincremental form of the pressure-correction method. A higher value of r represents a higher-order scheme where it uses pressure values from the previous time step; namely, p^t , p^{t-1} and so forth.

Also, the time derivative term can be formulated in the following way where a higher q represents a higher order of time accuracy.

$$\frac{\partial (u_i)}{\partial t} = \frac{1}{\Delta t}\left(\beta_q u_i*^{t+1} - \sum_{j=0}^{q-1} \beta_j u_i^{t-j}\right) \tag{15}$$

where β_q and β_j represents the coefficients of each velocity term for the different time steps (t, t-1 and so forth) and the expansion of the above equation will reveal that

$$\frac{\partial(u_i)}{\partial t} = \begin{cases} \frac{1}{\Delta t}(u_i{*}^{t+1} - u_i^{t}) & if\ q = 1 \\ \frac{1}{\Delta t}\left(\frac{3}{2}u_i{*}^{t+1} - 2u_i^{t} + \frac{1}{2}u_i^{t-1}\right) & if\ q = 2 \end{cases}$$

Following the same procedure as applied to Equation (9), the second step of the projection method is given below. The coefficient β_q comes from the coefficient of a higher-order method for the time derivative.

$$\frac{\beta_q}{\Delta t}(u_i^{t+1} - u_i{*}^{t+1}) + \frac{\partial \phi^{t+1}}{\partial x_i} = 0 \qquad (16)$$

where

$$\phi^{t+1} = p^{t+1} - p{*}^{t+1} + \chi\nu\frac{\partial u_i{*}^{t+1}}{\partial x_i} \qquad (17)$$

The coefficient χ may equal to 1 or 0. $\chi = 0$ yields the standard form while $\chi = 1$ yields the rotational form. Note that as mentioned previously, φ is a "pseudo pressure" and not the real thermodynamic pressure. The same is applicable for the p term in Equation (9).

Then, taking the divergence of Equation (16) and invoking the divergence-free condition, we obtain the following equation:

$$\left(\frac{\partial^2 \phi}{\partial x_i^2}\right)^{t+1} = \frac{\beta_q}{\Delta t}\left(\frac{\partial u_i{*}^{t+1}}{\partial x_i}\right) \qquad (18)$$

The velocity at the next time step can be simply calculated by rearranging Equation (16):

$$u_i^{t+1} = u_i{*}^{t+1} + \frac{\Delta t}{\beta_q}\frac{\partial \phi^{t+1}}{\partial x_i} \qquad (19)$$

Therefore, the general formulations can be coded and tested using different r, q, and χ parameters which represent variations of pressure-correction projection methods.

Boundary conditions for the intermediate auxiliary variable can be found by applying Equation (13a),

$$\frac{\beta_q}{\Delta t}(u_i^{t+1} - u_i{*}^{t+1}) + \frac{\partial \phi^{t+1}}{\partial x_i} = 0 \qquad (16)$$

Rearranging Equation (16) we obtain,

$$u_i{*}^{t+1} = (u_i^{t+1}) + \frac{\Delta t}{\beta_q}\frac{\partial \phi^{t+1}}{\partial x_i} \qquad (20)$$

Following Rempfer (2006), a Neumann boundary condition for the Poisson equation, Equation (18) is selected to be:

$$\frac{\partial \phi}{\partial n} = 0 \qquad (21)$$

This selection has a first-order accuracy and should not be confused with the one for Pressure Poisson Equation (PPE). Also, because ϕ is not the real "thermodynamic" pressure but instead it is a "pseudo pressure", the quantities that are used for the calculations are the gradient and not the real-absolute values.

2.1.2. *Simple algorithm*

An iterative pressure correction technique is the main crux in the SIMPLE algorithm (Patankar & Spalding, 1972). The grids are staggered where the velocities are located at the interface and the pressure is located at the cell centre.

The iterative process is started by guessing the pressure field which is denoted as p*. Then, the values of p* are used to solve for the u and v velocities from the momentum equations, Equation (2a). These velocities are denoted as u* and v*.

Then, lastly, the pressure correction, denoting p' is constructed from the continuity equation, Equation (1)

The pressure correction is added to the initial guessed pressure field resulting in the correct pressure,

$$p = p^* + \alpha_p p' \qquad (22)$$

Similarly, the corresponding velocity corrections are obtained from p' and are added to the initial guessed velocity field to yield the correct velocities.

$$u = u^* + \alpha_v u' \qquad (23)$$

$$v = v^* + \alpha_v v' \qquad (24)$$

Lastly, the value of p is used to replace the value of p* and the process is repeated until the velocity field satisfies the continuity equation. The number of iterations required depends upon the time step and convergence criteria. Here α_p and α_v are the relaxation factors to under relax the velocities and pressure corrections.

2.2. *Discretization*

Discretization of the convective and diffusive terms plays a major role in the convergence and accuracy of the solutions. A co-located grid is used in this work whereby the fluxes are located and calculated at the cell faces. All the velocity and pressure variables are stored at the center of the cell while the contra variant velocities and stresses are stored at the cell faces.

2.2.1. *Discretizing the convective term*

The descritization of the convective term can be treated two different ways, namely central differencing and upwinding. The derivatives are computed using the cell faces and therefore the scalar variables will be interpolated.

Following is the central difference discretization of the convective term in the x-momentum equation:

$$\frac{\partial(u^C u)}{\partial \xi} + \frac{\partial(v^C u)}{\partial \eta} = \frac{u^C{}_{i,j,k} * u_{i+\frac{1}{2},j,k} - u^C{}_{i-1,j,k} * u_{i-\frac{1}{2},j,k}}{\Delta \xi} + \frac{v^C{}_{i,j,k} * u_{i,j+\frac{1}{2},k} - v^C{}_{i,j-1,k} * u_{i,j-\frac{1}{2},k}}{\Delta \eta} \qquad (25)$$

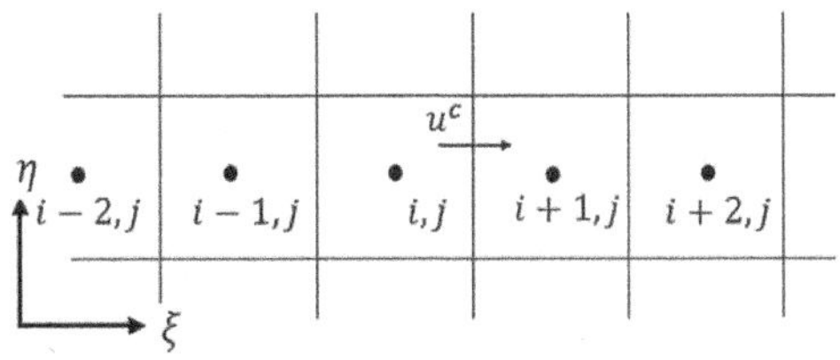

Figure 1. A 2-D Schematic for the selection of the grid points for upwinding scheme in the x direction.

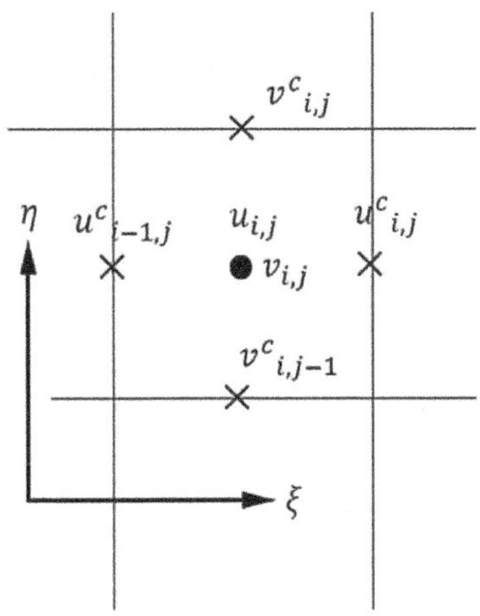

Figure 2. A 2-D Schematic of the placement of the contravariant velocities at the grid.

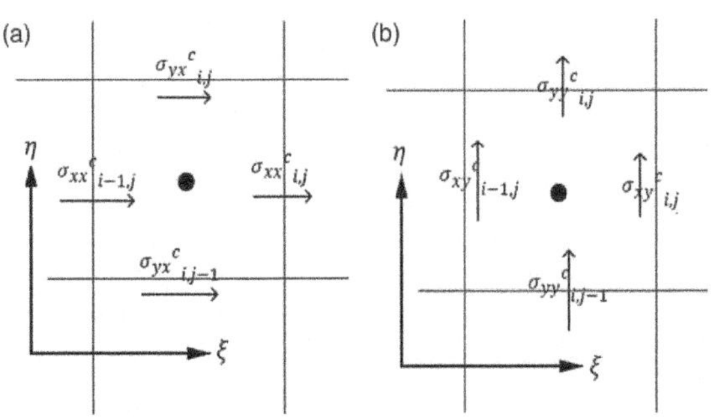

Figure 3. A 2-D schematic of the placement of the contravariant stress at the grid. First index of the stress term indicates the location while the second implies the direction of the tensor.

Where $u_{i+\frac{1}{2},j,k} = \frac{1}{2}(u_{i+1,j,k} + u_{i,j,k})$

For upwinding, the derivatives will be computed using second-order left or right sided derivatives based on the sign of the contravariant velocities. Figure 1 shows the selection of the grid points that are used for calculations in the x direction. For example, if $u^C > 0$, then the flow is moving in the right direction and the variables at point (i,j) , $(i-1,j)$, and $(i-2,j)$ will be used in computing the derivatives.

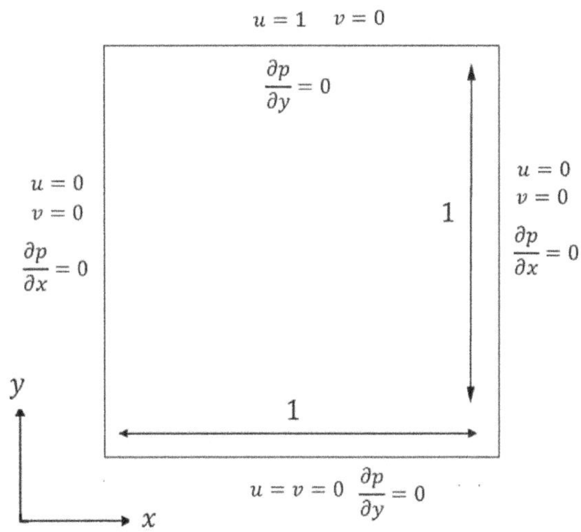

Figure 4. Geometry, dimensions and boundary conditions for the driven cavity flow.

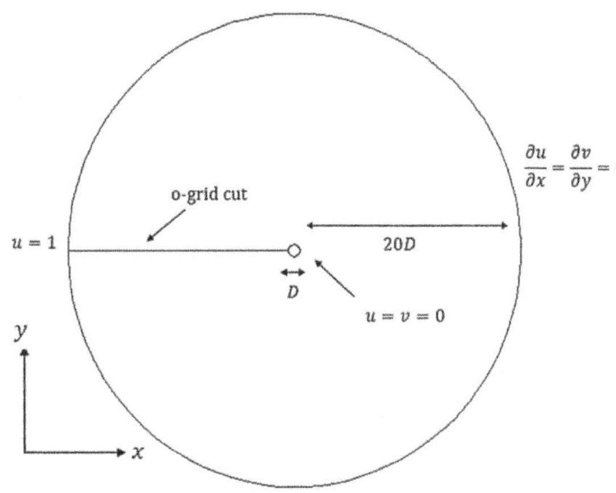

Figure 5. Geometry, dimensions and boundary conditions for flow over cylinder.

Therefore the discretization of the convective term in the x-momentum equation using upwinding is as follows:

$$
\begin{aligned}
&\frac{\partial(u^C u)}{\partial \xi} + \frac{\partial(v^C u)}{\partial \eta} = \\
&max\left(\frac{u^C}{|u^C|}, 0\right)(u^C{}_{i,j,k})\left(\frac{3u_{i,j,k} - 4u_{i-1,j,k} + u_{i-2,j,k}}{2\Delta\xi}\right) \\
&+ min\left(\frac{u^C}{|u^C|}, 0\right)(u^C{}_{i,j,k})\left(\frac{-3u_{i,j,k} + 4u_{i+1,j,k} - u_{i+2,j,k}}{2\Delta\xi}\right) \\
&+ max\left(\frac{v^C}{|v^C|}, 0\right)(v^C{}_{i,j,k})\left(\frac{3v_{i,j,k} - 4v_{i,j-1,k} + v_{i,j-2,k}}{2\Delta\eta}\right) \\
&+ min\left(\frac{v^C}{|v^C|}, 0\right)(v^C{}_{i,j,k})\left(\frac{-3v_{i,j,k} + 4v_{i,j+1,k} - v_{i,j+2,k}}{2\Delta\eta}\right)
\end{aligned} \tag{26}
$$

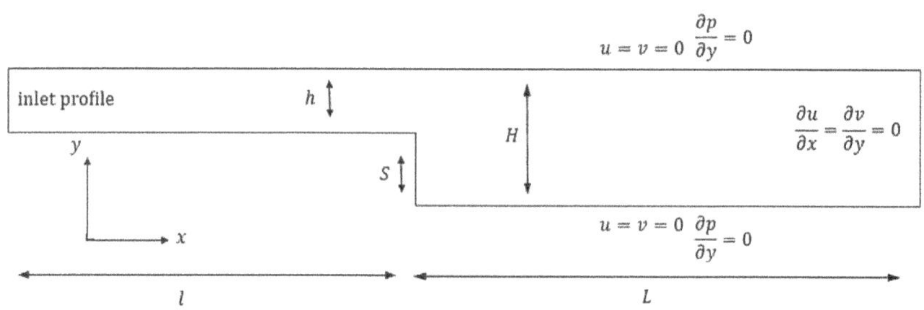

Figure 6. Geometry, dimensions and boundary conditions for backward facing step flow.

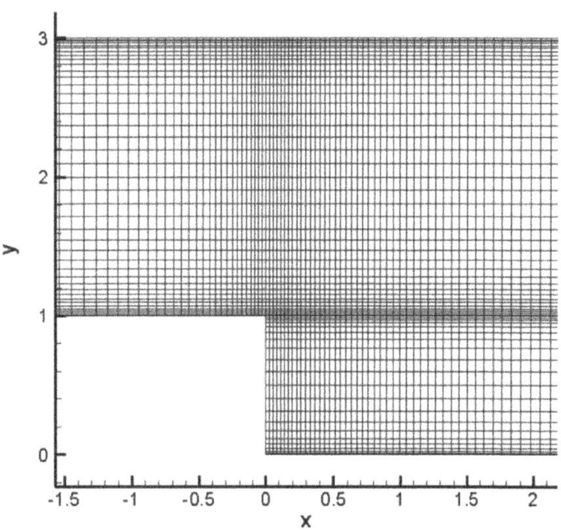

Figure 7. Grid used for the backward facing step benchmark case of Re = 5540.

2.2.2. *Discretization of the diffusive term*

Referring to Equation (4), the diffusive term is formulated using the stress notation. The contra variant velocities and stresses (Equation (5) and (6)) are calculated at the cell's face and not at the cell's center. Figures 2 and 3 show the placement of the contra variant velocities and stresses in the cell. It should be noted that that all the velocities and pressure variables as well as the metric terms are stored at the cell center.

For Newtonian fluids, the stresses are defined as:

$$\sigma_{ij} = -p\delta_{ij} + 2\nu S_{ij} \tag{27}$$

where $\delta_{ij} = \begin{cases} 1 \; if \;\; i = j \\ 0 \; if \;\; i \neq j \end{cases}$ and $S_{ij} = \frac{1}{2}\left(\frac{\partial u_i}{\partial x_j} + \frac{\partial u_j}{\partial x_i}\right)$

The pressure from the diffusion term is grouped together with the static pressure in the momentum equations. Therefore, Equation (27) simplifies to:

$$\sigma_{ij} = 2\nu S_{ij} = \nu\left(\frac{\partial u_i}{\partial x_j} + \frac{\partial u_j}{\partial x_i}\right) \tag{28}$$

Also, ν will be replaced by $\nu + \nu_T$ for turbulent flows where ν_T is the turbulent viscosity.

The derivatives of the velocities in Equation (28) are calculated using simple central differencing across the cell interfaces where the velocities and metric term variables are interpolated at the interfaces. The same interpolation is needed for calculation of the contra variant stresses.

2.3. ***Stability analysis***

Due to the complexity and the nonlinear characteristics of the Navier-Stokes equations, it is difficult to perform a standard analytical stability analysis such as von Neumann and discrete perturbation method. Therefore, for the numerical simulation using explicit time schemes, a commonly used stability analysis is the CFL stability condition. This stability condition is a necessary condition for solving partial differential equations by finite difference methods such as the one in this work.

In a two-dimensional case, the CFL condition is given as;

$$C = \frac{u\Delta t}{\Delta x} + \frac{v\Delta t}{\Delta y} \leq C_{max} \tag{29}$$

where u and v are the velocities in the x and y axes respectively, Δt is the time step and Δx and Δy are the grid spacing in their respective directions and C is the Courant number. The maximum Courant number is an upper bound criterion for the selection of time step in ensuring that the solution converges and provides accurate results. The Courant number depends heavily on the types of the schemes and the discretization being used. For this

Table 1. Comparison of different schemes for lid-driven cavity for Re = 1000.

Scheme	Time step (s)	Total time (s)	Iterations	Residuals	Maximum Residuals	Time Consumed (s)
Projection Method	0.001	35	1	$1.96 * 10^{-6}$	$5.04 * 10^{-5}$	538
SIMPLE	0.005	35	20	$5.50 * 10^{-7}$	$3.26 * 10^{-6}$	214

Table 2. Comparison of different schemes for Backward Facing Step for Re = 5540.

Scheme	Time step (s)	Total time (s)	Iterations	Residuals	Maximum Residuals	Time Consumed (s)
Projection Method	0.001	70	1	$3.12 * 10^{-8}$	$5.42 * 10^{-8}$	2945
SIMPLE	0.001	70	5	$1.06 * 10^{-7}$	$7.46 * 10^{-7}$	4090

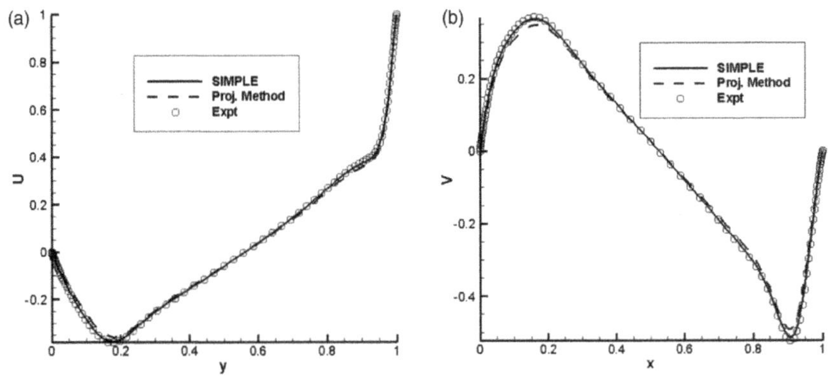

Figure 8. Comparison of velocity plots at x/L = 0.5 and y/L = 0.5 planes.

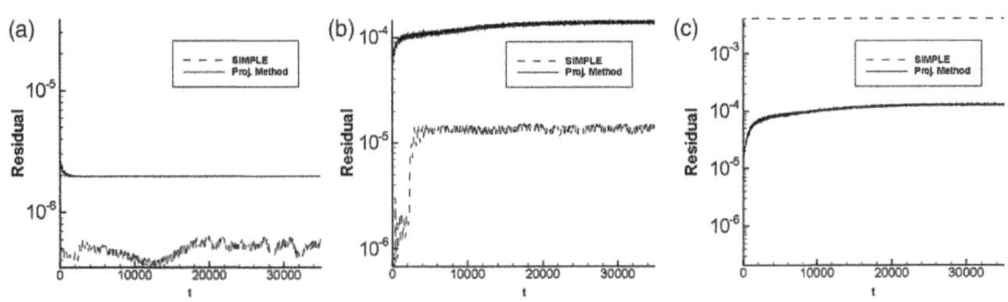

Figure 9. Plots of residuals for continuity (left), x-momentum (center) and y-momentum (right).

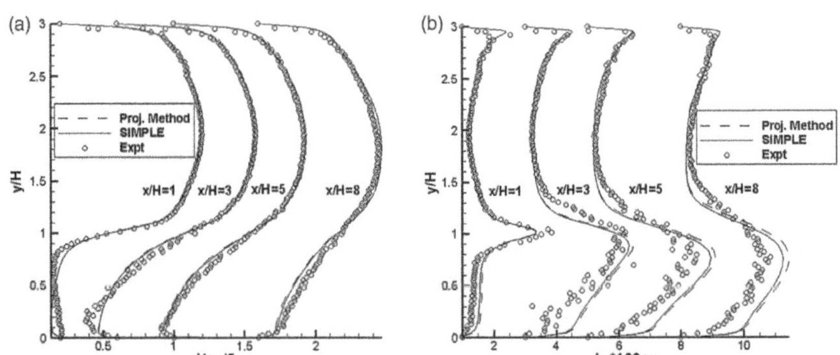

Figure 10. Comparison of velocity and turbulent kinetic energy plots at x/H = 1,3,5 and 8.

work, the Courant number will be calculated separately and compared for the different pressure-incremental projection method schemes. Referring to Equation (29), due to the nature of the flow and to simplify the problem, the Courant number is calculated to be:

$$C = \frac{\Delta t}{\Delta x} \leq C_{max} \quad (30)$$

where Δx is the smallest grid spacing in the domain and the characteristics velocity, u is selected to be 1. The maximum Courant number will be used as the stability criterion and can be represented as the slope of the Δt versus Δx plot.

3. Benchmark cases

This section describes the benchmark cases solved using the algorithms discussed in section 2 including geometry and boundary conditions.

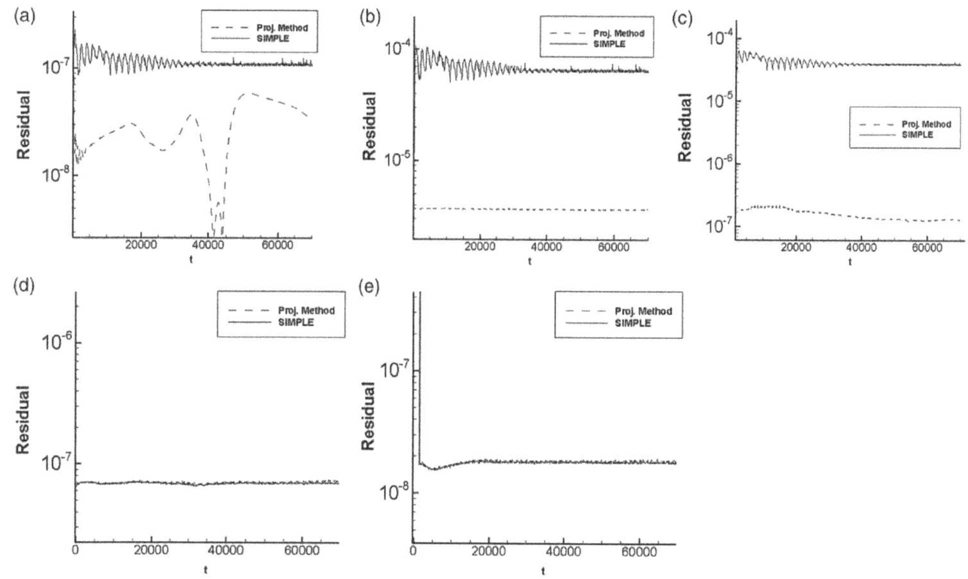

Figure 11. Plots of residuals for continuity (top left), x-momentum (top center), y-momentum (top right), turbulent kinetic energy (bottom left) and turbulent dissipation (bottom right)

Table 3. Comparison of the value of reattachment length with Kasagi et al. (1992).

Scheme	Reattachment Length
SIMPLE	6.2 H
Projection Methods	6.2 H
Experimental	6.51 H

3.1. Driven cavity flow at Re = 1000

A laminar-driven cavity flow is compared to the results obtained by Ghia, Ghia, and Shin (1982) for a Reynolds number of 1000. Figure 4 describes the geometry, dimensions and boundary conditions for the driven cavity flow. The grid size is 81 × 81 with the clustering near the walls. No-slip boundary conditions are applied for all velocities

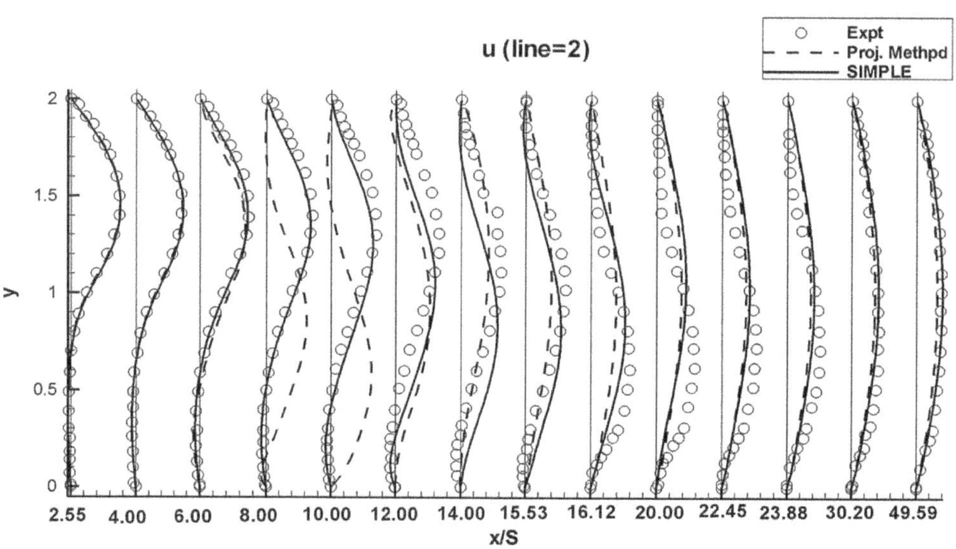

Figure 12. Comparison of velocity plots at x/S along the downstream. The spacing between the grid lines represents the magnitude of 2 for the u velocity. The maximum and average inlet velocity are scaled to 1.5 and 1, respectively.

Table 4. Comparison of different schemes for Backward Facing Step for Re = 1000.

Scheme	Time step (s)	Total time (s)	Iterations	Residuals	Maximum Residuals	Time Consumed (s)
Projection Method	0.003	900	1	$2.81 * 10^{-7}$	$1.96 * 10^{-6}$	18567
SIMPLE	0.003	900	5	$2.34 * 10^{-6}$	$2.58 * 10^{-5}$	29021

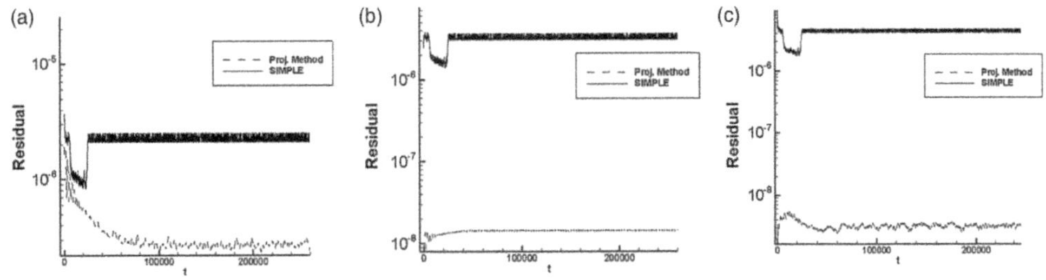

Figure 13. Plots of residuals for continuity (left), x-momentum (center) and y-momentum (right).

Table 5. Comparison of the values of different reattachment lengths based on convention used by Armaly, Durst, Pereira, and Schönung (1983).

	Reattachment Length				
Scheme	X1/S	X2/S	X3/S	X4/S	X5/S
Armaly (Numerical)	7.60	11.00	13.50	7.00	10.80
Armaly (Experimental)	17.00	-	-	13.00	20.00
SIMPLE	13.50	-	-	11.20	17.70
Projection Methods	8.50	12.20	15.20	6.60	13.10

except at the top. At the top the velocity is set to be u = 1, v = 0. All the dimensions are normalized to unity. A Neumann boundary condition of $\frac{\partial p}{\partial n} = 0$ is prescribed on all sides of the cavity for pressure.

3.2. *Flow past a cylinder at Re = 200*

A laminar flow past a cylinder is compared to the results of Qu, Norberg, Davidson, Peng, and Wang (2013) at a Reynolds number of 200. Figure 5 describes the geometry, dimensions and boundary conditions for flow past a cylinder. The grid size is 201 × 201 with the grid clustering being concentrated downstream. The o-grid outer boundary is set at 20D. No-slip boundary conditions are applied for velocities at the cylinder wall. Inlet boundary conditions are being applied at the outer boundary upstream where u = 1, v = 0 and convective boundary conditions are applied downstream. All the dimensions are normalized to a unit of 1. Neumann Boundary conditions are prescribed at all boundaries for pressure, namely $\frac{\partial p}{\partial n} = 0$.

3.3. *Backward facing step flow*

Flow in a backward facing step has also been solved in this work. Figure 6 describes the geometry, dimensions and boundary conditions for the backward facing step flow. The tested benchmark cases are transitional flow over a backward facing step established by Armaly et al. (1983) at a Reynolds number of 1000, and turbulent flow of Kasagi et al. (1992) at a Reynolds number of 5540. For the first case, the grid size is 41 × 51 upstream and 201 × 101 downstream. The dimensions are as follows: S = 1, h = 1, H = 2, l = 10 and L = 55 also shown in Figure 6. For the second case, the grid size is 71 × 31 upstream and 141 × 61 downstream. The dimensions are as follows: S = 1, h = 1,

Table 6. Comparison of different schemes for lid-driven cavity for Re = 1000.

Q	R	Rotational	Divergence Residuals	Maximum Residuals	Location of Maximum Residuals	Stability
1	0	No	$1.96 * 10^{-6}$	$5.04 * 10^{-5}$	Top left corner	Stable
1	0	Yes	$1.93 * 10^{-6}$	$4.95 * 10^{-5}$	Top left corner	Stable
1	1	No	$3.24 * 10^{-11}$	$9.55 * 10^{-11}$	Middle of top wall	Stable
1	1	Yes	$3.23 * 10^{-11}$	$9.36 * 10^{-11}$	Middle of top wall	Stable
2	0	No	$1.52 * 10^{-6}$	$3.89 * 10^{-5}$	Top left corner	Stable
2	0	Yes	$1.51 * 10^{-6}$	$3.84 * 10^{-5}$	Top left corner	Stable
2	1	No	$2.13 * 10^{-11}$	$6.30 * 10^{-11}$	Middle of top wall	Stable
2	1	Yes	$2.12 * 10^{-11}$	$6.23 * 10^{-11}$	Middle of top wall	Stable
2	2	No	-	-	-	Unstable
2	2	Yes	-	-	-	Unstable

Table 7. Comparison of different schemes for flow past cylinder.

Q	R	Rotational	Divergence Residuals	Maximum residuals	Location of maximum residuals	Stability	Cd (1.07)	St (0.1958)
1	0	No	$3.71 * 10^{-7}$	$4.57 * 10^{-6}$	Downstream near outlet	Stable	1.03	0.173
1	0	Yes	$3.69 * 10^{-7}$	$4.56 * 10^{-6}$	Downstream near outlet	Stable	1.03	0.173
1	1	Yes	$3.07 * 10^{-7}$	$3.76 * 10^{-6}$	Downstream near outlet	Stable*	1.02	0.175
1	1	No	$3.07 * 10^{-7}$	$3.76 * 10^{-6}$	Downstream near outlet	Stable*	1.02	0.175
2	0	No	$3.58 * 10^{-7}$	$4.45 * 10^{-6}$	Downstream near outlet	Stable	1.02	0.175
2	0	Yes	$3.57 * 10^{-7}$	$4.45 * 10^{-6}$	Downstream near outlet	Stable	1.02	0.175
2	1	No	$2.82 * 10^{-7}$	$3.52 * 10^{-7}$	Downstream near outlet	Stable*	1.04	0.177
2	1	Yes	$2.82 * 10^{-7}$	$3.52 * 10^{-7}$	Downstream near outlet	Stable*	1.04	0.177
2	2	No	-	-	-	Unstable	-	-
2	2	Yes	-	-	-	Unstable	-	-

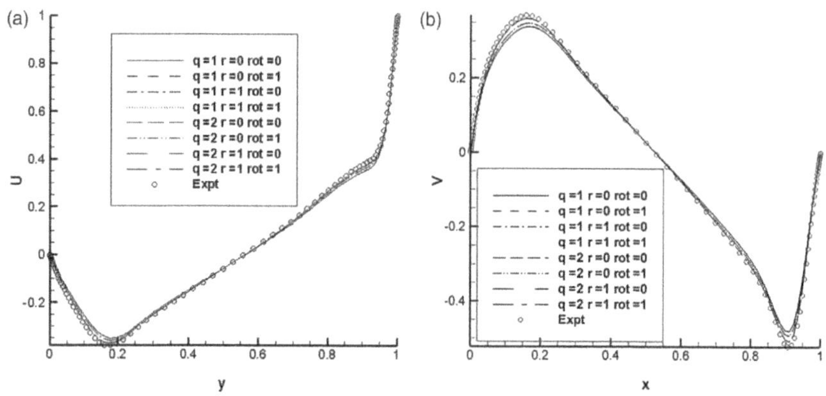

Figure 14. Comparison of velocity plots at x/L = 0.5 and y/L = 0.5 planes for different variant of projection method schemes.

H = 3, l = 10 and L = 20. Inlet velocity boundary conditions are applied upstream using the profile of fully developed flow in the channel. Downstream, convective boundary conditions are applied at the outlet. A no-slip boundary condition is applied at the upper and bottom walls while the Neumann Boundary condition $\frac{\partial p}{\partial n} = 0$ is prescribed at all boundaries for pressure. The mesh for the case at a Reynolds Number of Re = 5540 is shown in Figure 7. A standard k-ϵ model was used to calculate the turbulent viscosity for the turbulent flow at a Reynolds number of 5540.

3.4. *Calculations of residuals*

The L^2 and L^∞ norms of the residual are calculated to monitor convergence using the following definitions:

$$L^2 : \varepsilon_2 = \sum_{i=1}^{N} \frac{{\varepsilon_i}^2}{N} \quad (31)$$

$$L^\infty : \varepsilon_\infty = \max(\varepsilon_i) \quad (32)$$

where N is the number of grid points and ε_i is the residuals of the respective equations at each grid point. The residuals are computed by substituting the obtained velocities and pressure value of the current time step into the continuity and momentum equations, Equations (3) and (4).

The solution is considered to have converged when the L^2 norm value is below 10^{-6}.

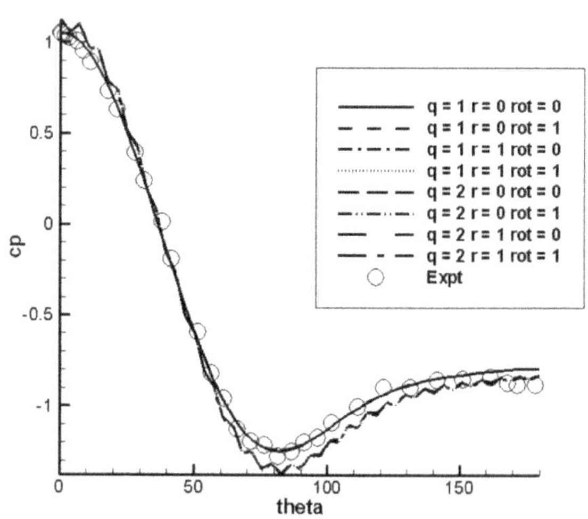

Figure 15. Comparisonofpressure coefficient along the cylinder for different variant of projection method schemes.

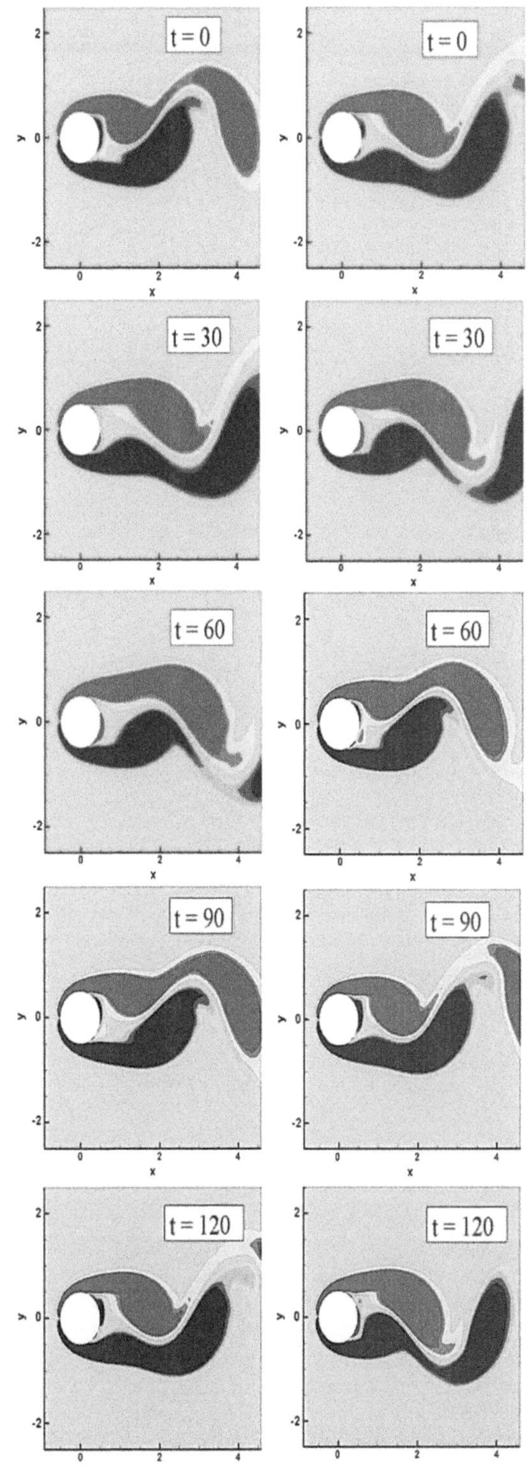

Figure 16. Comparisons of vorticity plot for r = 0 (left) and r = 1 (right) for a full shedding cycle where t = 0 is the start of the cycle.

4. Results and discussion

4.1. Projection method vs SIMPLE

Comparisons have been made between the projection method and SIMPLE algorithm for the lid-driven and backward step benchmark test cases for the same grid sizes and clustering. The projection method used here is the standard nonincremental scheme (i.e., r = 0 and q = 0). The differences are summarized in Table 1 and Table 2 for both lid-driven and backward step flow cases. The relaxation coefficients for the pressure and momentum equations in the SIMPLE algorithm are selected based on conventional default values proposed by previous researchers (see Patankar and Spalding (1972)). The nondimensional time step used for the lid-driven cavity is 0.005 while that for backward step flow it is 0.001. Comparison of velocity plots at x/L = 0.5 and y/L = 0.5 planes are shown in Figure 8 and the comparison of the residuals for the continuity, x-momentum and y-momentum are shown in Figure 9. Overall, the SIMPLE algorithm was found to be more accurate and faster in the laminar lid-driven cavity flow scenario.

The comparisons are different for the backward facing step flow. For the unsteady turbulent flow at Re = 5540, it can be concluded that there are no major differences between both schemes. This can be seen from the velocity and turbulent kinetic energy plots along the downstream shown in Figure 10. Figure 11 compares the residuals for the continuity, x-momentum, y-momentum as well as turbulent kinetic energy and turbulent dissipation. Also, the reattachment lengths obtained from both schemes are relatively the same as shown in Table 3. The time step chosen is the same for both schemes, but SIMPLE requires more iterations and therefore consumes more time. However, for the transitional case of Re = 1000, SIMPLE provides more accurate results as can be seen by comparison with the experimental data of Armaly et al. (1983) in Figure 12 although the simulation consumes more time (Table 4). Figure 12 compares the velocity plots for both schemes with the experimental data at many locations along the downstream. The corresponding comparisons of residuals are plotted in Figure 13. Additionally, the reattachment lengths obtained by the SIMPLE algorithm are closer to the experimental results (Armaly et al., 1983) as listed in Table 5.

For both lid-driven cavity and backward facing step flow, the number of iterations per time step of the Projection Method is selected to be 1 but not for the SIMPLE algorithm. This is because the Projection Method is fully explicit in time and does not require internal iterations within the same time step.

Lastly, the stability of the schemes depends on the maximum allowable time step for convergence. The SIMPLE algorithm holds the advantages of a larger time step for laminar flow since it is semi-implicit unlike the projection method which is fully explicit. However, both schemes are

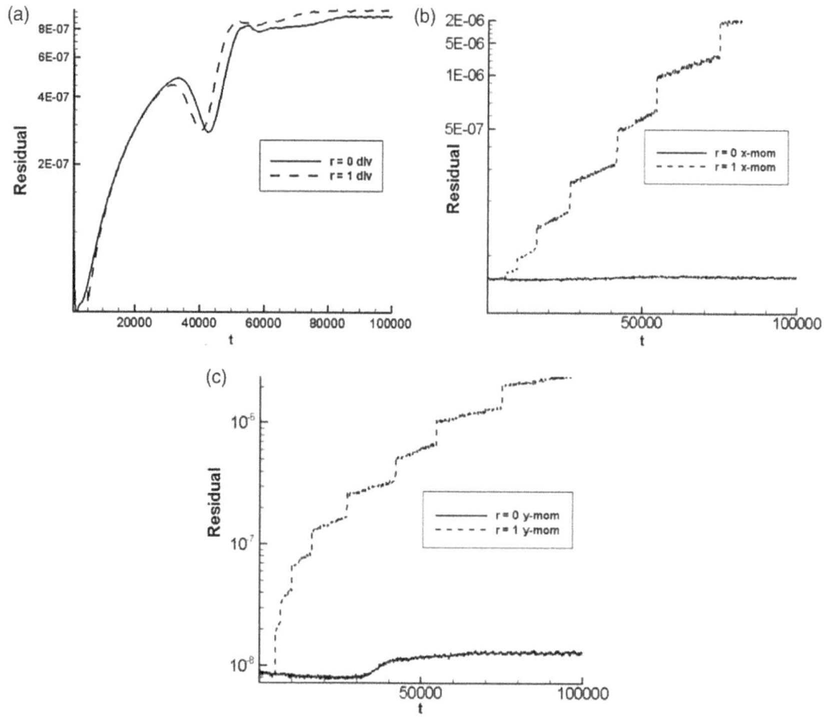

Figure 17. Plots of residuals for continuity (top left), x-momentum (top right) and y-momentum (bottom) for flow past a cylinder.

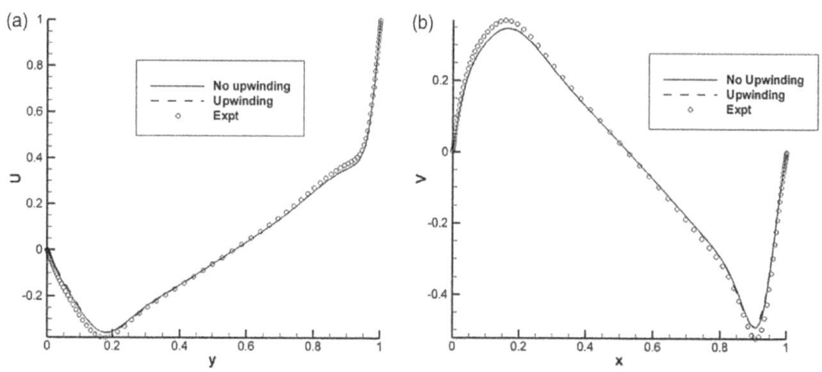

Figure 18. Comparisons of the velocity plot at x/L = 0.5 and y/L = 0.5 for no upwinding and second-order upwinding for flow past a cylinder.

restricted to the same time step for both transitional and turbulent flows.

4.2. *Variants of the projection method*

Table 6 and 7 show the comparison between different projection methods for the lid-driven cavity flow and flow over a cylinder. The location of maximum residuals is important in analyzing and determining the sources of instability.

Also, the simulation results are compared with benchmark DNS and experimental data. Figure 14 shows the velocity plots at x/L = 0.5 and y/L = 0.5 planes for the lid-driven cavity flow. Meanwhile, figures 15 and 16 show the distribution of the pressure coefficient along the surface of the cylinder and the vorticity plot for a typical vortex shedding cycle with t = 0 representing the beginning of the cycle for flow over a cylinder. Figure 17 compares the residuals of continuity, x-momentum and y-momentum for flow over a cylinder. The drag coefficient and Strouhal number are listed in Table 7.

The schemes are stable for various combinations of parameters except when the order of the pressure equation

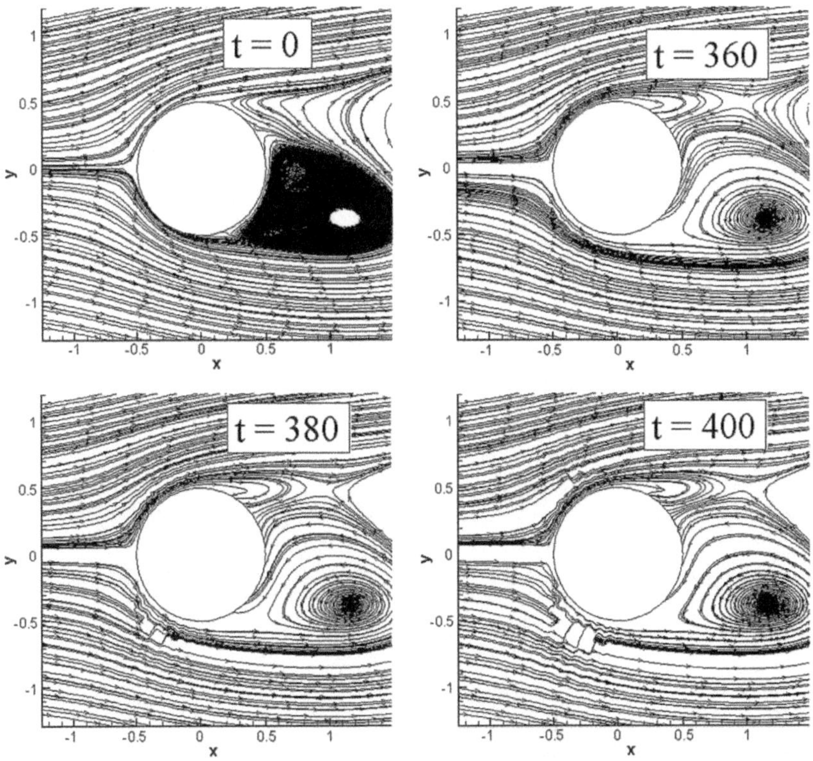

Figure 19. Progression of the streamlines around the cylinder when the simulation is switched into second-order upwinding

is r = 2 (Equation 10) and for time order q = 2 (Equation 20). It is hypothesized that the causes of this instability comes from the term $p*^{t+1} = 2p^t - p^{t-1}$. Also the instability holds, regardless of how small the time step and the grid size, and the exact source of the instability needs further investigation.

It is observed that the scheme with r = 1 allows the residuals of the divergence to go down to 10^{-11} as compared to r = 0 which give a value of 10^{-6} . This is also noticeable in the velocity plots shown in Figure 14 where increasing the order of the pressure equation produces more accurate results.

For flow over a cylinder, the scheme r = 1 is found to be stable, but the pressure coefficient plot shows oscillations especially after the separation region in Figure 15 and consequently the residuals are also observed to have an increasing trend in Figure 17.

4.3. Discretization

4.3.1. Convective term

The second-order upwinding scheme is unstable for flow over a cylinder and provides no improvement for lid-driven cavity flow as illustrated in Figures 18, 19 and 20. Figure 18 shows the velocity plot at the x/L =0.5 and y/L =0.5 planes for the lid-driven cavity where there is no differences between the results obtained using no upwinding and the second-order upwinding scheme. Figure 19 and 20 show the progression of the streamline and vorticity for the flow over a cylinder, where t=0 represents the time when the scheme is switched from no upwinding to second-order upwinding in the simulation. The instability for the flow over a cylinder (Figure 19) is observed to start at around 30 degrees from the stagnation point and is related to the upwinding discretization of the convective term as the central difference discretization does not lead to any instability.

4.3.2. Diffusive term

Comparisons have been made to evaluate the differences between storing the contravariant stresses at the cell interfaces versus storing them at the grid points. Using the contravariant stress values from the grid points results in a checkerboard velocity field as can be seen from Figure 21. The figure shows the checkerboard velocity field upstream and downstream of the cylinder. However, using the contravariant stress values from the cell interfaces yields a stable solution.

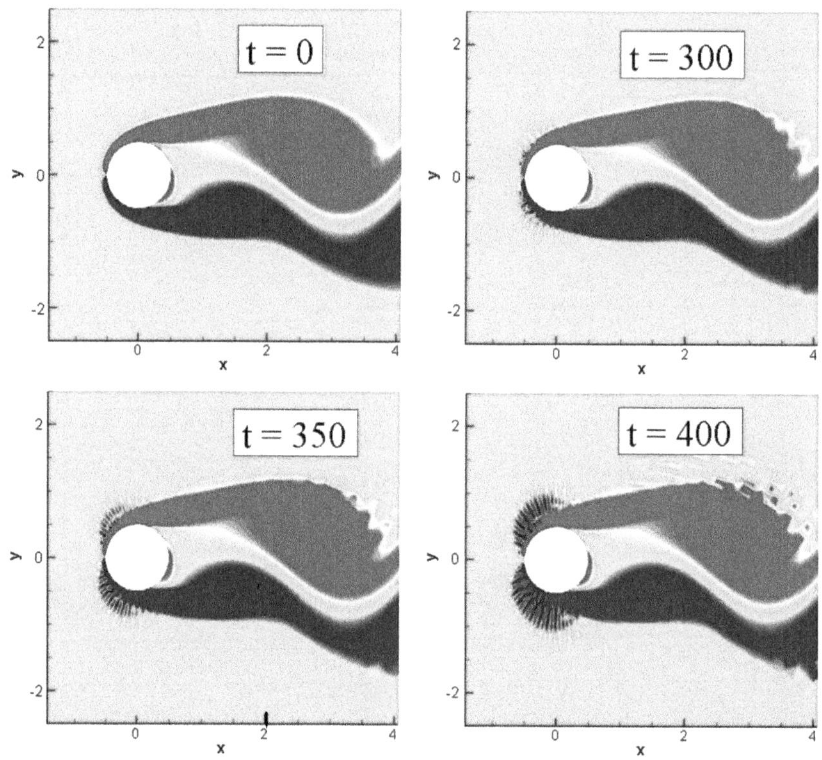

Figure 20. The progression of the vorticity when the simulation is switched into second-order upwinding.

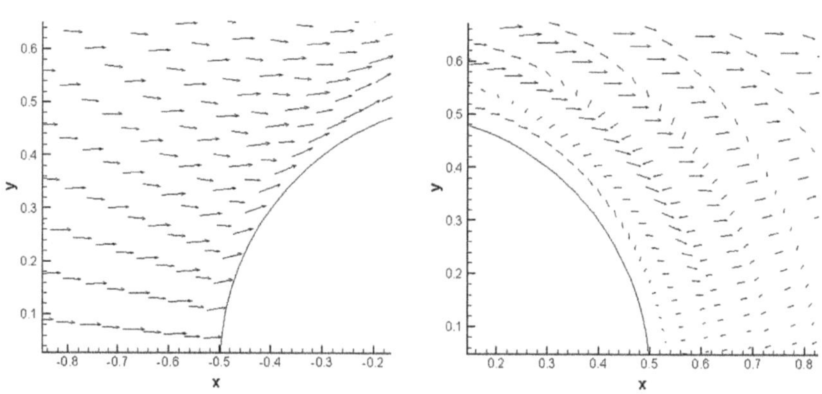

Figure 21. Checkerboard velocity vector field upstream and downstream of the cylinder.

4.4. Stability analysis

Stability analysis has been performed to determine the influence of the minimum grid spacing between any two grid points and the maximum allowable time step and is represented in Figure 22 for the schemes r = 0 and r = 1. Figure 22 plots the relationships between the minimum grid spacing and maximum allowable time step for both schemes for flow past a cylinder. The scheme r = 1 has a stricter time step requirement compared to r = 0. This is reflected in Figure 17 where the residuals shows an increasing trend and in Figure 15 which exhibits an oscillating behavior of the pressure coefficient. The scheme r = 2 was found to be unstable (Section 4.2).

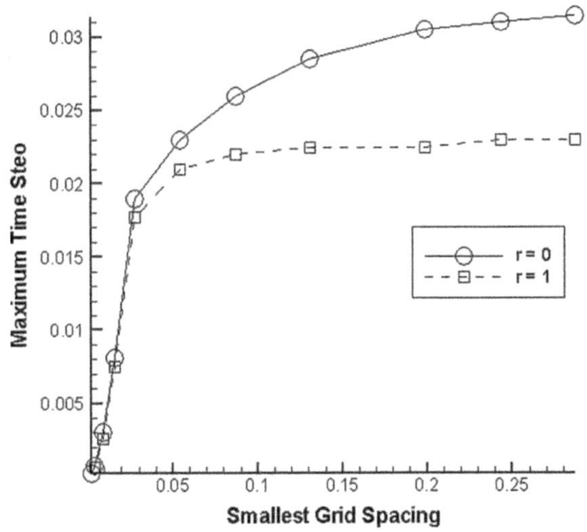

Figure 22. Plot of maximum time step versus smallest grid spacing for both r = 0 and r = 1 schemes for flow past a cylinder.

5. Conclusions

In conclusion, it is observed that the SIMPLE algorithm performs better for the laminar flow cases while the projection methods perform better for transitional and fully turbulent flows. Furthermore, comparisons of the different higher-order pressure-correction schemes show that the scheme r = 2 is unstable and the scheme r = 1 produces oscillations of the pressure coefficient and shows an increasing trend of the residuals for flow over a cylinder. The projection methods with r = 0 has been observed to be the most stable one. The second-order upwinding of the convective terms is unstable when using generalized curvilinear grids and does not show any substantial improvement over the central difference convective scheme. The diffusive terms involving the contravariant stresses are more stable when the stress values are taken from the cell interfaces as compared to taking values from the grid points. Lastly, stability analysis shows that the scheme r = 1 has a stricter time step requirement compared to r = 0 which is a more stable scheme overall.

The limitations of this study are that the benchmark cases are performed on a flow on simple geometries. Also, the stability analysis performed is unable to pinpoint to roots or real causes of the instability of the higher-order projection method schemes. In the future, it is suggested that the schemes should be used in a flow involving more complicated and complex 3-dimensional geometry close to real-world applications. Further investigations need to be done in order to better understand the sources of instability.

Acknowledgements

This work was carried out under the auspices of the Computational Mechanics Laboratory, Department of Mechanical Engineering, IIUM. We would also like to acknowledge the Ministry of Higher Education, Malaysia (MOHE) for providing financial support.

References

Abbasi, R., Ashrafizadeh, A., & Shadaram, A. (2013). A comparative study of finite volume pressure-correction projection methods on co-located grid arrangements. *Computers & Fluids*, *81*, 68–84.

Armaly, B. F., Durst, F., Pereira, J. C. F., & Schönung, B. (1983). Experimental and theoretical investigation of backward-facing step flow. *Journal of Fluid Mechanics*, *127*, 473–496.

Baniasadi, E., Aydin, M., Dincer, I., & Naterer, G. F. (2013). Computational aerodynamic study of automotive cooling fan in blocked conditions. *Engineering Applications of Computational Fluid Mechanics*, *7*(1), 66–73.

Chorin, A. (1968). Numerical solution of the Navier-Stokes equations. *Mathematics of Computation*, *22*(104), 745–762.

Fereidoon, A., Saedodin, S., Hemmet, E. M., & Noroozi, M. J. (2013). Evaluation of mixed convection in inclined square lid driven cavity filled with Al2O3/water nanofluid. *Engineering Applications of Computational Fluid Mechanics*, *7*(1), 55–65.

Ghia, U., Ghia, K., & Shin, C. (1982). High-Re solutions for incompressible flow using the Navier-Stokes equations and a multigrid method. *Journal of Computational Physics*, *411*, 387–411.

Goda, K. (1979). A multistep technique with implicit difference schemes for calculating two- or three-dimensional cavity flows. *Journal of Computational Physics*, *30*(1), 76–95.

Guermond, J. L., Minev, P., & Shen, J. (2006). An overview of projection methods for incompressible flows. *Computer Methods in Applied Mechanics and Engineering*, *195*(44–47), 6011–6045.

Guermond, J. L., & Shen, J. (2003). On the error estimates for the rotational pressure-correction projection methods. *Mathematics of Computation*, *73*(248), 1719–1738.

Kasagi, N., Horiuti, K., Miyake, Y., Miyauchi, T., & Nagano, Y. (1992). *Establishment of the direct numerical simulation data base of turbulent transport phenomena*. Co-operative Research No. 02302043, University of Tokyo, Japan.

Liu, J. G., Liu, J., & Pego, R. L. (2010). Stable and accurate pressure approximation for unsteady incompressible viscous flow. *Journal of Computational Physics*, *229*(9), 3428–3453.

Patankar, S., & Spalding, D. (1972). A calculation procedure for heat, mass and momentum transfer in three-dimensional parabolic flows. *International Journal of Heat and Mass Transfer*, *15*(10), 1787–1806.

Qu, L., Norberg, C., Davidson, L., Peng, S. H., & Wang, F. (2013). Quantitative numerical analysis of flow past a circular cylinder at Reynolds number between 50 and 200. *Journal of Fluids and Structures*, *39*, 347–370.

Rempfer, D. (2006). On boundary conditions for incompressible Navier-stokes problems. *Applied Mechanics Reviews*, *59*(3), 107–125.

Ren, Y., Jiang, Y., Liu, M., & Zhang, H. (2005). A class of fully third-order accurate projection methods for solving the incompressible Navier-Stokes equations. *Acta Mechanica Sinica*, *21*(6), 542–549.

Sun, H., He, Y., & Feng, X. (2011). On error estimates of the pressure-correction projection methods for the time-dependent Navier–Stokes equations. *International Journal of Numerical Analysis & Modeling*, *8*(1), 70–85.

Temam, R. (1969). Sur l'approximation de la solution des equations de Navier-Stokes par la methode des pas fractionnaires (I). *Archive for Rational Mechanics and Analysis*, *33*(5), 377–385.

Timmermans, L. J. P., Minev, P. D., & Van De Vosse, F. N. (1996). An approximate projection scheme for incompressible flow using spectral elements. *International Journal for Numerical Methods in Fluids*, *22*(7), 673–688.

Van Kan, J. (1986). A second-order accurate pressure-correction scheme for viscous incompressible flow. *SIAM Journal on Scientific and Statistical Computing*, *7*(3), 870–891.

2

Simulations and measurements of automotive turbocharger compressor whoosh noise

A. Broatch[a], J. Galindo[a], R. Navarro[a]*, J. García-Tíscar[a], A. Daglish[b] and R. K. Sharma[b]

[a]CMT - Motores Térmicos, Universitat Politècnica de València, Camino de Vera S/N, 46022 Valencia, Spain; [b]Jaguar Land Rover Limited, Abbey Road, Whitley, Coventry CV3 4LF, UK

Turbocharger noise has become a major concern in downsized automotive engine development. In this paper, the analysis is focused on the *whoosh noise* produced by the compressor when it is working near surge. A centrifugal compressor has been acoustically characterized on a turbocharger test rig mounted on an anechoic chamber. Three in-duct pressure signals forming a linear array are registered in order to obtain pressure components. In this way, meaningful pressure spectra and sound intensity level (SIL) compressor maps are obtained, showing an increase of SIL in the frequency window corresponding to whoosh noise. Besides, detached eddy simulations (DES) of the centrifugal compressor flow in two operating conditions near surge are performed. Good agreement is found between the experimental measurements and the CFD solutions in terms of predicted pressure spectra. Flow analysis is used to identify patterns responsible for the different features of the pressure spectra. At the simulated conditions, rotating instabilities in the compressor diffuser and inducer cause pressure oscillations in the frequency range of whoosh noise.

Keywords: CFD; aeroacoustics; rotating stall; centrifugal compressor; DES

1. Introduction

Turbocharging has been one of the enablers of automotive powerplants improvement in the last two decades. Nowadays almost all automotive diesel engines and more and more petrol engines are boosted with a turbocharger. Turbocharging, initially conceived as a means to increase the fresh charge within the cylinders, is now the responsible to reduce fuel consumption and pollutant emissions through downsizing and downspeeding (Knecht, 2008). This means that a smaller engine working at increased load and reduced speed can lead to meaningful improvements in consumption and emissions. However, this can only be carried out by increasing the boosting pressure with a turbocharger. Furthermore, the downsizing and downspeeding trends lead the turbomachinery used toward tighter requirements, particularly, in terms of flow range. In order to enlarge the flow range in the compressor from lower surge limit to higher choked flows, several techniques have emerged to reduce the surge limit flow such as ported shroud (Semlitsch et al., 2014) or improved compressor entry designs (Serrano, Margot, Tiseira, & García-Cuevas, 2013). However, the stable working area gained in this process results to be quite noisy (Evans & Ward, 2006). The issue, particularly for high-end automotive brands, is not only the overall acoustic output of turbochargers, but its quality. In particular, an undesirable noise identified as "whoosh" is detected in regions of near surge operation. Whoosh noise is usually identified as a broadband elevation in the medium frequency range between 1 and 3 kHz.

Evans and Ward (2006) found that whoosh noise is most likely to be encountered by the driver during high load acceleration typically below 2500 rpm or during transient tip-out or tip-in maneuvers, associated with operation of the turbocharger compressor close to the surge line. However, whoosh noise was not encountered at the area where operating line is closest to the surge line but at the region of the compressor map where, for a constant speed, the pressure ratio across the compressor increases with increasing mass flow rate. Particularly, the compressor outlet hose was found to be the main source of radiated noise. Teng and Homco (2009) investigated radiated noise on a powertrain dyno in a semi-anechoic cell measured at full load conditions. Improvement of the compressor surge margin was sought so as to reduce whoosh noise, either changing compressor trim or using pre-whirl devices. Additionally, the effects of various resonators at different locations of the intake line on whoosh noise attenuation were studied.

Mendonça, Baris, and Capon (2012) conducted a numerical study regarding flow-induced acoustics of an automotive centrifugal compressor. Sound pressure level (SPL) spectra were obtained for points in both inlet and outlet ducts. Two main features can be observed in the spectra: the blade passing frequency (BPF) tonal noise and a narrow band noise at a frequency about 70% of

*Corresponding author. Email: ronagar1@mot.upv.es

rotational speed. A spiral mode propagating upstream from the compressor impeller is detected for the narrow band noise aforementioned. Velocity vectors on the impeller flow-field revealed leading-edge separation and stalled passages, whereas velocity contours showed a low momentum region that rotates at a slower speed than the wheel. Therefore, inducer rotating stall was found to be the source of the narrow band noise.

Figurella et al. (2012) studied the acoustic and performance characteristics of a turbocharger compressor on a steady-flow turbocharger test bench. Near choke, the compressor noise was dominated by the rotor-order frequency and its harmonics (including BPF). As the flow rate was reduced (thus decreasing the slope of the speed lines) the compressor exhibited a broadband elevation of noise in the 4–12 kHz band. When the mass flow rate is less than a critical value, the temperature near the inducer tips increased sharply, suggesting local flow reversal, and the total SPL in the 4–12 kHz range suddenly reduced.

The bibliographic review shows that whoosh noise is commonly described as a broadband noise either in the 1 kHz–3 kHz band (Gaudé et al., 2008; Lee, Selamet, Kim, Kim, & Kim, 2009; Sevginer, Arslan, Sonmez, & Yilmaz, 2007) or in frequencies ranging from 4 kHz to 12 kHz (Figurella et al., 2012; Karim, Miazgowicz, Lizotte, & Zouani, 2013; Teng & Homco, 2009). Figurella et al. (2012) pointed out that the former could be related to Diesel engines whereas the latter correspond to turbochargers installed in gasoline engines.

In the literature, approaches to the understanding of the mechanism appear to be either purely experimental (Evans & Ward, 2006; Figurella et al., 2012; Teng & Homco, 2009) or essentially computational (Karim, Miazgowicz, Lizotte, & Zouani, 2013; Lee, Selamet, Kim, Kim, & Kim, 2009). The objective of this paper is to combine both techniques to attain a greater understanding of this particular phenomenon.

2. Experimental tests

The first stage of the present study consisted in the experimental characterization of the turbocharger aeroacoustic output across a representative region of the operation maps. To do so, the turbocharger was fitted in a turbocharger test rig that was built inside a large anechoic chamber, as depicted in Figure 1. The turbocharger test rig is described in detail by Galindo, Serrano, Guardiola, and Cervelló (2006). The rig is powered by a large Diesel engine capable of producing up to 250 kW. The engine powers a screw compressor, which is used to supercharge the engine itself and also provide pressurized air to a turbocharger turbine located in an adjacent anechoic chamber. The turbine drives the centrifugal compressor that is analyzed in this paper. This anechoic chamber is of the Wedge type, rated with a low cut-off frequency of 100 Hz, thus guaranteeing acoustical free-field conditions above this frequency, which is well below the frequency window of interest (1 kHz to 3 kHz).

Desired operation condition of the turbocharger was achieved by regulating a backpressure valve in the outlet pipe and adjusting the power output of the rig engine, which controls the air mass flow, pressure and temperature of the flow provided to the turbine. However, the operation of this rig is restricted to steady points, lacking the ability to run preprogrammed routines. This limitation was overcome during this study through the use of numeric cubic interpolation between all the recorded data points to obtain a fitted model for

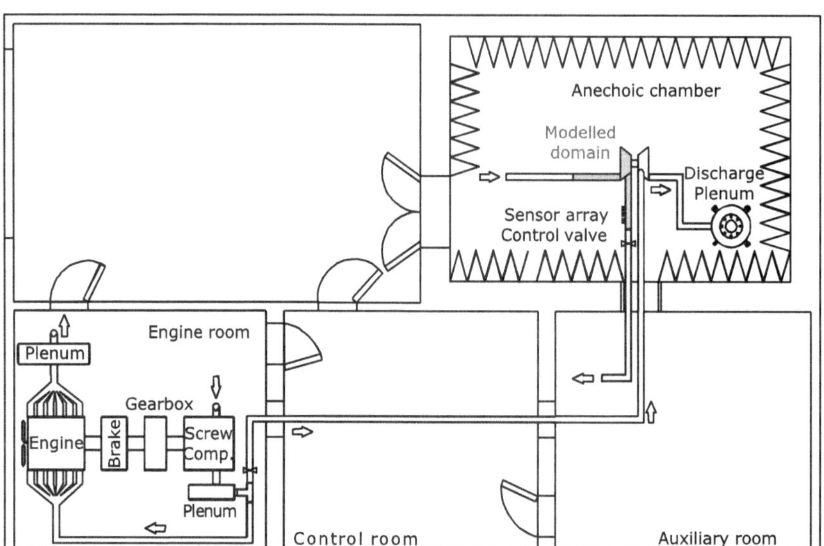

Figure 1. Schematic of the experimental installation.

the whole region. Then, this model can be applied to simulate routines such as a mass flow sweep at constant speed or even the lug line of the actual vehicle application.

Turbocharger status was constantly monitored through pressure, temperature, rotational speed and mass flow sensors, connected to a custom control interface. Using this interface, the operating condition (total-to-total pressure ratio and air mass flow) of the turbocharger could be adjusted and registered.

Besides, an array of three piezoelectric pressure sensors was flush mounted on the outlet pipe of the turbocharger, enabling the attainment of both direct pressure spectra and pressure wave decomposition. Instrumentation *inside* the compressor with pressure transducers as performed by Raitor and Neise (2008) was not considered due to the reduced size of the turbocharger compressor studied. A sketch of the compressor piping is shown in Figure 2, including the linear array. The sensor array was sampled by a standalone Yokogawa DAQ system at 100 kHz using independent channels.

A LCMV beamforming wave decomposition algorithm presented by Piñero, Vergara, Desantes, and Broatch (2000) was used to isolate the components of pressure signal. Then, in-duct sound intensity can be computed as described by Morfey (1971):

$$I = \frac{1}{\rho a}\left(|p_f|^2(1+M)^2 - |p_b|^2(1-M)^2\right) \qquad (1)$$

where ρ is the mean density of the flow, a is the mean sound speed, p_f and p_b are the forwards and backwards pressure spectra, and M is the mean Mach number. This expression enables the calculation of the in-duct sound intensity taking into account the often neglected effect of the mean flow.

However, the beamforming decomposition algorithm is not without limitations. One of the most restrictive is spatial aliasing. The sensor spacing (32 mm; see Figure 2) defines a Nyquist-type upper frequency cut-off that is below the onset of first higher-order acoustic mode due to spatial aliasing effects, as the distance between them implies a time delay, which has an associated limiting frequency. The closer the sensors are placed to each other the higher this limit becomes, but the differences of long-wavelength pressure components between two consecutive sensors diminish, falling eventually below the sensitivity of the sensor, thus imposing a low frequency cut-off as well. For our study, a safe window between 100 Hz and 6 kHz was considered when performing wave decomposition.

Intensity spectra were calculated using Equation 1 and integrated along the frequency band of interest (1–3 kHz) to obtain the overall intensity level of each of the tested operation points. Interpolating these data, it was possible to compute a "noise map" of the compressor, present in Figure 3, similar to others available in the literature Gaudé et al. (2008). Noise level increases with decreasing mass flow, with a steeper gradient (contour lines are close together) near the points with maximum compression ratio at each speed (Evans & Ward, 2006). These points, which present zero isospeed slope, are plotted in the dashed line of Figure 3.

Moreover, spectra of the data points recorded at the 160 krpm isospeed line were also interpolated in order to reconstruct the pseudo spectrogram included in Figure 3, analyzing not only the overall intensity, but its frequency distribution as well. It was found that the so-called whoosh phenomena appeared between 1 kHz and 2.5 kHz at much higher mass flow than the start of the low-frequency phenomena associated to the onset of surge.

Then, two points at different mass flows (59 g/s and 77 g/s) were considered. The latter corresponds to the maximum pressure ratio point at 160 krpm, whereas the former is closer to surge. Analyzing their respective sound intensity spectra in the bottom part of Figure 3, it is obvious that whoosh noise is more prominent at 77 g/s point than at 59 g/s, even though its acoustic intensity integrated in the 1 kHz–2.5 kHz band is lower. This may be the reason why the so-called whoosh noise is more noticeable for the customer in automobile applications in operating conditions

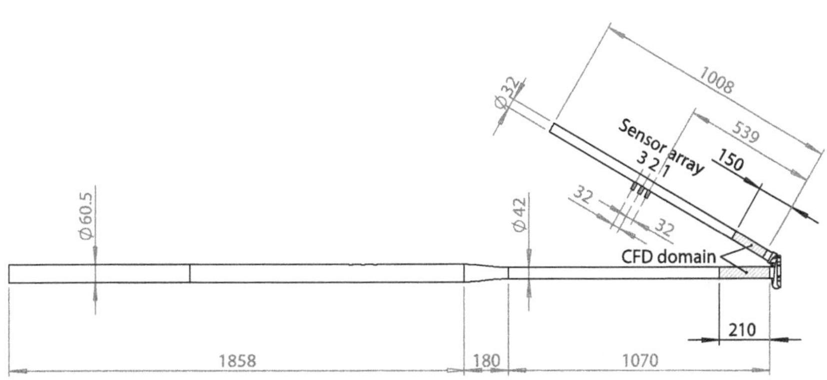

Figure 2. Sketch of compressor piping, highlighting the CFD domain (dimensions in mm).

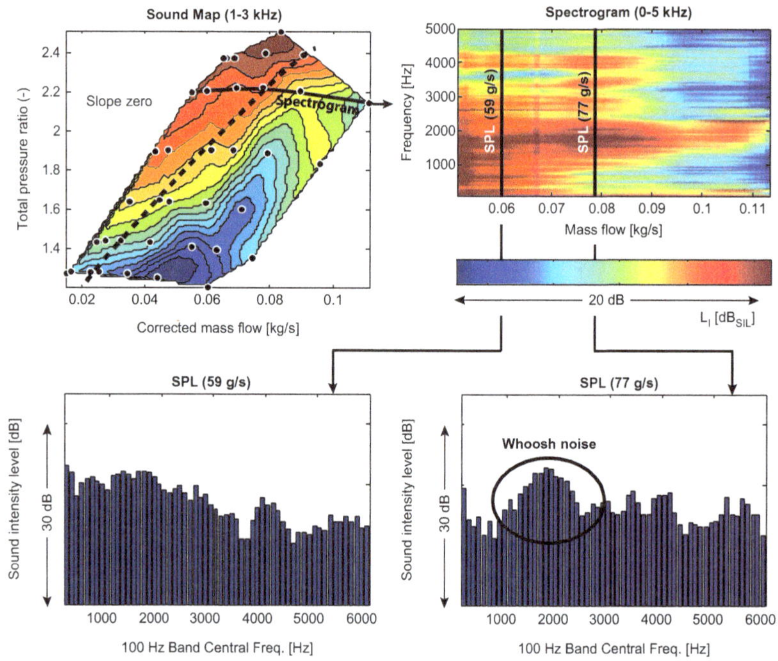

Figure 3. Noise map of the compressor and spectrogram of the 160 krpm isospeed line, including the spectra of two points of interest.

not so close to surge (Evans & Ward, 2006), although the overall noise is not necessarily high.

3. Numerical simulations

Measurements described in Section 2 show that, for a speed line, whoosh noise is more relevant for operating conditions corresponding to zero pressure ratio slope than for working points closer to the surge line, as pointed out by Evans and Ward (2006). 3-dimensional CFD simulations using StarCCM+ (CD-adapco, 2013) are performed to investigate the flow at the two operating conditions studied in Section 2.

Numerical simulations are configured following the criteria used by Mendonça et al. (2012). Detached eddy simulations with underlying k-ω SST turbulence model were performed using the rigid body motion approach. Time-step size was selected for each operating condition causing an impeller rotation of 1° per time step.

The modeled domain is depicted in Figure 4. It includes the vaneless compressor and 5-diameters-long straight inlet and outlet ducts, which were employed to reduce the effect of boundary conditions. Backplate region and tip clearance have been modeled as well. Even though tip gap is expected to decrease from "cold" conditions to actual compressor working conditions, CAD clearance is used in the impeller because a sensitivity analysis carried out by Galindo, Tiseira, Navarro, and López (2015) proves that compressor acoustic signature does not depend on tip clearance ratio.

Figure 4. Modeled domain.

Polyhedral grids are expected to perform with higher accuracy than hexahedral meshes when secondary flows are relevant (Tritthart & Gutknecht, 2013), thus being a common choice for turbomachinery flows (Galindo, Hoyas, Fajardo, & Navarro, 2013; Baris & Mendonça, 2011). Prism layer meshing increases near wall grid resolution so as to obtain y^+ values close to the unity in the impeller (see Figure 5). Nine and half million polyhedral cells were used to mesh the domain.

Since the experimental rig described in Section 2 is not entirely modeled for the sake of computational cost (see Figure 2), raw comparison between spectra of numerical

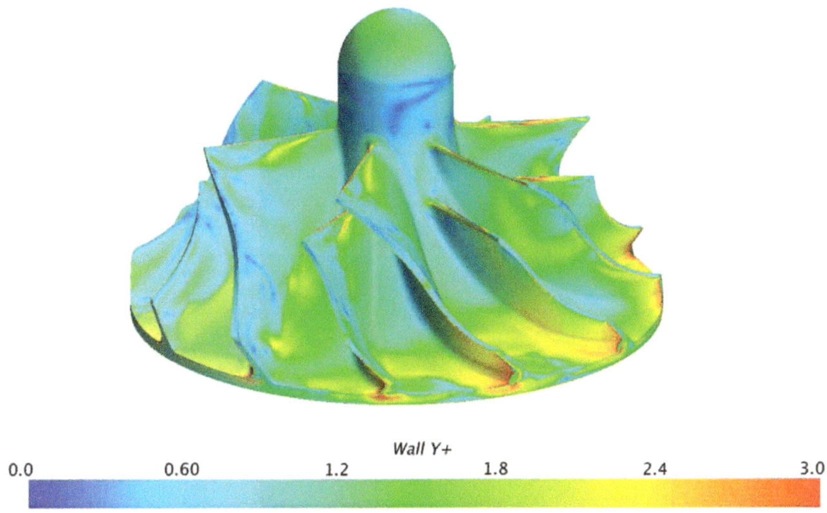

Figure 5. Wall y$^+$ contours at 59 g/s mass flow rate.

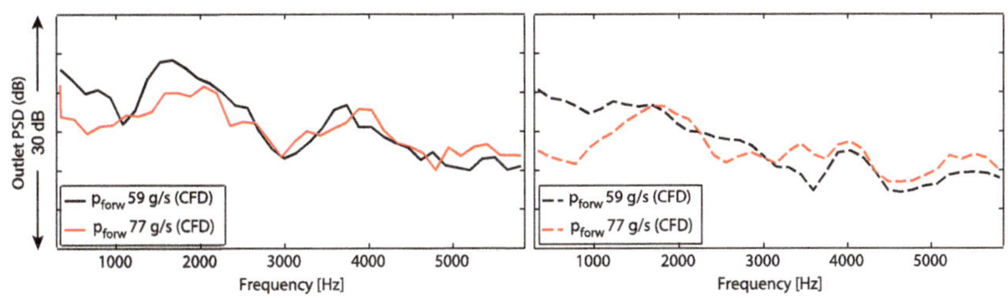

Figure 6. PSD of experimental (left) and numerical (right) forward pressure at outlet duct at 59 g/s (blue) and 77 g/s (red).

and measured pressure signals is not advisable (Broatch, Galindo, Navarro, & García-Tíscar, 2014). Pressure wave decomposition is performed in this study in order to better isolate the spectral information generated specifically by the aeroacoustic sources of the compressor. A pressure record captured in a sole point does not only contain this information, but also includes reflection from the domain boundaries and can even present undesired information from spurious standing waves. Therefore, an array of three pressure probes is placed at inlet and outlet ducts of turbocharger test rig to decompose experimental pressure signals, following again the work of Piñero et al. (2000).

For numerical simulations, pressure decomposition is performed using:

$$p_f = p_0\left[\frac{1}{2}\left(1+\left(\frac{p}{p_0}\right)^{\frac{\gamma-1}{2\gamma}}\left(1+\frac{\gamma-1}{2}\frac{u}{a}\right)\right)\right]^{\frac{2\gamma}{\gamma-1}}$$

$$p_b = p_0\left[\frac{1}{2}\left(1+\left(\frac{p}{p_0}\right)^{\frac{\gamma-1}{2\gamma}}\left(1-\frac{\gamma-1}{2}\frac{u}{a}\right)\right)\right]^{\frac{2\gamma}{\gamma-1}} \tag{2}$$

which are derived in the spirit of the Method of Characteristics, following the same notation as Torregrosa, Fajardo, Gil, and Navarro (2012), i.e., u is the axial velocity, γ is the ratio of specific heats and p and p_0 are the pressure and reference pressure, respectively. The methodology developed by Broatch et al. (2014) is used as follows: temporal traces of pressure, axial velocity and speed of sound cross-section average are extracted from the simulations, calculating then the pressure components by means of Equation 2.

Power spectral density (PSD) of the forward component of pressure obtained experimentally and predicted by CFD at the outlet duct for both operating conditions are compared in Figure 6. The agreement between measured and modeled signals is quite good. Mean relative error is 2.8% for the 59 g/s case and 2.5% for the 77 g/s case, with the maximum relative error for a particular frequency of about 10% at 3600 Hz for 59 g/s mass flow rate (see Figure 7). Whoosh noise is captured by numerical simulations, i.e., the predominance of pressure amplitude at the range of 1–2.5 kHz is well predicted.

Figure 6 shows that the experimental spectrum monotonously decays from low frequencies until 3.5 kHz for 59 g/s mass flow rate. Then, a narrow band is present from 3.5 kHz to 4.5 kHz and a broadband noise exists until 6 kHz. Instead, the simulation predicts a broadband elevation from 1 kHz to 3 kHz (whoosh noise), even though the decay rate is similar to the measured one. The rest of

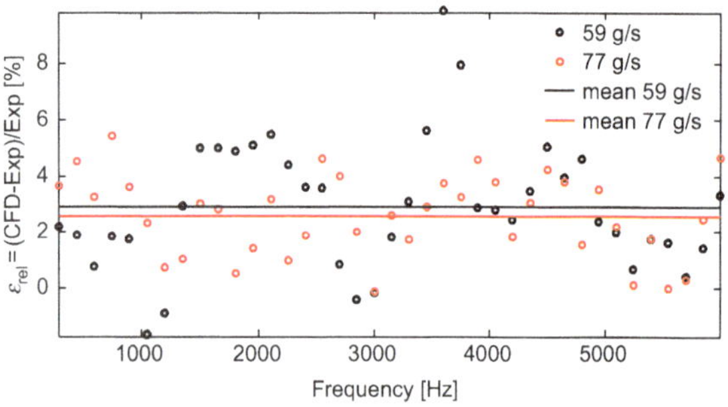

Figure 7. Relative error of forward pressure PSD at outlet duct at 59 g/s (blue) and 77 g/s (red).

the spectrum is similar to the experimental one. For 77 g/s, numerical and measured spectra are very much alike. Whoosh noise is clearly present between 1 and 2.5 kHz, and a broadband noise exists at higher frequencies

Flow is analyzed to detect phenomena occurring at the frequency range of whoosh noise. Figure 8 shows three snapshots of the transient behavior of flow structures at 59 g/s (left side) and 77 g/s (right side). The frames are obtained at three consecutive blade-passing periods. At the impeller inducer, contours of pressure are plotted at two different cross-sections, depicting regions of low pressure (particularly, $p < 85$ kPa). At diffuser midspan, radial velocity magnitude contour is presented, highlighting the level of 100 m/s to facilitate the identification of flow structures.

Flow behavior at the diffuser is quite similar for both operating conditions. The jet-wake structure at the impeller trailing edge is emphasized by the stall cells generated at the blade suction side shown at Figure 9, in which meridional velocity at 50% span is represented using a solid black line to indicate reverse flow. These regions with low momentum are not axisymmetrically distributed. A stalled channel can recover when its low-momentum structure is split and left behind (a close-up of this phenomenon is represented in Figure 10), causing the blockage of subsequent channel. The so-called rotating stall turns at

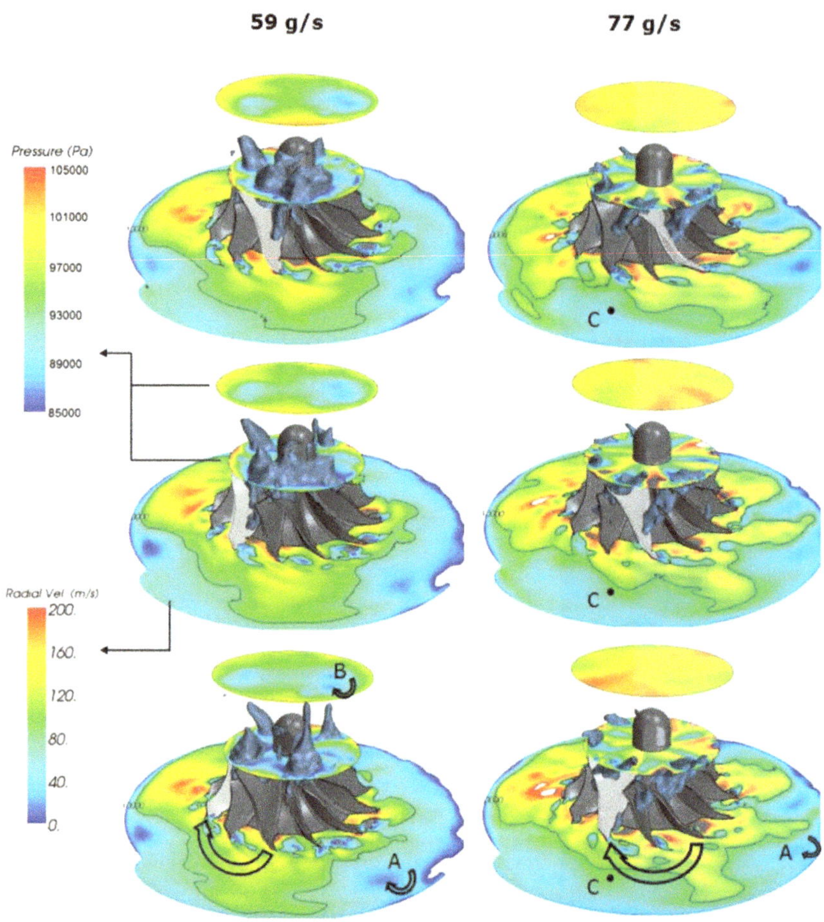

Figure 8. Sequence of snapshots of pressure contours at inducer planes and radial velocity contours at mid-span diffuser for 59 g/s (left) and 77 g/s (right) operating conditions.

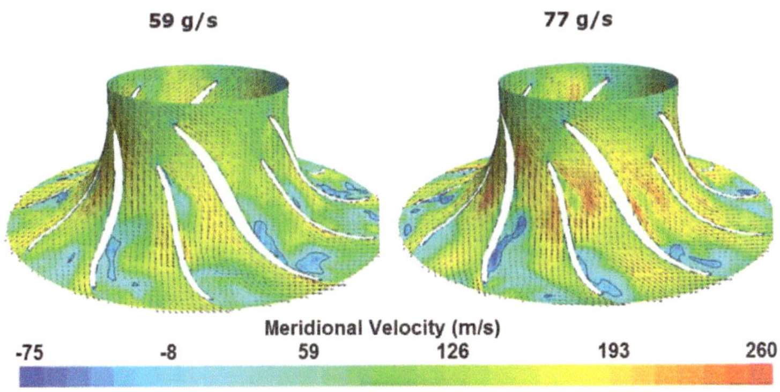

Figure 9. Meridional velocity at 50% span surfaces, at 59 g/s (left) and 77 g/s (right).

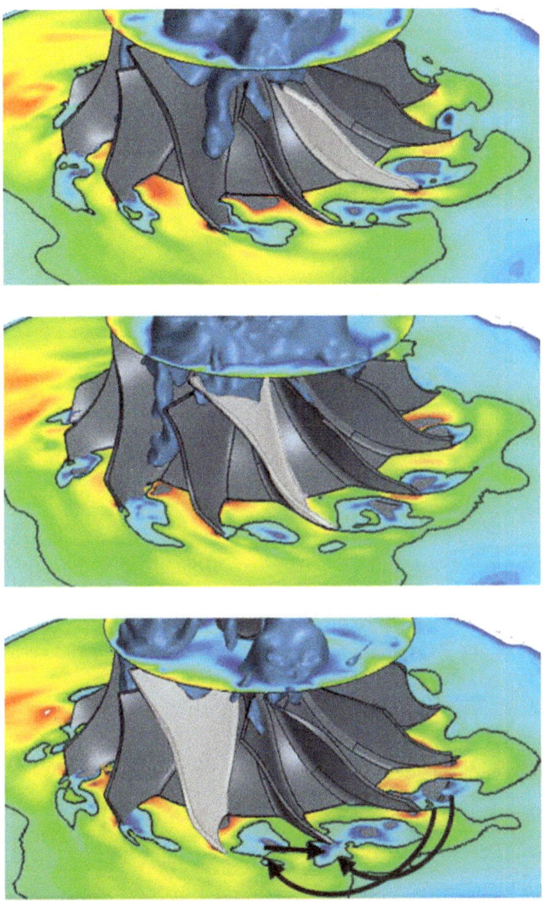

Figure 10. Rotating stall process at impeller trailing edge at 59 g/s.

approximately half the impeller speed (corresponding to 2600 Hz). This phenomenon is responsible for the rotating structures appearing at the diffuser outlet (see A point in Figure 8), which eventually will cause a fluctuating discharge into the volute and outlet duct. So far, diffuser rotating stall is the only phenomenon occurring at the frequency range of whoosh noise, so it may be responsible for this noise.

Several authors (Evans & Ward, 2006; Gaudé et al., 2008; Teng et al., 2009) attributed whoosh noise to turbulence, without going into greater detail. The phenomenon may be related to the so-called tip clearance noise described by Raitor and Neise (2008), but Galindo et al. (2015) found no influence of tip clearance ratio on whoosh noise. Simulations performed by Mendonça et al. (2012) predicted a broadband noise at 70% of impeller rotation frequency, which corresponds to the whoosh noise frequency range. Mendonça et al. ascribed the broadband feature to rotating stall, although their analysis was focused on the inducer flow behavior.

Examination of inducer flow in Figure 8 reveals some differences between operating conditions. Low-pressure regions appear due to flow detachment at the suction side, caused by a bad incidence angle at the blade leading edge. This negative incidence is cause by the low axial component of velocity compared to the rotational component (the operating conditions are close to surge).

At 59 g/s, these low-pressure bubbles are transferred between channels in a similar fashion as their diffuser counterparts. Inducer rotating stall can also be appreciated at the upstream-located cross-section (point B in Figure 8). At 77 g/s, there still exist inducer rotating stall, but with a reduced intensity. In fact, pressure at the upstream plane is quite homogeneous. It should be noted that a difference of 3 kPa exists at this plane between operating conditions because pressure is set as a boundary condition at the outlet, so predicted compression ratio dictates inlet pressure.

Diffuser rotating stall is more intense than inducer rotating stall at operating conditions not so close to surge (77 g/s), where whoosh noise is more prominent, therefore supporting the attribution of compressor whoosh noise phenomenon to diffuser rotating stall. This is in agreement with the observation made by Evans and Ward (2006): the outlet hose contribution to whoosh noise radiation is larger than that corresponding to the inlet duct.

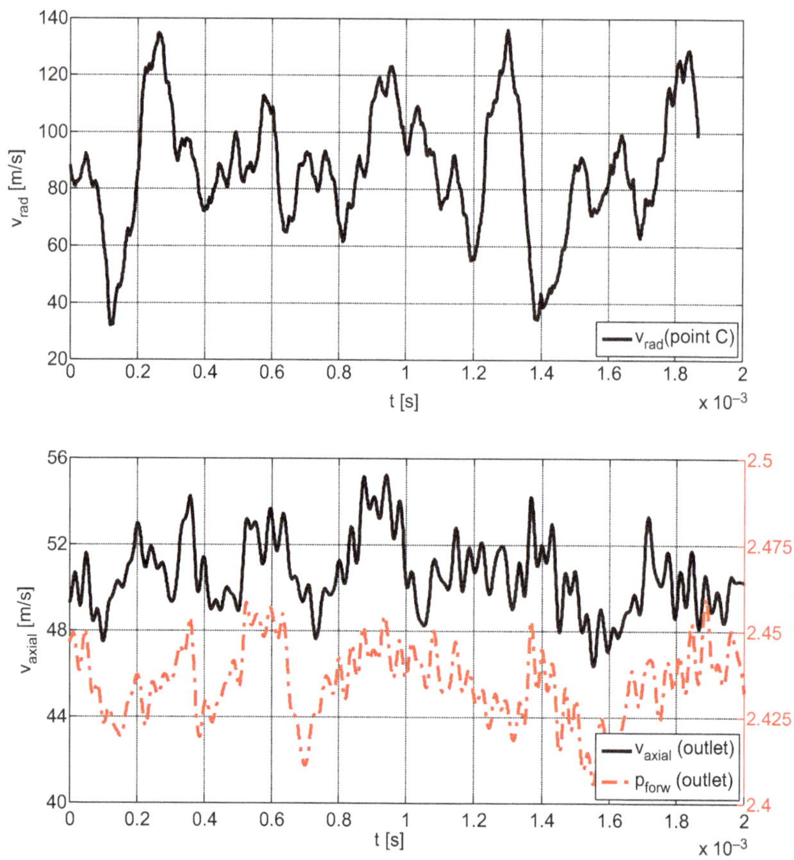

Figure 11. Temporal evolution of radial velocity at point C (top) and axial velocity and forward pressure at outlet cross section (bottom) at 77 g/s.

In order to correlate the fluctuation of radial velocity to the oscillation of pressure in the outlet duct, a point probe is placed at diffuser midspan for 77 g/s mass flow rate case (point C in Figure 8). Figure 11 (top) depicts the evolution of radial velocity at point C, covering a timespan of 2 ms after the compressor reaches a steadily cyclic state. Large fluctuations in the frequency range of whoosh noise can be appreciated. Pressure trace at point C is not representative because it includes all reflections produced at the diffuser. Figure 11 (bottom) is produced considering the outlet cross-sectional surface used in Figure 6 instead. Axial velocity presents low oscillations as in point C, but with a reduced magnitude. Forward pressure is coherent with such fluctuations, which confirms the relation between diffuser rotating stall and whoosh noise.

4. Conclusions

In this paper experiments and CFD calculations have been carried out in order to relate flow phenomena to the acoustic signature of a centrifugal compressor operating close to surge. Whoosh noise, which extends from 1 kHz to 2.5 kHz, can be more *relevant* at operating conditions not so close to surge, according to experimental measurements. Two operating conditions at same compressor speed have been examined through calculation: one near the maximum pressure ratio at 160 krpm and the other point closer to surge. Detached eddy simulations of these operating points have been performed, obtaining numerical pressure spectra in agreement with experiments.

The numerical model is used to analyze the flow at these operating conditions, with the objective of identifying aeroacoustic phenomena in the frequency range of whoosh noise (1–3 kHz). Inducer rotating stall is detected, although it is more noticeable when the compressor is working close to surge. Diffuser rotating stall, on the contrary, presents similar intensity for both simulated points, therefore it seems to be the main phenomenon to explain the existence of whoosh noise. Moreover, diffuser rotating stall is located closer to the outlet hose, in which the intensity of whoosh noise is greater. Observation of temporal traces at virtual monitors confirms the coherence between the velocity oscillations at the diffuser due to rotating stall and pressure fluctuations related to whoosh noise in the outlet duct.

The flow behavior during diffuser rotating stall is analyzed, finding flow detachment at the blade suction side, which promote the classical jet-wake structure at the impeller trailing edge. The passages are not evenly loaded, and stall cells travel in the impeller rotation direction at subsynchronous speeds corresponding to the whoosh noise frequency range.

Acknowledgement

The equipment used in this work has been partially supported by the Spanish Ministerio de Economía y Competitividad through grant no. TRA2012-36954 and by FEDER project funds "Dotación de infraestructuras científico técnicas para el Centro Integral de Mejora Energética y Medioambiental de Sistemas de Transporte (CiMeT), (FEDER-ICTS-2012-06)" framed in the operational program of unique scientific and technical infrastructure of the Spanish Ministerio de Economía y Competitividad.

References

Baris, O., & Mendonça, F. (2011, Jun 6–10). *Automotive turbocharger compressor CFD and extension towards incorporating installation effects*. Proceedings of ASME Turbo Expo 2011: Power for Land, Sea and Air, Vancouver, Canada.

Broatch, A., Galindo, J., Navarro, R., & García-Tíscar, J. (2014). Methodology for experimental validation of a CFD model for predicting noise generation in centrifugal compressors. *International Journal of Heat and Fluid Flow*, *50*, 134–144.

CD-adapco. (2013). *STAR-CCM+ 8.02*, Retrieved from http://www.cd-adapco.com

Evans, D., & Ward, A. (2006, May 17–18). *The reduction of turbocharger whoosh noise*. Proceedings of the 8th International Conference on Turbochargers and Turbocharging, London, UK, pp. 29–42.

Figurella, N., Dehner, R., Selamet, A., Tallio, K., Miazgowicz, K., & Wade, R. (2012, August 19–22). *Noise at the mid to high flow range of a turbocharger compressor*. Proceedings of the 41st International Congress and Exposition on Noise Control Engineering 2012, New York, pp. 786–797.

Galindo, J., Hoyas, S., Fajardo, P., & Navarro, R. (2013). Set-up analysis and optimization of CFD simulation for radial turbines. *Engineering Applications of Computational Fluid Mechanics*, *7*(4), 441–460.

Galindo, J., Tiseira, A., Navarro, R., & López, M. (2015). Influence of tip clearance on flow behaviour and noise generation of centrifugal compressors in near-surge conditions. *International Journal of Heat and Fluid Flow*, *52*, 129–139.

Galindo, J., Serrano, J. R., Guardiola, C., & Cervelló, C. (2006). Surge limit definition in a specific test bench for the characterization of automotive turbochargers. *Experimental Thermal and Fluid Science*, *30*(5), 449–462.

Gaudé, G., Lefèvre, T., Tanna, R., Jin, K., McKitterick, T. J. B., & Armenio, S. (2008, October 26–29). *Experimental and computational challenges in the quantification of turbocharger vibro-acoustic sources*. Proceedings of the 37th International Congress and Exposition on Noise Control Engineering 2008, Shanghai, China, pp. 5598–5611.

Karim, A., Miazgowicz, K., Lizotte, B., & Zouani, A. (2013). Computational aero-acoustics simulation of compressor whoosh noise in automotive turbochargers. SAE Paper 2013-01-1880.

Knecht, W. (2008). Diesel engine development in view of reduced emission standards. *Energy*, *33*(2), 264–271.

Lee, I., Selamet, A., Kim, H., Kim, T., & Kim, J. (2009). Design of a multi-chamber silencer for turbocharger noise. *SAE International Journal of Passenger Cars – Mechanical Systems*, *2*(1), 1339–1344.

Mendonça, F., Baris, O., & Capon, G. (2012, June 11–15). *Simulation of radial compressor aeroacoustics using CFD*. Proceedings of ASME Turbo Expo 2012, Copenhaguen, Denmark.

Morfey, C. (1971). Sound transmission and generation in ducts with flow. *Journal of Sound Vibration*, *14*(1), 37–55.

Piñero, G., Vergara, L., Desantes, J. M., & Broatch, A. (2000). Estimation of vellocity fluctuation in internal combustion engine exhaust systems through beamforming techniques. *Measurement Science & Technology*, *11*, 1585–1595.

Raitor, T., & Neise, W. (2008). Sound generation in centrifugal compressors. *Journal of Sound and Vibration*, *314*, 738–756.

Semlitsch, B., JyothishKumar, V., Mihaescu, M., Fuchs, L., Gutmark, E., & Gancedo, M. (2014). Numerical flow analysis of a centrifugal compressor with ported and without ported shroud. *SAE Paper* 2014-01-1655.

Serrano, J. R., Margot, X., Tiseira, A., & García-Cuevas, L. M. (2013). Optimization of the inlet air line of an automotive turbocharger. *International Journal of Engineering Research*, *14*(1), 92–104.

Sevginer, C., Arslan, M. O., Sonmez, N., & Yilmaz, S. I. (2007, August 28–31). *Investigation of turbocharger related whoosh and air blow noise in a diesel powertrain*. Proceedings of the 36th International Congress and Exposition on Noise Control Engineering 2007, Instanbul, Turkey, pp. 476–485.

Teng, C., & Homco, S. (2009). Investigation of compressor whoosh noise in automotive turbochargers. *SAE International Journal of Passenger Cars – Mechanical Systems*, *2*(1), 1345–1351.

Torregrosa, A. J., Fajardo, P., Gil, A., & Navarro, R. (2012). Development of a non-reflecting boundary condition for its application in 3D computational fluid dynamic codes. *Engineering Applications of Computational Fluid Mechanics*, *6*(3), 447–460.

Tritthart, M., & Gutknecht, D. (2013). Three-dimensional simulations of free-surface flows using polyhedral finite volumes. *Engineering Applications of Computational Fluid Mechanics*, *1*(1), 1–14.

3

Evaluating wave forces on groups of three and nine cylinders using a 3D numerical wave tank

Arun Kamath*, Mayilvahanan Alagan Chella, Hans Bihs and Øivind A. Arntsen

Department of Civil and Transport Engineering, Norwegian University of Science and Technology, Trondheim, Norway.

The evaluation of the complex wave regime due to wave interaction with a large group of cylinders placed in proximity requires an efficient and accurate numerical model. This paper presents the application of a two-phase Computational Fluid Dynamics (CFD) model to carry out a detailed investigation of wave forces and flow around vertical circular cylinders placed in groups of different configurations at low Keulegan-Carpenter (KC) numbers. The 3D numerical wave tank is validated by comparing the numerical results with experimental data. Further, the hydrodynamic effects associated with three cylinders placed in tandem, side by side and in a 3 × 3 square array of nine cylinders are investigated. Wave forces are seen to reduce along the row in a tandem array. In a side-by-side arrangement, the central cylinder experiences the highest force. A combination of these effects is seen in the 3 × 3 square array. The variation of the wave forces on the cylinders in the array for different center-to-center distances and incident wavelengths is evaluated and the results show that the wave forces are the highest on the cylinders when the center-to-center distance is slightly less than half the incident wavelength.

Keywords: CFD; wave forces; multiple cylinders; numerical wave tank; level set method

1. Introduction

Cylindrical structures are commonly seen in marine environments as support structures for oil and gas platforms and offshore wind turbines. The study of wave interaction with a group of cylinders and the accurate assessment of the wave forces acting on each of the cylinders is important for design considerations. The ratio of the diameter of a single cylinder to the wavelength of the incident waves leads to different wave force regimes on the cylinder, which is determined by the Keulegan-Carpenter number $KC = UT/D$, where U is the amplitude of the horizontal water particle velocity, T is the wave period and D is the diameter of the cylinder. When $KC > 2$, the total force on a single cylinder is calculated using the Morison formula (Morison, O'Brien, Johnson, & Schaaf, 1950) as a sum of the inertia and drag forces. The formula uses experimentally determined coefficients to calculate the total wave force on the cylinder. When $KC < 2$, the wave forces are inertia-dominated and the MacCamy-Fuchs theory (MacCamy & Fuchs, 1954) is used to determine the total wave force on a single cylinder. In this force regime, the effect of wave diffraction is important and the incident waves are scattered by the cylinder. In the case of multiple vertical cylinders in a diffraction regime, the incident waves and the diffracted waves are scattered multiple times, leading to complex wave interactions in the vicinity of the cylinders. This phenomenon of multiple scattering from the different cylinders influences the wave forces on each of the cylinders in the group. Ohkusu (1974) proposed an iterative method to evaluate the successive scattering from floating structures in a group to determine the wave force experienced by each structure. The velocity potential functions used to evaluate the successive scattering become complicated as the number of structures increases. Spring and Monkmeyer (1974) presented a method with all the boundary conditions imposed at once to obtain a matrix equation to solve the problem. Linton and Evans (1990) improved on Spring and Monkmeyer by proposing a method based on potential theory to calculate the wave forces on each of the cylinders placed in a group. This method is employed in this paper for the comparison of numerical results in the case of multiple cylinders.

Several studies have presented numerical wave tanks based on boundary integral equations (Isaacson 1982; Kim, Liu, & Ligget, 1983; Longuet-Higgins & Cokelet, 1976) and Boussinesq equations (Nwogu, 1993; Wei & Kirby, 1995). The numerical wave tanks based on boundary integral equations are limited in their application due to being single phase and due to the assumption of the irrotational flow. Boussinesq equations are depth-averaged, two-dimensional and valid only in intermediate and shallow waters. The Navier-Stokes equations include three-dimensional flow features, the effects of rotational flow, viscosity and higher order non-linear wave

*Corresponding author. Email: arun.kamath@ntnu.no

interactions between the cylinders. Computational Fluid Dynamics (CFD) simulations provide a great amount of detail concerning the free surface around obstacles in the flow, the pressure and forces acting on an object, and the velocity of fluid particles. With an accurate representation of the free surface, the various flow features can be studied, including wave diffraction and the evolution of the diffracted and reflected waves and their influence on the wave forces experienced by the objects. In the current literature, CFD methods have been used to investigate wave propagation in coastal regions (Jacobsen, Fuhrman, & Fredsøe, 2011; Morgan et al., 2010), wave interaction with coastal structures (del Jesus et al., 2012; Higuera, Lara, & Losada, 2014) and modeling of breaking waves (Alagan Chella, Bihs, Myrhaug, & Muskulus, 2015). Two-phase models are essential for obtaining a good representation of free surface flows due to the interaction of air and water at the interface (Liu & Yang, 2014). Numerical methods such as Smoothed Particle Hydrodynamics (SPH) have also been used to model the interaction of fluids with structures (Pu, Shao, Huang, & Hussain, 2013). Three-dimensional numerical modeling of flow and transport over large domains can provide insights into the large-scale phenomena which can have a significant impact on the engineering solutions in the field of river, estuarine and coastal engineering and management (Chau & Jiang, 2001, 2004).

The objective of this paper is to obtain insights into the physical processes and hydrodynamic features associated with wave interaction with multiple cylinders placed in groups in different configurations using the open-source CFD model REEF3D (Alagan Chella, Bihs, Myrhaug, & Muskulus, 2015). CFD can deliver deep insight into the underlying flow physics and wave kinematics, and produce accurate and detailed results without the assumptions and simplifications of typical wave models. The evaluation of the complex wave diffraction regime in the case of multiple cylinders placed in proximity is challenging due to the multiple wave reflection and interference effects involved. The center-to-center distance between the cylinders is also an important factor influencing the wave forces, and lower-order methods using potential theory cannot account for the hydrodynamics when the cylinders are spaced very close together.

The model is validated by comparing the numerical results with data from the large-scale experiments carried out in the Large Wave Flume (GWK) in Hannover, Germany (Mo, Irschik, Oumeraci, & Liu, 2007) for the wave kinematics and the wave forces on a single cylinder. Further, numerical simulations are carried out with three cylinders in tandem and three cylinders placed side by side. The final case is a combination of the two arrangements, with nine cylinders placed in a 3 × 3 square array. The effect of the center-to-center distance and the incident wavelength on the wave forces experienced by each of the cylinders in the array is studied.

2. Numerical model: REEF3D

The incompressible Reynolds-Averaged Navier-Stokes (RANS) equations are used to solve the fluid flow problem:

$$\frac{\partial U_i}{\partial x_i} = 0 \tag{1}$$

$$\frac{\partial U_i}{\partial t} + U_j \frac{\partial U_i}{\partial x_j} = -\frac{1}{\rho}\frac{\partial P}{\partial x_i} + \frac{\partial}{\partial x_j}\left[(\nu + \nu_t)\left(\frac{\partial U_i}{\partial x_j} + \frac{\partial U_j}{\partial x_i}\right)\right] + g_i \tag{2}$$

where U_i is the time averaged velocity, ρ is the density of water, P is the pressure, ν is the kinematic viscosity, ν_t is the eddy viscosity, t is time and g is the acceleration due to gravity. The pressure is treated using Chorin's projection method (Chorin, 1968) and the resulting Possion pressure equation is solved using a preconditioned BiCGStab solver (van der Vorst, 1992). Turbulence modeling is carried out using the two-equation $k-\omega$ model proposed by Wilcox (1994). Eddy viscosity, ν_t, is bounded to avoid unphysical overproduction of turbulence in strained flow by using a stress limiter in the definition of eddy viscosity (Bradshaw, Ferriss, & Atwell, 1967) as shown by Durbin (2009). The large difference in density of air and water in a two-phase model leads to an overproduction of turbulence at the interface due to the large strain. Free surface turbulence damping using a limiter around the interface for the source terms as shown by Naot and Rodi (1982) is carried out to avoid the unphysical overproduction of turbulence at the interface. The damping is carried out only around the interface using the smoothed Dirac delta function.

The fifth-order conservative finite difference Weighted Essentially Non-Oscillatory (WENO) scheme proposed by Jiang and Shu (1996) is used for the discretization of convective terms for the velocity U_i, the level set function ϕ, the turbulent kinetic energy k and the specific turbulent dissipation rate ω. A Total Variation Diminishing (TVD) third-order Runge-Kutta explicit time scheme developed by Harten (1983) is employed for time discretization in the model. This is a three-step scheme and involves the calculation of the spatial derivatives three times per time step. This scheme is used for the time advancement of the level set function and the re-initialization equation. An adaptive time stepping approach is used to maintain the time step in accordance with the Courant-Friedrichs-Lewy (CFL) criterion for numerical stability (Griebel, Dornsheifer, & Neunhoeffer, 1998).

The free surface is obtained using the level set method where the zero level set of a signed distance function, $\phi(\vec{x}, t)$ is used to represent the interface between air and water (Osher & Senthian, 1988). Moving away from the interface, the level set function gives the closest distance of the point from the interface. The sign of the function represents the two fluids across the interface, as shown in

Equation (3):

$$\phi(\vec{x},t)\begin{cases} >0 & \text{if } \vec{x} \text{ is in air} \\ =0 & \text{if } \vec{x} \text{ is at the interface} \\ <0 & \text{if } \vec{x} \text{ is in water} \end{cases} \quad (3)$$

The velocity field in the simulation moves the level set function using the convection equation:

$$\frac{\partial \phi}{\partial t} + U_j \frac{\partial \phi}{\partial x_j} = 0 \quad (4)$$

The level set method is re-initialized after every iteration using a partial differential equation (PDE) based re-initialization procedure presented by Sussman et al. (1994):

$$\frac{\partial \phi}{\partial t} + S(\phi)\left(\left|\frac{\partial \phi}{\partial x_j}\right| - 1\right) = 0 \quad (5)$$

where $S(\phi)$ is the smoothed sign function by Peng, Merriman, Osher, Zhao, and Kang (1999) to retain its signed distance property after convection:

$$S(\phi) = \frac{\phi}{\sqrt{\phi^2 + \left|\frac{\partial \phi}{\partial x_j}\right|^2 (\Delta x)^2}} \quad (6)$$

The jump in density and viscosity at the interface is handled with the regularized Heaviside function over a thickness of $2.1\Delta x$ around the interface.

A uniform Cartesian grid is used in the numerical model for spatial discretization. The Immersed Boundary Method (IBM; Peskin, 1972) is used to incorporate the boundary conditions for complex geometries. In the current study, a local directional ghost cell IBM presented by Berthelsen and Faltinsen (2008) is implemented using object-oriented programming techniques, where the ghost cell values can be updated from multiple directions (Bihs, 2011). The ghost cells store multiple values and return a particular value when called from the respective direction.

The relaxation method (Mayer, Garapon, & Sørensen, 1998) is used for wave generation and absorption in the numerical wave tank. This method uses a wave generation zone and a wave absorption zone to moderate the computational values and the analytical values given by wave theory. The relaxation function proposed by Jacobsen et al. (2012) presented in Equation (7) is used in this study:

$$\Gamma(x) = 1 - \frac{\exp(x^{3.5}) - 1}{\exp(1) - 1} \text{ for } x \in [0;1] \quad (7)$$

where x is the coordinate along the x-axis scaled to the length of the relaxation zone. In the wave generation zone, the values for the free surface elevation and the velocity given by wave theory are prescribed to the computational values using the relaxation function $\Gamma(x)$, as shown in Equation (8), and the generated waves are released into the working zone of the wave tank. In the wave absorption zone, the relaxation function is operated as $\Gamma(1-x)$. The computational values of velocity and free surface from the working zone are reduced to zero and the wave energy is smoothly removed from the computational domain.

$$\begin{aligned} U_{\text{relaxed}} &= \Gamma(x)\, U_{\text{analytical}} + (1 - \Gamma(x))\, U_{\text{computational}} \\ \phi_{\text{relaxed}} &= \Gamma(x)\, \phi_{\text{analytical}} + (1 - \Gamma(x))\, \phi_{\text{computational}} \end{aligned} \quad (8)$$

The free surface, horizontal and vertical water particle velocities for linear wave theory are given by Equations (9), (10) and (11) respectively:

$$\eta = A \sin(\omega t - \kappa x) \quad (9)$$

$$u = \omega A \frac{\cosh \kappa(z+d)}{\sinh \kappa d} \sin(\omega t - \kappa x) \quad (10)$$

$$w = \omega A \frac{\sinh \kappa(z+d)}{\sinh \kappa d} \cos(\omega t - \kappa x) \quad (11)$$

where A is the wave amplitude, κ is the wave number $= 2\pi/L$, L is the wavelength, ω is the angular wave frequency and d is the water depth. Typically, the wave generation zone is one wavelength long and the absorption zone is two wavelengths long. The computational efficiency of the program is increased by parallelizing the code using a Message Passing Interface (MPI). Here the domain is decomposed into smaller pieces and each assigned to a processor. So the program runs separately on each processor and the values between the processes are communicated using the MPI library.

3. Calculation of wave forces

3.1. *Calculation of wave forces in the numerical model*

The wave force on an object is calculated as the integral of the pressure and the shear stress around the surface of the object as follows:

$$F = \int_{\Omega} (-\vec{n}\mathrm{P} + \vec{n} \cdot \tau)\, d\Omega \quad (12)$$

where $\vec{n}$ and τ are a unit normal vector and a viscous stress tensor, respectively. This is easily accomplished in the numerical model as the values for the pressure at every point in the domain are available from the solution of the Poisson pressure equation. No-slip boundary conditions are imposed on the boundary between the object and the fluid domain.

3.2. *Analytical formula for calculating wave forces on multiple cylinders*

Wave forces on cylinders in an inertia-dominated force regime ($KC < 2$) can be calculated using the MacCamy-Fuchs formula (MacCamy & Fuchs, 1954), which is based on potential theory:

$$F_0 = \left| \frac{4\rho giA\tanh(\kappa d)}{\kappa^2 H_1'(\kappa a)} \right| \quad (13)$$

where F_0 is the inline force on a single cylinder, $i = \sqrt{-1}$, d is the water depth, a is the radius of the cylinder and H_1' is the derivative of the Hankel function of the first kind.

Linton and Evans (1990) presented a theory to evaluate the wave forces experienced by each of the N cylinders placed in an array by solving a system of $N(2M + 1)$ unknowns, where M is order of the solution:

$$A_m^k + \sum_{j=1, j\neq k}^{N} \sum_{n=-M}^{M} A_n^j Z_n^j e^{i(n-m)\alpha_{jk}} H_{n-m}(\kappa R_{jk}) = -I_k e^{im(\pi/2-\beta)},\ k = 1,\ldots,N, m = -M,\ldots,M. \quad (14)$$

where A_m^k and A_n^j are unknown coefficients of the scattered radiating wave from the kth and the jth cylinders, $Z_n^j = J_n'(\kappa a)/H_n'(\kappa a)$, where κ is the wave number, J' is the derivative of the Bessel function of the first kind, α_{jk} is the angle between the x-axis and the line connecting the centers of cylinders j and k, R_{jk} is the center-to-center distance of the cylinders j and k, I_k is the incident wave potential on cylinder k and β is the angle of wave incidence with the x-axis.

$$\left| \frac{F^j}{F_0} \right| = \frac{1}{2} |A_{-1}^j \pm A_1^j| \quad (15)$$

where F^j is the force on the jth cylinder. The subtraction of the coefficients in Equation (15) gives the force on the cylinder along the x-axis and the addition of the coefficients gives the force along the y-axis.

4. Results and discussion

4.1. *Validation of wave force calculation*

A numerical simulation is carried out to determine the wave force on a single cylinder due to a regular wave. The wave parameters from the Large Wave Flume (GWK) experiments (Mo et al., 2007) are used in this simulation. The numerical wave tank is 132 m long, 5 m wide and 8 m high. A regular wave of height 1.2 m and wave period $T = 4.0$ s is generated in a water depth of $d = 4.76$ m. A cylinder of diameter $D = 0.7$ m is placed 44 m from the wave generation zone to provide sufficient length for the waves to propagate (Figure 1) and avoid the influence of waves reflected from the cylinder on the incident waves. The grid resolution required to represent the cylinder geometry in the domain and accurately compute the wave forces is quantified by the number of cells contained in a square with a side equal to the diameter of the cylinder. By this definition, a grid size of 0.1 m is used with a density of 49 cells/D^2 and 5.28 million cells in the domain. Wall boundary conditions are enforced on the side walls and bottom of the numerical domain and the symmetry plane boundary condition is applied at the top.

The free surface elevation is measured near the wall along the front line of the cylinder (WG 1), in front of the cylinder (WG 2), at the side (WG 3) and behind the cylinder (WG 4), as shown in Figure 1. The water particle velocities near the wall along the frontline of the cylinder are measured at the depths of $z = -0.93$ m, -1.53 m and -2.73 m in experiments using Acoustic Doppler Velocimeters (ADVs).

The wave forces acting on the cylinder are calculated and compared to the experimental results (Figure 2(a)) and the numerical results are seen to match the experimental observations very well. A grid refinement study is carried out for the wave forces and simulations are carried out with 49 cells/D^2, 22 cells/D^2 and 12 cells/D^2 (Figure 2(b)) and it is seen that the numerical model calculates the wave forces accurately from 49 cells/D^2 onwards. Mo et al. (2007) reported that the measured wave forces corresponded to the force calculated from the inertia part of the Morison equation. So, this case is inertia-dominated with low influence from the viscous boundary layer. The numerical model employs wall layer turbulence modeling and this is

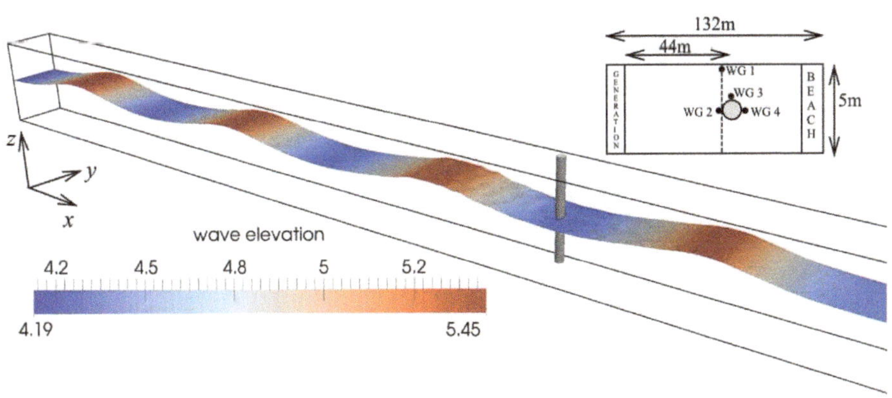

Figure 1. Set-up for calculating forces on a single cylinder.

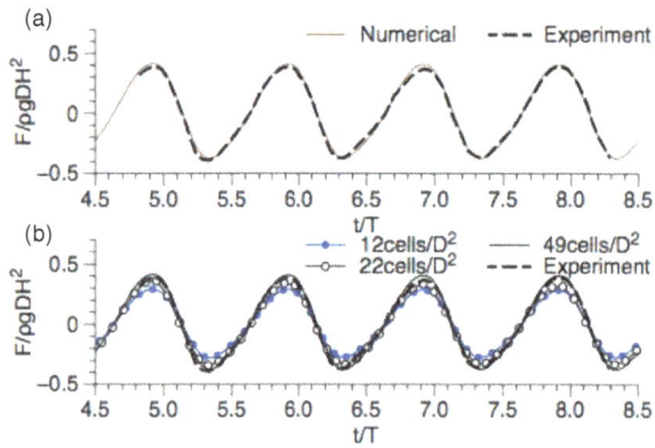

Figure 2. Wave forces on a single cylinder: (a) comparison of experimental and numerical results; (b) grid convergence study for wave forces.

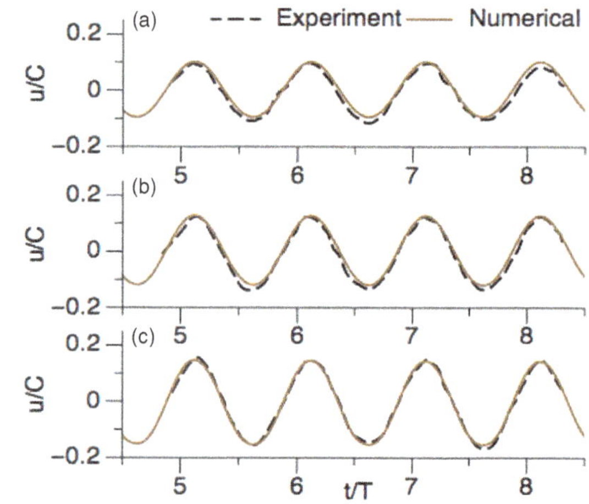

Figure 4. Comparison of experimental and numerical results for water particle velocities in front of the cylinder: (a) at $z = -0.93$ m; (b) at $z = -1.53$ m; (c) at $z = -2.73$ m.

sufficient to account for the hydrodynamics close to the surface of the cylinder in this case. This is also the case for the simulations carried out in the following sections, where the forces are inertia-dominated due to the low KC numbers.

The free surface elevation near the wall along the front line of the cylinder is compared with the experimental data (Figure 3(a)). The amplitude of the wave near the wall along the front line of the cylinder is taken as $\eta_{max,wall}$. The comparisons of the free surface in front, at the side and behind the cylinder are presented in Figure 3(b), 3(c) and 3(d), respectively. The rise in the water level in front of the cylinder recorded in the experiments is seen in the numerical results as well. The water elevations beside and behind the cylinder also show a good match with the experimental observations. Further, the fluid velocity computed by the numerical model is compared with the experimental results. The values are scaled with wave celerity C and are presented in Figure 4; a good agreement is seen between the numerical and experimental results. Thus, it is concluded that the numerical model produces a good representation of the wave elevation, water particle velocity and wave force.

4.2. *Wave interaction with multiple cylinders*

4.2.1. *Wave interaction with three cylinders placed in tandem*

Three cylinders of diameter $D = 0.26$ m are placed in tandem with a center-to-center distance $S = 0.8$ m (Figure 5). Linear waves of amplitude $A = 0.03$ m, wavelength $L = 2.0$ m and wave period $T = 1.18$ s ($KC = 0.79$) in a water depth $d = 0.5$ m are generated in a wave tank 15 m long, 5 m wide and 1 m high, with a grid resolution of 108 cells/D^2 around each cylinder, resulting in 4.8 million cells. The width of the tank is selected such that reflections from the side wall do not reach the region around the cylinders by ensuring a distance of $9D$ between the wall and the cylinder closest to the wall. The clear distance between the surfaces of the neighboring cylinders in the set-up is $0.27L$ and the wave diffraction and interference effects over a quarter of the incident wavelength are studied. The computed wave forces on the three cylinders are compared with the predictions from the analytical formula (Equation 15)

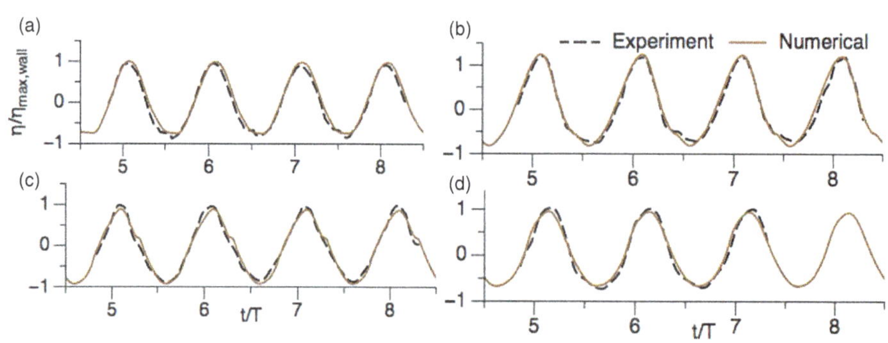

Figure 3. Comparison of experimental and numerical results for wave elevations around the cylinder: (a) at WG 1; (b) at WG 2; (c) at WG 3; d) at WG 4.

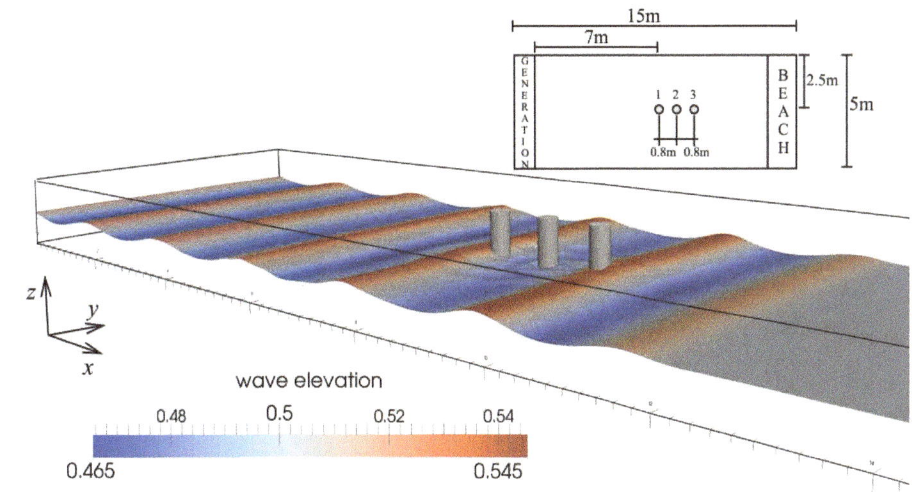

Figure 5. Set-up for three cylinders placed in tandem.

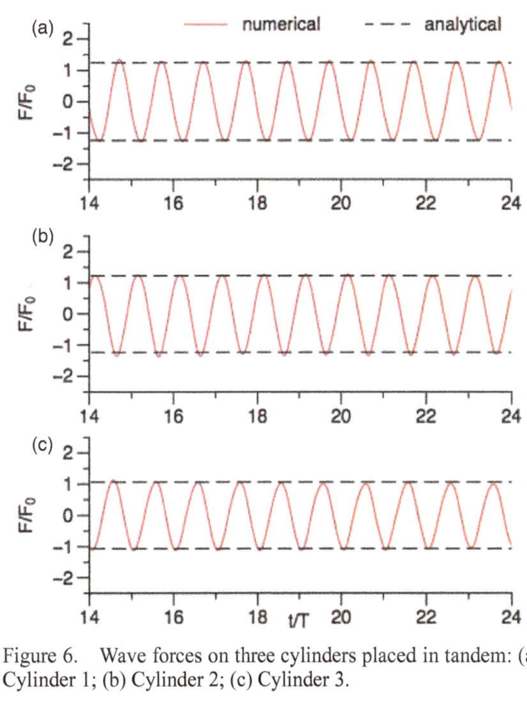

Figure 6. Wave forces on three cylinders placed in tandem: (a) Cylinder 1; (b) Cylinder 2; (c) Cylinder 3.

and a good agreement is seen in Figure 6. Cylinders 1 and 2 experience a similar force of $1.24F_0$ and $1.23F_0$ and the downstream cylinder experiences a force of $1.06F_0$. The interaction of the incident waves and the diffracted waves from the cylinders results in the wave pattern seen in Figure 7 for $t/T = 9.0$, 9.25, 9.50 and 9.75. A circular diffracted wave pattern is noticed in the vicinity of Cylinder 1, which decays as it propagates away from the cylinder. In an inertia-dominated force regime, the pressure around the cylinders influences the forces acting on the cylinders, which can be studied using the free surface elevations. In Figure 7(a), Cylinder 1 is directly exposed to the incident wave crest in the front and a diffracted wave trough is behind the cylinder. This leads to a larger net difference in the pressure around the cylinder, resulting in a higher force compared to a single cylinder. The constructive interference seen in Figure 7(b) results in a larger wave elevation behind Cylinder 1, when a wave trough is incident on it in the front. Thus, Cylinder 1 is always subject to large differences in pressure and it experiences a force that is larger than the force experienced by a single cylinder. Cylinder 2 is exposed to scattered waves from both the upstream Cylinder 1 and the downstream Cylinder 3. This counteracts the shadowing of the incident waves by Cylinder 1, and Cylinder 2 experiences a force higher than the force on a single cylinder. A similar interaction is noted between Cylinders 2 and 3, but Cylinder 3 experiences a lower force of $1.06F_0$ because it is largely exposed only to the scattered waves from upstream Cylinders 1 and 2. It is also clearly seen from Figure 7(c) and 7(d) that the diffracted waves from Cylinder 1 radiate away along the y-axis and the diffracted pattern along x-axis is extended until the vicinity of Cylinder 3. The diffracted waves from the cylinders upstream interact with each other and the diffracted waves from Cylinder 3, resulting in a slightly higher force than on a single cylinder on Cylinder 3. The total inline wave forces are seen to decrease along the direction of wave propagation. As the wave crest is incident on Cylinder 1, the flow is obstructed and the waves are scattered around the cylinder, resulting in a downstream pressure gradient. A diffracted wave is radiated from Cylinder 1, which propagates further downstream and interacts with the diffracted waves radiated from Cylinder 2. This complex interaction between the cylinders leads to higher forces on both Cylinders 1 and 2 compared to the forces on a single cylinder.

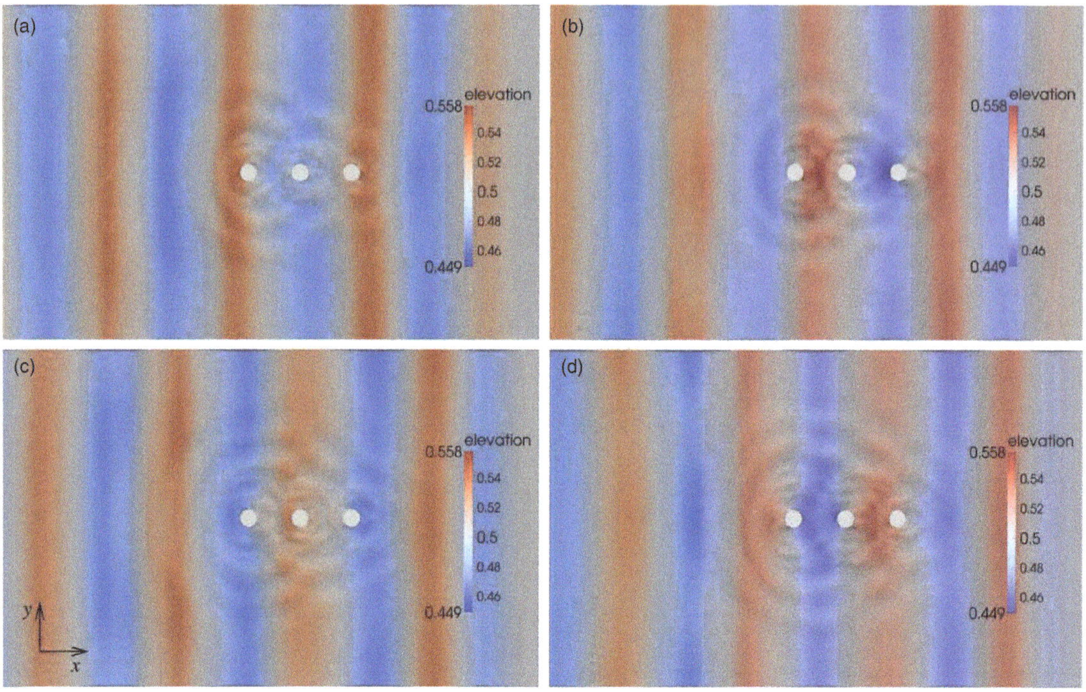

Figure 7. Wave elevations around the cylinders showing wave diffraction patterns: (a) $t/T = 9.0$; (b) $t/T = 9.25$; (c) $t/T = 9.50$; (d) $t/T = 9.75$.

4.2.2. *Wave interaction with three cylinders placed side by side*

In this case, three cylinders of diameter $D = 0.26$ m are placed next to each other with a center-to-center distance of $S = 0.3$ m. A small center-to-center distance is chosen to magnify the effects of flow obstruction caused by the cylinders. Regular linear waves of amplitude $A = 0.03$ m, wavelength $L = 2.0$ m and wave period $T = 1.18$ s ($KC = 0.79$) in a water depth $d = 0.5$ m are generated in a wave tank 15 m long, 7 m wide and 1 m high, with a grid resolution of 108 cells/D^2 around each cylinder, resulting in 6.72 million cells.

The cylinders on either side experience the same force of $1.31F_0$, due to the symmetrical arrangement of the cylinders. The central cylinder experiences the maximum force of $1.58F_0$ in the array and the numerical results in Figure 8 show a good match between the computed wave forces and the results from the analytical solution. The contraction effect due to the small distance of separation $S = 0.3$ m between the centers of the cylinders seen in Figure 9 results in a large pressure difference in front and behind the cylinders. This is seen through the increase in the water level in front of the cylinders and lower water elevations behind the cylinders. The distinct increase in water elevation in front of Cylinder 2 during the incidence of the crest and the propagation of the elevated water level through the space around the cylinder is clearly seen in Figure 9. The circular diffracted waves radiating from the group of cylinders is also observed. The result of the flow obstruction is a large pressure in front of the cylinders and a much lower pressure behind them. This large difference in the net pressure leads to a 31% larger force on Cylinders 1 and 3 (Figure 9(b) and 9(f)) and 58% larger force on

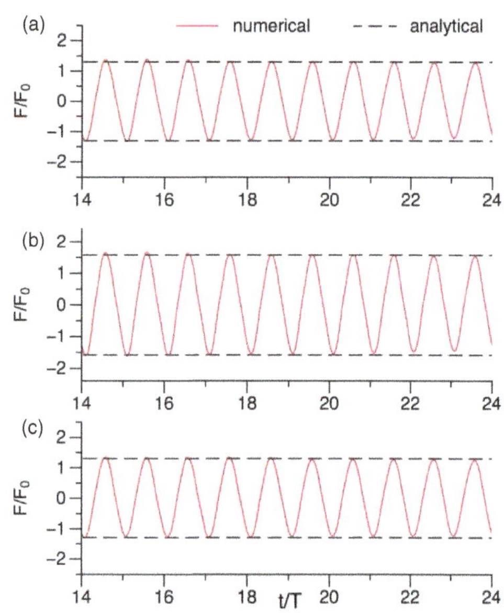

Figure 8. Wave forces on cylinders placed side-by-side: (a) Cylinder 1; (b) Cylinder 2; (c) Cylinder 3.

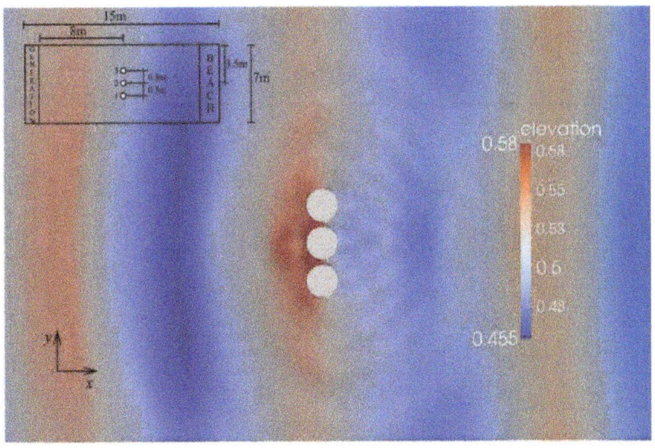

Figure 9. Set-up for three cylinders placed side by side and wave elevation contours around the cylinders.

Cylinder 2 (Figure 9(d)) in comparison to the total force on a single cylinder.

4.2.3. *Wave interaction with a 3 × 3 square array of nine cylinders*

In this case, a 3 × 3 square array of cylinders of diameter $D = 0.26$ m (Figure 10) was placed in a wave tank 16 m long, 7 m wide and 1 m high to study the wave interaction with a large array of cylinders (Figure 11). Waves of amplitude $A = 0.03$ m, wavelength $L = 2.0$ m and wave period $T = 1.18$ s ($KC = 0.79$) in a water depth of $d = 0.5$ m are incident on the cylinder array. The computational domain consists of a grid resolution of 64 cells/D^2 around each cylinder with $S = 0.8$ m. Figure 11 shows the wave forces on cylinders and wave surface elevation at $t/T = 11$, 11.25 and 11.5 (Figure 11(a)–11(c)) in the part of the computational domain around the cylinder array. The computed wave forces on the cylinders are in good agreement with the results from the analytical formula. The wave forces on Cylinders 1, 2 and 3 are the same as the wave forces on Cylinders 7, 8 and 9 due to the symmetrical arrangement. Figure 11(a)–11(c) illustrates the interaction of the waves with a 3 × 3 square array of cylinders and the resulting wave patterns due to the incident, diffracted and reflected waves in the computational domain. An increase in the free surface elevation in front of each of the columns of the arrangement is seen in Figure 11(a)–11(c) similar to the increase seen in the case of a single column in section 4.2.2. It is also observed that the increase in the water elevation in front of the cylinders is reduced as the wave progresses downstream. The formation of diffracted waves of large radii is seen behind the first upstream column, showing that the strongest diffracted waves propagate between the first and the second columns in the array. This is similar to the phenomenon observed in section 4.2.1 where the strongest diffracted waves were seen to propagate between the first and second tandem cylinders.

The radiating diffracted waves from cylinders influence the wave forces on the cylinders besides and diagonally across from each cylinder. The central cylinder (Cylinder 4) in the front row experiences the highest force in the array of $1.46F_0$, due to the combined influence of the cylinders beside and behind it (Figure 11(g)). The other two cylinders in the front row (Cylinders 1 and 7) experience a force of $1.34F_0$ (Figure 11(d)). The results are consistent with the case of three cylinders side by side, presented in section 4.2.2, where the central cylinder experienced the maximum force in the column and the two cylinders on either side had lower forces acting on them. Cylinders 2 and 8 in the

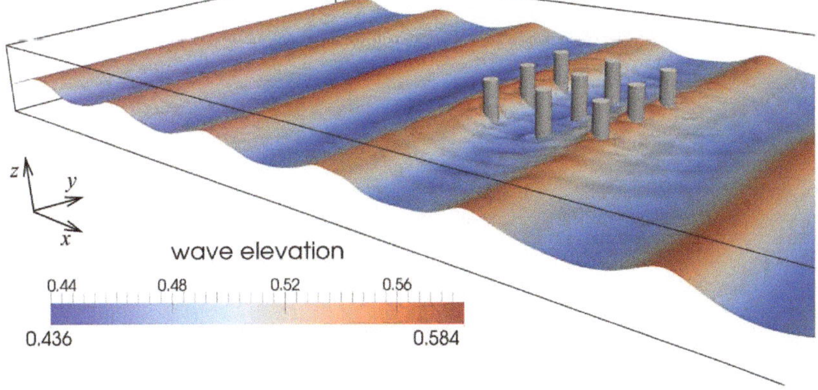

Figure 10. Set-up for nine cylinders placed in a 3 × 3 square array.

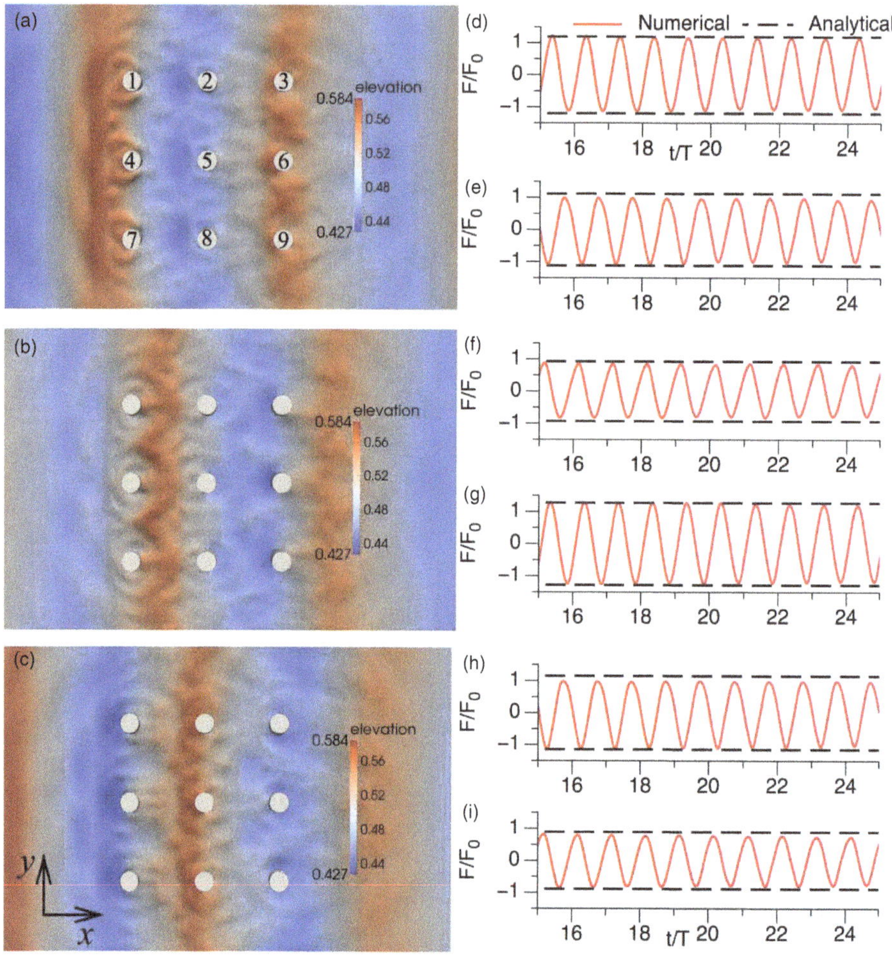

Figure 11. Wave elevations around the cylinders: (a) at $t/T = 11.0$; (b) at $t/T = 11.25$; (c) at $t/T = 11.50$; and wave forces on (d) Cylinder 1; (e) Cylinder 2; (f) Cylinder 3; (g) Cylinder 4; (h) Cylinder 5; (i) Cylinder 6.

second row experience a force of $1.15F_0$ (Figure 11(e)) due to the diffraction effects from the cylinders upstream and downstream. Cylinders 3, 6 and 9 in the downstream row experience the lowest forces in the array as they are protected from the direct incident waves by the upstream cylinders and do not have cylinders placed behind them. The wave forces on the cylinders reduce along the row in the direction of wave propagation, as seen in the case of three tandem cylinders in section 4.2.1. Cylinders 3 and 9 experience a force of $0.92F_0$ (Figure 11(f)) and Cylinder 6 a force of $0.96F_0$ (Figure 11(i)). The central cylinder in the array is exposed to diffracted and reflected waves from upstream and downstream cylinders. The central position of Cylinder 5 results in diffraction effects from all directions and the cylinder experiences a force of $1.26F_0$ (Figure 11(h)).

In order to study the influence of the center-to-center distance between the cylinders, simulations with $S = 0.6$ m and $S = 1.0$ m were carried out with the incident wavelength kept at $L = 2$ m. The numerically determined amplitude of the wave forces on each of the cylinders in the different cases is compared and presented in Figure 12. It is seen that for $S = 0.6$ m, the cylinders in the first column (Cylinders 1, 4 and 7) experience the same force, lower than the force on a single cylinder, F_0. The cylinders in the second column (Cylinders 2, 5 and 8) experience the highest forces in the arrangement and the cylinders in the last column (Cylinders 3, 6 and 9) experience forces higher than F_0. In this case, a different diffraction regime arises from the center-to-center distance, being much less than half of the incident wavelength compared to $S = 0.8$ m. The cylinders in the second row experience higher forces than the cylinders in the first row because of the destructive interference of the diffracted waves around the first column and constructive interference around the second column in this arrangement.

On the other hand, at $S = 1.0$ m, the cylinders in the first column experience the same force as that on a single

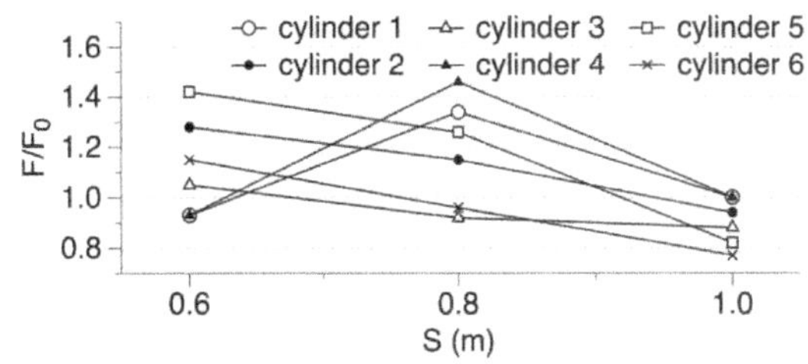

Figure 12. Variation of the wave forces on the cylinders with a distance of separation S.

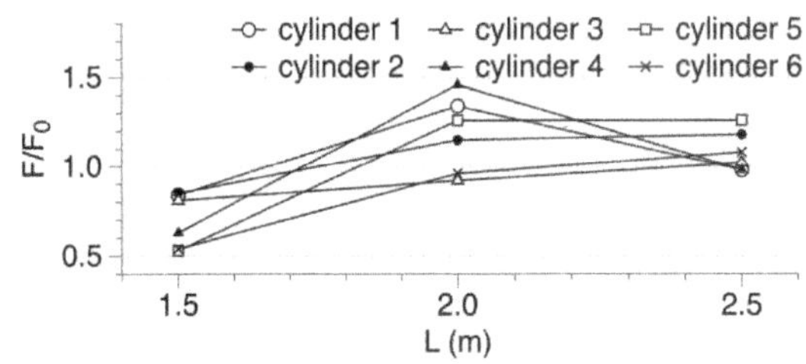

Figure 13. Variation of the wave forces on the cylinders with incident wavelength L.

cylinder, F_0. The cylinders in the second and third columns experience forces lower than F_0. This is justified by the fact that the cylinders are placed wide apart at $S = 1.0$ m and the wave forces experienced by the cylinders in the first column are the same as those experienced by a single cylinder. As the wave propagates through the array, the interaction with the cylinders reduces the wave energy carried by it and successive columns experience lower forces. The central cylinder (Cylinder 5) benefits from cancelation effects from two sides and experiences a lower force than Cylinder 3 in the third column.

The effect of incident wavelength on the wave forces experienced by the cylinders in the arrangement is studied by carrying out simulations with $L = 1.5$ m and $L = 2.5$ m. The center-to-center distance between the cylinders is kept at $S = 0.8$ m. The variation of the wave forces on the cylinders in the array with the incident wavelength is presented in Figure 13. A general trend of the wave forces increasing with increasing incident wavelength is seen. It is also observed that the change in incident wavelength does not greatly affect the wave forces on Cylinders 3, 6 and 9 in the last column. The central cylinder in the first column, Cylinder 4, undergoes the most change in the wave forces with change in the incident wavelength. The center-to-center distance of $S = 0.8$ m is almost equal to half the incident wavelength for $L = 1.5$ m, but the wave forces on all the cylinders are seen to be lower than the force on a single cylinder. The highest increase in the forces is seen when the incident wavelength is increased from $L = 1.5$ m to $L = 2$ m. This is further evidence that the strong diffraction and interference effects occur between the cylinders when the center-to-center distance is slightly lower than but not equal to half the incident wavelength.

5. Conclusion

The open-source CFD model REEF3D was used to model the wave interaction with a single vertical cylinder, three cylinders placed in tandem, three cylinders placed side by side and a 3×3 square array of nine cylinders at low KC numbers. The numerical model was validated by comparing the numerical results with the experimental data from the large-scale experiments performed by Mo et al. (2007). The computed wave surface elevation, water particle velocity and wave forces on a single cylinder agree well with the experimental data.

The wave interaction with cylinder groups in different configurations was investigated. Circular diffracted waves were observed radiating from the cylinders with constructive and destructive wave interference influencing the wave forces on the cylinders. The following observations were made in each of the cases simulated:

Three cylinders in a tandem arrangement:

- The computed wave forces on the first two cylinders were about 24% larger than the wave forces on a single cylinder due to the strong interaction of the diffracted waves between them. The last cylinder in the array experienced only a 6% higher force than a single cylinder due to the absence of neighbors downstream from the cylinder.

Three cylinders in a side-by-side arrangement:

- An increase in the free surface elevation in front of the column of cylinders was observed due to the flow obstruction caused by the three closely-spaced cylinders.
- The central cylinder experienced a 57% higher force than a single cylinder and the cylinders beside it experienced a similar force that was 30% higher than the force experienced by a single cylinder.

Nine cylinders in a 3×3 square array arrangement:

- The circular radiating waves from each of the cylinders had an influence on all the cylinders in the arrangement.
- The cylinders experience the highest forces when the center-to-center distance is slightly less than half the incident wavelength due to the constructive wave interference around the cylinders.
- The central cylinder in the first column experiences the largest change in the wave force on changing the incident wavelength.

The paper presents the results for three cylinders arranged in tandem and side by side. Further a combination of these two cases, a 3 × 3 square array of cylinders, was simulated and the effect of incident wavelength and center-to-center distance on the wave forces was investigated. The results obtained in this study represent a few combinations of cylinder arrangements. Further studies can be carried out to explore the influence of the center-to-center distance and the incident wavelength on the wave forces. The variation of the force regime for high steepness waves and larger cylinders can also be investigated.

Disclosure statement

No potential conflict of interest was reported by the author(s).

Funding

This work was supported by the OWCBW Project funded by the Research Council of Norway (grant number 217622/E20). This work was supported in part with computational resources provided by the Norwegian University of Science and Technology (NTNU) by NOTUR, http://www.notur.no.

References

Alagan Chella, M., Bihs, H., Myrhaug, D., & Muskulus, M. (2015). Breaking characteristics and geometric properties of spilling breakers over slopes. *Coastal Engineering*, *95*, 4–19.

Berthelsen, P. A., & Faltinsen, O. M. (2008). A local directional ghost cell approach for incompressible viscous flow problems with irregular boundaries. *Journal of Computational Physics*, *227*, 4354–4397.

Bihs, H. (2011). *Three-dimensional numerical modeling of local scouring in open channel flow* (PhD thesis). Department of Hydraulic and Environmental Engineering, NTNU Trondheim.

Bradshaw, P., Ferriss, D. H., & Atwell, N. P. (1967). Calculation of boundary-layer development using the turbulent energy equation. *Journal of Fluid Mechanics*, *28*, 593–616.

Chau, K. W., & Jiang, Y. W. (2001). 3D numerical model for Pearl river estuary. *Journal of Hydraulic Engineering*, *127*, 72–82.

Chau, K. W., & Jiang, Y. W. (2004). A three-dimensional pollutant transport model in orthogonal curvilinear and sigma coordinate system for Pearl river estuary. *International Journal of Environment and Pollution*, *21*(2), 188–198.

Chorin, A. (1968). Numerical solution of the Navier-Stokes equations. *Mathematics of Computation*, *22*, 745–762.

del Jesus, M., Lara, J. L., & Losada, I. J. (2012). Three-dimensional interaction of waves and porous coastal structures: Part I: Numerical model formulation. *Coastal Engineering*, *64*, 57–72.

Durbin, P. A. (2009). Limiters and wall treatments in applied turbulence modeling. *Fluid Dynamics Research*, *41*, 1–18.

Griebel, M., Dornsheifer, T., & Neunhoeffer, T. (1998). *Numerical simulation in fluid dynamics: A practical introduction*. Philadelphia, PA. Society for Industrial and Applied Mathematics.

Harten, A. (1983). High resolution schemes for hyperbolic conservation laws. *Journal of Computational Physics*, *49*, 357–393.

Higuera, P., Lara, J. L., & Losada, I. J. (2014). Three-dimensional interactions of waves and porous coastal structures using OpenFOAM: Part I: Formulation and validation. *Coastal Engineering*, *83*, 253–258.

Isaacson, M de St Q. (1982). Nonlinear-wave effects on fixed and floating bodies. *Journal of Fluid Mechanics*, *120*, 267–281.

Jacobsen, N. G., Fuhrman, D. R., & Fredsøe, J. (2012). A wave generation toolbox for the open-source CFD library: OpenFOAM. *International Journal for Numerical Methods in Fluids*, *70*, 1073–1088.

Jiang, G. S., & Shu, C. W. (1996). Efficient implementation of weighted ENO schemes. *Journal of Computational Physics*, *126*, 202–228.

Kim, S. K., Liu, P. L-F., & Ligget, J. A. (1983). Boundary integral equation solutions for solitary wave generation, propagation and run-up. *Coastal Engineering*, *7*, 299–317.

Linton, C. M., & Evans, D. V. (1990). The interaction of waves with arrays of vertical circular cylinders. *Journal of Fluid Mechanics*, *215*, 549–569.

Liu, T., & Yang, J. (2014). Three-dimensional computations of water-air flow in a bottom spillway during gate opening. *Engineering Applications of Computational Fluid Mechanics*, *8*(1), 104–115.

Longuet-Higgins, M. S., & Cokelet, E. D. (1976). The deformation of steep surface waves on water I – A numerical method of computation. *Proceedings of the Royal Society of London. Series A, Mathematical and Physical Sciences*, *350*, 1–26.

MacCamy, R., & Fuchs, R. (1954). *Wave forces on piles: A diffraction theory* (Technical Memorandum No. 69). Washington, DC: Beach Erosion Board, Coastal Engineering Research Center, US Army.

Mayer, S., Garapon, A., & Sørensen, L. S. (1998). A fractional step method for unsteady free-surface flow with applications to non-linear wave dynamics. *International Journal for Numerical Methods in Fluids*, *28*(2), 293–315.

Mo, W., Irschik, K., Oumeraci, H., & Liu, P. (2007). A 3D numerical model for computing non-breaking wave forces on slender piles. *Journal of Engineering Mathematics*, *58*, 19–30.

Morgan, G. C. J., Zang, J., Greaves, D., Heath, A., Whtilow, C. D., & Young, J. R. (2010). Using the rasInterFoam CFD model for wave transformation and coastal modeling. *Proceedings of the Coastal Engineering Conference*, Shanghai, China, 1–9.

Morison, J. R., O'Brien, M. P., Johnson, J. W., & Schaaf, S. A. (1950). Force exerted by surface waves on piles. *Journal of Petroleum Technology*, *2*, 149–154.

Naot, D., & Rodi, W. (1982). Calculation of secondary currents in channel flow. *Journal of the Hydraulic Division. Proceedings of the American Society of Civil Engineers*, *108*(HY 8), 948–968.

Nwogu, O. (1993). Alternative form of Boussinesq equations for nearshore wave propagation. *Journal of Waterway, Port, Coastal, and Ocean Engineering*, *119*(6), 618–638.

Ohkusu, M. (1974). Hydrodynamic forces on multiple cylinders in waves. *Proceedings of the International Symposium on Dynamics of Marine Vehicles and Structures in Waves*. London, 107–112.

Osher, S., & Sethian, J. A. (1988). Fronts propagating with curvature-dependent speed: algorithms based on Hamilton-Jacobi formulations. *Journal of computational physics*, *79*(1), 12–49.

Peng, D., Merriman, B., Osher, S., Zhao, H., & Kang, M. (1999). A PDE-based fast local level set method. *Journal of Computational Physics*, *155*, 410–438.

Peskin, C. S. (1972). Flow patterns around the heart valves: A numerical method. *Journal of Computational Physics*, *10*, 252–271.

Pu, J. H., Shao, S., Huang, Y., & Hussain, K. (2013). Evaluations of SWEs and SPH numerical modelling techniques for dam break flows. *Engineering Applications of Computational Fluid Mechanics*, *7*(4), 544–563.

Spring, B., & Monkmeyer, P. L. (1974). Interaction of plane waves with vertical cylinders. *Proceedings of the International Conference on Coastal Engineering, ASCE*, Copenhagen, 1828–1847.

Sussman, M., Smereka, P., & Osher, S. (1994). A level set approach for computing solutions to incompressible two-phase flow. *Journal of Computational physics*, *114*(1), 146–159.

van der Vorst, H. (1992). Bi-CGStab: A fast and smoothly converging variant of Bi-CG for the solution of non symmetric linear systems. *SIAM Journal on Scientific and Statistical Computing*, *13*, 631–644.

Wei, G., & Kirby, J. T. (1995). Time-dependent numerical code for extended Boussinesq equations. *Journal of Waterway, Port, Coastal and Ocean Engineering*, *121*(5), 251–261.

Wilcox, D. C. (1994). *Turbulence modeling for CFD*. La Canada, California: DCW Industries Inc.

4

Lattice Boltzmann simulations on the role of channel structure for reactive capillary infiltration

Danilo Sergi*, Loris Grossi, Tiziano Leidi and Alberto Ortona

The iCIMSI Research Institute, The University of Applied Sciences and Arts of Southern Switzerland, Manno, Switzerland

It is widely recognized that the structure of porous media is of relevance for a variety of mechanical and physical phenomena. The focus of the present work is on capillarity, a pore-scale process occurring at the micron scale. We attempt to characterize the influence of pore shape for capillary infiltration by means of Lattice Boltzmann simulations in 2D with reactive boundaries leading to surface growth and ultimately to pore closure. The systems under investigation consist of single channels with different simplified morphologies: namely, periodic profiles with sinusoidal, step-shaped and zigzag walls, as well as constrictions and expansions with rectangular, convex and concave steps. This is a useful way to break the complexity of typical porous media down into basic structures. The simulations show that the minimum radius alone fails to properly characterize the infiltration dynamics. The structure of the channels emerges as the dominant property controlling the process. A factor responsible for this behavior is identified as being the occurrence of the pinning of the contact line. It turns out that the optimal configuration for the pore structure arises from the packing of large particles with round shapes. In this case, the probability of having wide and straight flow paths is higher. Faceted surfaces with sharp edges should be avoided because of the phenomenon of pinning near narrow-to-wide parts. This study is motivated by the infiltration of molten metals into carbon preforms. This is a manufacturing technique for ceramic components devised for advanced applications. Guidelines for experimental work are discussed.

Keywords: porosity; microstructure; capillary infiltration; Lattice Boltzmann simulations; liquid silicon infiltration

1. Introduction

A porous medium is a material presenting some void content in the solid matrix (Dullien, 1992). The fraction of void space defines the porosity ε. Common examples of porous materials include wood (Jeje & Zimmermann, 1979; Ota et al., 1995; Studart, Gonzenbach, Tervoort, & Gauckler, 2006) and building materials (Martys & Hagedorn, 2002). For the latter, the porosity is exploited in order to trap air with the purpose of achieving better thermal insulation (Dullien, 1992). Soil science is another fruitful research area in the field (Pan, Hilpert, & Miller, 2004; Schaap, Porter, Christensen, & Wildenschild, 2007; Sukop & Or, 2004). In materials engineering, the porosity can also be a key factor in order to meet the highest demands in a variety of applications (Clyne, Golosnoy, Tan, & Markaki, 2006; Furler, Scheffe, Gorbar, et al., 2012; Furler, Scheffe, & Steinfeld, 2012; Goodall & Mortensen, 2013; Roberts & Garboczi, 2000; Trimis, Pickenäcker, & Wawrzinek, 2006; Zeschky et al., 2005). In general, the pore structure is recognized as having a prominent role for the operating behavior of the material.

In the present work we are concerned with the capillary properties of porous materials. Capillarity is the phenomenon responsible for the spontaneous infiltration of a liquid into a porous medium. It is a pore-scale process arising from the adhesive forces between the liquid and solid phases. Computational fluid dynamics (CFD) relying on the finite element method (FEM) allows the problem to be addressed (Chibbaro, Costa, et al., 2009; Svihla & Xu, 2006; Xu & Guetari, 2004). Our study is based on simulations using the Lattice Boltzmann (LB) method in 2D (Benzi, Succi, & Vergassola, 1992; Chen & Doolen, 1998; Succi, 2009; Sukop & Thorne, 2010; Wolf-Gladrow, 2005). This approach has been gaining popularity for handling hydrodynamic systems out of equilibrium involving complex boundaries and interfacial phenomena. More generally, many applications can be figured out in the incompressible limit, i.e., at low Mach numbers (Ghosh, Patil, Mishra, Das, & Das, 2012; Guiet, Reggio, & Teyssedou, 2011; Haghani, Rahimian, & Taghilou, 2013).

The motivating problem for our investigation is the reactive infiltration of molten silicon (Si) into carbon (C) (Bougiouri, Voytovych, Rojo-Calderon, Narciso, & Eustathopoulos, 2006; Dezellus & Eustathopoulos, 2010; Dezellus, Hodaj, & Eustathopoulos, 2003; Einset, 1996, 1998; Eustathopoulos, Nicholas, & Drevet, 1999; Hillig, Mehan, Morelock, DeCarlo, & Laskow, 1975; Israel et al., 2010; Liu, Muolo, Valenza, & Passerone, 2010; Messner

*Corresponding author. Email: danilo.sergi@icimsi.ch

& Chiang, 1990; Mortensen, Drevet, & Eustathopoulos, 1997; Voytovych, Bougiouri, Calderon, Narciso, & Eustathopoulos, 2008). This process is of particular relevance to industrial practices which involve the processing of carbon and graphite materials (Gadow, 2000; Gadow & Speicher, 2000; Krenkel, 2005; Paik et al., 2002; Salamone, Karandikar, Marshall, Marchant, & Sennett, 2008). Specifically, ceramization through liquid Si infiltration, or impregnation, is necessary for high-temperature applications such as tribology (Gadow, 2000). In this process, the reaction between Si and C to form silicon carbide (SiC) has been the subject of intense research activity, since it coincides with a wetting transition (Bougiouri et al., 2006; Dezellus et al., 2003; Dezellus & Eustathopoulos, 2010; Einset, 1996, 1998; Eustathopoulos et al., 1999; Hillig et al., 1975; Israel et al., 2010; Liu et al., 2010; Messner & Chiang, 1990; Mortensen et al., 1997; Voytovych et al., 2008). In other words, without this reaction, infiltration would not occur (Bougiouri et al., 2006). Importantly, the formation of SiC can lead to a thickening of the surface behind the invading front (contact line) (Bougiouri et al., 2006; Einset, 1996, 1998; Israel et al., 2010). Surface growth can result in retardation effects and ultimately hinder the impregnation because of pore obstruction (Bougiouri et al., 2006; Einset, 1996, 1998; Israel et al., 2010; Messner & Chiang, 1990). In the LB framework, the surface reaction is treated as a precipitation process (Kang, Lichtner, & Zhang, 2007; Kang, Zhang, & Chen, 2003; Kang, Zhang, Chen, & He, 2002; Kang, Zhang, Lichtner, & Tsimpanogiannis, 2004; Lu, DePaolo, Kang, & Zhang, 2009; Sergi, Grossi, Leidi, & Ortona, 2014). In this model, the driving mechanism for surface growth is exhaustively compliant with the requirement of mass conservation (Kang et al., 2007). Among the pioneering works for this subject, the article by Miller and Succi (2002) deserves attention.

A porous medium can be regarded as a network of capillary channels. For the sake of clarity, we shall still limit ourselves to single capillaries. In general, the analysis of more complex geometries is based on the theoretical results for these types of system. In a previous work (Sergi et al., 2014), we investigated the retardation effects induced by surface growth for the linear Washburn law for interstices of uniform radii. The linear time dependence for the infiltration depth is characteristic of reactive Si infiltration (Israel et al., 2010; Voytovych et al., 2008). It is important to note that the employed LB models (Chibbaro, 2008; Chibbaro, Biferale, Diotallevi, & Succi, 2009; Diotallevi, Biferale, Chibbaro, Lamura, et al., 2009; Diotallevi, Biferale, Chibbaro, Pontrelli, et al., 2009) only reproduce the macroscopic behavior observed in experiments without providing any explanation for its origin (Israel et al., 2010; Voytovych et al., 2008).

Here, attention is paid to simplified channel structures consisting of single capillaries with different geometric attributes. Namely, we consider periodic profiles (Gern & Kochendörfer, 1997; Patro, Bhattacharyya, & Jayram, 2007) with varying degrees of angularity and tortuosity (Duda, Koza, & Matyka, 2011; Matyka & Koza, 2012). These patterns could arise for example from the juxtaposition of medium-sized C grains. The patterns resulting from the juxtaposition of larger C grains are instead modeled as capillary systems with constrictions or expansions (Einset, 1996). Also in this case, the results for different morphologies and degrees of tortuosity are analyzed.

Our simulations confirm previous results for a uniform channel (Sergi et al., 2014), pointing out that the infiltration velocity marginally affects the process of surface growth and pore closure. The main result of the present study consists in a system of rules for the evaluation of the response of microstructures of different characteristics to capillary infiltration and surface growth. For ceramic products (Gadow, 2000; Gadow & Speicher, 2000; Krenkel, 2005; Paik et al., 2002; Salamone et al., 2008), these are competing phenomena which are both important in order to obtain complete densification and optimize the final properties. As explained before, the complexity of fully-developed porous systems is decomposed into basic structures. In the following, a comparative analysis is proposed in order to understand the role of the various structural features. Specifically, we shall address issues such as front dynamics, flow retardation, thickening of the surface, pore closure, analysis of characteristic radii and tortuosity. Notably, it arises that the structure of the channels can significantly affect the resulting effective radius. It appears that our findings provide useful inputs for preform preparation. The optimal configuration for the porosity is one which contains the larger number of constrictions, since the effect of surface growth is weaker. This configuration could realize a percolating network which allows the flow to be guided into the porous preforms. Another advantage of this configuration is that it requires coarse particles and so it should be easier to obtain the desired arrangement. Of course, these guidelines need to be verified and improved by more simulation work before proceeding to testing by experiments.

2. LB models

The LB method can simulate hydrodynamic phenomena at a mesoscopic level; streaming and collision processes are reminiscent of the particulate nature of liquids, demanding nevertheless for a statistical treatment. More precisely, the velocity space is discretized and distribution functions account for the particles moving along a given direction (Benzi et al., 1992; Chen & Doolen, 1998; Succi, 2009; Sukop & Thorne, 2010; Wolf-Gladrow; 2005). In the present work we employ the same models as in Sergi et al. (2014) with the same notation and terminology. More details can be found in the related literature. Namely, fluid flow is treated by means of multicomponent models (Chibbaro, 2008; Chibbaro, Biferale, Diotallevi, & Succi, 2009; Diotallevi, Biferale, Chibbaro, Lamura, et al., 2009; Diotallevi, Biferale, Chibbaro, Pontrelli, et al., 2009). For solute transport and reactive boundaries we closely follow Kang et al. (2007). For similar works the interested reader is addressed to the articles by Kang, Zhang, Chen, and He (2002), Kang et al. (2003, 2004) and Lu et al. (2009).

In the sequel, we shall express the simulation results in model units. The basic units are those of mass (mu), length (lu) and time (ts) (Sukop & Thorne, 2010). Ordinary units are obtained after suitable transformations (Gross, Varnik, Raabe, & Steinbach, 2010; Lu et al., 2009). Of course, comparative analyses with experiments based on dimensionless parameters (Reynolds, Damkohler, Peclet, capillary, Bond, Weber numbers, etc.) are preferable. These numbers allow for the establishment of equivalences between systems (Landau & Lifshitz, 2008) and have the same value in the common systems for units like the SI (international system of units).

3. Capillary systems and simulation settings

Wetting and infiltration studies have highlighted the relevance of surface and pore microstructures (Alava, Dube, & Rost, 2004; Grzelakowski et al., 2009; Quéré, 2008). The importance of pore characteristics has also gained general acceptance for the processing route of C/SiC composites (Gadow & Speicher, 2000; Israel et al., 2010; Paik

Figure 1. Periodic profiles for the upper wall; that is, sinusoidal profile, rectangular step and zigzag profile. The last figure shows the case of a sinusoidal capillary with the lower wall misaligned with respect to the upper one. A is the amplitude of the profiles and Δx denotes the displacement inducing misalignment. Similar definitions also apply for the other geometries.

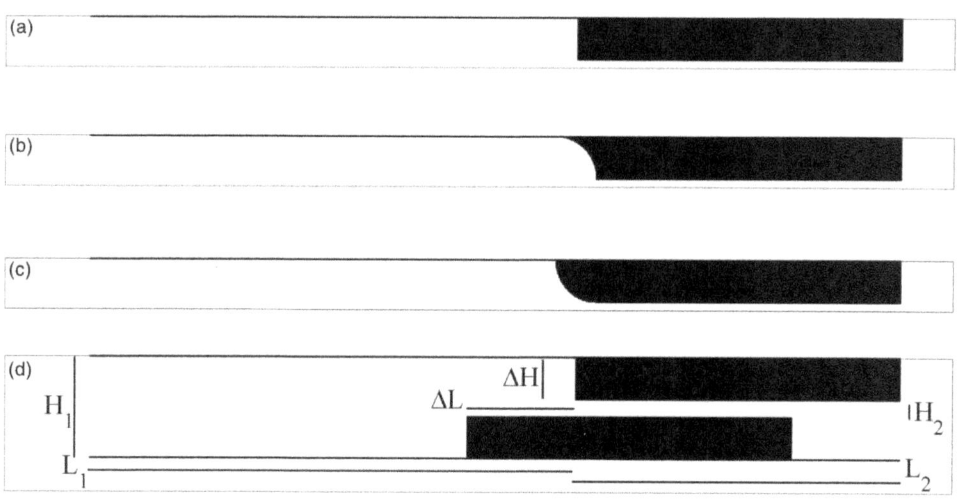

Figure 2. Constricted channel with rectangular, convex and concave junctions (upper wall). In the last panel, $L = L_1 + L_2$ is the overall length of the interstice, H_1 is the radius of the first segment, and H_2 that of the second one, $\Delta H = (H_1 - H_2)/2$ the width of the solid phase and ΔL the displacement introducing misalignment between the upper and lower walls. Similar definitions also hold for the capillaries with expansion.

et al., 2002; Salamone et al., 2008). One way to control the pore properties of the matrix is via enrichment by ceramic powders. Our interest resides in the optimization of the infiltration process taking into account surface reactivity and subsequent thickening. We investigate basic, simplified capillary geometries that could arise from intraparticle porosity. In a general porous medium, the pores do not have a constant cross-section. Furthermore, it is possible to identify larger void spaces. To start with, we thus consider capillaries presenting the alternation between wide and narrow portions. Aligned periodic walls reproduce structures with such properties (Gern & Kochendörfer, 1997; Patro et al., 2007). Our focus is on the sinusoidal profile. Other geometries are also considered, that is, rectangular steps and zigzag walls. The degree of tortuosity is modified by introducing a misalignment between the upper and lower walls. Figure 1 illustrates some representative geometries for this type of system. Another complication arises from the fact that the pore size distribution is generally to some extent broad. In order to understand how this aspect can affect capillary impregnation, simulations are also performed for capillaries with constrictions or expansions. The step at the junction can be rectangular, convex or concave. Also in this case, the tortuosity of the systems is varied. Examples for this type of system are shown in Figure 2. Loosely speaking, a low degree of angularity is for example representative of directly synthesized powders, while a higher degree of angularity can be observed for milled powders.

For the implementation of the various capillary systems we proceed as follows. The simulation domains are N_x long and N_y wide. The length of the capillaries is indicated by L and their height by H. For all systems we have $L=N_x/2$. The left extremity of the capillary always coincides with the point $x_s=2N_x/5$. The periodic profiles are defined by the period T, the amplitude A and the displacement Δx controlling the misalignment between the walls. Since the minimum of the profiles

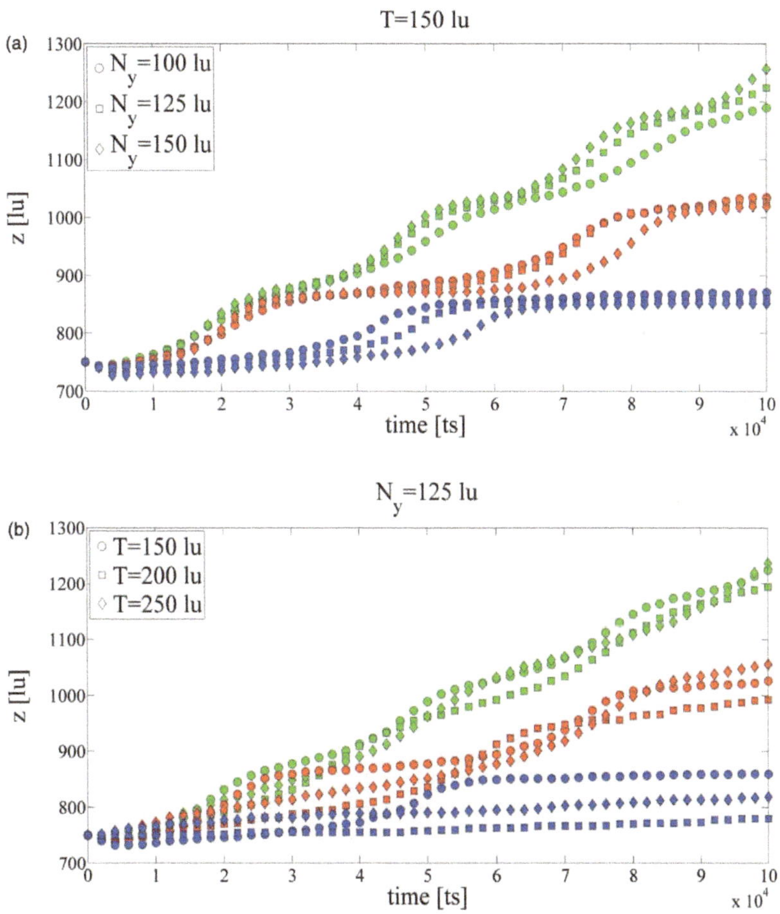

Figure 3. Centerline position z of the advancing front over the course of time for sinusoidal capillaries of length $L=750$ lu without surface reaction. Color code based on the amplitude A. We use green, red and blue for increasing amplitude $A=i(N_y-25)/16$, where $i=2,3,4$.

corresponds to the starting point of the solid phase, the amplitude A is equal to half the maximal initial width of the solid phase (see Figure 1). As an example, the lower wall of the sinusoidal capillary is described by the function $f(x) = A\sin(\omega\tilde{x} + 3\pi/2) + A$ with $\omega = 2\pi/T$ and $\tilde{x} = x - x_s - \Delta x$. These parameters can take on the following values: L = 750,1000,1250 lu, N_y = 100,125,150 lu, T = 150,200,250 lu, $A = (N_y - 25)/8, 3(N_y - 25)/16, (N_y - 25)/4$, $\Delta x = 0, T/5, 2\,T/5$. The choice for the amplitude A guarantees that the width of the necks is at least 25 lu.

For constricted channels and expansions, the overall length of the capillary is still denoted by L. The length of the first segment, wider or narrower, is indicated by L_1 and that of the second one by L_2; of course, $L = L_1 + L_2$. The other parameters characterizing these systems are the width of the first segment H_1, that of the second part H_2, the width of the step $\Delta H = (H_1 - H_2)/2$ (for constrictions) and the displacement ΔL introducing a misalignment between the two walls (see Figure 2). The convex and concave shapes for the step are defined using circles of center (x_c, y_c) and radius $r = \Delta H$. For the lower wall, the coordinates of the center are given by $x_c = 2N_x/5 + L_1 \mp \Delta H/2 - 1$ and $y_c = \Delta H$ (the signs refer to the convex and concave shapes respectively). The various parameters can vary as follows: L = 750,1000,1250 lu, H_1 = 100,125,150 lu, $L_1 = 2L/5, 3L/5, 4L/5$, $\Delta H = (N_y - 25)/4, 3(N_y - 25)/8, (N_y - 25)/2$, $\Delta L = 0, L_2/3, 2L_2/3$ with $\Delta L < L_1$ (for constrictions). Furthermore, the step at the junction can have a rectangular, convex or concave shape (see Figure 2).

Our systems consist of binary mixtures (Chibbaro, 2008; Chibbaro, Biferale, Diotallevi, & Succi, 2009). In the initial condition, the first component always occupies half of the simulation domain. This means that the first fluid fills the capillary up to one tenth of its length. The density values are chosen so that $\rho_0 = \rho_1 + \rho_2 = 2$ mu/lu^2 and $\rho_1/\rho_2 = 2.5\%$, where the first fluid is the main component; elsewhere the ratio between the densities of the

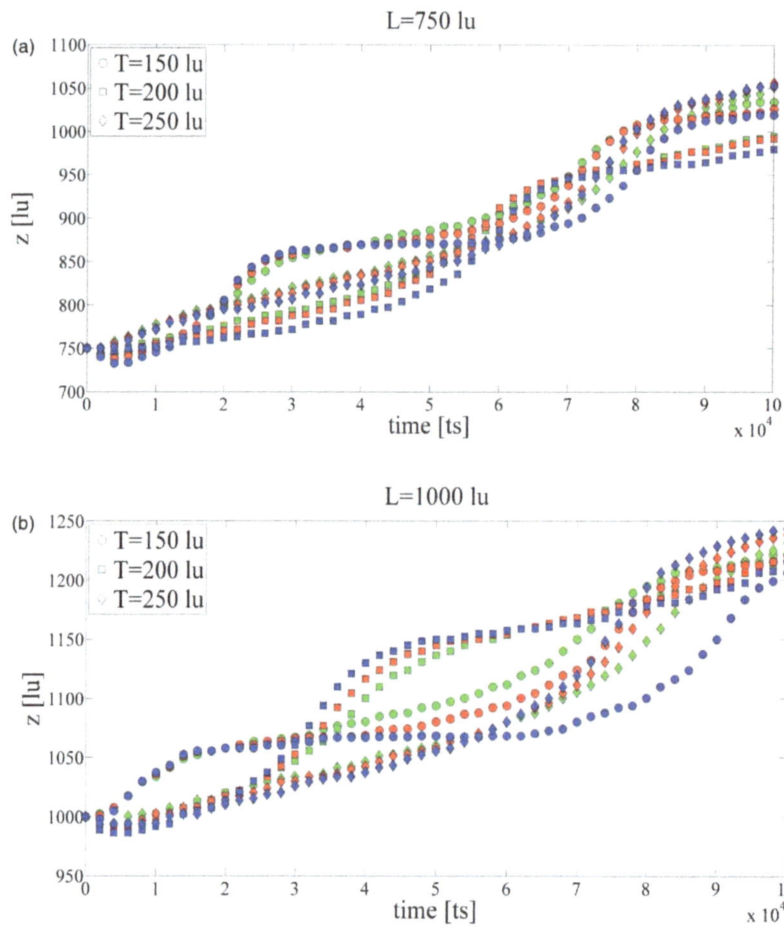

Figure 4. Time dependence of the centerline position z of the invading front in the absence of reaction for sinusoidal profiles. Color code based on the domain width N_y: green, red and blue for increasing values N_y = 100,125,150 lu. The systems have the average minimum height $< H_{min} > = 50$ lu, determined using the amplitude $A = 3(N_y - 25)/16$.

two components is inversed. Unless stated otherwise, the parameter for fluid-fluid interactions (cohesive forces) is $G_c = 0.9$ lu/mu/ts^2, for which the surface tension turns out to be $\gamma = 0.16403$ lu·mu/ts^2 (Sergi et al., 2014). Solid-fluid interactions are instead determined by the parameters $G_{ads,1} = -G_{ads,2} = -0.35$ lu/ts^2. In so doing, the first component wets the solid phase while the second is a non-wetting fluid. These settings (Sergi et al., 2014) reproduce an equilibrium contact angle around 30°, typical for droplets of molten Si on SiC substrates (Bougiouri et al., 2006; Voytovych et al., 2008). Regarding solute transport, we proceed as Sergi et al. (2014). In the initial configuration, in the region filled with the first fluid component, the wetting one, the solute concentration is $C_1 = 10^{-2}$ mu/lu^2; elsewhere the solute concentration is $C_2 = 2 \cdot 10^{-3}$ mu/lu^2. The parameters for the function introducing an interface for solute are assumed to be $G_s = -4.875 \cdot 10^{-3}$ mu/lu/ts^2, $\varphi_0 = 1$ and $C_0 = 4.9 \cdot 10^{-3}$ mu/lu^2 (Sergi et al., 2014). For the saturated concentration we choose $C_s = 5 \cdot 10^{-3}$ mu/lu^2. At the beginning, the mass deposited on solid boundaries is $b_0 = 2 \cdot 10^{-3}$ mu; the threshold value determining surface growth is $b_{max} = 10^{-2}$ mu. The relaxation time for solute transport is set to $\tau_s = 1$ ts, so the diffusion coefficient D is kept fixed. The reaction-rate constant is varied according to the rule $k_r = (8/5)/I$, where $I = 1000, 3000$ ts. We recall that the Damkohler number is defined as $Da = k_r N/D$ (Kang et al., 2003, 2004; Lu et al., 2009), where N is a characteristic length for the system. Last, the evolution of every system amounts to 100,000 ts. For the analysis, 50 evenly-spaced frames are collected. The interfaces are tracked as explained in Sergi et al. (2014). Here the interested reader can also find more details about the motivations for the simulation settings.

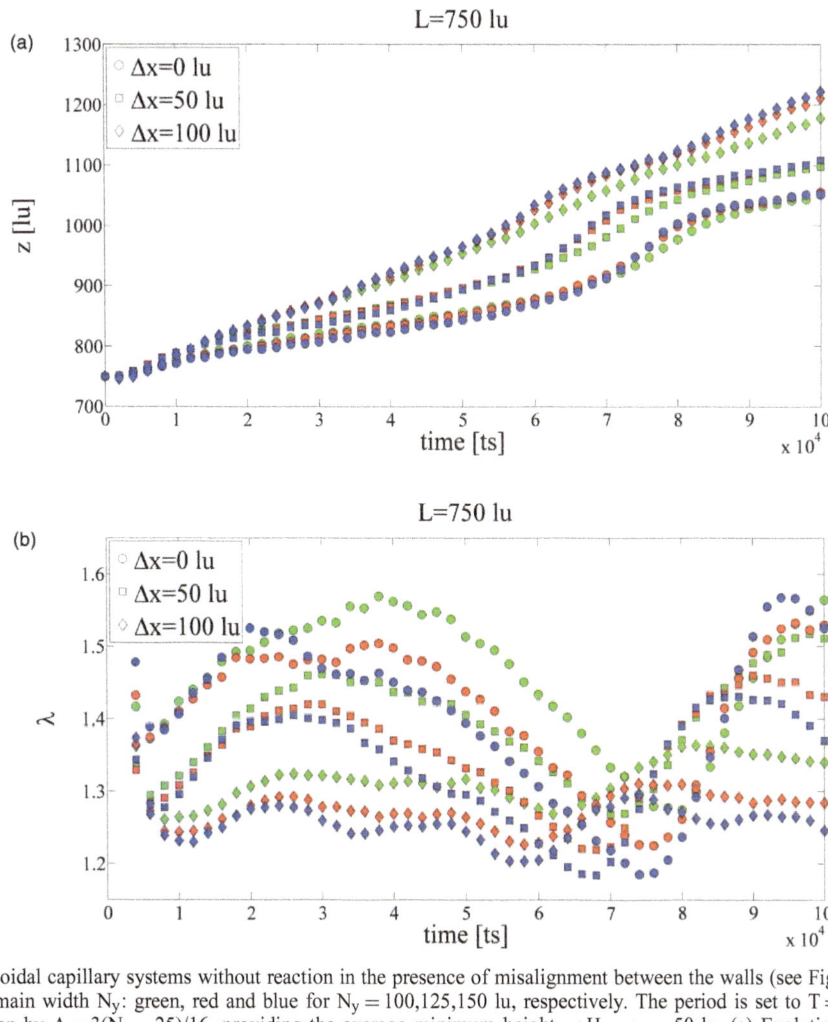

Figure 5. Sinusoidal capillary systems without reaction in the presence of misalignment between the walls (see Figure 1). Color code based on the domain width N_y: green, red and blue for $N_y = 100, 125, 150$ lu, respectively. The period is set to $T = 250$ lu, while the amplitude is given by $A = 3(N_y - 25)/16$, providing the average minimum height $< H_{min} > = 50$ lu. (a) Evolution of the centerline position z of the meniscus. (b) Evolution of the tortuosity λ.

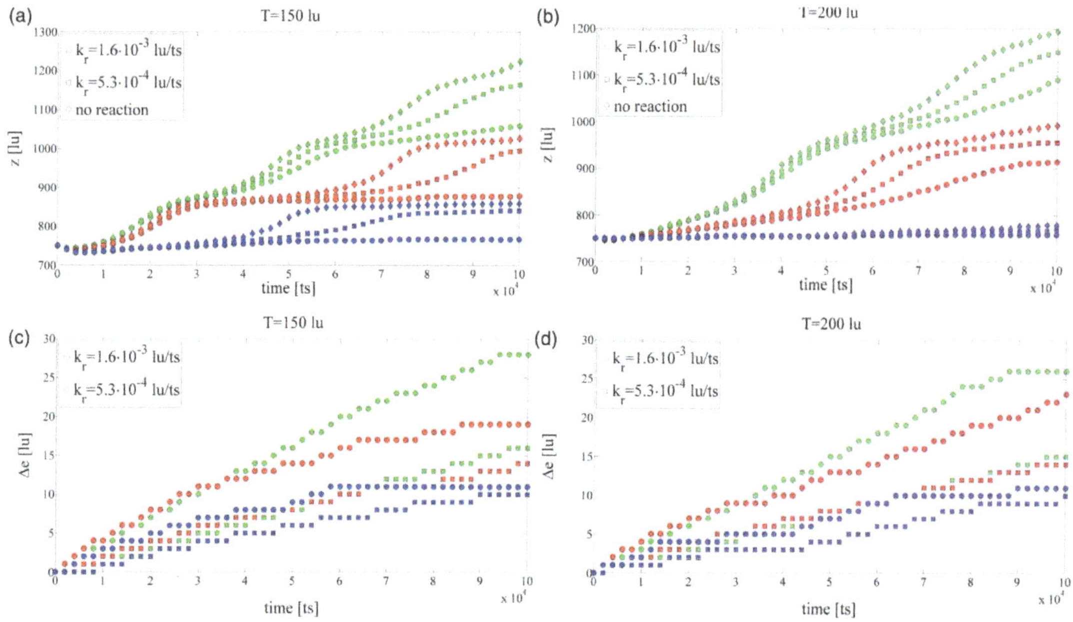

Figure 6. Results for sinusoidal capillary systems of length $L = 750$ lu with $N_y = 125$ lu in the presence of surface reaction, controlled by the reaction-rate constant k_r. Color code based on the amplitude $A = i(N_y - 25)/16$ with $i = 2,3,4$, for green, red and blue, respectively. Top: Time dependence of the invading front. Bottom: Maximal thickening Δe of the solid surface over the course of time.

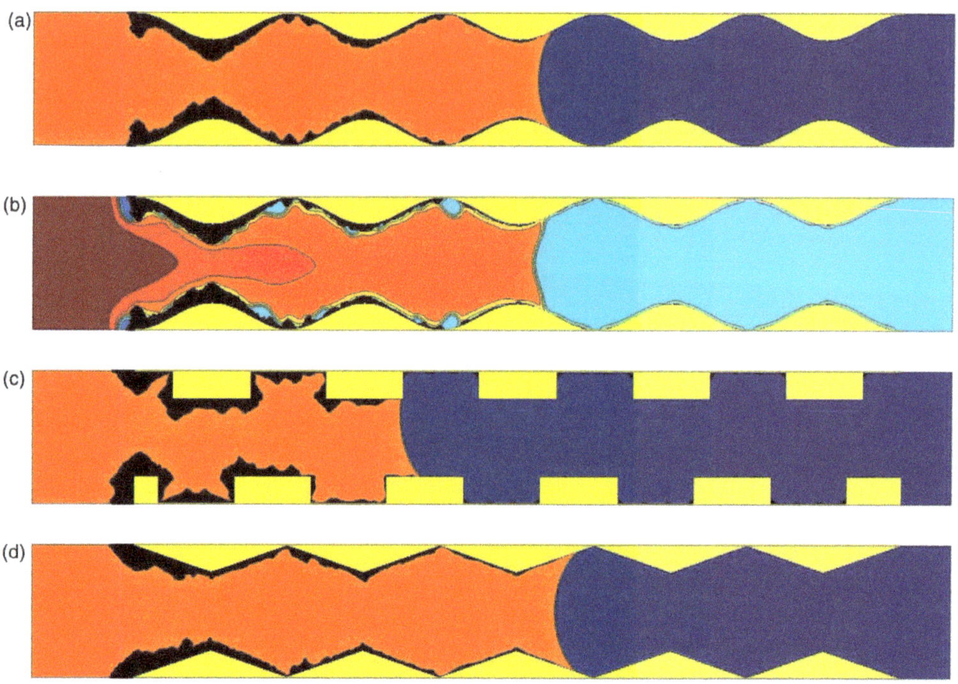

Figure 7. Periodic systems after 60,000 timesteps with channel width $N_y = 125$ lu, capillary length $L = 750$ lu, period $T = 150$ lu, amplitude $A = (N_y - 25)/8$, corresponding to $H_{min} = 75$ lu, and reaction-rate constant $k_r = 1.6 \cdot 10^{-3}$ lu/ts. Unless stated otherwise, the figures represent fluid flow with color code based on the density of the wetting component. Dark points for the phase resulting from surface growth and yellow points for the initial solid phase. (a) Sinusoidal profile. (b) Sinusoidal profile. Representation of solute transport with color code based on the solute concentration. (c) Step-shaped profile with misalignment given by $\Delta x = 100$ lu. (d) Zigzag profile.

4. Results and discussion

4.1. Periodic profiles

Periodic profiles give rise to capillaries with a rich variety of morphology characteristics (see Figure 1). Simply put, with T variations the pore chambers and necks become longer, for increasing values. Increases in amplitude A lead to deeper pore chambers (wider stomachs) and narrower necks (closer tips). With increasing values of Δx the

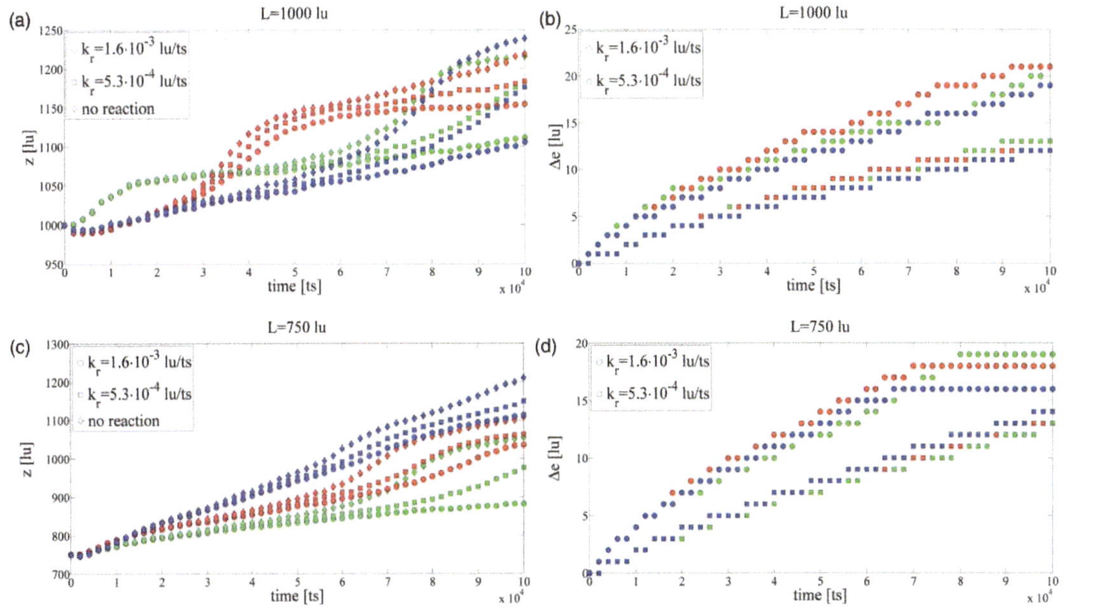

Figure 8. Sinusoidal capillary systems with reactivity enabled, adjusted via the reaction-rate constant k_r. The domain width is set to $N_y = 125$ lu and the amplitude is kept fixed at $A = 3(N_y - 25)/16$, yielding the minimum height $H_{min} = 50$ lu. Top: Color code based on the period T: green, red and blue for $T = 150,200,250$ lu, respectively. Bottom: Color code based on the parameter Δx controlling the misalignment between the walls: green, red and blue for $\Delta x = 0,50,100$ lu. The period is $T = 250$ lu.

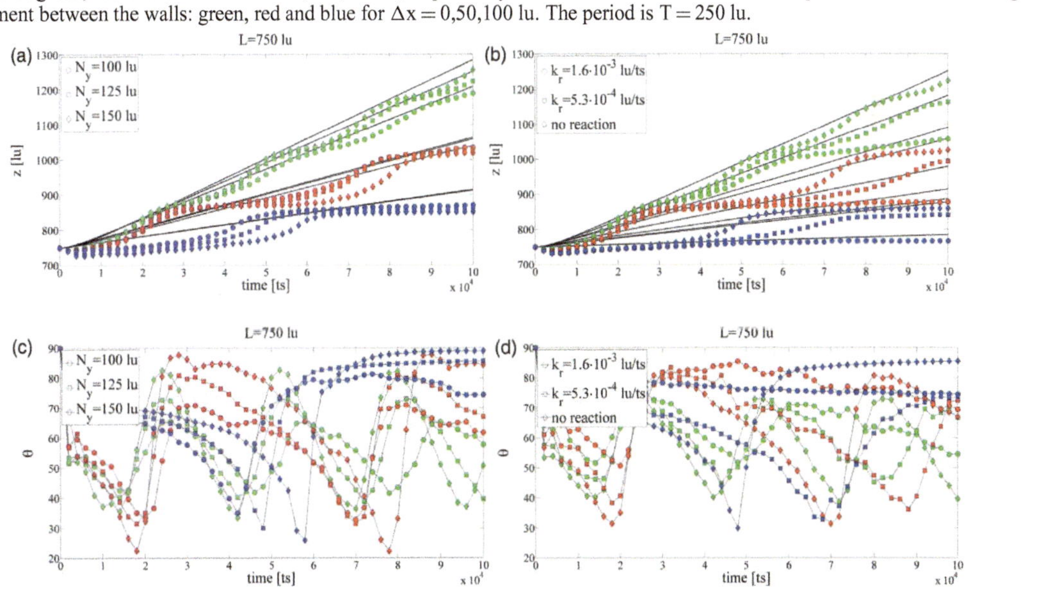

Figure 9. Sinusoidal capillary systems of period $T = 150$ lu. Points for simulation results; the solid lines are determined using Eq. 1. (a) The solid surface is inert. Color code based on the amplitude A: green, red and blue for $A = i(N_y - 25)/16$ with $i = 2,3,4$, respectively. (b) The surface reaction is controlled via the reaction-rate constant k_r. The domain width is $N_y = 125$ lu. Color code based on the amplitude $A = i(N_y - 25)/16$, with $i = 2,3,4$, corresponding to green, red and blue, respectively. (c) Dynamic contact angle for the conditions of a. (d) Dynamic contact angle for the systems of b.

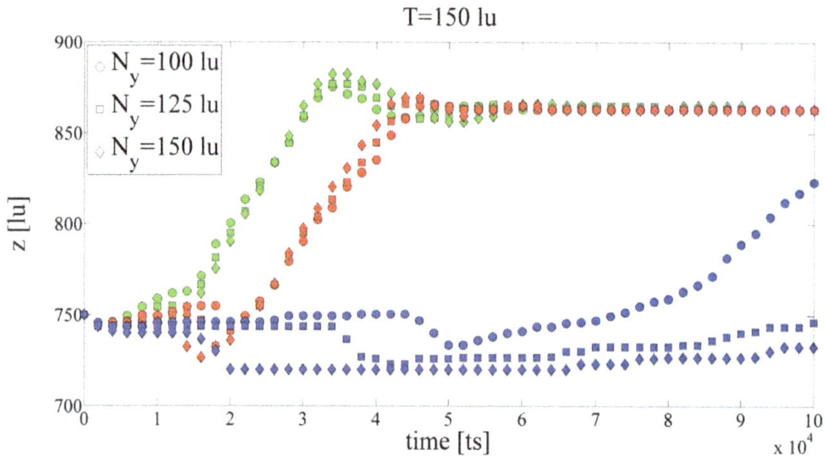

Figure 10. Front displacement over the course of time for step-walled capillaries without reaction. The length of the capillary is L = 750 lu. Color code based on the amplitude A. It is given by $A = i(N_y - 25)/16$ with i = 2,3,4, corresponding to the colors green, red and blue, respectively.

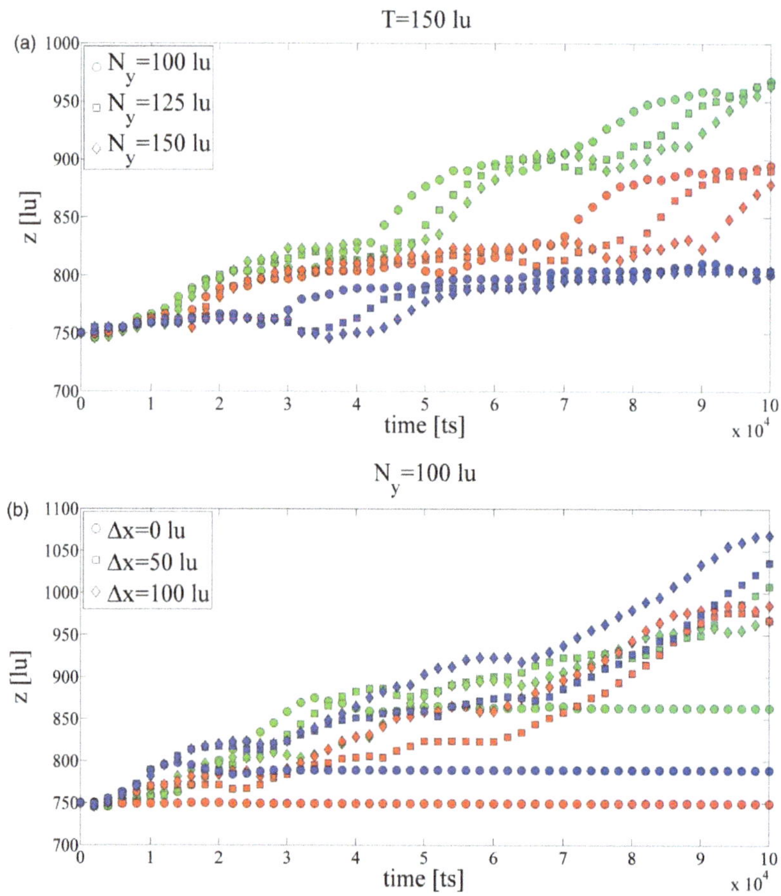

Figure 11. Centerline position of the invading front over the course of time for capillary systems with periodic step-shaped walls in the absence of surface reaction. The capillaries have length L = 750 lu. (a) The misalignment between the walls is Δx = 100 lu. Color code based on the amplitude $A = i(N_y - 25)/16$ with i = 2,3,4 for green, red and blue, respectively. (b) The amplitude is fixed to $A = (N_y - 25)/8$, determining the minimum height $H_{min} = 62$ lu. Color code based on the period: T = 150,200,250 lu for green, red and blue.

pore chambers become narrower and the necks are shorter. Moreover, the path connecting adjacent maxima and minima is shorter for the zigzag and sinusoidal profiles while it is longer for the rectangular steps.

4.1.1. Sinusoidal profiles

To start with, let us consider sinusoidal profiles in the absence of surface reaction. Figure 3 shows the behavior of the invading front over the course of time for varying channel width N_y, period T and amplitude A. In general, it is possible to distinguish phases of pronounced acceleration followed by phases of weaker acceleration eventually leading to plateaus. This phenomenon is usually referred to as the pinning of the interface (Blow, Kusumaatmaja, & Yeomans, 2009; Chibbaro, Biferale, Binder, et al., 2009; Chibbaro, Costa, et al., 2009; Kusumaatmaja, Pooley, Girardo, Pisignano, & Yeomans, 2008; Mognetti & Yeomans, 2009; Wiklund & Uesaka, 2012, 2013). Basically, pinning occurs in proximity to the necks, i.e., close to narrow-to-wide structures. This phenomenon is of course more marked for capillaries with narrower necks. The role of the period is not so clear. It seems that for narrower necks pinning occurs for the smaller period. Increasing periods mainly tend to weaken the capillary forces without enhancing the phenomenon of pinning (results not shown for brevity), as we could verify with longer simulations of up to 500,000 timesteps. It should be noted that for these systems the minimum and maximum radii are the same while the average and hydraulic radii are comparable. The hydraulic radius is defined as $r_h = (1/2)\text{Area/Perimeter}$ (Dullien, 1992). The plots of Figure 3 also indicate that the various curves are associated with three distinct infiltration behaviors. It is important to note that the minimum radius is not the same within a group of capillaries. What seems to really matter is the ratio of the amplitude to the domain width A/N_y. It follows that to first approximation the infiltration process depends on the structure of the capillary. For the sake of clarity, we consider the quantity $R = \langle (V - \langle V \rangle)^2 \rangle / \langle V \rangle$, where V is a variable and

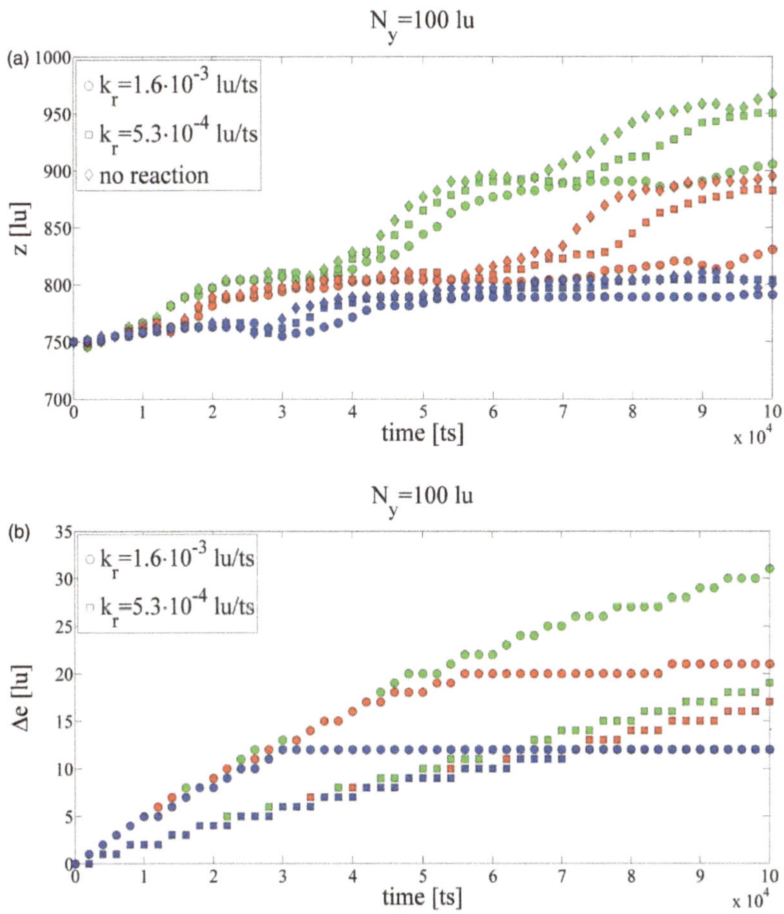

Figure 12. Capillary systems with periodic step-shaped walls in the presence of surface reaction. The period is fixed to T = 150 lu and the misalignment is set to Δx = 100 lu. Color code based on the amplitude: green, red and blue for $A = i(N_y - 25)/16$ when i = 2,3,4. (a) Front displacement over the course of time. (b) Evolution of the maximal thickening of the solid surface.

the brackets indicate an average. Let us consider the data for which N_y varies: if $V = A/N_y$, it is found that R is at least 1000 times smaller than for $V = r_{min}$. Thus, A/N_y better characterizes the infiltration behavior. It really has to be noted that if T = 150 lu and $A = (N_y - 25)/8$, the minimum heights are $H_{min} = 62,75$, 87 lu for $N_y = 100,125$, 150 lu, respectively. For $N_y = 150$ lu the minimum radius is bigger, but this is not sufficient in order to ease penetration, since the pore chamber becomes deeper (stronger drag force).

Figure 4 refers to capillary systems with varying channel width N_y and period T. It clearly arises that, for a medium value of the average minimum height $< H_{min} >$, the phenomenon of pinning is accentuated with shorter periods and longer capillary lengths, but this does not result in a strong retardation for capillary infiltration. If we assume that the infiltration velocity is proportional to 1/L, as for smooth capillaries (Chibbaro, Biferale, Diotallevi, & Succi, 2009), for the infiltration velocity we can write $v(L + \Delta L)/v(L) \approx (1 - \Delta L/L)$. In passing from L = 750 lu to L = 1000 lu, this ratio becomes $v(L + \Delta L)/v(L) = 0.67$. From Figure 4 it appears that the invading fronts reach almost the same depth. Specifically, on average we find that $v(L + \Delta L)/v(L) = 0.82$; the infiltration velocities are determined by applying the method of least squares. If we restrict our attention to the data for T = 200 lu, it even turns out that $v(L + \Delta L)/v(L) = 0.96$. In this case, if we overcome the obstacle of estimating the velocity by considering z as a function of L for the time fixed at t = 100,000 lu, this phenomenon is clearer. It is found that $z(L + \Delta L)/z(L) = 0.91$ for L = 750 lu and $z(L + \Delta L)/z(L) = 0.88$ for L = 1000 lu. As a result, in the case of sinusoidal profiles, the length of the capillaries affects the infiltration process to a weaker extent. The discussion on the results of Figure 3 is also valid for longer capillaries. Unless stated otherwise, in the sequel the conclusions can be assumed to hold for all capillary lengths.

The results for capillary systems with misalignment between the upper and lower walls (see Figure 1) are

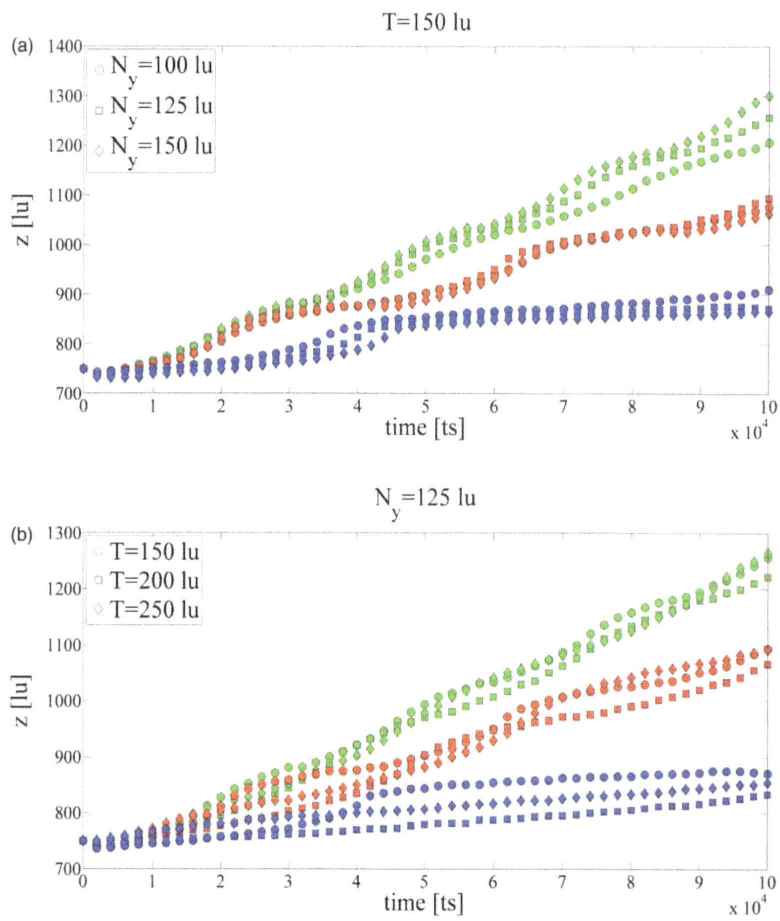

Figure 13. Front dynamics for zigzag walls in the absence of reactivity. The length of the capillary is L = 750 lu. Color code based on the amplitude A: green, red and blue are associated with $A = i(N_y - 25)/16$ where i = 2,3,4.

shown in Figure 5. The infiltration process is faster for increasing misalignment of the walls. In this respect, it is important to note that for a given value of Δx the minimum and maximum radii vary with N_y within a group of systems displaying similar behavior. In general, it appears that the effect of pinning is weaker for increasing misalignment of the walls. Furthermore, for higher misalignment, it is found that pinning is clearly stronger in the case of narrower necks and smaller periods (results not shown for brevity). We also consider the tortuosity (Duda et al., 2011; Matyka & Koza, 2012). The tortuosity is a measure of the departure of fluid flow from straight pathways. The calculations are made by means of the formula $\lambda = <u>/<u_x>$ (Duda et al., 2011; Matyka & Koza, 2012), u being the overall fluid velocity (Kang, Zhang, & Chen, 2002; Shan & Chen, 1993; Shan & Doolen, 1995). It should be kept in mind that at the interface there are spurious currents (Wagner, 2003). The results shown in Figure 5 indicate that the tortuosity is smaller for higher misalignment of the walls. This means that it is more important to have wider pathways rather than only narrow, straight ones. The curves are grouped approximately according to the misalignment of the walls.

Figure 6 has to do with sinusoidal capillaries in the presence of reaction. Without reaction, the increase in period T results in a slightly lower penetration depth for $H_{min} = 75$ and 50 lu. When the reactivity is enabled, no drastic difference for the penetration depth can be observed. In the absence of reaction, in passing from T = 150 lu to T = 200 lu, flow slows down significantly for $H_{min} = 25$ lu (cf. discussion on Figure 3). Again, this points out that the structure of the capillary is more important than the minimum radius (cf. comments on Figure 3). With surface reaction, the results for the maximal thickening Δe of the solid surface (Sergi et al., 2014) indicates that the process of surface growth is not strongly affected by fluid behavior. The morphology presented in Figure 7 supports this thesis. This consequence is expected since

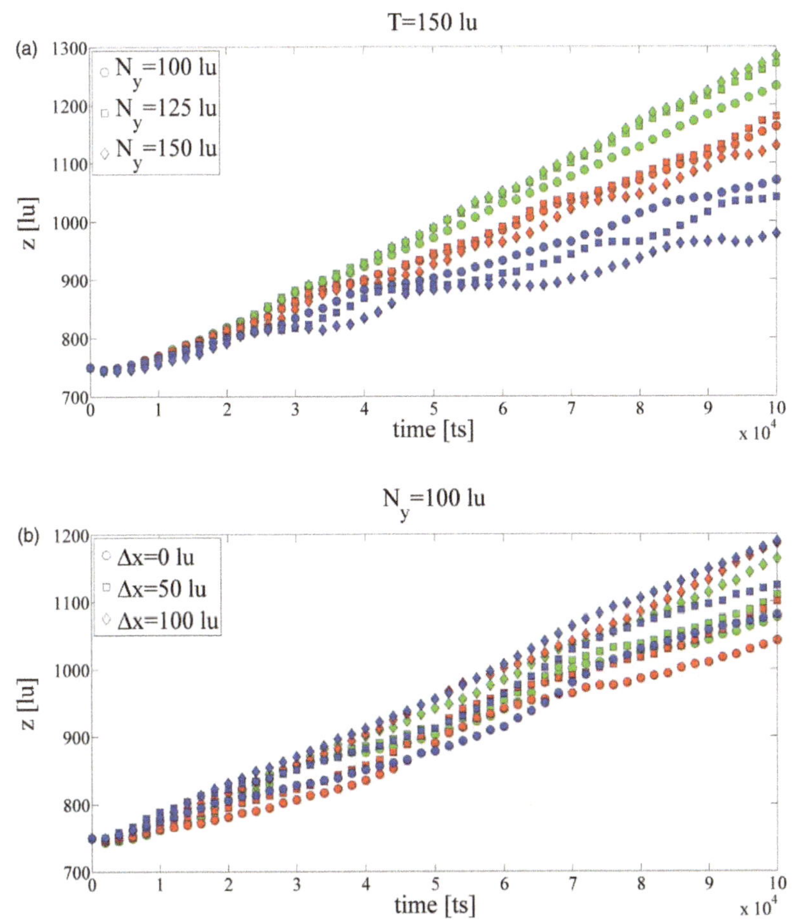

Figure 14. Motion of the invading front for zigzag walls without reactivity. The capillaries have length L = 750 lu. (a) The misalignment between the walls is set to $\Delta x = 100$ lu. Color code based on the amplitude $A = i(N_y - 25)/16$ with i = 2,3,4 for green, red and blue. (b) The amplitude is fixed to $A = 3(N_y - 25)/16$. Color code based on the period T = 150,200,250 lu for green, red and blue.

we know that the infiltration velocity was proven to have a marginal role on the kinetics of pore closure (Sergi et al., 2014). There appears to be an exception for the system with $H_{min} = 50$ lu and $k_r = 1.6 \cdot 10^{-3}$ lu/ts. Visual inspection of the dynamics reveals that this difference is chiefly due to an asymmetry in the thickening of the surface near the neck. For other N_y values the curves for Δe have a similar behavior.

In Figure 8 we analyze the effects of pinning for surface growth. When the period T varies, it is possible to distinguish different infiltration behaviors due to pinning. What is important to note is that the maximal width of the growing surface Δe is not affected by the hydrodynamic behavior. The evolution of Δe depends essentially on the reaction-rate constant k_r. It is observed that the infiltration process tends to be retarded to a larger extent by surface growth with smaller periods as the plateaus become longer (i.e., pinning more marked). Concerning the results for different tortuosities, from Figure 8 it can be seen that the various curves can be considered as grouped according to the degree of misalignment. On the other hand, the quantity Δe depends mainly on the reaction-rate constant k_r. These findings give further evidence for the prominence of the capillary structure, while the hydrodynamic behavior is of secondary importance for the occurrence of pore closure. The infiltration process is slowed down by surface growth more significantly for aligned walls.

We recall that for a uniform capillary, in the absence of surface reaction, the process of capillary infiltration is described by the following equation (Chibbaro, Biferale, Diotallevi, & Succi, 2009):

$$z(t) = \frac{V_{cap} H \cos\theta}{6L} t_d[\exp(-t/t_d) + t/t_d - 1] + z_0 \,, \quad (1)$$

where $V_{cap} = \gamma/\mu$ and $t_d = H^2/12\mu$. By z we designate the centerline position of the invading front, γ is the surface tension, μ the dynamic viscosity, ρ the density, H the capillary height, L the length, and θ the contact angle.

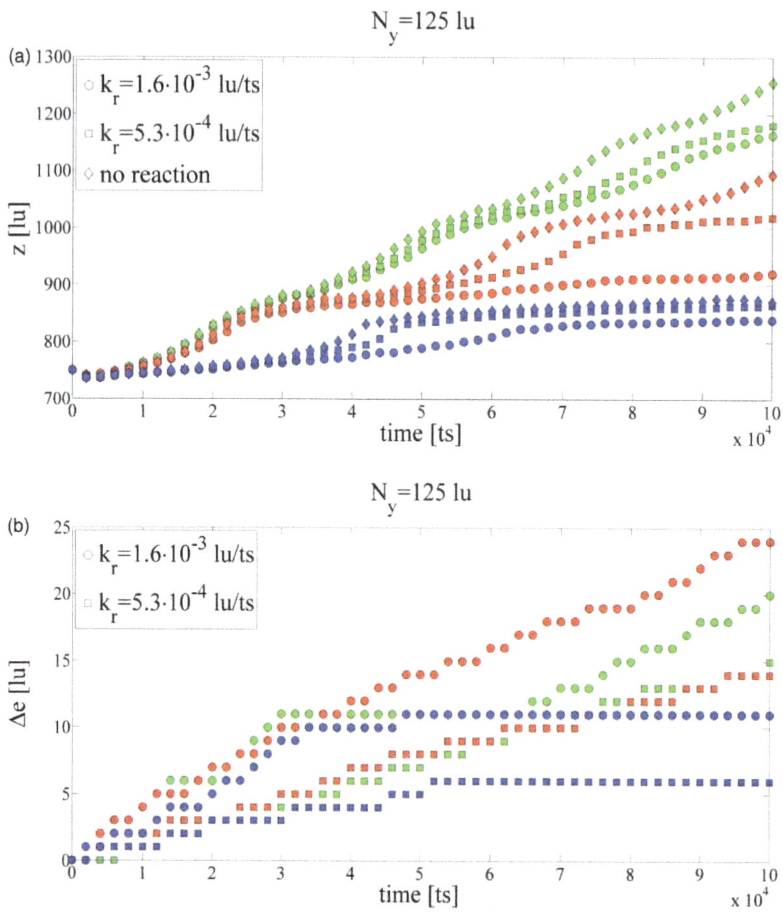

Figure 15. Capillary systems with zigzag walls in the presence of surface reaction. The length of the capillaries is L = 750 lu and the period is set to T = 150 lu. Color code based on the amplitude $A = i(N_y - 25)/16$ with i = 2,3,4 for green, red and blue. (a) Time dependence of the position of the invading front. (b) Evolution of the maximal thickening of the growing surface.

The quantity z has a linear dependence with time. We want to use this formula in order to extract approximations for the effective radius r_{eff}. The well-known result for the capillary pressure $\Delta P = 2\gamma \cos\theta / H$ allows the fitting of the simulation data to a straight line with a slope given by $H^2 \Delta P / 12\mu L$. By using a suitable average for ΔP, an estimate for the effective radius can be obtained. From this result, we can then calculate the dynamic contact angle using the capillary pressure (Joos, Van Remoortere, & Bracke, 1990; Van Remoortere & Joos, 1993). Finally, by taking an average for the contact angle it is possible to employ Eq. (1), as shown in Figure 9. Deviations from the behavior predicted by Eq. (1) are more marked when the phenomenon of pinning occurs. The reactivity turns out to enhance the effect of pinning. At the beginning of every plateau the dynamic contact angle θ reaches the local minimum value. When the infiltration accelerates, θ increases sharply. When pinning is more significant, θ remains quite large and then decreases more slowly. This phenomenon is clearer from the results with reaction, for which pinning is stronger. It is found that the effective radius is better approximated by the average radius, with the exception when pinning becomes more significant. In this case r_{min} is closer to r_{eff}. It is important to recall that pinning depends mainly on the structure: r_{min} is a parameter among others defining it, as seen before for the discussion related to Figure 3.

4.1.2. Step-shaped profiles

For periodic step-shaped walls, we limit ourselves to capillaries of length $L = 750$ lu. From Figure 10 it emerges that the fluid does not advance any further once the meniscus has reached the zones of enlargement at $x = 865$ lu. This occurs because at the corners the contact line remains pinned. This behavior was ascertained by repeating the simulations for $N_y = 100{,}125$ lu with 500,000 timesteps. The same conclusions also hold for longer periods. As seen

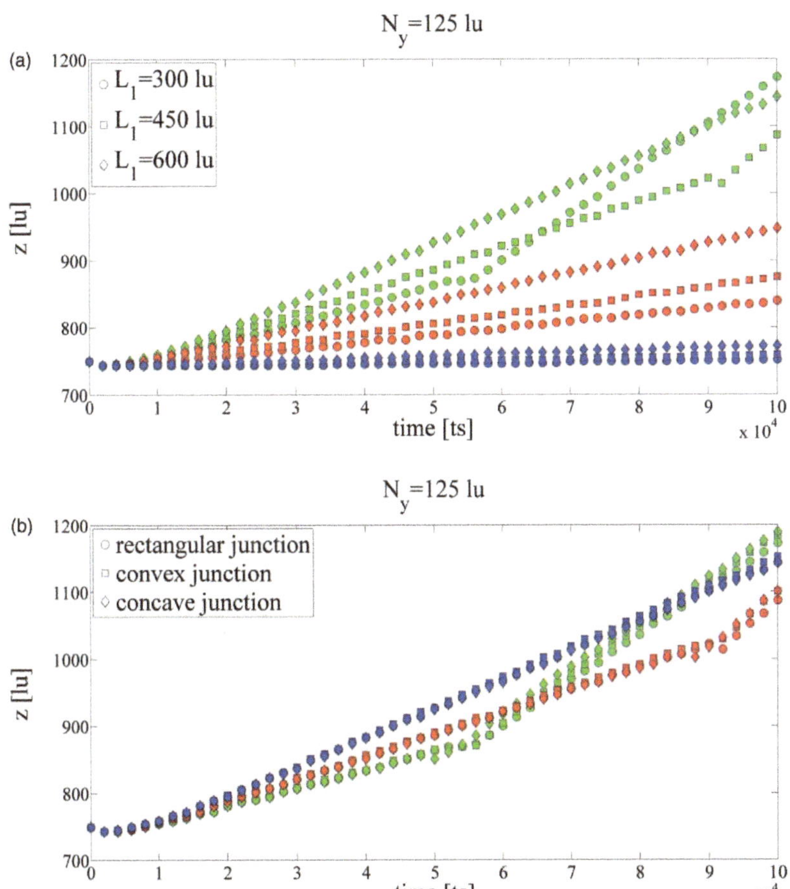

Figure 16. Time dependence of the front position for constrictions without reactivity. The length of the capillary is $L = 750$ lu. (a) The junction has a rectangular shape. Color code based on the width $\Delta H = 25{,}38{,}50$ lu associated with green, red and blue, respectively. The corresponding minimum heights are $H_{min} = 75{,}50{,}25$ lu. (b) The width of the step is set to $\Delta H = 25$ lu. Green, red and blue are used for the length of the wider segment given by $L_1 = 300{,}450{,}600$ lu.

before, an increase in capillary forces could be obtained by considering smaller ratios A/N_y (Mognetti & Yeomans, 2009). A detailed analysis of the conditions for depinning is outside the scope of the present study: the interested reader is directed to the existing literature (Blow et al., 2009; Chibbaro, Biferale, Binder, et al., 2009; Chibbaro, Costa, et al., 2009; Kusumaatmaja, et al., 2008; Mognetti & Yeomans, 2009).

Figure 11 presents the results in the case of misalignment between the upper and lower walls. The effect of misalignment is that the pinning barrier can be overcome and the meniscus does not remain stuck. The curves are grouped as for sinusoidal profiles (see Figure 3). Different degrees of misalignment do not separate the infiltration behaviors neatly. Moreover, the dependence on the period is not so marked. The infiltration is slightly faster for T = 250 lu, the largest period. The results for the other capillary systems lead to the same conclusions. Of course, the infiltration process is slower for step-shaped walls than for sinusoidal profiles. For the considered number of timesteps, the difference amounts to about 100 lu. Furthermore, the systems with the period T = 150 lu have the same minimum radius of the sinusoidal capillaries of Figure 9. It thus follows that the presence of corners quite significantly lowers the effective radius.

The results with reactivity are also in line with previous findings. A typical morphology is shown in Figure 7. Again, it is clear that the quantity Δe, controlling pore closure, does not depend on the flow velocity, as it comes out from Figure 12. The surface growth opposite to fluid flow seems to have a marginal influence on the infiltration process. The remarks presented in the previous paragraph are material to this statement. The curves of Figure 12 give further support to this thesis since their behavior is related. Similar results are obtained for the other capillary systems. From a comparison with the equivalent sinusoidal capillary

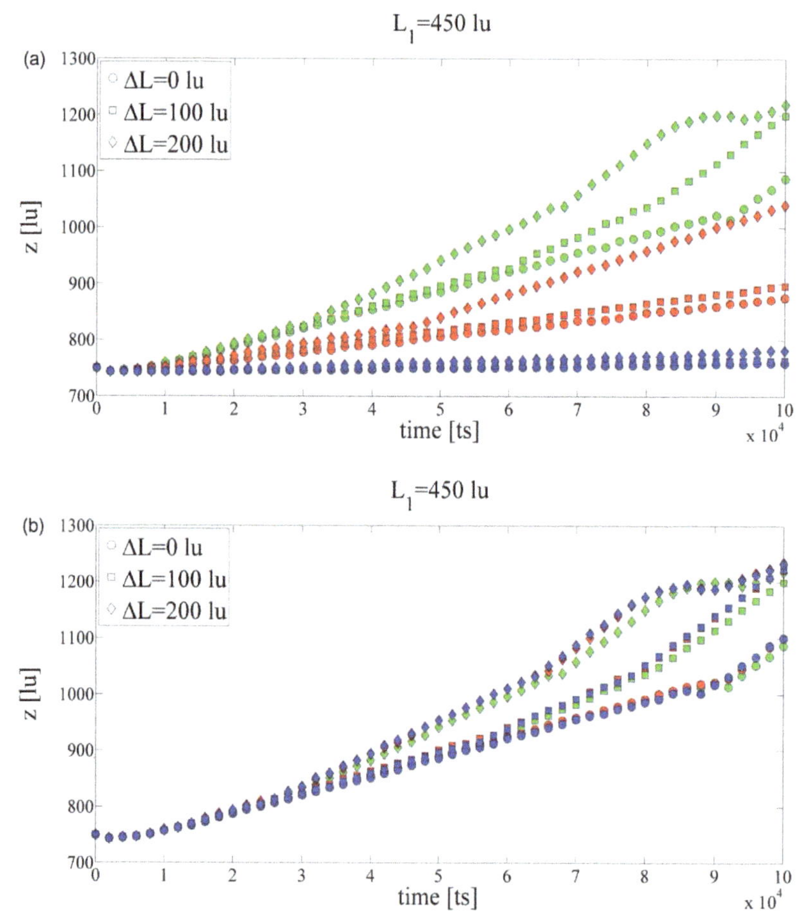

Figure 17. Front dynamics for constrictions with inert solid surface. The length of the capillary is L = 750 lu; the width of the simulation domain is N_y = 125 lu. (a) Color code based on the width of the solid phase ΔH: green, red and blue for ΔH = 25,38,50 lu, leading to the minimum heights H_{min} = 75,50,25 lu. (b) The width of the solid phase is ΔH = 25 lu (i.e., H_{min} = 75 lu). Color code based on the shape of the junction: green, red and blue for rectangular, convex and concave.

systems it turns out that the process of pore closure occurs sooner with step-shaped walls. Corners are thus more detrimental. Interestingly, it follows that the phenomenon of pore closure is actually sensitive to the wall structure.

4.1.3. Zigzag profiles

For zigzag profiles we also restrict the simulations to the case of capillaries of length 750 lu. In general, the effect of pinning appears to be weaker for this geometry. As can be seen in Figure 13, the dependence of z on time is almost linear, with the exception for narrower necks. The curves are grouped as for sinusoidal profiles. But the infiltration process turns out to be faster for zigzag profiles. The period T does not seem to have a prominent role. For narrower necks, the same conclusions reached for sinusoidal profiles hold (see Figure 3), which was also verified up to 500,000 timesteps in this case. Namely, the infiltration depth is almost the same but for larger periods there are less evident plateaus.

Misalignment between walls has the effect of speeding up capillary infiltration, especially for narrower necks, as shown in Figure 14. For wider necks, the advantage of misalignment is comparable to that for the other geometries, that is within 150 lu (see Figure 5). It is found that the dynamics is slightly faster for larger periods (not shown for brevity); the corresponding characteristic radii (e.g., minimum and hydraulic) are almost the same. The other capillary systems lead to the same conclusions.

From Figure 15 it can be seen that, for the same reaction conditions, Δe does not display approximately the same behavior until pore closure occurs, as for the other profiles (see Figures 6, 8 and 12). The difference can be ascribed to the fact that the surface does not grow principally near the maximum, i.e., the peak (see Figure 7). By varying the width of the channels, this phenomenon takes place for

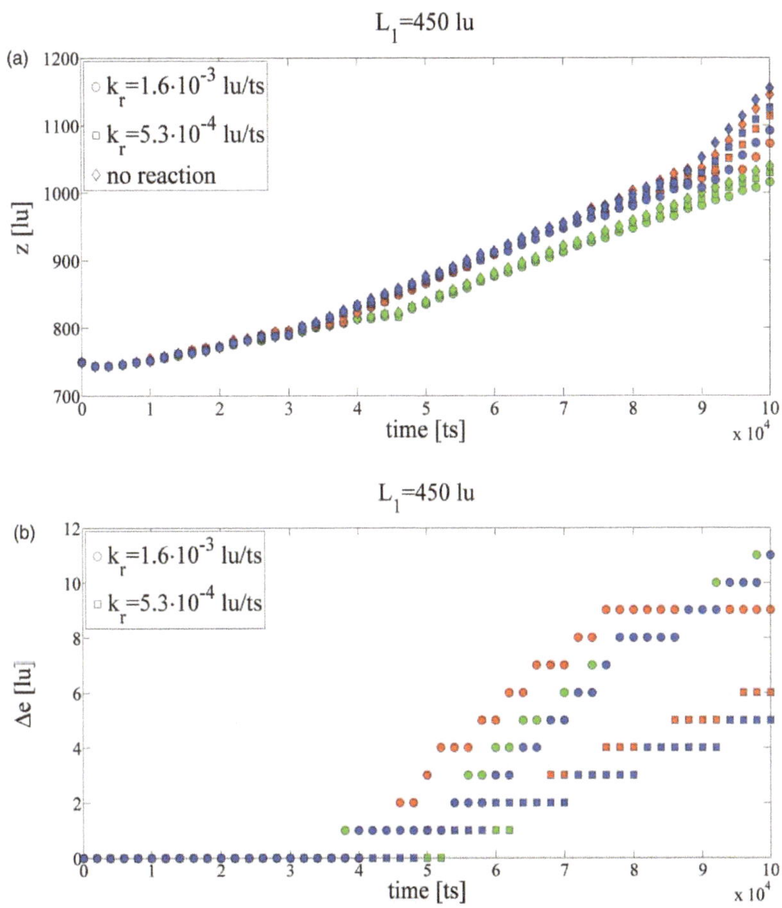

Figure 18. Results for constrictions in the presence of surface reaction. The length of the capillaries is set to $L = 750$ lu, the width of the simulation domain is $N_y = 125$ lu and the misalignment is given by $\Delta L = 200$ lu. The initial width of the solid surface is given by $\Delta H = 38$ lu, corresponding to a minimum height of $H_{min} = 50$ lu. Color code based on the morphology of the junction: green, red and blue for rectangular, convex and concave. (a) Front position over the course of time. (b) Time dependence of the maximal thickening of the growing surface in the neighborhood of the junction.

other combinations of the other parameters. It is thus difficult to identify the causes with precision. In any case, we can conclude that the process of pore closure is slower for zigzag profiles, which is due to their structural properties (sharp peaks).

Since the infiltration process appears to be slightly faster for zigzag profiles than for sinusoidal ones, it turns out that the effective radius for this geometry is higher (see Figure 9), as verified. Presumably, the reason resides in the fact that the role of pinning is less significant. However, in general this phenomenon is not captured exhaustively by characteristic radii. Although all periodic profiles have the same minimum radius, in the absence of misalignment the flow cannot proceed for the step-shaped structure. Of course, the tortuosity is expected to be smaller for zigzag profiles. As an example, for the equivalent systems of Figure 5, it is found that the first maxima are slightly lower. For completeness, for the equivalent systems with step-shaped structures, the first maxima are around 3 for the highest misalignment of the walls.

4.2. *Two-pore systems*

Constrictions and expansions are characterized by constant radii (see Figure 2). In our systems, L_1 determines the relative length of the two segments. ΔH controls the width of the narrower segment. As ΔL increases, the necks become shorter while the main cavity becomes smaller. For the rectangular step, at the junction there are two corners. For the convex and concave shapes there is only one corner, but the profile of the junction is longer.

4.2.1. *Constrictions*

The length of the capillary is also fixed to $L = 750$ lu for constrictions. Furthermore, we consider for the width of

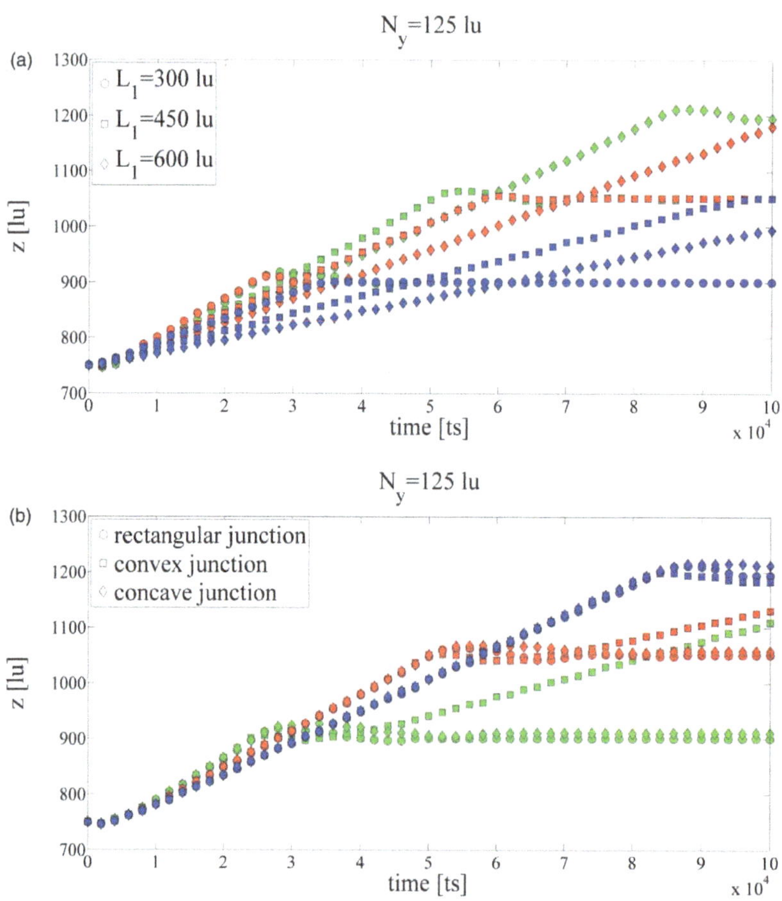

Figure 19. Front position over the course of time for capillary systems with expansions. The surface reaction is disregarded. The length of the capillaries is $L = 750$ lu. (a) The shape of the junction is rectangular. Color code based on the width of the solid phase $\Delta H = 25,38,50$ lu for green, red and blue, respectively. These values determine the minimum heights $H_{min} = 75,50,25$ lu. (b) The width of the step is set to $\Delta H = 25$ lu, i.e., $H_{min} = 75$ lu. Color code based on the length of the narrower segment $L_1 = 300,450,600$ lu, corresponding to green, red and blue.

the simulation domain only the case $N_y = 125$ lu. From Figure 16 it is interesting to see that for the channels with a higher minimum radius, when the capillary shrinks, the flow accelerates. Otherwise, the infiltration is clearly faster for the capillaries with longer segments of large width. The shape of the junction has no noticeable effect. The same conclusions can be drawn from the other capillary systems. Comparison with the periodic profiles indicates that the infiltration process is slower (see Figures 3 and 13) despite the absence of pinning.

Figure 17 shows the results in the presence of misalignment between the upper and lower walls. Misalignment can speed up the infiltration process. As the minimum radius increases – that is, the step ΔH decreases – the effect of misalignment becomes more significant. For smaller segments of the wider part, it is found that misalignment is associated with a weaker resistance to flow (results not shown for brevity). For higher misalignment, when the channels widen, the front remains pinned, creating a plateau. Again, the shape of the junction has no appreciable consequence. We start seeing a clear separation of the curves for smaller minimum radii and higher misalignment: the rectangular shape leads to a slower dynamics. For increasing L_1, it also turns out that the convex shape is associated with a faster infiltration process (not shown for brevity). For periodic profiles, the misalignment of the walls leads to a slightly faster dynamics (see Figures 5 and 14).

For these capillary systems the reaction has a less marked effect. In Figure 18 it is instructive to notice that the process of surface growth starts becoming detrimental especially when the front reaches the minimum radius. This observation holds also for the other structures. As a result, the reaction has a higher impact for smaller segments with a large radius. The analysis proposed for sinusoidal profiles in order to estimate the effective radius

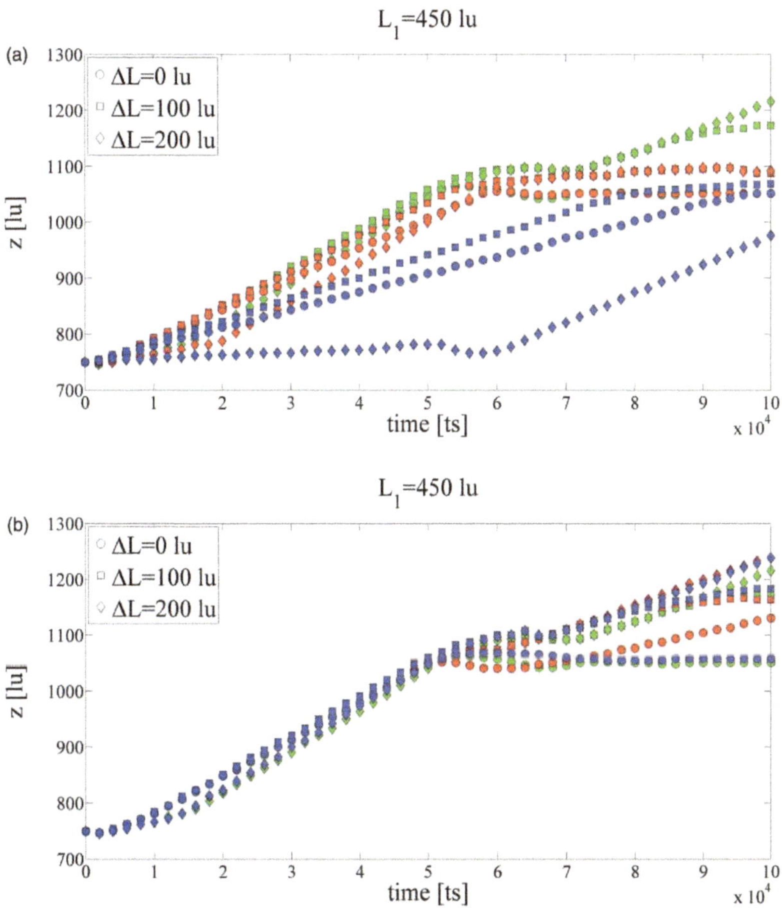

Figure 20. Infiltration dynamics for capillary systems with expansions in the presence of misalignment between the walls. The surface reaction is not taken into account. The length of the capillaries is fixed to $L = 750$ lu and the width of the simulation domain is $N_y = 125$ lu. (a) The shape of the junction is rectangular. Color code based on the width of the solid phase $\Delta H = 25,38,50$ lu corresponding to green, red and blue. The minimum heights are thus $H_{min} = 75,50,25$ lu. (b) The width of the step is set to $\Delta H = 25$ lu, leading to $H_{min} = 75$ lu. Color code based on the shape of the junction: green, red and blue for rectangular, convex and concave, respectively.

was also performed for the systems of Figures 17 and 18. It arises that the average radius characterizes the infiltration process better for large minimum radii and high misalignment. Departure from these conditions leads to effective radii closer to the minimum one. For the equivalent sinusoidal profiles the dynamics is faster. Their effective radius is higher. The results for the tortuosity also indicate that the resistance to flow tends to be stronger for constrictions (results not shown for brevity).

4.2.2. Expansions

In this case we also restrict ourselves to lengths of 750 lu and widths of $N_y = 125$ lu for the channels. The results for the channels presenting expansions are very interesting for the problem of pinning because it might be presumed that the drag force is weaker while the capillary forces increase because of longer flat walls. From Figure 19 it can be seen that, for the rectangular shape of the junction, the pinning barrier cannot be overcome in any case. For the largest minimum height $H_{min} = 75$ lu, the front can proceed for the convex geometry. This is easier for the smaller value of L_1. It might seem surprising that the concave geometry can be less favorable. We point out that, for the convex shape, the front has higher momentum when it encounters the sharp corner. For the data of Figure 19, with $L_1 = 450$ lu, the average over the last three frames before a negative momentum is recorded yields 724 mu·lu/ts for the convex shape and 333 mu·lu/ts for the concave shape. Furthermore, because of the convex geometry, the contact line can slide along the border of the junction and again reach the uniform part of the channel. The penetration for the convex geometry is slightly slower than for constrictions (see Figure 16). For the sake of completeness, simulations up to 500,000 ts proves that the pinning barrier is overcome only for the convex geometry

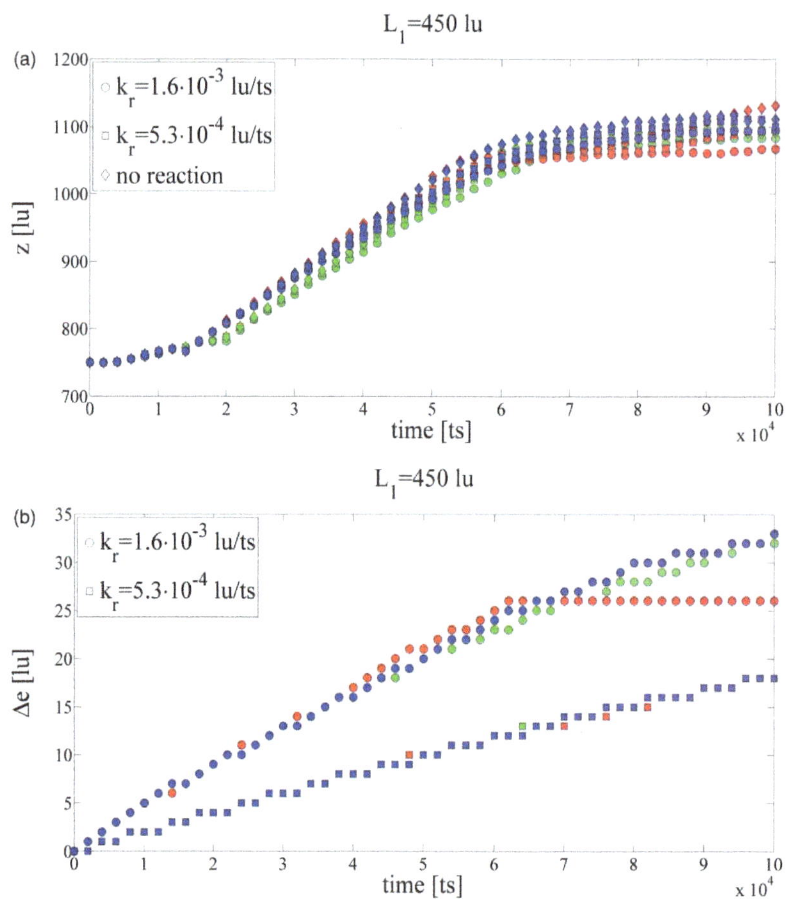

Figure 21. Results for capillary systems with expansions in the presence of surface reaction. The length of the capillaries is $L = 750$ lu, the width of the simulation domain is $N_y = 125$ lu and the misalignment between the walls is set to $\Delta L = 200$ lu. The initial width of the solid step is given by $\Delta H = 38$ lu, determining the minimum height $H_{min} = 50$ lu. Color code based on the shape of the junction: green, red and blue for a rectangular, convex and concave morphology, respectively. (a) Front position over the course of time. (b) Time dependence of the maximal thickening of the growing surface.

with $L_l = 300,450$ lu and $\Delta H = 25$ lu, corresponding to $H_{min} = 75$ lu.

Figure 20 shows the results in the presence of misalignment between the walls. For a rectangular shape of the junction, it is interesting to remark that the misalignment seems to have no effect for $H_{min} = 50$ lu. The centerline position of the front remains almost unchanged, but the contact line can proceed inside the widening part of the channel. We observe this behavior during a shorter time interval for smaller L_l (results not shown for brevity). In this case, the capillary forces are expected to be stronger. In the light of the considerations of the previous paragraph, this configuration can also be assumed to be one that to a larger extent eases the infiltration. At the beginning, the infiltration process is particularly slow for $H_{min} = 25$ lu and higher misalignment because the orifice is not behind the contact line. From Figure 20 we also see that the shape of the junction has no particular influence for $H_{min} = 75$ lu with misalignment. For smaller minimum heights, we distinguish no privileged geometry. In general, the effect of misalignment is weaker for longer narrow segments and larger minimum heights. Comparison with constrictions indicates that for wider channels the results are similar (see Figure 17). Otherwise, constrictions are more interesting for the absence of the phenomenon of pinning. In any case, constrictions can be regarded as expansions for flow in the other sense. So, in a random porous medium, both types of channel are present.

The results with reaction provide a compelling example for the independence of surface growth from the infiltration velocity (see Figure 21). Comparison with the outcome for constrictions allows the possibility of making this statement more precise (see Figure 18). We see that this is true in particular when the structural features responsible for the retardation of the flow are already behind the contact line. For an extended porous medium, this means that the flow is interrupted because of pore obstruction at the surface of the solid with a relation linear with time. Now, the infiltration velocity is controlled by the pore characteristics. Our finding does not imply that the porous structure is irrelevant. Indeed, a larger porosity should require more time before the flow stops and, on the basis of previous experiments, it might be possible to make predictions. We verified that the curves for z separate more neatly with reactivity for the other values of L_l and ΔL, since at the beginning both solid steps are behind the contact line for these configurations. Furthermore, the effect of surface growth is more marked for decreasing minimum radii. In general, we recognize no particular influence of the shape of the junction for surface growth.

The infiltration process starts with a strong acceleration near the converging part of the channel, followed by a phase where the front (centerline position) is almost at rest (see Figure 21). As a consequence, it is more difficult to fit the data to a linear relation. In any case, it follows that the results for expansions are in part comparable with those for constrictions and periodic profiles (sinusoidal and zigzag) and even slightly faster (see Figures 9 and 18). The analysis for tortuosity reveals that expansions are less advantageous than constrictions. It also appears that the concave geometry favors infiltration more than the rectangular shape (results not shown for brevity). As an example, for higher misalignment and $H_{min} = 50$ lu, it is found that the tortuosity is relatively low, around 1.35, with a peak in the range 2–3, when the flow meets the narrowing part of the channel. With reaction, the peaks can reach 5 with stronger reactivity.

5. Conclusion

Our work deals with LB simulations in 2D for capillary penetration into single channels with walls characterized by different geometric properties. The focus is on the effects on fluid flow of surface growth from a supersaturated solution (Kang et al., 2003, 2004, 2007; Kang, Zhang, Chen, & He, 2002; Lu et al., 2009). The study is motivated by the problem of Si infiltration into C preforms (Bougiouri et al., 2006; Dezellus et al., 2003; Dezellus & Eustathopoulos, 2010; Einset, 1996, 1998; Eustathopoulos et al., 1999; Hillig et al., 1975; Israel et al., 2010; Liu et al., 2010; Messner & Chiang, 1990; Mortensen et al., 1997; Voytovych et al., 2008). Several inconsistencies exist between the experiments and the models, as well as the simulation conditions. For example, in the LB models the liquid and vapor phases have the same density. But this choice allows for the reproduction of a linear dynamics for the Washburn infiltration (Chibbaro, 2008; Chibbaro, Biferale, Diotallevi, & Succi, 2009), as observed in experiments (Israel et al., 2010; Voytovych et al., 2008). However, the accordance with experimental results is poor since in our simulations the Reynolds number is overestimated (Sergi et al., 2014). This shortcoming could be limited by considering longer capillaries with more iterations. This is not necessary for our purposes, as the process of surface growth exhibits a relatively weak dependence on the infiltration velocity. This is expected to be the case, especially for smaller Reynolds numbers. Furthermore, for interconnected porous systems smaller Reynolds numbers could be attained. Another point is that in our simulations the reaction-rate constant k_r varies, while in real systems this quantity is fixed to a single value. From experimental data the values of the parameter k_r could be chosen more accurately on the basis of the morphology of the growing solid phase at the surface. This kind of analysis is important because a proper comparison with the experimental results should be based on dimensionless, characteristic numbers involving the time set by the process of pore closure and the corresponding infiltrated distance. Indeed, the flow behavior is expected to depend on the relative effects between the intervening mechanical forces and the speed of reactivity. In this way, it would be possible to have good accordance for the full infiltration dynamics resulting from simulations and experiments. This is easy to understand because otherwise it is imposed only the condition to realize the last stage of the process. Finally, it is worth noting that other phenomena are completely disregarded in our description, such as thermal effects (Sangsuwan, Tewari, Gatica, Singh, & Dickerson, 1999) and the transition to wetting (i.e., the reaction at the contact line). As a consequence, our investigation provides no explanation for the emergence of the linear Washburn behavior (Bougiouri et al., 2006; Israel et al., 2010; Voytovych et al., 2008).

Our modeling approach consists in breaking intraparticle porosity down into basic, simplified structures. Their function for capillary infiltration and surface growth is assessed. Inputs for preform preparation can be singled out. In general, it is found that the structure of the channels is more important than the minimum radius. More specifically, for periodic profiles, the simulations indicate that the ratio of the amplitude to the channel width, A/N_y, mainly affects the infiltration speed. Smaller ratios are more advantageous. The misalignment between the walls further speeds up the infiltration process. The period does not appear to be a significant parameter. Of course, smaller periods are associated with a higher drag force. For the sinusoidal profiles, the length of the capillaries is not so important, as opposed to uniform channels. Faster infiltration occurs for zigzag profiles. The results for the sinusoidal profiles are similar. Step-walled capillaries lead to a slower dynamics. In this case, the phenomenon of pinning is more marked.

These considerations have interesting consequences for the selection of ceramic powders and the preform preparation. Small A/N_y ratios can be realized via wide pathways with weakly pronounced radius variations. It follows that powders containing larger particles are good candidates. The results for the misalignment of the walls suggest that an excessive compaction of the grains should be avoided, especially in the direction normal to the infiltration. The use of particles with rounded edges is highly recommended. In particular, the results for step-walled capillaries show that powders with fine grains are detrimental.

Constrictions and expansions can arise by the arrangement of quite coarse particles. By inverting the sense of fluid motion the other configuration is obtained. The danger of pinning associated with expansions should be minor in random porous media given the typical size of the grains. It is interesting to note that the effect of reaction is weaker for constrictions and expansions. An important consequence is that the flow paths which allow fluid to be guided into the porous structures should be created by means of large aggregates. Small, faceted grains could instead be added in order to enhance the process of surface growth. We recall that in C/SiC composites the SiC formation is desired for the mechanical properties.

To summarize, in order to ensure the impregnation by Si it is advisable to start with preforms presenting pore pathways as wide as possible, straight and with round morphologies. Preferably, this could be achieved by using large ceramic particles. The residual Si in the ceramics could be limited by adding carbon powders containing smaller grains. Particles of small size are likely to slow down the infiltration significantly, particularly if the surface is characterized by corners. Previous investigations have already highlighted the importance of the particle size, the packing properties and the microstructure for the preform to be impregnated (Gadow & Speicher, 2000; Israel et al., 2010; Paik et al., 2002; Salamone et al., 2008). Another finding of relevance is that, especially at the solid surface of the preform, the growth kinetics are expected to depend marginally on the infiltration velocity. As explained, this finding can have practical consequences for the processing conditions (see Section 4.2.2). As suggested by the analysis for misalignment, the porosity alone does not exhaustively characterize the whole infiltration process. Our investigation still leaves open the question of the role of the flow pattern inside real porous media for reactive infiltration. Further progress can be made by considering more complex models for the microstructure, including interconnected pore channels. Needless to say that this problem is critical for a better determination of the optimal configuration of the porous preform for the manufacturing of ceramic components devised for advanced applications. In this respect, the LB method seems to have an advantage over other numerical schemes for hydrodynamics.

Disclosure statement

No potential conflict of interest was reported by the authors.

Funding

This research was funded by the European Union Seventh Framework Programme (FP7/2007-2013) under grant agreement no. 280464, 'High-frequency Electro-Magnetic technologies for the advanced processing of ceramic matrix composites and graphite expansion' (HELM).

References

Alava, M., Dube, M., & Rost, M. (2004). Imbibition in disordered media. *Advances in Physics*, *53*, 83–175.

Benzi, R., Succi, S., & Vergassola, M. (1992). The lattice Boltzmann equation: Theory and applications. *Physics Reports*, *222*, 145–197.

Blow, M. L., Kusumaatmaja, H., & Yeomans, J. M. (2009). Imbibition through an array of triangular posts. *Journal of Physics: Condensed Matter*, *21*, 464125.

Bougiouri, V., Voytovych, R., Rojo-Calderon, N., Narciso, J., & Eustathopoulos, N. (2006). The role of the chemical reaction in the infiltration of porous carbon by NiSi alloys. *Scripta Materialia*, *54*, 1875–1878.

Chen, S., & Doolen, G. D. (1998). Lattice Boltzmann method for fluid flows. *Annual Review of Fluid Mechanics*, *30*, 329–364.

Chibbaro, S. (2008). Capillary filling with pseudo-potential binary Lattice-Boltzmann model. *The European Physical Journal E*, *27*, 99–106.

Chibbaro, S., Biferale, L., Binder, K., Dimitrov, D., Diotallevi, F., Milchev, A., & Succi, S. (2009). Hydrokinetic simulations of nanoscopic precursor films in rough channels. *Journal of Statistical Mechanics: Theory and Experiments*, *2009*, P06007.

Chibbaro, S., Biferale, L., Diotallevi, F., & Succi, S. (2009). Capillary filling for multicomponent fluid using the pseudo-potential Lattice Boltzmann method. *The European Physical Journal Special Topics*, *171*, 223–228.

Chibbaro, S., Costa, E., Dimitrov, D. I., Diotallevi, F., Milchev, A., Palmieri, D., ... Succi, S. (2009). Capillary filling in microchannels with wall corrugations: A comparative study of the Concus-Finn criterion by continuum, kinetic, and atomistic approaches. *Langmuir*, *25*, 12653–12660.

Clyne, T. W., Golosnoy, I. O., Tan, J. C., & Markaki, A. E. (2006). Porous material for thermal management under extreme conditions. *Philosophical Transactions of the Royal Society A*, *364*, 125–146.

Dezellus, O., & Eustathopoulos, N. (2010). Fundamental issues of reactive wetting by liquid metals. *Journal of Materials Science*, *45*, 4256–4264.

Dezellus, O., Hodaj, F., & Eustathopoulos, N. (2003). Progress in modelling of chemical-reaction limited wetting. *Journal of the European Ceramic Society*, *23*, 2797–2803.

Diotallevi, F., Biferale, L., Chibbaro, S., Lamura, A., Pontrelli, G., Sbragaglia, M., ... Toschi, T. (2009). Capillary filling using lattice Boltzmann equations: The case of multi-phase flows. *The European Physical Journal Special Topics*, *166*, 111–116.

Diotallevi, F., Biferale, L., Chibbaro, S., Pontrelli, G., Toschi, F., & Succi, S. (2009). Lattice Boltzmann simulations of capillary filling: Finite vapour density effects. *The European Physical Journal Special Topics*, *171*, 237–243.

Duda, A., Koza, Z., & Matyka, M. (2011). Hydraulic tortuosity in arbitrary porous media flow. *Physical Review E*, *84*, 36319–36326.

Dullien, F. A. L. (1992). *Porous media: Fluid transport and pore structure*. New York, NY: Academic Press.

Einset, E. O. (1996). Capillary infiltration rates into porous media with application to silcomp processing. *Journal of the American Ceramic Society*, *79*, 333–338.

Einset, E. O. (1998). Analysis of reactive melt infiltration in the processing of ceramics and ceramic composites. *Chemical Engineering Science*, *53*, 1027–1039.

Eustathopoulos, N., Nicholas, M. G., & Drevet, B. (1999). *Wettability at high temperatures*. Oxford: Pergamon Press.

Furler, P., Scheffe, J., Gorbar, M., Moes, L., Vogt, U., & Steinfeld, A. (2012). Solar thermochemical CO_2 splitting utilizing a reticulated porous ceria redox system. *Energy and Fuels*, *26*, 7051–7059.

Furler, P., Scheffe, J. R., & Steinfeld, A. (2012). Syngas production by simultaneous splitting of H_2O and CO_2 via ceria redox reactions in a high-temperature solar reactor. *Energy and Environmental Science*, *5*, 6098–6103.

Gadow, R. (2000). Current status and future prospects of CMC brake components and their manufacturing technologies. In T. Jessen and E. Ustundag (Eds.), *24th Annual Conference on Composites, Advanced Ceramics, Materials, and Structures – A* (pp. 15–29). Hoboken, NJ: John Wiley & Sons.

Gadow, R., & Speicher, M. (2000). Optimized morphological design for silicon infiltrated microporous carbon preforms. In T. Jessen and E. Ustundag (Eds.), *24th Annual Conference on Composites, Advanced Ceramics, Materials, and Structures – A* (pp. 485–492). Hoboken, NJ: John Wiley & Sons.

Gern, F. H., & Kochendörfer, R. (1997). Liquid silicon infiltration: Description of infiltration dynamics and silicon carbide formation. *Composites Part A: Applied Science and Manufacturing*, *28*, 355–364.

Ghosh, S., Patil, P., Mishra, S. C., Das, A. K., & Das, P. K. (2012). 3-D lattice Boltzmann model for asymmetric Taylor bubble and Taylor drop in inclined channel. *Engineering Applications of Computational Fluid Mechanics*, *6*, 383–394.

Goodall, R., & Mortensen, A. (2013). *Porous metals: Physical metallurgy*. Philadelphia, PA: Elsevier.

Gross, M., Varnik, F., Raabe, D., & Steinbach, I. (2010). Small droplets on superhydrophobic substrates. *Physical Review E*, *81*, 51606–51619.

Grzelakowski, C., Jazia, D. B., Lebeau, B., Vonna, L., Dupuis, D., & Haidara, H. (2009). On the influence of pore structure on the free-imbibition of sessile drops into nanoporous substrates. *Langmuir*, *25*, 5855–5860.

Guiet, J., Reggio, M., & Teyssedou, A. (2011). Implementation and application of the lattice Boltzmann method using MATLAB. *Engineering Applications of Computational Fluid Mechanics*, *5*, 117–126.

Haghani, R., Rahimian, M. H., & Taghilou, M. (2013). LBM simulation of a droplet dripping down a hole. *Engineering Applications of Computational Fluid Mechanics*, *7*, 461–470.

Hillig, W. B., Mehan, R. L., Morelock, C. R., DeCarlo, V. J., & Laskow, W. (1975). Silicon/silicon carbide composites. *American Ceramic Society Bulletin*, *54*, 1054–1056.

Israel, R., Voytovych, R., Protsenko, P., Drevet, B., Camel, D., & Eustathopoulos, N. (2010). Capillary interaction between molten silicon and porous graphite. *Journal of Materials Science*, *45*, 2210–2217.

Jeje, A. A., & Zimmermann, M. H. (1979). Resistance to water flow in xylem vessels. *Journal of Experimental Botany*, *30*, 817–827.

Joos, P., Van Remoortere, P., & Bracke, M. (1990). The kinetics of wetting in a capillary. *Journal of Colloid and Interface Science*, *136*, 189–197.

Kang, Q., Lichtner, P. C., & Zhang, D. (2007). An improved lattice Boltzmann model for multicomponent reactive transport in porous media at the pore scale. *Water Resources Research*, *43*, 5551–5562.

Kang, Q., Zhang, D., & Chen, S. (2002). Displacement of a two-dimensional immiscible droplet in a channel. *Physics of Fluids*, *14*, 3203–3214.

Kang, Q., Zhang, D., & Chen, S. (2003). Simulation of dissolution and precipitation in porous media. *Journal of Geophysical Research*, *108*, 2504–2513.

Kang, Q., Zhang, D., Chen, S., & He, X. (2002). Lattice Boltzmann simulation of chemical dissolution in porous media. *Physical Review E*, *65*, 36318–36325.

Kang, Q., Zhang, D., Lichtner, P. C., & Tsimpanogiannis, I. N. (2004). Lattice Boltzmann model for crystal growth from supersaturated solution. *Geophysical Research Letters*, *31*, 21107–21111.

Krenkel, W. (2005). Carbon fibre reinforced silicon carbide composites (C/SiC, C/C-SiC). In N. P. Bansal (Ed.), *Handbook of ceramic composites* (pp. 117–148). New York, NY: Kluwer Academic.

Kusumaatmaja, H., Pooley, C. M., Girardo, S., Pisignano, D., & Yeomans, J. M. (2008). Capillary filling in pattern channels. *Physical Review E*, *77*, 67301–67304.

Landau, L. D., & Lifshitz, E. M. (2008). *Fluid mechanics*. Cambridge, MA: Elsevier.

Liu, G. W., Muolo, M. L., Valenza, F., & Passerone, A. (2010). Survey on wetting of SiC by molten metals. *Ceramics International, 36*, 1177–1188.

Lu, G., DePaolo, D. J., Kang, Q., & Zhang, D. (2009). Lattice Boltzmann simulation of snow crystal growth in clouds. *Journal of Geophysical Research, 114*, 11087–11100.

Martys, N. S., & Hagedorn, J. G. (2002). Multiscale modeling of fluid transport in heterogeneous materials using discrete Boltzmann methods. *Materials and Structures, 35*, 650–659.

Matyka, M., & Koza, Z. (2012). How to calculate tortuosity easily? In K Vafai (Ed.), *Porous Media and Its Applications in Science, Engineering and Industry: Fourth International Conference* (pp. 17–22). Melville, NY: AIP.

Messner, R. P., & Chiang, Y. M. (1990). Liquid-phase reaction-bonding of silicon carbide using alloyed silicon-molybdenum melts. *Journal of the American Ceramic Society, 73*, 1193–1200.

Miller, W., & Succi, S. (2002). A lattice Boltzmann model for anisotropic crystal growth from melt. *Journal of Statistical Physics, 107*(1–2), 173–186.

Mognetti, B. M., & Yeomans, J. M. (2009). Capillary filling in microchannels patterned by posts. *Physical Review E, 80*, 56309–56316.

Mortensen, A., Drevet, B., & Eustathopoulos, N. (1997). Kinetics of diffusion-limited spreading of sessile drops in reactive wetting. *Scripta Materialia, 36*, 645–651.

Ota, T., Takahashi, M., Hibi, T., Ozawa, M., Suzuki, S., Hikichi, Y., & Suzuki, H. (1995). Biomimetic process for producing SiC 'Wood'. *Journal of the American Ceramic Society, 78*, 3409–3411.

Paik, U., Park, H. C., Choi, S. C., Ha, C. G., Kim, J. W., & Jung, Y. G. (2002). Effect of particle dispersion on microstructure and strength of reaction-bonded silicon carbide. *Materials Science and Engineering: A, 334*, 267–274.

Pan, C., Hilpert, M., & Miller, C. T. (2004). Lattice-Boltzmann simulation of two-phase flow in porous media. *Water Resources Research, 40*, W01501.

Patro, D., Bhattacharyya, S., & Jayram, V. (2007). Flow kinetics in porous ceramics: Understanding with non-uniform capillary models. *Journal of the American Ceramic Society, 90*, 3040–3046.

Quéré, D. (2008). Wetting and roughness. *Annual Review of Materials Research, 38*, 71–99.

Roberts, A. P., & Garboczi, E. J. (2000). Elastic properties of model porous ceramics. *Journal of the American Ceramic Society, 83*, 3041–3048.

Salamone, S., Karandikar, P., Marshall, A., Marchant, D. D., & Sennett, M. (2008). Effects of SI:SiC ratio and SiC grain size on properties for RBSC. In E. Lara-Curzio, J. Salem, & D. Zhu (Eds.), *Mechanical properties and performance of engineering ceramics and composites III* (pp. 101–109). Hoboken, NJ: Wiley.

Sangsuwan, P., Tewari, S. N., Gatica, J. E., Singh, M., & Dickerson, R. (1999). Reactive infiltration of silicon melt through microporous amorphous carbon preforms. *Metallurgical and Materials Transaction B, 30*, 933–944.

Schaap, M. G., Porter, M. L., Christensen, B. S. B., & Wildenschild, D. (2007). Comparison of pressure-saturation characteristics derived from computed tomography and lattice Boltzmann simulations. *Water Resources Research, 43*, W12S06.

Sergi, D., Grossi, L., Leidi, T., & Ortona, A. (2014). Surface growth effects on reactive capillary-driven flow: Lattice Boltzmann investigation. *Engineering Applications of Computational Fluid Mechanics, 8*, 549–561.

Shan, X., & Chen, H. (1993). Lattice Boltzmann model for simulating flows with multiple phases and components. *Physical Review E, 47*, 1815–1819.

Shan, X., & Doolen, G. D. (1995). Multicomponent lattice-Boltzmann model with interparticle interaction. *Journal of Statistical Physics, 81*, 379–393.

Studart, A. R., Gonzenbach, U. T., Tervoort, E., & Gauckler, L. J. (2006). Processing routes to macroporous ceramics: A review. *Journal of the American Ceramic Society, 89*, 1771–1789.

Succi, S. (2009). *The Lattice Boltzmann equation for fluid dynamics and beyond*. Oxford: Oxford University Press.

Sukop, M. C., & Or, D. (2004). Lattice Boltzmann method for modeling liquid-vapor interface configurations in porous media. *Water Resources Research, 40, W01509*.

Sukop, M. C., & Thorne, D. T., Jr. (2010). *Lattice Boltzmann modeling: An introduction for geoscientists and engineers*. New York, NY: Springer.

Svihla, C. K., & Xu, H. (2006, May). Simulation of free surface flows with surface tension with ANSYS CFX. Paper presented at the 2006 International ANSYS Users Conference & Exhibition, Pittsburgh, PA, USA.

Trimis, D., Pickenäcker, O., & Wawrzinek, K. (2006). Porous burners. In M. Scheffler & P. Colombo (Eds.), *Cellular ceramics: Structure, manufacturing, properties and applications* (pp. 484–508). Hoboken, NJ: John Wiley & Sons.

Van Remoortere, P., & Joos, P. (1993). About the kinetics of partial wetting. *Journal of Colloid and Interface Science, 160*, 387–396.

Voytovych, R., Bougiouri, V., Calderon, N. R., Narciso, J., & Eustathopoulos, N. (2008). Reactive infiltration of porous graphite by NiSi alloys. *Acta Materialia, 56*, 2237–2246.

Wagner, A. J. (2003). The origin of spurious velocities in lattice Boltzmann. *International Journal of Modern Physics B, 17*, 193–196.

Wiklund, H. S., & Uesaka, T. (2012). Edge-wicking: Microfluidics of two-dimensional liquid penetration into porous structures. *Nordic Pulp and Paper Research Journal, 27*, 403–408.

Wiklund, H. S., & Uesaka, T. (2013). Microfluidics of imbibition in random porous media. *Physical Review E, 87*, 23006–23014.

Wolf-Gladrow, D. A. (2005). *Lattice-gas cellular automata and lattice Boltzmann models – an introduction*. New York, NY: Springer.

Xu, H., & Guetari, C. (2004, May). The use of CFD to simulate capillary rise and comparison to experimental data. Paper presented at the 2004 International ANSYS Users Conference & Exhibition, Pittsburgh, PA, USA.

Zeschky, J., Höfner, T., Arnold, C., Weissmann, R., Bahloul-Hourlier, D., Scheffler, M., & Greil, P. (2005). Polysilsesqui oxane derived ceramic foams with gradient porosity. *Acta Materialia, 53*, 927–937.

5

On the applicability of 2D URANS and SST k - ω turbulence model to the fluid-structure interaction of rectangular cylinders

F. Nieto[a]*, D.M. Hargreaves[b], J.S. Owen[b] and S. Hernández[a]

[a]*School of Civil Engineering, University of A Coruña, Spain;* [b]*Faculty of Engineering, University of Nottingham, UK*

In this work the practical applicability of a 2D URANS approach adopting a block structured mesh and Menter's SST k-ω turbulence model in fluid-structure interaction (FSI) problems is studied using as a test case a ratio $B/H = 4$ rectangular cylinder. The vortex-induced vibration (VIV) and torsional flutter phenomena are analyzed based on the computation of the out-of-phase and in-phase components of the forced frequency component of lift and moment coefficients when the section is forced to periodically oscillate both in heave and pitch degrees of freedom. Also the flutter derivatives are evaluated numerically from the same forced oscillation simulations. A good general agreement has been found with both experimental and numerical data reported in the literature. This highlights the benefits of this relatively simple and straightforward approach. These methods, once their feasibility has been checked, are ready to use in parametric design of bridge deck sections and, at a later stage, in the shape optimization of deck girders considering aeroelastic constraints.

Keywords: computational fluid dynamics; URANS; bluff body aerodynamics; vortex-induced vibration; torsional flutter; flutter derivatives; $B/H = 4$ rectangular cylinder

1. Introduction

The rectangular cylinder is a classic example of a bluff body and has been extensively studied for decades by the scientific community. There are several reasons for this interest: their aerodynamic response is governed by the separation of the shear layers from the wind-ward edges and the flow patterns are dramatically different depending on the B/H ratio (B is the cross-section width while H is the section depth). Thus rectangular prisms are prone to vortex-induced vibration (VIV), torsional flutter, galloping and coupled flutter (Takeuchi & Matsumoto, 1992) and they are also often found in the built environment, for instance as the main structural members in arch bridges. Besides this, rectangular cylinders can be regarded as simplified geometry cases for studying the aerodynamic and aeroelastic response of bridge decks and buildings, which makes them particularly appealing as test cases for the validation of new methods of analysis.

From an experimental perspective, the number of published works on the aerodynamic and aeroelastic response of rectangular prisms is enormous, and covers all the different aerodynamic phenomena. It is not possible to make a comprehensive summary here, therefore just as a small and nonsystematic sample, the following are mentioned: Nakaguchi, Hashimoto, and Muto (1968), Ogawa, Sakai, and Sakai (1988), Norberg (1993), Matsumoto (1996) or Le, Tamura, and Matsumoto (2011).

Rectangular cylinders have also been extensively studied by means of numerical simulations. Due to the intrinsic complexity of the CFD problem and the associated computer power demand, the first applications were confined to the analysis of the aerodynamics of static rectangular cylinders. A detailed summary of the main contributions in the 1990s can be found in Shimada and Ishihara (2002). Some more recent references on static rectangular cylinders are Sohankar (2006), Kuroda, Tamura, and Suzuki (2007) and Sohankar (2012), along with the research carried out in the frame of the BARC (Benchmark on the Aerodynamics of a Rectangular 5:1 Cylinder) initiative such as Mannini, Soda, and Schewe (2011) and Bruno, Coste, and Fransos (2012). The current state of the art, shows that 3D LES (Large Eddy Simulation) or DES (Detached Eddy Simulation) are generally used in order to study the complex three-dimensional flow around static rectangular cross-sections both in isolation and in combination with other bluff bodies. It must be pointed out that in these models, the spanwise length of the body influences the length scales of turbulence in the spanwise direction.

However, the number of references in the literature dealing with the numerical simulation of FSI (Fluid–Structure Interaction) problems for rectangular cross-section cylinders is more sparse, and in some cases the research has been carried out as a preliminary study prior to a further application in bridge deck aerodynamics

*Corresponding author. Email: felix.nieto@udc.es

problems. An early application is Murakami, Mochida, and Sakamoto (1997), where the results of forced oscillations and wind-induced free oscillation of a square cylinder using a 3D LES approach are reported. Piperno (1988), obtained with reasonable accuracy the in-phase and out-of-phase components of a ratio 4:1 rectangular cylinder under forced oscillation without considering a turbulence model. The works by Tamura and Itoh (1999); Tamura (1999), Tamura and Ono (2003) on the oscillations of rectangular cylinders and turbulence effects must also be highlighted. Also, in Mendes and Branco (1999), the computed values for the H_1^* and A_2^* flutter derivatives of the $B/H = 4$ rectangular cylinder are reported employing a 2D approach with no turbulence model. Braun and Awruch (2003), computed the response of a $B/H = 5$ rectangular cylinder displaying wind-induced free oscillation in heave and pitch degrees of freedom. A more recent application can be found in Sarwar, Ishihara, Shimada, Yamasaki, and Ikeda (2008), where the flutter derivatives of rectangular cylinders with aspect ratios 10:1 and 20:1 were computed as part of a numerical study for obtaining the flutter derivatives of a box girder deck. In Sun, Owen, Wright, and Liaw (2008), the numerical results for the A_2^* flutter derivative of the $B/H = 4$ rectangular cylinder using a 3D LES approach are reported. Later, Sun and co-workers (Sun, Owen, & Wright, 2009) published the complete set of 18 flutter derivatives for the $B/H = 4$ rectangular cylinder employing a 2D URANS (Unsteady Reynolds Averaged Navier-Stokes) modeling and a k-ω turbulence model; also the effect of the incoming level of turbulence was studied. Finally, in Shimada and Ishihara (2012), the responses of 2:1 and 4:1 ratio rectangular prisms subject to forced and free wind-induced oscillations were computed and compared with experimental tests and other numerical results available in the literature finding a good agreement. The authors in the last reference adopted a 2D URANS approach and a modified k-ε turbulence model (Shimada & Ishihara, 2002) to study the aerodynamic instabilities of interest avoiding the burdensome 3D numerical simulations.

In this work a $B/H = 4$ rectangular cylinder has been selected as a study case since it belongs to the category of permanent reattached flow (Bruno, Salvetti, & Ricciardelli, 2014) and therefore it can be considered as a basic bridge deck geometry. Furthermore, rectangular cylinders, as bridge decks, present fixed separation points instead of moving ones, characteristic of circular cylinders (Wu & Kareem, 2012). From the existing literature on CFD applications, two references cited above dealing specifically with the FSI response of the $B/H = 4$ rectangular cylinder must be highlighted. In Sun et al. (2009) the flutter derivatives of the 4:1 rectangular prism were computed. However, the VIV and torsional flutter problems were not addressed and therefore, the reduced velocities ($U_R = U/(fB)$) considered were in the range (3, 6), outside the VIV zone close to $U_R = 1.7$. On the other hand, in Shimada and Ishihara (2012) the focus was put on the VIV, galloping and torsional flutter of rectangular prisms. The range of reduced velocities studied for the $B/H = 4$ case was ample (approximately between 0.75 and 7.5); nevertheless no attempt was made to compute the flutter derivatives. Consequently, to the authors' knowledge the fundamental FSI phenomena in bridges, VIV response and computation of flutter derivatives required for the identification of the critical flutter wind speed, have not been simultaneously addressed in the literature.

Furthermore, in Sun et al. (2009) it is stated that the most feasible turbulence model depends on the type of aeroelastic phenomena being modeled. Since both flutter and VIV are narrow band processes, it is argued that the same basic requirement holds for the two phenomena. That is the correct simulation of the average shear layer aerodynamics induced by the deck motion. The motivation of the current piece of research is to explore the applicability of a 2D URANS approach in combination with a low Reynolds strategy in the proximity of the bluff body, in order to obtain the VIV, the torsional flutter response and the flutter derivatives for the 4:1 rectangular cylinder. A key point, which is discussed in the next section, is the selection of a turbulence model which must be suitable for modeling the aforementioned aeroelastic phenomena.

Furthermore, in this investigation a block structured mesh has been employed since it is particularly suited to the automatic grid generation in the context of parametric and optimum design problems. In addition, the time required for solving unstructured meshes in CFD problems is often higher than for the structured ones (Zhang, Jia, Wang, & Altinakar, 2013). The solver of choice has been OpenFOAM, which is a free, open source CFD software package based on the finite volume method.

The success of the described strategy would allow the application of this methodology in the parametric design of bridge box decks. Its implementation in the frame of the numerical shape optimization of bridge deck sections considering aeroelastic responses has already been proposed by some of the authors of this work (Hernández, Nieto, Jurado, & Pérez, 2012).

In the first part of the paper the available numerical approaches and turbulence models are discussed and the choice of the SST k-ω is explained. Then the basic numerical formulation is introduced and the formulation for computing the in-phase and out-of-phase components of lift and moment coefficients for bluff bodies under forced oscillation is briefly described, along with the extraction of the flutter derivatives. In the next section the flow modeling and computational approach adopted in the $B/H = 4$ rectangular cylinder case study is explained and the results of the verification analyses and validation for the static 0° angle of incidence case are described. Concerning the FSI problems, the results for the VIV and torsional flutter simulations are presented and compared with experimental and numerical data in the literature. Furthermore, the flutter derivatives are extracted and also the current results are compared with computer simulations and wind tunnel test data. Finally, the main conclusions from this piece of research are summarized.

2. The choice of turbulence model

It is well known that turbulence is a time-dependent three-dimensional process whose simulation by means of computational techniques is particularly challenging. Several turbulence models, comprising modified versions have been proposed, which differ on their physical foundations, their degree of complexity and computer power demands.

A lot of effort has been devoted to the development of reliable general-use turbulence models for solving the closure problem in the Reynolds-averaged Navier-Stokes (RANS) and Large Eddy Simulation (LES) approaches. In the following some of the available models are mentioned and their expected behavior on the type of problems considered in the present piece of research is commented.

Amongst the one-equation models, the Spalart-Allmaras turbulence model (Spalart & Allmaras, 1992) must be highlighted. This model applies the Boussinesq approximation and it is based on the proposal of a transport equation for the eddy viscosity (Wilcox, 2006). This turbulence model has proved to be effective in airfoil and wing applications, as well as free shear flows. However, the model does not provide good results for massively separated flows (Mannini, Soda, & Schewe, 2010a) but it is of interest in the frame of Dettached Eddy Simulation (SA-DES). Focusing on its performance in the aerodynamics of 2D rectangular cylinders, Mannini and co-workers found that for a static $B/H = 5$ rectangular cylinder at 0° angle of attack and $Re_H = 2 \times 10^4$, the model predicted an almost steady solution, which is in disagreement with experimental evidence. In the same manner, the recirculation region at the leeward side of the prism was overestimated and the mean values of the pressure coefficient were not particularly accurate.

Perhaps the most commonly used turbulence models in RANS applications are the two-equation type based on the Boussinesq assumption. They are the k-ε and the k-ω models with their different versions. The k-ε turbulence model has enjoyed great popularity, particularly in industrial applications. However, according to Wilcox (2006), it has proved to be inaccurate for separated flows, it is extremely difficult to integrate through the viscous sublayer and very often requires a certain degree of tuning to suit each particular application. Consequently, the k-ω turbulence model has gained popularity since it appears to be more accurate for modelling 2D boundary layers with both adverse and favorable pressure gradients and it can be integrated through the viscous sublayer without any special viscous correction. In this respect, some of the authors of this work (Nieto, Kusano, Hernández, & Jurado, 2010) have found poor performance of the k-ε model simulating the vortex shedding from a twin box deck. In contrast, the k-ω model provided better results for the same problem. Menter (2009) remarked on the sensitivity of the k-ω model with the free stream turbulent characteristics. As an alternative that avoids this problem, the SST version of the k-ω turbulence model was proposed by Menter (1994), and a number of modifications and improvements have been later published by Menter and co-workers. Following Menter (2009), the basic idea consists in using the k-ω model in the near wall regions and the k-ε model in the remaining flow domain, introducing blending functions that join the two models in a single formulation along with the JK model. The eddy viscosity is limited to improve the model's behavior for adverse pressure gradients and in the wake region. Also, a limiter is introduced in the production term of the kinetic energy (Blazek, 2005; Mannini et al., 2010a).

A more sophisticated approach consists in providing alternatives to the Boussinesq eddy-viscosity assumption, introducing the anisotropy tensor for modeling the components of the Reynolds stress tensor. These models, known as Explicit Algebraic Reynolds Stress Models are a subclass of the algebraic stress models (Wilcox, 2006). Mannini and co-workers (Mannini et al., 2010a; Mannini, Soda, Voß, & Schewe, 2010b) have applied the linearized explicit algebraic version of the model, coupled with Wilcox's standard k-ω model. This model has shown accurate results in 2D simulations of a static $B/H = 5$ rectangular cylinder, as well as qualitatively capturing the Reynolds number dependency of the response for a 4° angle of attack. These kinds of models are particular useful for flows over curved surfaces, since the Boussinesq eddy-viscosity approximation generally fails. The potential of the model for bluff bodies with curved edges has been exploited in Mannini et al. (2010b) studying the effect of corner sharpness of the cross-section of a bridge deck.

When unsteady RANS models are applied considering a three-dimensional flow domain, they typically produce single-mode large-scale unsteady structures without resolving any of the details of the turbulence (Menter, 2009) and the integral results are practically identical to those obtained from two-dimensional simulations (Sun ct al., 2009). On the other hand, the application of Reynolds Stress models in three-dimensional flow domains allows for the capture of three-dimensional features in the flow. In the application reported in Mannini et al. (2010a) it was found that for the static 5:1 rectangular cylinder studied, the three-dimensionality of the flow was weak. There were no remarkable differences from the two-dimensional simulation, apart from a 50% increment in the standard deviation of the drag coefficient. In fact, the authors elaborate on the difficulties in setting an adequate span-wise dimension of the domain and in triggering the three-dimensionality of the flow.

Nowadays the most accurate, but also the most computationally burdensome, model is the Large Eddy

Simulation (LES) approach applied in a 3D domain. In fact, the high cost of LES in wall boundary layers has restricted its application in industrial applications (Menter, 2009). A less computationally onerous alternative is the so-called Dettached Eddy Simulation (DES), where different versions of RANS models, such as Spalart-Allmaras, are applied in boundary layers. Focusing on LES, the basic idea consists in computing the large eddies in the flow, while the smallest eddies are modeled. These small scales of the turbulence are solved by means of the subgrid-scale model, which simulates the energy transfer between the large eddies and the subgrid scales. General references describing this model are, for instance, Blazek (2005) or Sagaut (2001) who focuses on incompressible flows. Concerning the high computer cost associated with the method, some computer times are provided in Bruno et al. (2012) for a static 5:1 rectangular cylinder when different space discretization is considered in the span-wise direction.

Taking into account that the primary goal of this piece of research is the CFD-based computation of vortex-induced response and flutter derivatives of a ratio $B/H = 4$ rectangular cylinder by means of forced oscillations, 3D simulations have been discarded. Since several reduced velocities must be considered for two different degrees of freedom and the simulations must be extended for tens of nondimensional time steps, the computer cost linked with DES or LES approaches could not be afforded by the authors. In fact, Shimada and Ishihara (2012) make a similar reflection in order to justify their k-ε approach. Furthermore, the application of URANS models, or even the Explicit Algebraic Reynolds Stress model, in three-dimensional flows, while being computationally expensive, does not provide significantly better results than a 2D simulation, as it has been noted above.

Focusing on 2D URANS, the drawbacks of the Spalart-Allmaras model for massively separated flows have already been noted. Reynolds Stress models, while providing good results for 2D simulations of rectangular cylinders, offer better results in problems dealing with bluff bodies with curved surfaces. Since the case considered herein has sharp edges, it is was not clear a priori that the more sophisticated Reynolds Stress model would provide better results while being robust in the forced oscillation simulations to be conducted. Amongst the remaining alternatives, the general SST version of the k-ω model seemed to offer the right balance between computer cost, robustness and expected accuracy. Therefore, it has been chosen by the authors for this study.

3. Numerical formulation

The time averaging of the equations for conservation of mass and momentum gives the Reynolds averaged equations of motion in conservation form (Wilcox, 2006).

$$\frac{\partial U_i}{\partial x_i} = 0 \tag{1.a}$$

$$\rho\frac{\partial U_i}{\partial t} + \rho U_j\frac{\partial U_i}{\partial x_j} = -\frac{\partial P}{\partial x_i} + \frac{\partial}{\partial x_j}(2\mu S_{ij} - \rho\overline{u_i' u_j'}) \tag{1.b}$$

where U_i is the mean velocity vector, x_i is the position vector, t is the time, ρ is the fluid density, u_i' is the fluctuating velocity vector and the over-bar represents the time average, P is the mean pressure, μ is the fluid viscosity, S_{ij} is the mean strain-rate tensor. From Equation 1b, the specific Reynolds stress tensor is defined as:

$$\tau_{ij} = -\overline{u_i' u_j'} \tag{2}$$

which is an additional unknown to be modeled based on the Boussinesq assumption for one and two equation turbulence models (Wilcox, 2006).

$$\tau_{ij} = 2\nu_T S_{ij} - \frac{2}{3}k\delta_{ij} \tag{3}$$

where ν_T is the kinematic eddy viscosity and k is the kinetic energy per unit mass of the turbulent fluctuations.

In this work the closure problem is solved applying Menter's k-ω SST model for incompressible flows (Menter & Esch, 2001).

For the simulations where oscillations of the bluff body have been imposed, the Arbitrary Lagrangian Eulerian (ALE) formulation has been applied. The conservation of mass and momentum equations are written as follows (Bai, Sun, & Lin, 2010; Sarkic, Fisch, Hoffer, & Bletzinger, 2012):

$$\frac{\partial(U_i - U_{gi})}{\partial x_i} = 0 \tag{4.a}$$

$$\rho\frac{\partial U_i}{\partial t} + \rho U_j\frac{\partial(U_i - U_{gi})}{\partial x_j} = -\frac{\partial P}{\partial x_i} + \frac{\partial}{\partial x_j}(2\mu S_{ij} - \rho\overline{u_i' u_j'}) \tag{4.b}$$

where U_{gi} is the grid velocity in the i-th direction.

The forced displacement is imposed at the bluff-body boundary and the mesh control is achieved computing the motion of the grid points solving the Laplace equation with variable diffusivity using a quadratic distance-based method.

4. Forced vibration

Forced oscillation of a bluff body allows its vortex-induced response to be analyzed as well as torsional flutter. Next, a brief summary of the fundamental formulation is going to be presented. More comprehensive explanations can be found in Washizu, Ohya, Otsuki, and Fujii (1978), or more recently, in Shimada and Ishihara (2012).

Forced displacement is imposed on a bluff body, in a single degree of freedom, according to the following expression for heave:

$$h(t) = h_0 \sin(\omega_m t) \tag{5}$$

where $h(t)$, positive upwards, represents the forced oscillation in heave, h_0 is the displacement amplitude, ω_m is the forced vibration angular frequency and t is the time.

For the pitch oscillation, the equivalent expression for the forced motion is:

$$\alpha(t) = \alpha_0 \sin(\omega_m t) \tag{6}$$

with $\alpha(t)$ being the pitch rotation, positive in the clock-wise direction, and α_0 is the amplitude of the forced vibration.

The main components of the unsteady wind force acting on the oscillating bluff-body are the vortex-shedding frequency component and the forced frequency component. In the results that are going to be presented below, the initial transient region in the simulations has been excluded and the data analyses have been conducted for time histories that can be considered steady and stationary. In Figure 1 the simulated time history of the lift coefficient is presented along with its frequency decomposition for a reduced velocity $U_R = U/(fB) = 4.0$ for the 4:1 rectangular cylinder (where $f = \omega_m/2\pi$ the frequency of oscillation in Hz). The first peak, at 2.5 Hz, corresponds to the frequency of oscillation while the second one, at 5.7 Hz, identifies the vortex shedding frequency. The component of the unsteady lift force (or moment) at the frequency of the forced oscillation can be written as:

$$L_m(t) = L_0 \sin(\omega_m t + \beta) \tag{7}$$

Where L_0 is the amplitude of the unsteady lift at the excitation frequency, and β is the phase shift with respect to the forced oscillation.

Applying the Fourier decomposition of the unsteady lift force per unit of span length, L_0 and β can be obtained:

$$[a_m,\ b_m] = \frac{1}{T}\int_{-T}^{T} L(t)[\cos\omega_m t,\ \sin\omega_m t]dt \tag{8.a}$$

$$L_0 = \sqrt{a_m^2 + b_m^2} \tag{8.b}$$

$$\beta = \tan^{-1}(b_m/a_m) \tag{8.c}$$

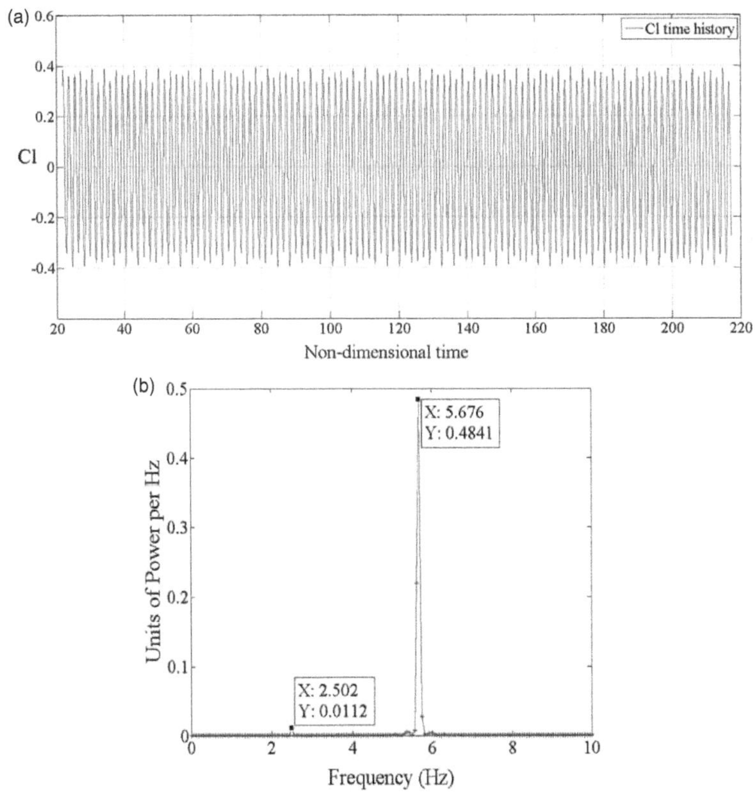

Figure 1. Ratio 4:1 rectangular cylinder (a) time history and (b) spectral content of the lift coefficient at forced oscillation in the heave degree of freedom.

And the forced frequency component of the unsteady lift force acting on the bluff-body is:

$$L_m(t) = L_0 \cos\beta \sin(\omega_m t) + L_0 \sin\beta \cos(\omega_m t) \quad (9)$$

Where $L_{mR} = L_0 \cos\beta$ is the in-phase component and $L_{mI} = L_0 \sin\beta$ is the out-of-phase component of the forced frequency term of the unsteady lift force. The out-of-phase component is of great importance since it plays the role of the aerodynamic damping, while the in-phase component plays the role of the aerodynamic stiffness. When the out-of-phase component is positive, it acts as negative aerodynamic damping indicating that self-excited oscillation may take place (see Shimada and Ishihara (2012) and Sarwar and Ishihara (2010), for a more detailed explanation).

The in-phase and out-of-phase components of the forced frequency component of the unsteady lift force can be expressed as nondimensional coefficients:

$$C_{LR} = \frac{L_{mR}}{\frac{1}{2}\rho U^2 B} \quad (10.a)$$

$$C_{LI} = \frac{L_{mI}}{\frac{1}{2}\rho U^2 B} \quad (10.b)$$

The same principles can be applied to the forced frequency component of the unsteady moment for the pitch degree of freedom. In this case, the nondimensional coefficients are written as follows:

$$C_{MR} = \frac{M_{mR}}{\frac{1}{2}\rho U^2 B^2} \quad (11.a)$$

$$C_{MI} = \frac{M_{mI}}{\frac{1}{2}\rho U^2 B^2} \quad (11.b)$$

where M_{mR} is the in-phase component and M_{mI} is the out-of-phase terms of the forced frequency component of the unsteady moment.

5. Flutter derivatives computation by means of forced oscillation simulations

Flutter derivatives are nonanalytical parameters which relate motion-induced forces and the velocities and movements of the structure. As a consequence, these parameters have been traditionally identified using wind tunnel tests, and more recently from numerical-based simulations.

According to Sarkar, Caracoglia, Haan, Sato, and Murakoshi (2009) and Simiu and Scanlan (1996), the aeroelastic forces on a bridge deck, considering two degrees of freedom (heave and pitch) can be written as follows, using Scanlan's formulation: [

$$L_{ae}^S(t) = \frac{1}{2}\rho U^2 B \left[KH_1^* \frac{\dot{h}(t)}{U} + KH_2^* \frac{B\dot{\alpha}(t)}{U} + K^2 H_3^* \alpha(t) + K^2 H_4^* \frac{h(t)}{B} \right] \quad (12.a)$$

$$M_{ae}^S(t) = \frac{1}{2}\rho U^2 B^2 \left[KA_1^* \frac{\dot{h}(t)}{U} + KA_2^* \frac{B\dot{\alpha}(t)}{U} + K^2 A_3^* \alpha(t) + K^2 A_4^* \frac{h(t)}{B} \right] \quad (12.b)$$

Where $L_{ae}^S(t)$ is the aeroelastic lift force per unit of span length, $M_{ae}^S(t)$ is the aeroelastic moment per unit of span length,$K = (B\omega)/U$ is the reduced frequency, $h(t)$ is the heave oscillation and $\dot{h}(t)$ is its time derivative, $\alpha(t)$ in the torsional rotation and $\dot{\alpha}(t)$ its time derivative, H_i^* and $A_i^*(i = 1, \ldots, 4)$ are the flutter derivatives.

Assuming prescribed harmonic forced oscillations $h = h_0 \sin(\omega_h t)$ and $\alpha = \alpha_0 \sin(\omega_\alpha t)$, where h_0 and α_0 are the amplitudes of the oscillations, and also that motion-induced forces are linear functions of the movement; after some manipulation, the following expressions are obtained for the identification of the flutter derivatives:

$$H_1^* = \frac{-L_0 \sin\phi_L}{qK^2 h_0} \quad (13.a)$$

$$H_2^* = \frac{-L_0 \sin\phi_L}{qBK^2 \alpha_0} \quad (13.b)$$

$$H_3^* = \frac{L_0 \cos\phi_L}{qBK^2 \alpha_0} \quad (13.c)$$

$$H_4^* = \frac{L_0 \cos\phi_L}{qK^2 h_0} \quad (13.d)$$

$$A_1^* = \frac{-M_0 \sin\phi_M}{qBK^2 h_0} \quad (13.e)$$

$$A_2^* = \frac{-M_0 \sin\phi_M}{qB^2K^2 \alpha_0} \quad (13.f)$$

$$A_3^* = \frac{M_0 \cos\phi_M}{qB^2K^2 \alpha_0} \quad (13.g)$$

$$A_4^* = \frac{M_0 \cos\phi_M}{qBK^2 h_0} \quad (13.h)$$

Where L_0 and M_0 are the amplitudes of the fluctuating lift and moment acting on the bluff-body, ϕ_L and ϕ_M are the phase lags of the aeroelastic lift or moment with respect to the prescribed oscillation and q is the dynamic pressure.

The adopted sign criterion has been the one reported in Sarkar et al. (2009) that is aeroelastic lift force and heave displacement positive downwards, moment and pitch rotation positive in the clock-wise direction, for a flow coming from the left side.

The aerodynamic force representation employed by Washizu and co-workers (Washizu, Ohya, Otsuki, & Fujii,

1978; 1980) is equivalent to the flutter derivatives formulation proposed by Scanlan. In this respect, in Sarkar et al. (2009), the expressions relating the out-of-phase and in-phase components of the lift and moment coefficients with the direct flutter derivatives H_1^*, H_4^*, A_2^* and A_3^* are derived.

6. Geometry and computer modeling

As it has been previously noted, a rectangular section with a width to depth ratio $B/H = 4$ and sharp edges has been chosen as the case study for computing its forced oscillation response.

In the FSI simulations the rectangular cylinder, which is considered as rigid, is forced to oscillate at a prescribed frequency in heave or pitch degrees of freedom. In the adopted approach no structural solver is involved and the mesh movement is handled by the solver using the ALE method.

6.1. *Boundary conditions*

The flow domain size adopted in all the simulation reported herein is 41B by 30.25B, similar to the size employed in successful simulations by other researchers (Bruno, Fransos, Coste, & Bosco, 2010; Fransos & Bruno, 2010) and bigger than flow domains in Sun et al. (2008), Sarwar and Ishihara (2010) or Arslan, Pettersen, and Andersson (2011). The flow domain size has been chosen in order to guarantee that flow conditions near the bluff body are not influenced by the distance to the boundaries.

As boundary conditions, a constant velocity inlet has been set at the left side (see Figure 2) while a pressure outlet at atmospheric pressure has been imposed at the right side. The upper and lower boundaries have been defined as slip walls, that is, neglecting viscous effects caused by the wall surface. Taking into account that in the region close to these boundaries the flow is practically undisturbed, given the distance of the upper and lower boundaries from the bluff body, this is similar to imposing the uniform value of the inlet velocity along these boundaries as in Fransos and Bruno (2010) or Sarwar and Ishihara (2010). The walls of the rectangular cylinder surface are modeled as no-slip, taking into account that in the forced oscillation simulations the kinematic requirement that no flow can cross the wall is enforced (Donea, Huerta, Ponthot, & Rodríguez-Ferrán, 2004). In the solver, the resultant velocity field around the rectangular cylinder wall is corrected, imposing that the velocity at the boundary is equal to the mesh velocity and therefore no flux across the wall takes place.

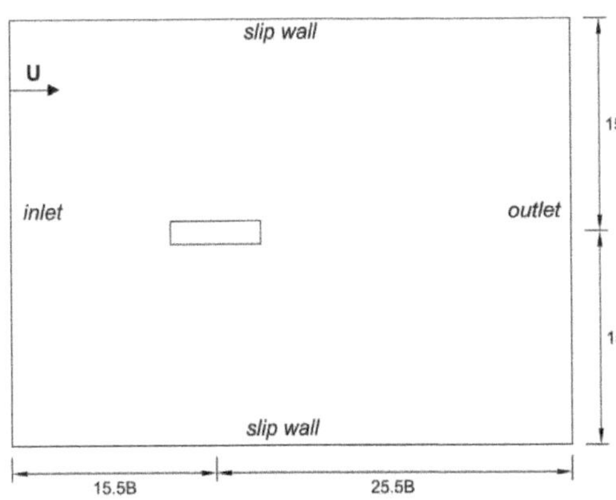

Figure 2. Flow domain definition and boundary conditions (rectangular cylinder out of scale).

A turbulence intensity of 1% has been chosen along with a 0.1B turbulent length scale for the incoming flow, as in Ribeiro (2011). This has the same order of magnitude as the 3% turbulence intensity and 0.082B turbulent length scale values in Sarkic et al. (2012).

A 2D block structured regular mesh has been generated taking special care in the definition of the refined grid around the deck cross-section in order to obtain target values for the nondimensional first grid height ($y^+ = (\delta_1 u_*)/\nu$, where δ_1 is the height of the first prismatic grid layer around the deck and u_* is the friction velocity) close to 1. In this manner, no wall functions are required.

The forced oscillation simulations have been conducted using a transient solver for incompressible flow of Newtonian fluids on a moving mesh by means of the PIMPLE (merged PISO-SIMPLE) algorithm distributed with OpenFOAM (2014a). The interpolation of values from the cell centers to face centers is done using a linear scheme and the gradient terms are discretized using the cell limited version of the Gauss discretization scheme. In the same manner, the divergence terms are discretized by means of the Gauss schemes. For the Laplacian terms, the Gauss scheme with a linear interpolation scheme for the diffusion coefficient and a limited surface normal gradient scheme has been chosen. The first-order time derivative is discretized using the Euler implicit scheme. Concerning the temporal discretization, the PISO scheme discretizes the momentum equations in an implicit manner, while the pressure gradient is explicit (OpenFOAM, 2014b, 2014c). The segregated operator splitting in PISO results in a solution that is highly sensitive to Courant number, although certain aspects of the discretization are implicit. Thus, Courant numbers below 1 are generally required to maintain stability.

6.2. *Mesh control*

The computer implementation of the ALE formulation requires a mesh-update method that assigns mesh-node velocities or displacements at each calculation time step (Donea et al., 2004). The computational grid must be kept as regular as possible, avoiding distortions and the squeezing of mesh elements, and therefore decreasing numerical errors.

In the fluid structure interaction simulations conducted in this piece of research the boundary motion is defined by the prescribed forced oscillations of the rectangular cylinder, which follows a sinusoidal law with given frequency and amplitude. On the other hand, the exterior boundaries of the fluid domain are fixed along the simulations.

Amongst the available mesh movement algorithms a Laplacian smoothing technique for each component of the node-mesh position has been chosen (Oliver, 2009). According to Jasak and Rusche (2009), the Laplace equation can be expressed as:

$$\nabla \cdot k \nabla \boldsymbol{u} = 0 \tag{14}$$

where **u** is the node-mesh displacement vector and k is the diffusion coefficient.

In this work the diffusivity of the field is computed based on the quadratic inverse distance from the oscillating boundary. This prevents the distortion of the smallest elements around the rectangular cylinder (Löhner, 2008).

6.3. *Verification studies*

A grid independence study has been conducted for three different grids, namely Coarse, Medium and Fine grids, for the static rectangular cylinder with 0° angle of incidence. The thickness of the high-density layer attached to the rectangular cylinder is *B*/8. The mesh comprises 80 rows of elements in this zone and the height of the first element around the cross-section is the same for the three meshes considered in the verification study, and is defined as $\delta_1/B = 2.44 \times 10^{-4}$. Besides this, the expansion factor (ratio between the dimension of the furthest and the closest elements to the wall) is 20. As a consequence, the mean value of y^+ around the rectangular cylinder is about 0.9–0.95 and a maximum value, limited to the windward corners, is approximately 5.25, well below the maximum value of 8 reported in Sarkic et al. (2012) for a streamlined deck cross-section (see Figure 3 for images of the Medium grid). The Reynolds number of the simulations reported herein is $Re = (UB)/\nu = 1.06 \times 10^5$ In Table 1 the data

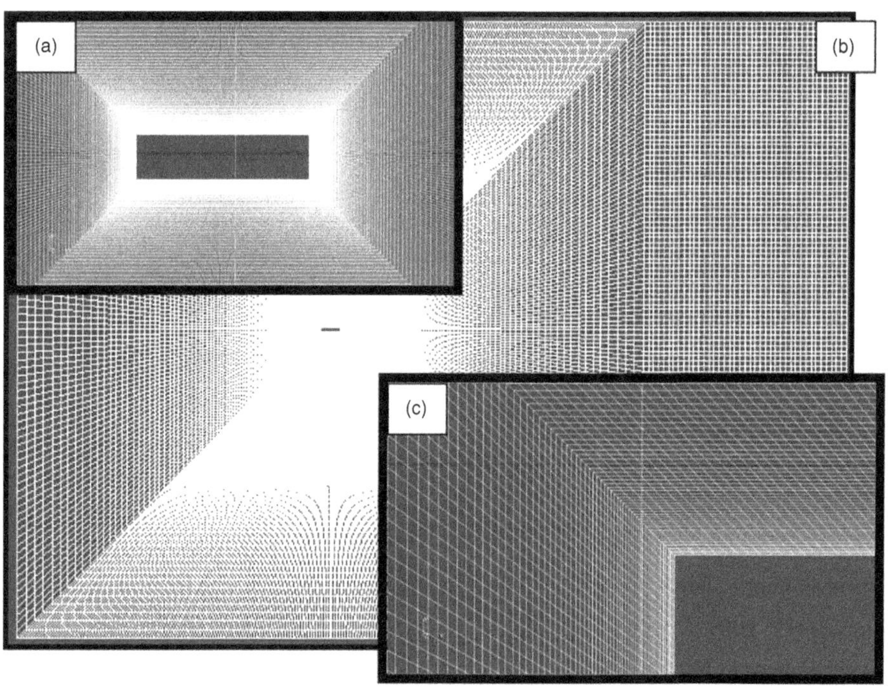

Figure 3. Block-structured grid: (a) close-up of the rectangular cylinder; (b) a larger region around the cylinder and (c) detail around the corner.

Table 1. Properties and results of the grid-refinement study.

Grid	Total cells	Cells around body	S_t	C_D	C'_D	C'_L	C'_M
Coarse	104000	560	0.145	0.31	0.009	0.27	0.046
Medium	147600	680	0.145	0.31	0.012	0.28	0.046
Fine	235200	880	0.145	0.31	0.011	0.29	0.045

Table 2. Results of the time-refinement study.

Max. Co. numb.	$\overline{\Delta s}$	S_t	C_D	C'_D	C'_L	C'_M
1	3.e-4	0.145	0.31	0.012	0.28	0.046
0.5	1.5e-4	0.143	0.31	0.013	0.28	0.047

of these three grids, along with the Strouhal number and the mean values and standard deviations of the force coefficients are reported. All the computations have been carried out imposing a maximum Courant number of 1.

The definition of the force coefficients in this work is the following:

$$C_D = \frac{D}{\frac{1}{2}\rho U^2 B} \quad C_L = \frac{L}{\frac{1}{2}\rho U^2 B} \quad C_M = \frac{M}{\frac{1}{2}\rho U^2 B^2} \tag{15}$$

In the former expressions, D is the drag force per span length, positive windward, L is the lift force per span length, positive upwards, and M is the twist moment per unit of span length, positive in the clock-wise direction. The reference dimension for the three coefficients is B. The standard deviation of the force coefficients is identified with the prime symbol in the following.

Table 1 shows very similar results for the three grids considered, which highlights the level of independence of the solution from the mesh density. For the simulations hereafter the Medium grid has been retained since it offers similar accuracy to the Fine mesh with lower computational cost and its extra spatial resolution with respect to the Coarse mesh offers additional guaranties for the more demanding fluid-structure interaction simulations to be addressed in the present research.

Two different maximum Courant numbers have been considered for studying the stability of the numerical simulation depending on the time step discretization: 1 and 0.5. The corresponding mean nondimensional time-steps, have been: $\Delta s = (\Delta t U)/B \approx 3.0 \times 10^{-4}$ and 1.5×10^{-4}. In Table 2 the results obtained for the Medium mesh are reported. It has been found that both maximum Courant numbers offer very similar results, thus the higher has been retained hereafter.

Table 3. $B/H = 4$ rectangular cylinder: results and validation.

	S_t	C_D	C'_D	C'_L	C'_M
Present simulation	0.145	0.31	0.012	0.28	0.046
Sun et al. (2009) - 2D URANS	0.15	0.33	0.024	0.351	0.053
Vairo (2003) - 2D URANS	0.149	0.356	0.065	0.25	0.071
Okajima (1982); Nakaguchi et al. (1968) - EXP.	0.135	0.30			
Vairo (2003) - EXP. by CSTB	0.159	0.348	0.081	0.289	0.054

In the simulations which are going to be presented next, the nondimensional time steps are roughly between 3.0×10^{-4} and 2.5×10^{-4}.

The FSI problems have been solved in a High Performance Computer Cluster which mainly comprises nodes with 2 AMD 8 cores processors. The forced oscillations in pitch simulations have been computed in parallel using 8 cores for each simulation. The heave-related problems, which demanded lower computational power, were solved using 4 cores in parallel. The execution time per time step has been about 4.5 seconds, depending obviously on the specific node load during the simulation.

7. Results and discussion

7.1. *Flow simulation around the static rectangular cylinder*

In Figure 4 the time history of the force coefficients for the static ratio 4:1 rectangular cylinder can be seen.

As a first step in the validation of the computational results reported in this work, the main aerodynamic integral parameters are presented and compared with available experimental and numerical results in the literature. The Strouhal number is computed from the dominant frequency in the spectrum of the lift coefficient. In Table 3 the aforementioned results are summarized along with the standard deviation of the force coefficients and compared

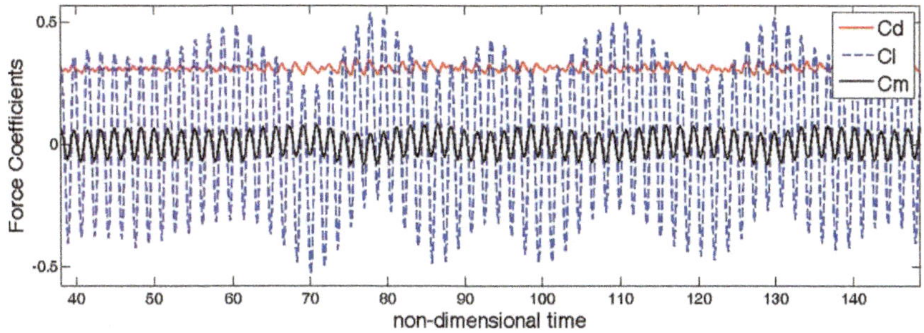

Figure 4. Time history of force coefficients for the $B/H = 4$ static rectangular cylinder.

with numerical and experimental results reported in the literature.

It has been found that both mean drag force coefficient and the Strouhal number predicted in this study fall inside the range of the experimental data available in the literature. Also the results are similar to other numerical simulations carried out using a 2D URANS approach (Sun et al., 2009; Vairo, 2003). Some differences have been found in the standard deviation of the force coefficients with respect to both experimental and numerical publications; in fact, Bruno and co-workers, have pointed out the existing scattering in the literature for standard deviation of the lift coefficient for rectangular cylinders with high aspect ratios (Bruno et al., 2010). It has been found that the results obtained in this work for the standard deviation of the lift and moment coefficients are very close to the experimental values reported in Vairo (2003), while the standard deviation of the drag coefficient is lower, although all the available results collected agree in pointing towards a low value for the standard deviation of the drag coefficient, well below 0.1.

7.2. One degree of freedom heave forced oscillation

With the aim of exploring the feasibility of the 2D URANS approach using Menter's SST k-ω turbulence model in fluid-structure interaction problems, the forced oscillation response of the $B/H = 4$ rectangular cylinder has been computed. The amplitude of the oscillation is $h_0/D = 0.02$ as in the experimental tests conducted by Washizu et al. (1978). These forced oscillation simulations with constant flow speed and at different frequencies allow the identification of the reduced velocity regions where vortex-induced vibration can take place (Sarwar & Ishihara, 2010). The change in the out-of-phase component of the unsteady lift force from negative to positive values indicates that vortex-induced vibrations will occur due to the negative aerodynamic damping introduced in the system for positive values of C_{LI}.

The current numerical results are compared with the experimental ones in the former reference by Washizu and co-workers, along with the computational results reported in Sarwar and Ishihara (2010) and Shimada and Ishihara (2012). In the first reference concerning CFD, a 3D approach with a LES turbulence model is employed; while in the latter one, the authors successfully applied a 2D URANS approach using a two-layer k-ε turbulence model modified in the production term. In Figures 5(a) and 5(b) the out-of-phase and in-phase components of the lift coefficient are reported. Also, in Figures 6(a) and 6(b) the amplitude of the unsteady lift coefficient at the frequency of oscillation and the phase difference between the forced displacement in the heave degree of freedom and the forced oscillation component of the lift force are shown. It must be noted that in Sarwar and Ishihara (2010) and Shimada and Ishihara (2012) the values of C_{LR} and C_{L0} are not reported, however they have been computed for this study from the available C_{LI} and β data in order to provide the complete picture in Figures 5(b) and 6(a). From a qualitative perspective, the CFD results obtained in the frame of the current study agree well with the experimental data reported by Washizu et al. (1978): a region showing positive values of the out-of-phase component (C_{LI}) has been identified and the phase difference between the forced displacement and the forced oscillation component of the lift coefficient, for different reduced velocities, has also been obtained. However, a closer look at these results allows a further understanding of the turbulence model performance. As it has already been mentioned, the out-of-phase component of the lift coefficient is of utmost importance since it allows the identification of reduced velocity regions prone to VIV. From Figure 5(a), the V-shaped branch in the range of reduced velocities (1.0, 1.7) has been captured as well as the inverted V-shaped branch between reduced velocities (1.7, 2.5). In the same manner, a curved branch in the interval (2.5, 4.0) and the nearly horizontal branch in the range (4.0, 8.0) have been correctly identified. Nevertheless, differences between the numerical values in the CFD simulations and the experimental tests are apparent. In fact the numerical simulations reported herein overestimate the magnitude of the peak of the V-shaped branch at $U_R = 1.6$. Also the peak in the inverted

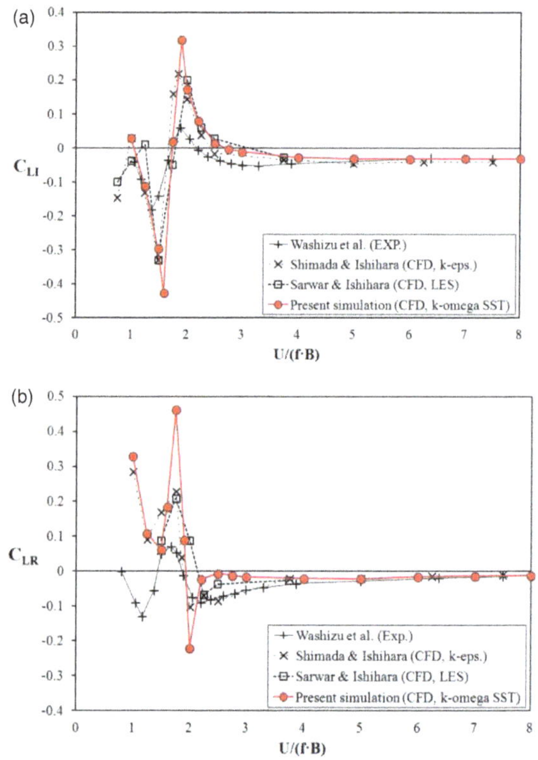

Figure 5. $B/H = 4$ rectangular cylinder lift coefficient (a) Out-of-phase component; (b) In-phase component.

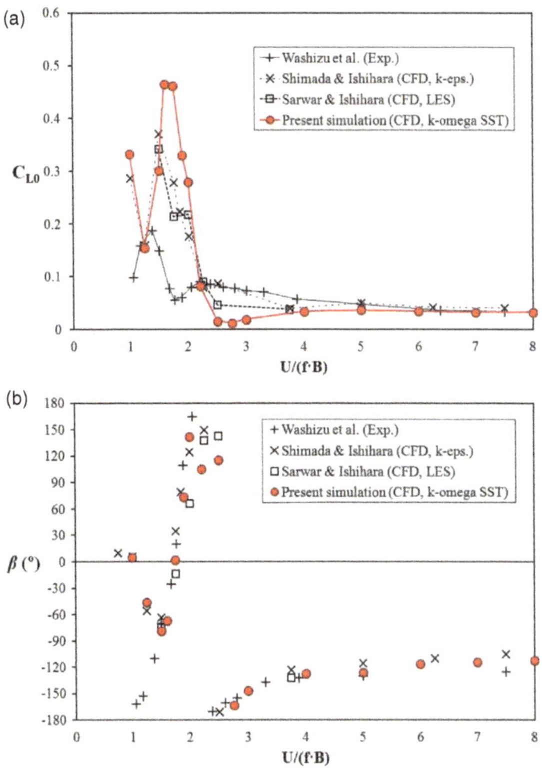

Figure 6. $B/H = 4$ rectangular cylinder lift coefficient (a) Amplitude of the forced frequency component of the lift coefficient; (b) Phase angle.

V-shaped branch for $U_R = 1.9$ is significantly higher than the experimental value in Washizu et al. (1978) (see Table 4).

These discrepancies can be explained as follows: The simulations have been obtained by means of 2D URANS simulations, therefore the predicted force coefficients show unrealistic smooth and periodic oscillations (Mannini et al., 2010b) since complex turbulence phenomena are not resolved. Furthermore, in Brusiani, de Miranda, Patruno, Ubertini, and Vaona (2013) it is noted that for RANS models based on the Boussinesq hypothesis, 2D simulations are equivalent to 3D simulations showing perfect correlation in the span-wise direction. As a consequence, the aerodynamic forces are generally overestimated. Consequently, in this simulation, the forced frequency component of the lift force is expected to be higher than in the physical experiments. At reduced velocities equal to 1.6 and 1.9 the frequency of vortex shedding is locked-in with the frequency of oscillation, therefore the out-of-phase component of the lift coefficient is strongly over-predicted. The same phenomenon takes place at reduced velocities of 1.75 and 2.0, which are the peaks in the in-phase component of the lift coefficient chart (see Figure 5(b). In the same manner, Figure 6(a) shows high values for the amplitude of the forced frequency component of the lift coefficient in the range of reduced velocities (1.5, 2.0), and these values are higher than in the other CFD-based references used for comparison.

Also a region of reduced velocities between 2.2 and 3.0 can be indentified which corresponds to the range of frequencies of oscillation below the frequency of vortex shedding and therefore this is the region where the lock-in must be surpassed. It has been found that the SST k-ω model has not been able to properly simulate this complex FSI region. This problem manifests itself in the underestimation of the in-phase component of the lift coefficient in that range of reduced frequencies, and the extension of the range of reduced velocities showing positive values for the out-of-phase component. Similar behavior can be identified in the amplitude C_{L0} (see Figure 6a).

Furthermore, the phase difference between the forced oscillation lift force component and the heave forced displacements has not been correctly obtained for $U_R = 2.2$ and $U_R = 2.5$ (see Figure 6b) which corresponds to phase angles close to 180° according to Washizu et al. (1978).

On the other hand, the turbulence model has offered results very close to the experimental ones in the range of reduced velocities (4.0, 8.0), which corresponds to the region where the complex VIV aerodynamic response has

Table 4. Discrepancies in the definition of the out-of-phase component of the forced frequency component of the lift coefficient using different turbulence models.

	EXP. (Washizu et al., 1978)	k-ε (Shimada & Ishihara, 2012)	LES (Sarwar & Ishihara, 2010)	SST k-ω
C_{LI}^{max}	0.06	0.22	0.2	0.3183
C_{LI}^{min}	– 0.18	– 0.33	– 0.33	– 0.4274
Rangepositive C_{LI}	1.765, 2.05	1.75, 2.25	2, 2.5	1.75, 2.5
ΔU_R^+	0.285	0.5	0.5	0.75
$e(C_{LI}^{max})$		2.67	2.33	4.30
$e(C_{LI}^{min})$		0.83	0.83	1.37
$e(\Delta U_R^+)$		0.75	0.75	1.63

been overcome. Similar issues can be identified, with lower severity, in Shimada and Ishihara (2012) and Sarwar and Ishihara (2010).

With the purpose of assessing the level of accuracy offered by the SST k-ω turbulence model and the k-ε and LES models, the main features in the definition of the outline of the C_{LI} component are compared with the wind tunnel data in Table 4. The following magnitudes have been considered: the maximum and minimum values of the out-of-phase component of the forced frequency term of the lift coefficient C_{LI}^{max} and C_{LI}^{min}; the range of reduced velocities with positive values of C_{LI} and the length of the corresponding interval ΔU_R^+. The discrepancies are estimated as:

$$e = \frac{|exp.value - num.value|}{|exp.value|} \tag{16}$$

It must be borne in mind that the number of reduced frequencies considered in the numerical simulations and the experimental tests are different, as they are also different the concrete values of the reduced velocities adopted in each work. Therefore the values reported in Table 4 must be considered as an approximation for the purpose of assessing the accuracy of the different models.

According to the data reported in Table 4, it can be concluded that in spite of the qualitative agreement of the SST k-ω turbulence model with the experimental data, its accuracy is lower than the k-ε and LES models, since it predicts higher peaks in the C_{LI} and the range of reduced frequencies showing positive values in C_{LI} is also overestimated (for practical applications this is in the safe side). It is remarkable the performance of the 2D k-ε model which at least matches the 3D LES results. In this respect, the lower overestimation of the peak values when compared with the SST k-ω model can be related with the behavior of the modified k-ε model, since in Shimada and Ishihara (2002) it is noted that the fluctuations in the lift force were considerably underestimated in some cases, as the 4:1 ratio rectangular cylinder.

In the above discussion the range of reduced velocities (4.0, 8.0) has not been considered since it is outside the VIV prone region, no significant discrepancies have been observed and that range will be specifically considered in the latter discussion on the flutter derivatives estimation.

7.3. *One degree of freedom pitch forced oscillation*

The engineering interest in analyzing the response of a bluff-body undergoing harmonic torsional oscillation under wind flow lies in identifying the possibility of torsional flutter taking place at a certain range of reduced velocities, when the out-of-phase component of the moment coefficient changes from negative to positive values.

For the $B/H = 4$ rectangular cylinder, the amplitude of the pitch oscillation in the current simulations has been $\alpha_0 = 3.82°$, as in the experiments reported in Washizu et al. (1980) and the numerical simulation in Shimada and Ishihara (2012). In Figure 7.a and 7.b the out-of-phase and in-phase components of the unsteady moment (C_{MI} and C_{MR}) are reported along with equivalent experimental and numerical results for validation.

The plot of the out-of-phase component of the moment coefficient as a function of the reduced velocity is qualitatively similar to the experimental values. The positive values of the out-of-phase component allow identifying the range of reduced velocities for which the cross-section is prone to torsional flutter. In the CFD simulation the reduced velocity for which C_{MI} is null is lower than in the wind tunnel tests, which is in the safe side for design applications. When the current simulation is compared with the one reported by Shimada and Ishihara, it can be concluded that the overall behavior is similar: both of them are shifted upwards in the vertical axis from $U_R = 3.0$, but without great differences with respect to the wind tunnel experiments.

For the sake of completeness, the amplitude of the forced frequency component of the moment coefficient and the phase angle with the forced oscillation in pitch degree

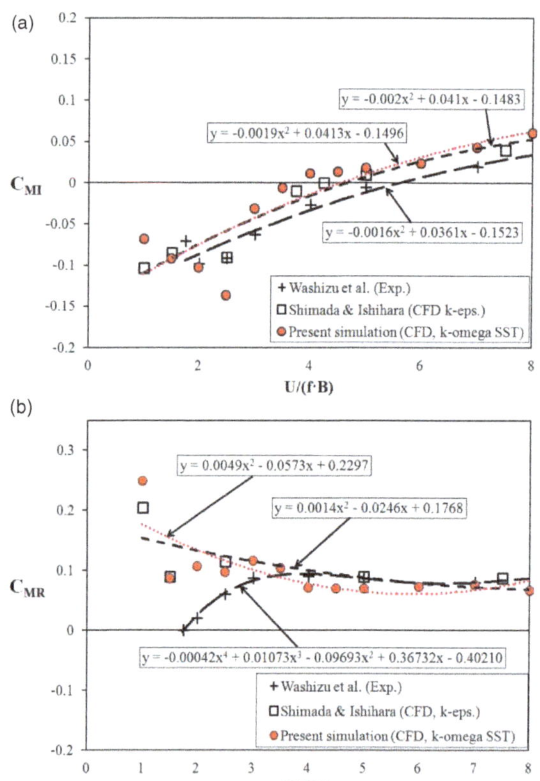

Figure 7. $B/H = 4$ rectangular cylinder moment coefficient (a) Out-of-phase component; (b) In-phase component.

of freedom are reported in Figures 8(a) and 8(b). The data labeled as Shimada and Ishihara (CFD, k-eps.) have been computed from the C_{MR} and C_{MI} data reported in Shimada and Ishihara (2012).

In this case, since the changes in the out-of-phase component of the forced frequency moment coefficient C_{MI} with the reduced velocity are smooth, an order two polynomial can be approximated (see Figure 7a) for the experimental values as well as the numerical values reported in the present work and in Shimada and Ishihara (2012). This allows the level of accuracy in the 2D URANS simulations to be judged when compared with the wind tunnel data. In order to provide a global assessment for the range of reduced velocities studied, the error is estimated

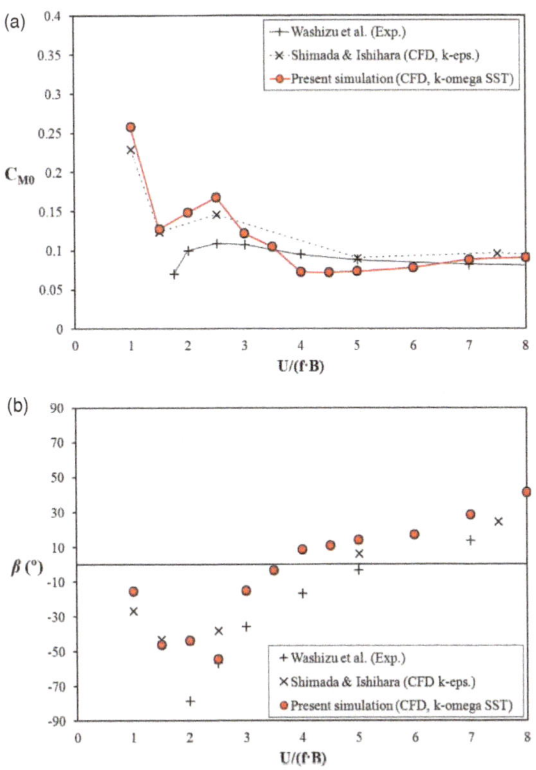

Figure 8. $B/H = 4$ rectangular cylinder moment coefficient (a) Amplitude of the forced frequency component of the moment coefficient; (b) Phase angle.

Table 5. Discrepancies in the definition of the out-of-phase and in-phase components of the forced frequency component of the moment coefficient using different turbulence models.

	k-ε (Shimada & Ishihara, 2012)	SST k-ω
$e(C_{MI})$	0.51	0.59
$e(C_{MR})$	0.27	0.33

from the following expression:

$$e = \frac{\int_{U_R^0}^{U_R^f} |exp.\ aprox. - num.aprox.| dU_R}{\int_{U_R^0}^{U_R^f} |exp.\ aprox| dU_R} \qquad (17)$$

Where *exp. aprox.* and *num. aprox.* are the order two polynomial approximations for the experimental and numerical data; and U_R^0 and U_R^f are the bounds of the considered interval.

In Table 5 the reported results for the out-of-phase component correspond to the interval of reduced velocities (1.75, 8.0), which is the one for which the results are available for the experimental test and the CFD simulations. It can be concluded that the SST k-ω offers slightly poorer accuracy when compared with the modified k-ε model in Shimada and Ishihara (2012).

Furthermore, the results for the in-phase component C_{MR} show a similar trend as in Shimada and Ishihara's paper, with overestimated values at $U_R < 3.5$, but particularly for reduced velocities in the interval (1.0, 2.0). In the former reference it is argued that this phenomenon, which was found also for rectangular cylinders of ratio 2:1 and 5:1, could be linked with the lack of turbulence diffusion in 2D models.

On the other hand, for reduced velocities above 3.5, both turbulence models offer results very close to the experimental ones. In this case a polynomial of order 4 has been approximated for the experimental data since it offers a very close match with the available points. The accuracy of the different numerical approaches is assessed using the expression in Equation (17) and the results are also presented in Table 5. The range of reduced velocities has been $U_R^0 = 1.75$ and $U_R^f = 8.0$. In this case also, the modified ε turbulence model has provided slightly better results than the SST k-ω model.

7.4. *Flutter derivatives computation*

The flutter derivatives of the $B/H = 4$ rectangular cylinder have been computed from the same forced oscillation simulations in heave and pitch degrees of freedom that have been employed to obtain the VIV and torsional flutter response presented in previous sections. In this way, always costly simulations have been avoided, which enhances the feasibility and industrial applicability of the computational-based approach proposed.

A VIV prone region around $U_R = 2$ was identified in section 6.2. Therefore for the heave-related flutter derivatives H_1^*, H_4^*, A_1^* *and* A_4^* the range of reduced velocities considered is (2.5, 8.0), while for the pitch-related H_2^*, H_3^*, A_2^* *and* A_3^* flutter derivatives the range of reduced velocities is (2.0, 8.0).

In Figure 9 the numerical results are presented along with the experimental data in Matsumoto, Yagi, Tamaki,

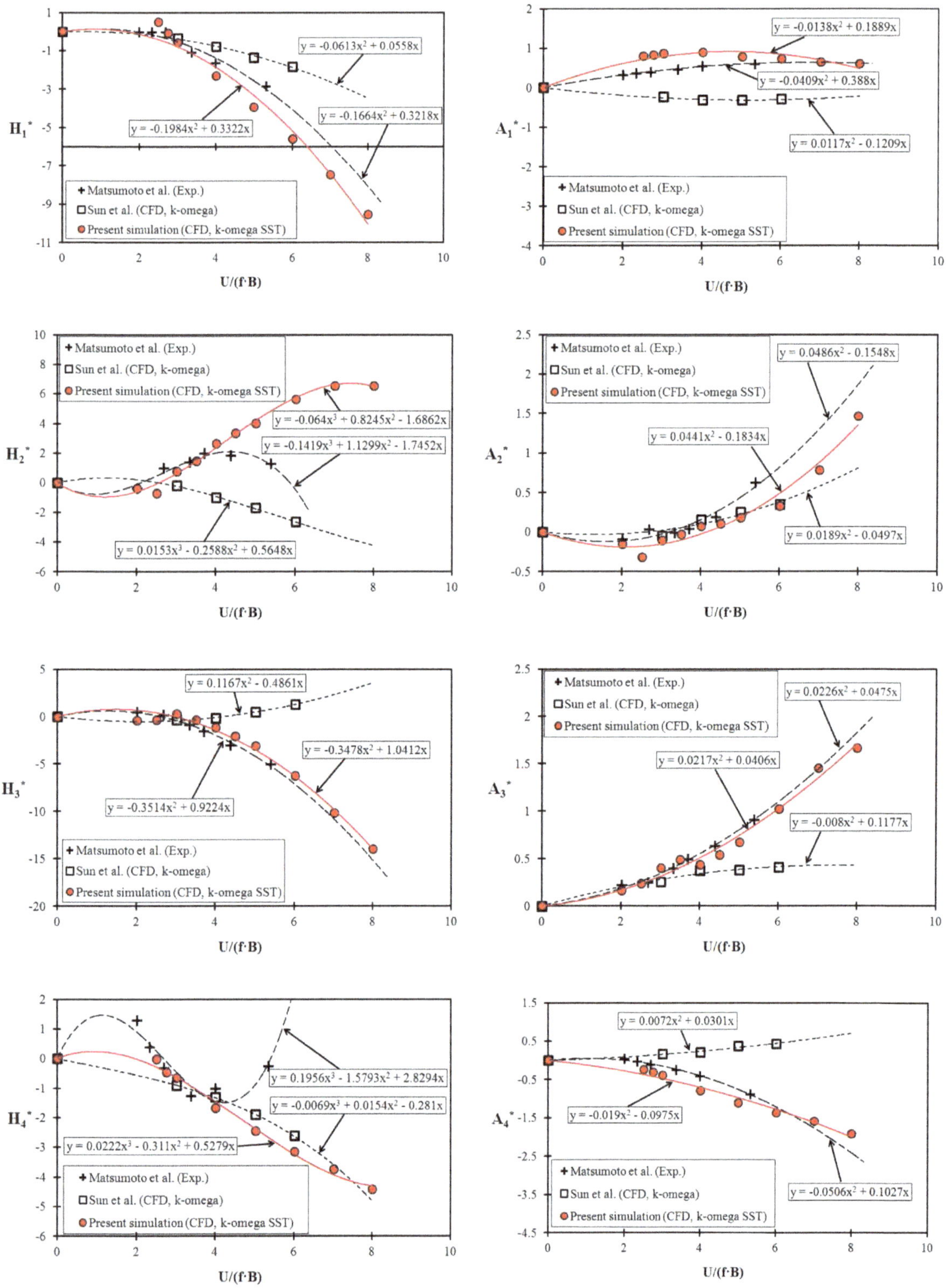

Figure 9. Flutter derivatives of the $B/H = 4$ rectangular cylinder: numerical results and comparison with experimental and numerical data.

Table 6. Discrepancies in the flutter derivatives for the 4:1 rectangular cylinder computed from the standard and the SST versions of the k-ω model.

	Standard k-ω (Sun et al., 2009)	SST k-ω
$e(H_1^*)$	0.54	0.28
$e(H_2^*)$	1.74	1.07
$e(H_3^*)$	1.20	0.14
$e(H_4^*)$	1.19	1.09
$e(A_1^*)$	1.34	0.34
$\varepsilon(A_2^*)$	0.55	0.57
$\varepsilon(A_3^*)$	0.58	0.07
$\varepsilon(A_4^*)$	1.38	0.30

and Tsubota (2008) and the computational ones reported in Sun et al. (2009). Second-order polynomial approximations are also included, which allow judging the general agreement between numerical simulations and wind tunnel tests, and assessing the accuracy of the numerical approach.

The comparison of the reported results with the wind tunnel test data show, in general terms, that they are very similar. In fact the only remarkable differences take place for the H_2^* and H_4^* flutter derivatives at reduced velocities above 4. On the other hand flutter derivatives H_3^* or A_3^* are extremely close to the experimental data.

A better general agreement for the current computations has been found than the numerical simulations in smooth flow in Sun et al. (2009). This is arguably due to the higher size of the flow domain which prevents blockage effects and the higher mesh density adopted in the current work. Also the block structured mesh must play a role in the improved results since unstructured meshes are more diffusive, as it has been noted in Mannini et al. (2010a). Furthermore, Menter's SST k-ω turbulence model seems to offer better performance than the standard k-ω employed by Sun and co-workers.

In Table 6 the accuracy of the numerical simulations with respect to the experimental data in Matsumoto et al. (2008) is assessed base on the second-order polynomial approximations. The error is estimated (Equation 17) in the range of reduced velocities (0, 8.0), except for the flutter derivatives H_2^* and H_4^*, where the reduced velocities considered are between 0 and 6.0 since the discrepancies are important above the latest value, and the maximum reduced velocity considered in Matsumoto et al. (2008) is 5.30. It can be remarked that in this case the SST k-ω model offers a remarkable agreement with the experimental data. In fact it has offered a better approximation for all the flutter derivatives considered in this work, with the exception of A_2^*, when compared with the standard k-ω turbulence model employed by Sun and co-workers.

8. Concluding remarks

In this work a 2D URANS approach using a block-structured mesh and Menter's SST k-ω turbulence model has been applied to the study of the fluid-structure interaction of a $B/H = 4$ rectangular cylinder. The aeroelastic phenomena of interest have been VIV, torsional flutter and the computation of flutter derivatives. A single set of heave and pitch forced oscillations has been considered for computing the responses of interest.

A study considering different mesh densities and time steps has been carried out, which has allowed guaranteeing the stability of the solution in terms of grid and time step refinement as well as identifying the most convenient mesh and time step in terms of better accuracy at lower computational cost. The results obtained for the fixed rectangular cylinder have been similar to the experimental ones available in the literature.

The SST k-ω turbulence model has been capable of identifying the region with positive values of the out-of-phase term of the forced frequency component of the lift coefficient as well as the phase difference between the forced heave displacement and the force frequency component of the lift force. As a consequence, the possibilities of the current approach, in the frame of industrial applications where the preliminary identification of VIV prone regions is the main goal, have been demonstrated. When the obtained results are compared with the ones computed by Shimada and Ishihara by means of a modified k-ε model, it is found that the SST k-ω models offers lower accuracy in the evaluation of the magnitude of the peak values or the extend of the region with positive values of C_{LI}.

In a similar fashion, the SST k-ω model has correctly predicted the torsional flutter prone region for the 4:1 rectangular cylinder. However, when the results are compared with the modified k-ε model in Shimada and Ishihara (2012), again the SST k-ω model has offered slightly lower accuracy.

Where the SST k-ω model offers a better performance is in the computation of the flutter derivatives. The agreement with the wind tunnel tests reported in Matsumoto et al. (2008) is remarkable. In this respect, the SST version offers higher accuracy than the standard k-ω turbulence model.

The current approach has proved its feasibility for identifying fluid-structure interaction responses, avoiding burdensome 3D LES simulations or the development of in-house sophisticated software. Therefore it is ready for being applied in the frame of parametric studies of bridge deck sections and, in the future, in the shape optimization of bridge deck considering aeroelastic constraints.

Acknowledgements

This work has been funded by the Spanish Ministry of Education, Culture and Sport under the Human Resources National Mobility Program of the R-D + i National Program 2008–2011, extended by agreement of the Cabinet Council on October 7th 2011. It has also been partially financed by the Galician Government (including FEDER funding) with reference GRC2013-056 and by the Spanish Minister of Economy and Competitiveness (MINECO) with reference BIA2013-41965-P. The authors fully acknowledge the received support.

The authors are grateful for access to the University of Nottingham High Performance Computing Facility and the Breogán Cluster at the University of La Coruña.

References

Arslan, T., Pettersen, B., & Andersson, H. I. (2011, July 10–15). *Calculations of the flow around rectangular shaped floating structures*. Proceedings of the 13th International Conference on Wind Engineering, Amsterdam, The Netherlands.

Bai, Y., Sun, D., & Lin, J. (2010). Three dimensional numerical simulations of long-span bridge aerodynamics using block-iterative coupling and DES. *Computers and Fluids*, *39*(9), 1549–1561.

Blazek, J. (2005). *Computational fluid dynamics: Principles and applications* (2nd ed.). Amsterdam: Elsevier.

Braun, A. L., & Awruch, A. M. (2003). Numerical simulation of the wind action on a long-span bridge deck. *Journal of the Brazilian Society of Mechanical Sciences and Engineering*, *25*(4), 352–363.

Bruno, L., Coste, N., & Fransos, D. (2012). Simulated flow around a rectangular 5:1 cylinder: Spanwise discretisation effects and emerging flow features. *Journal of Wind Engineering and Industrial Aerodynamics*, *104–106*, 203–215.

Bruno, L., Fransos, D., Coste, N., & Bosco, A. (2010). 3D flow around a rectangular cylinder: A computational study. *Journal of Wind Engineering and Industrial Aerodynamics*, *98*(6–7), 263–276.

Bruno, L., Salvetti, M. V., & Ricciardelli, F. (2014). Benchmark on the aerodynamics of a rectangular 5:1 cylinder: An overview after the first four years of activity. *Journal of Wind Engineering and Industrial Aerodynamics*, *126*, 87–106.

Brusiani, F., de Miranda, S., Patruno, L., Ubertini, F., & Vaona, P. (2013). On the evaluation of bridge deck flutter derivatives using RANS turbulence models. *Journal of Wind Engineering and Industrial Aerodynamics*, *119*, 39–47.

Donea, J., Huerta, A., Ponthot, J. Ph., & Rodríguez-Ferrán, A. (2004). Arbitray Lagrangian-Eulerian methods. *Encyclopedia of Computational Mechanics, Fundamentals, 1*, 413–437.

Fransos, D., & Bruno, L. (2010). Edge degree-of-sharpness and free-stream turbulence scale effects on the aerodynamics of a bridge deck. *Journal of Wind Engineering and Industrial Aerodynamics*, *98*(10–11), 661–671.

Hernández, S., Nieto, F., Jurado, J. Á, & Pérez, I. (2012, September 2–6). *Bluff body aerodynamics of simplified bridge decks for aeroelastic optimization*. Proceedings of the 7th International Colloquium on Bluff Body Aerodynamics and Applications, Shanghai, China.

Jasak, H., & Rusche, H. (2009, June 1–4). *Dynamic mesh handling in OpenFOAM*. Fourth OpenFOAM workshop, Montreal, Canada.

Kuroda, M., Tamura, T., & Suzuki, M. (2007). Applicability of LES to the turbulent wake of a rectangular cylinder – comparison with PIV data. *Journal of Wind Engineering and Industrial Aerodynamics*, *95*(9–11), 1242–1258.

Le, T. H., Tamura, Y., & Matsumoto, M. (2011). Spanwise pressure coherence on prisms using wavelet transform and spectral proper orthogonal decomposition based tools. *Journal of Wind Engineering and Industrial Aerodynamics*, *99*(4), 499–508.

Löhner, R. (2008). *Applied computational fluid dynamics techniques: An introduction based on finite element methods* (2nd ed). Chichester, UK: John Wiley.

Mannini, C., Soda, A., & Schewe, G. (2010a). Unsteady RAS modelling of flow past a rectangular cylinder: Investigation on Reynolds number effects. *Computers and Fluids*, *39*(9), 1609–1624.

Mannini, C., Soda, A., Voß, R., & Schewe, G. (2010b). Unsteady RANS simulations of flow around a bridge section. *Journal of Wind Engineering and Industrial Aerodynamics*, *98*(12), 742–753.

Mannini, C., Soda, A., & Schewe, G. (2011). Numerical investigation on the three-dimensional unsteady flow past a 5:1 rectangular cylinder. *Journal of Wind Engineering and Industrial Aerodynamics*, *99*(4), 469–482.

Matsumoto, M. (1996). Aerodynamic damping of prisms. *Journal of Wind Engineering and Industrial Aerodynamics*, *59*(2–3), 159–175.

Matsumoto, M., Yagi, T., Tamaki, H., & Tsubota, T. (2008). Vortex-induced vibration and its effect on torsional flutter instability in the case of B/D = 4 rectangular cylinder. *Journal of Wind Engineering and Industrial Aerodynamics*, *96*(6–7), 971–983.

Mendes, P. A., & Branco, F. A. (1999). Analysis of fluid-structure interaction by an arbitrary lagrangian-eulerian finite element formulation. *International Journal for Numerical Methods in Fluids*, *30*(7), 897–919.

Menter, F. R. (1994). Two-equation Eddy-viscosity turbulence models for engineering applications. *AIAA Journal*, *32*(8), 269–289.

Menter, F. R. (2009). Review of the shear-stress transport turbulence model experience from an industrial perspective. *International Journal of Computational Fluid Dynamics*, *23*(4), 305–316.

Menter, F., & Esch, T. (2001, November 26–30). *Elements of industrial heat transfer prediction*. Proceedings of the 16th Brazilian Congress of Mechanical Engineering, Uberlandia, Brasil.

Murakami, S., Mochida, A., & Sakamoto, S. (1997). CFD analysis of wind-structure interaction for oscillating square cylinders. *Journal of Wind Engineering and Industrial Aerodynamics*, *72*, 33–46.

Nakaguchi, H., Hashimoto, K., & Muto, S. (1968). An experimental study on aerodynamic drag of rectangular cylinders. *Journal of the Japan Society of Aeronautical and Space Sciences*, *16*, 1–5.

Nieto, F., Kusano, I., Hernández, S., & Jurado, J. Á. (2010, May 23–27). *CFD analysis of the vortex-shedding response of a twin-box deck cable-stayed bridge*. Proceedings of the 5th International Symposium on Computational Wind Engineering, Chapel Hill, NC.

Norberg, C. (1993). Flow around rectangular cylinders: pressure forces and wake frequencies. *Journal of Wind Engineering and Industrial Aerodynamics*, *49*(1–3), 187–196.

Ogawa, K., Sakai, Y., & Sakai, F. (1988). Aerodynamic device for suppressing wind-induced vibration of rectangular section structures. *Journal of Wind Engineering and Industrial Aerodynamics*, *28*(1–3), 391–400.

Okajima, A. (1982). Strouhal numbers of rectangular cylinders. *Journal of Fluid Mechanics*, *123*, 379–398.

Oliver, A. (2009). *Mesh motion alternatives in OpenFOAM*. Project report: PhD course in CFD with OpenSource

software [Internet]. University of Chalmers. Retrieved from: http://www.tfd.chalmers.se/ ~ hani/kurser/OS_CFD_2009/. [cited 2014 July 18].
OpenFOAM Foundation. (2014a). *Standard solvers* [Internet]. Retrieved from: http://www.openfoam.org/features/standard-solvers.php. [cited 2014 July 24].
OpenFOAM Foundation. (2014b). OpenFOAM User Guide.
OpenFOAM Foundation. (2014c). OpenFOAM Programmer's Guide.
Piperno, S. (1998). *Numerical simulation of aeroelastic instabilities of elementary bridge decks*. Rapport de recherche n° 3549. INRA (Institute National de Recherche en Informatique et en Automatique).
Ribeiro, A. F. P. (2011, July 10–15). *Unsteady RANS modeling of flow past a rectangular 5:1 cylinder: Investigation of edge sharpness effects*. Proceedings of the 13th International Conference on Wind Engineering, Amsterdam, The Netherlands.
Sagaut, P. (2001). *Large eddy simulation for incompressible flow: An introduction*. Berlin: Springer-Verlag.
Sarkar, P. P., Caracoglia, L., Haan, F. L. Jr., Sato, H., & Murakoshi, J. (2009). Comparative and sensitivity study of flutter derivatives of selected bridge deck sections, Part 1: Analysis of inter-laboratory experimental data. *Engineering Structures*, *31*(1), 158–169.
Sarkic, A., Fisch, R., Hoffer, R., & Bletzinger, K. (2012). Bridge flutter derivatives based on computed, validated pressure fields. *Journal of Wind Engineering and Industrial Aerodynamics*, *104–106*, 141–151.
Sarwar, M. W., & Ishihara, T. (2010). Numerical study on suppression of vortex-induced vibrations of box girder bridge section by aerodynamic countermeasures. *Journal of Wind Engineering and Industrial Aerodynamics*, *98*(12), 701–711.
Sarwar, M. W., Ishihara, T., Shimada, K., Yamasaki, Y., & Ikeda, T. (2008). Prediction of aerodynamic characteristics of a box girder bridge section using LES turbulence model. *Journal of Wind Engineering and Industrial Aerodynamics*, *96*(10–11), 1895–1911.
Shimada, K., & Ishihara, T. (2002). Application of a modified *k-ε* model to the prediction of aerodynamic characteristics of rectangular cross-section cylinders. *Journal of Fluids and Structures*, *16*(4), 465–485.
Shimada, K., & Ishihara, T. (2012). Predictability of unsteady two-dimensional *k-ε* model on the aerodynamic instabilities of some rectangular prisms. *Journal of Fluids and Structures*, *28*, 20–39.
Simiu, E., & Scanlan, R. H. (1996). *Wind effects on structures* (3rd ed.). New York, NY: John Wiley.
Sohankar, A. (2006). Flow over a bluff body from moderate to high Reynolds numbers using large eddy simulation. *Computers and Fluids*, *35*(10), 1154–1168.
Sohankar, A. (2012). A numerical investigation of the flow over a pair of identical square cylinders in a tandem arrangement. *International Journal for Numerical Methods in Fluids*, *70*(10), 1244–1257.
Spalart, P. R., & Allmaras, S. R. (1992). A one-equation turbulence model for aerodynamic flows. *AIAA paper*, 92–439.
Sun, D., Owen, J. S., & Wright, N. G. (2009). Application of the *k-ω* turbulence model for a wind-induced vibration study of 2D bluff bodies. *Journal of Wind Engineering and Industrial Aerodynamics*, *97*(2), 77–87.
Sun, D., Owen, J. S., Wright, N. G., & Liaw, K. F. (2008). Fluid-structure interaction of prismatic line-like structures, using LES and block-iterative coupling. *Journal of Wind Engineering and Industrial Aerodynamics*, *96*(6–7), 840–858.
Takeuchi, T., & Matsumoto, M. (1992). Aerodynamic response characteristics of rectangular cylinders in tandem arrangement. *Journal of Wind Engineering and Industrial Aerodynamics*, *41*(1–3), 565–575.
Tamura, T. (1999). Reliability on CFD estimation for wind-structure interaction problems. *Journal of Wind Engineering and Industrial Aerodynamics*, *81*(1–3), 117–143.
Tamura, T., & Itoh, Y. (1999). Unstable oscillation of rectangular cylinder at various mass ratios. *Journal of Aerospace Engineering*, *12*(4), 136–144.
Tamura, T., & Ono, Y. (2003). LES analysis on aeroelastic instability of prisms in turbulent flow. *Journal of Wind Engineering and Industrial Aerodynamics*, *91*(12–15), 1827–1846.
Vairo, G. (2003). A numerical model for wind loads simulation on long-span bridges. *Simulation Modelling Practice and Theory*, *11*(5–6), 315–351.
Washizu, K., Ohya, A., Otsuki, Y., & Fujii, K. (1978). Aeroelastic instability of rectangular cylinders in heaving mode. *Journal of Sound and Vibration*, *59*(2), 195–210.
Washizu, K., Ohya, A., Otsuki, Y., & Fujii, K. (1980). Aeroelastic instability of rectangular cylinders in a torsional mode due to a transverse wind. *Journal of Sound and Vibration*, *72*(4), 507–521.
Wilcox, D. C. (2006). Turbulence modeling for CFD (3rd ed.). La Cañada: DCW Industries.
Wu, T., & Kareem, A. (2012). An overview of vortex-induced vibration (VIV) of bridge decks. *Frontiers of Structural and Civil Engineering*, *6*(4), 335–347.
Zhang, Y., Jia, Y., Wang, S. S. Y., & Altinakar, M. (2013). Composite structured mesh generation with automatic domain decomposition in complex geometries. *Engineering Applications of Computational Fluid Mechanics*, *7*(1), 90–102.

6

Effect of reservoir side slopes on dam-break flood waves

Farhad Hooshyaripor and Ahmad Tahershamsi*

Faculty of Civil and Environment Engineering, Amirkabir University of Technology, Tehran, Iran

Dam failure causes catastrophic flooding and immense damage downstream. There are many factors that affect the magnitude of such flash floods. In this study the influence of reservoir side slopes on dam-break hydrographs and water levels was investigated using experimental and numerical methods. Ideal trapezoidal reservoirs with different side angles varying from 30° to 90° were examined in the laboratory, while the reservoir capacity and breach cross section were considered to be the same. Due to the experimental limitations for measuring the flow parameters at any required points, 3D-numerical modeling was used. The results showed that the dam-break flood from the reservoir with a 30° side slope has the highest peak-discharge, the sharpest rising and recession limbs of the hydrograph, and the earliest maximum water level and discharge. Based on the results, in view of risk assessment, the reservoirs with lower side slopes are expected to have more potential risk for damage in the event of dam break.

Keywords: dam break; reservoir side slope; experimental method; numerical simulation; outflow hydrograph

1. Introduction

Dam failure is a topic of great importance for hydraulic researchers because in reality this catastrophic phenomenon can result in immense economic damage and loss of lives. A variety of hydraulic, hydrological, geomorphological, geotechnical, and geometric factors affect the outflow characteristics and the consequent damage downstream; however, there are considerable uncertainties involved in the problem (Hooshyaripor, Tahershamsi, & Golian, 2014). It has been shown that the reservoir geometry can influence the dam-break outflow hydrograph. Tahershamsi, Shetty, and Ponce (2003) developed an empirical relationship to relate the peak outflow discharge to reservoir geometry. Feizi Khankandi, Tahershamsi, and Soares-Frazão (2012) investigated the effect of reservoir shape on outflow hydrographs at different points of the downstream channel and showed the considerable impact of this factor on the bell-shaped hydrograph, the value of peak-discharge, and time to peak. Pilotti, Tomirotti, Valerio, and Bacchi (2010) showed that peak outflow depends on the reservoir bathymetry and the breach area. They defined the valley's geometry where the dam is located as prismatic. This definition meets a wide range of valley cross sections, from rectangular and parabolic ('U' shape) to triangular ('V' shape). The results emphasized the effect of reservoir geometry on the outflow hydrograph. Despite the importance of this topic, to the authors' knowledge, very few works have considered the role of reservoir geometry. A literature review indicates that most of the former investigations have focused on the features of the channel downstream of the dam and the effect of the reservoir geometry has rarely been studied. Therefore, because of the possible influence of the reservoir geometry on the dam-break flow, it seems worthwhile to study the effect of the reservoir's features on the outflow hydrograph. Accordingly, this paper aims to carry out an in-depth investigation into the differences in outflow hydrographs caused by altering the reservoir side slope, and rout the hydrographs at the downstream channel. To this end, several experiments were conducted and the results were analyzed. The experimental set-up included a reservoir upstream, a rectangular channel downstream, and a gate between them which could be pulled up rapidly. One of the main challenges of this study was to determine the cross section of the reservoir. Feizi Khankandi et al. (2012) considered reservoirs with a rectangular cross section of different dimensions. Pilotti et al. (2010) collected the data for several real valleys of alpine mountain reservoirs and characterized the cross sections with parabolic functions. In this paper, the complicated cross section of real reservoirs was simplified as trapezoidal, in which the lateral diagonal faces can have different slopes. This simplified cross section can easily be constructed in the laboratory on the one hand, and on the other hand it is an acceptable representative for the geometry presented by Pilotti et al. (2010).

The other challenge was the limitation of the apparatus in data acquisition. It was impossible to measure the velocity variations at the gate point and far downstream locations where the low water level influences the tool's performance. Furthermore, recording the time evolution of

*Corresponding author. Email: tshamsi@aut.ac.ir

water level and velocities at every point along the reservoir and channel in the laboratory required a considerable amount of time, so in this study the key parameters were measured in a few chosen positions. To overcome these limitations, three-dimensional numerical modeling, which has been used in various water engineering problems (Chau & Jiang, 2001, 2004; Dargahi, 2006; Liu & Yang, 2014), was applied in order to study the temporal and spatial variations of the water level and velocities at any arbitrary location. Recently, two-dimensional shallow water equations (SWEs) have been utilized to provide weak solutions to the dam-break flow (Gallegos, Schubert, & Sanders, 2009; LaRocque, Imran, & Chaudhry, 2013; Liang, Lin, & Falconer, 2007; Pu, Shao, Huang, & Hussain, 2013). Liang (2012) showed that in the location of the dam, the vertical momentum is considerable where the flow changes abruptly, so the flow is no longer horizontal and the pressure distribution is no longer hydrostatic. Ozmen-Cagatay, Kocaman, and Guzel (2014) evaluated the Reynolds-averaged Navier-Stokes (RANS) equations and the SWEs in the modeling of a dam-break flood wave in a flume with a hump downstream and showed that the RANS equations have a slightly better performance. Consequently, the present study employs the full Navier-Stokes equations in studying dam-break phenomena near dam locations to better link the mathematics to the physical features of the studied flow. The rest of the paper is organized as follows. Section 2 describes the experimental apparatus

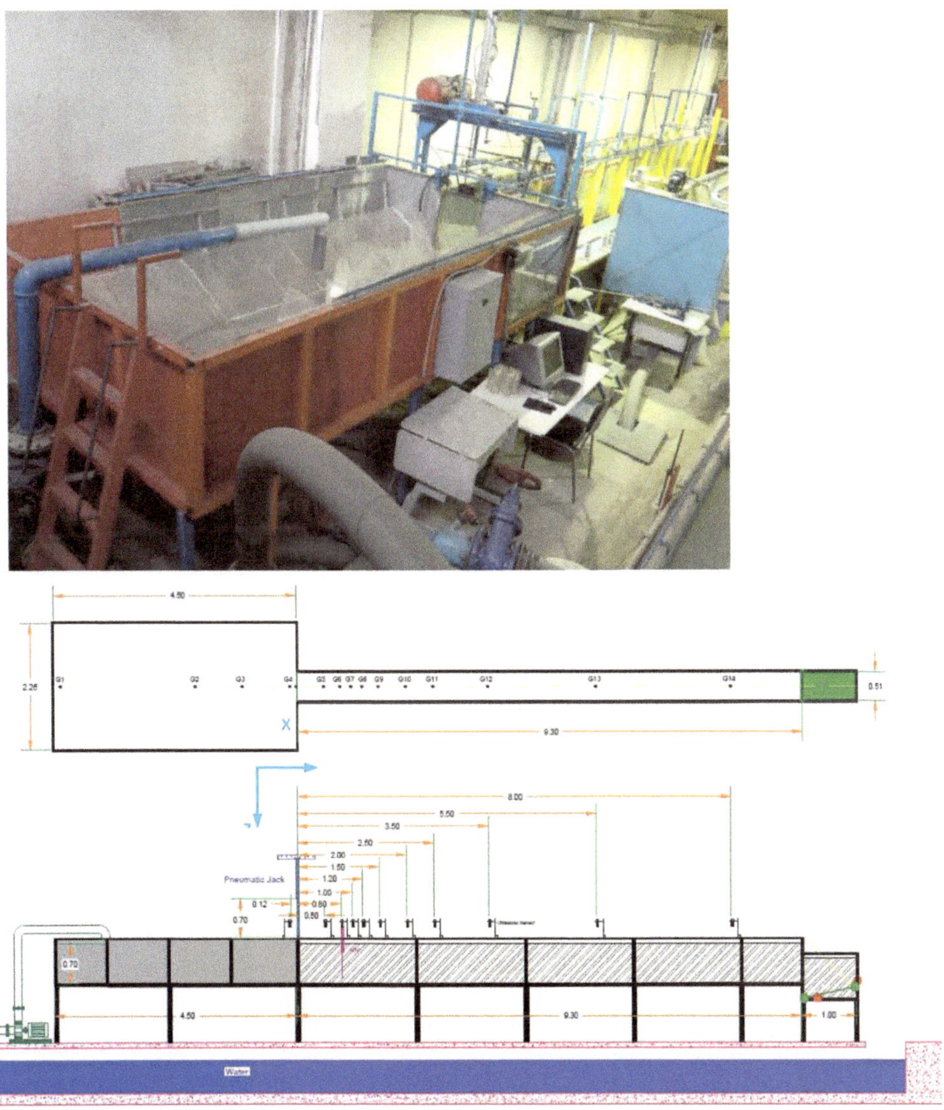

Figure 1. (a) Photograph and (b) sketch of the experimental apparatus.

and the modalities of the tests. Section 3 presents the calibration and verification steps of the numerical model. Section 4 discusses the experimental and numerical results, and Section 5 reports some conclusions.

2. Experiments

2.1. *Experimental set-up*

Figure 1 illustrates the experimental set-up, which was located in the Hydraulics Laboratory of the Civil Engineering Department, Amirkabir University of Technology, Iran. The upstream reservoir with a length of 4.5 m and a width of 2.25 m was made of galvanized sheet and fixed by nine steel columns to the laboratory floor. The dimensions of the rectangular flat and horizontal flume downstream are 0.51 m width, 0.7 m height, and 9.3 m length, and its Manning roughness was estimated under steady flow condition as 0.011 m/s$^{1/3}$. In the set-up there is a gate between the reservoir and the flume which has the ability to be opened instantaneously. The quick gate removal is accomplished with the aid of a pneumatic jack simulating a sudden dam break (Feizi Khankandi et al., 2012). The end of the downstream reach is open (Figure 1). To minimize the interference of the gate and water, the gate is made of Plexiglas plate of 10 mm thickness. The scale effect in a smooth prismatic and rectangular channel is insignificant, provided that the initial water depth against the gate $h_u \geq 0.3$ m and that the dam-break flow follows essentially the Froude similitude (Lauber & Hager, 1998). Therefore, h_u is 0.35 m above the flume bottom and the flume bed is initially dry in all tests. The gate opening time t_{op} is 0.14 s. Dam-break simulation can be considered as instantaneous if t_{op} is less than $t_{cr} = 1.25\sqrt{h_u/g}$ (Vischer & Hager, 1998), where g is the gravity acceleration (m/s^2). In this experiment, $t_{cr} = 0.24$ s.

2.2. *Flow measurement*

The x-coordinate of the measurement points are presented in Table 1. As illustrated in Figure 1(b), all the points are positioned on the centerline. The table also indicates the flow features which were measured at each point. The water-level evolution was measured by means of six ultrasonic sensors and the velocity components were acquired by an Acoustic Doppler Velocimeter (SonTec MicroADV). The ultrasonic sensors with response frequencies up to 100 Hz were placed at points G4 to G14 (see Figure 1). All sensors were installed above the flume on a movable device mounted on a rail at the top of the model. To reduce the side wall boundary-layer effects, the measurement points were positioned at the flume centerline. Water level measurement has an accuracy of ± 0.12 mm (Feizi Khankandi et al., 2012). The ADV has response frequencies up to 50 Hz and has been successfully used in dam-break studies (Feizi Khankandi et al., 2012; Soares-Frazão & Zech, 2007). The small diameter of the ADV probe (6 mm) causes local (Soares-Frazão & Zech, 2007) and negligible (Feizi Khankandi et al., 2012) perturbations in the flow.

Table 1. Location of the measurement points (for all points $z = 0.255$ m).

				Experimental model		Numerical model	
No.	Point	x (m)	$X = x/h_u$	Water level	Velocity	Water level	Velocity
1	G1	− 4.38	− 12.51	–	–	√	√
2	G2	− 1.88	− 5.371	–	–	√	√
3	G3	− 1.00	− 2.857	–	–	√	√
4	G4	− 0.12	− 0.343	√	–	√	√
5	Gate	0.00	0.000	–	–	√	√
6	G5	0.50	1.429	√	√	√	√
7	G6	0.80	2.286	√	√	√	√
8	G7	1.00	2.857	√	√	√	√
9	G8	1.20	3.429	√	√	√	√
10	G9	1.50	4.286	√	√	√	√
11	G10	2.00	5.714	√	√	√	√
12	G11	2.50	7.143	√	–	√	√
13	G12	3.50	10.000	√	–	√	√
14	G13	5.50	15.71	√	–	√	√
15	G14	8.00	22.86	√	–	√	√

2.3. *Program of the experiments*

In the laboratory, the cross section of the reservoirs is considered to be trapezoidal, in which the lateral diagonal faces are representative of the reservoir sides with slopes of *ZH:1V*. To assess the effect of this feature on dam-break outflow, reservoirs with different side slopes (*1/Z*) were examined, where $Z = 1.73$, 1.0, 0.58, and 0.0 ($\alpha = 30°$, 45°, 60°, and 90°, respectively; see Figure 2). The geometric characteristics of the reservoir are top reservoir width (w_d), bottom reservoir width (w_r), initial water depth (h_u), length (l), and capacity (v_r).

In the experiments, the reservoir capacity is considered to be the same in all the tests. In order to have a constant capacity after changing the side slopes in which the cross-sectional area of the reservoir A increases or decreases, the upstream end of the reservoir needed to be shifted downstream or upstream. Therefore, the reservoir length is a function of the side slope. According to this assumption, the reservoir of 90° side slope is the longest with a length of 4.5 m. Considering the 0.35 m initial water depth in this reservoir with $w_r = w_d = 0.51$ m, the reservoir capacity is 0.803 m^3 (Figure 3). Since the breach cross section (a) is unchanged in all the experiments while the reservoir cross section changes as the side slope varies, the value of breach ratio a/A defined by Pilotti et al. (2010) will depend on the reservoir side slopes.

Table 2 summarizes the reservoir dimensions in the experiments. Based on the number of the reservoirs, instrument limitations (just one ADV and six ultrasonic sensors) and retesting runs to ensure that the results from different

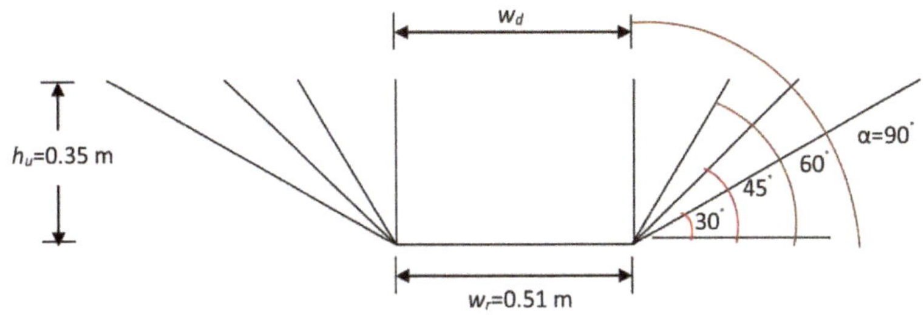

Figure 2. The reservoir cross section of the models.

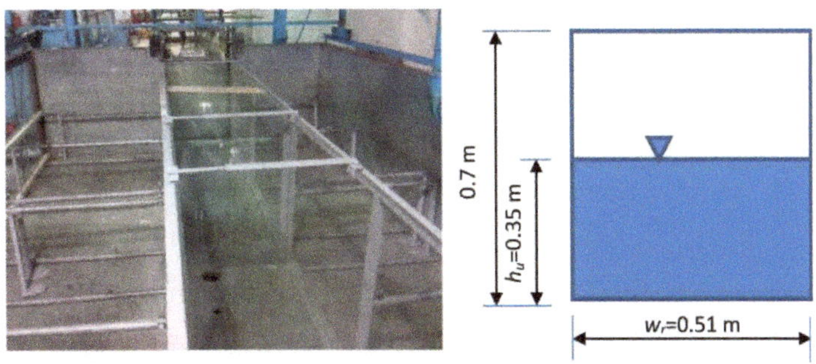

Figure 3. (a) View of the reservoir with a 90° side slope, and (b) detail of the reservoir's section.

Table 2. The reservoirs dimension in the experiments.

	α	Z	w_r	h_u	w_d	A	l
No.	(deg.)	(-)	(m)	(m)	(m)	(m^2)	(m)
1	30	1.73	0.51	0.35	1.73	0.391	2.05
2	45	1.00	0.51	0.35	1.05	0.301	2.67
3	60	0.58	0.51	0.35	0.92	0.249	3.23
4	90	0.00	0.51	0.35	0.51	0.179	4.50

runs could be combined, a total of 312 runs needed to be carried out.

2.4. *Observation from the experiments*

At time $t = 0$, the gate is removed instantaneously and the water that was at rest in the reservoir abruptly starts to flow to the downstream with a high turbulence. More perturbation in the outflow is seen next to the gate downstream when the side slope decreases. Figure 4(a) shows the generated cross waves downstream from the gate with a reservoir of 30° side slope. These cross waves cause variations of the water level in the downstream channel, both in the transverse and longitudinal directions (Figure 4(b)).

On the other hand, the flow pattern in the reservoir is influenced by the reservoir's side slopes. A one-dimensional flow pattern is seen if the side walls are vertical. In this case (90° side reservoir), as soon as the gate is opened, a negative wave propagates upstream, while in the other cases two other negative waves are generated which propagate to the side walls (Figure 5). The negative side waves meet the walls and then return to incident in the reservoir centerline. The wave interference continues sequentially until the reservoir depletion, so the resultant periodic perturbation propagates downstream the channel.

3. Numerical model

To determine the flow characteristics at the gate and improve the design of the process equipment, the 3D-module of the Computational Fluid Dynamics (CFD) package *FLUENT* (ver. 6.3.26) was used to simulate the experiments. *FLUENT* has previously been applied to and evaluated for dam-break simulation (LaRocque et al., 2013; Maghsoodi, Roozgar, & Sarkardeh, 2012) and modeling a flow below a spillway gate when the gate is opening (Liu & Yang, 2014). The *FLUENT* software uses the finite volume method to solve the Reynolds-averaged conservation equations for mass and momentum for an incompressible fluid as the governing equations

$$\frac{\partial \overline{u_i}}{\partial x_i} = 0 \tag{1}$$

$$\frac{\partial \rho \overline{u_i}}{\partial t} + \frac{\partial \rho \, \overline{u_j} \, \overline{u_i}}{\partial x_j} = -\frac{\partial \overline{P}}{\partial x_i} + \frac{\partial}{\partial x_j}\left[\mu\left(\frac{\partial \overline{u_i}}{\partial x_j} + \frac{\partial \overline{u_j}}{\partial x_i}\right)\right] + \frac{\partial}{\partial x_j}(-\rho\overline{u_i' u_j'}) \tag{2}$$

where $\overline{u}$ and $\overline{P}$ = Reynolds-averaged velocities and pressure, respectively, x = Cartesian coordinate axes,

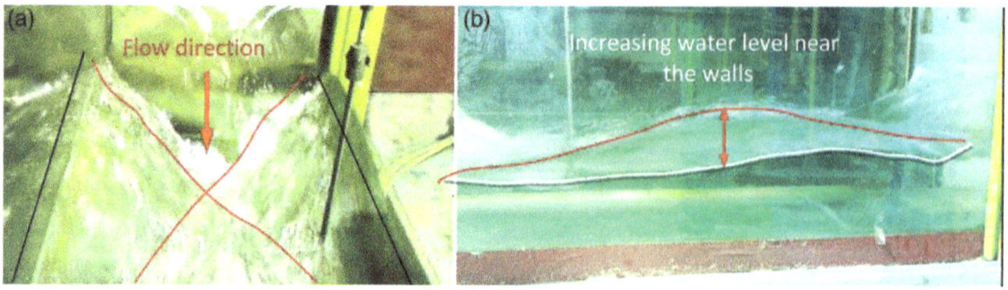

Figure 4. (a) Cross waves downstream from the gate at t = 1.0 s (30° reservoir) and (b) effect of the cross waves on the flow depth near the channel walls.

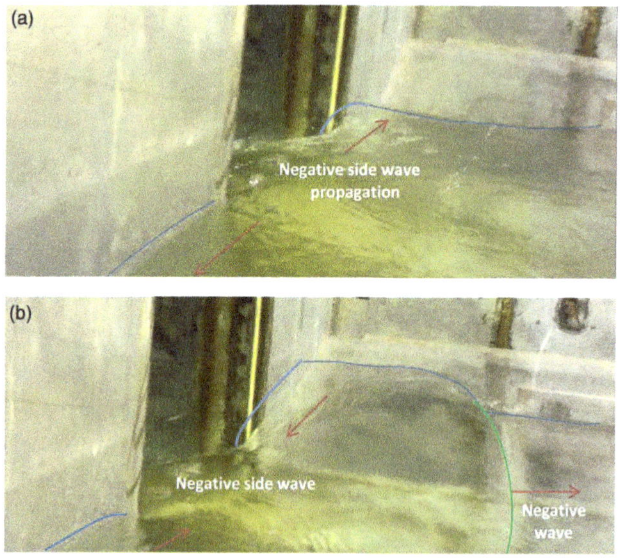

Figure 5. Negative waves in the 30° reservoir at (a) t = 1.0 s and (b) t = 2 s after gate removal.

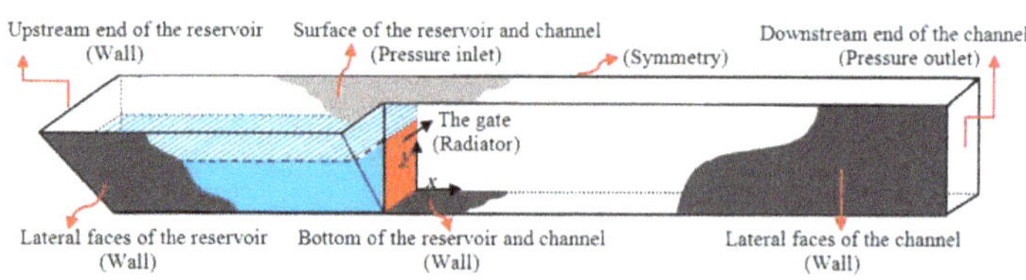

Figure 6. Boundary condition in the numerical model.

ρ = density of fluid, t = time, μ = viscosity of fluid, and $-\rho\overline{u_i'u_j'}$ is known as Reynolds stress.

In this study the so-called segregated solver was used, so the governing equations were solved sequentially, i.e., segregated from each other. As the governing equations are non-linear and coupled, a number of iterations of the solution loop must be performed to reach a converged solution. In this iterative scheme, all the equations are solved iteratively, for a given time step, until the convergence criteria are met. With this iterative scheme, non-linearity of the individual equations and inter-equation couplings are fully accounted for, eliminating the splitting error. In order to capture the free surface, the Volume of Fluid (VOF) scheme was applied. The back, lateral, and bottom faces were specified as the wall, top and end downstream boundaries respectively as pressure inlet and outlets, and a plane of symmetry was defined along the model (Figure 6). To simulate the initial conditions of dam failure, the area of water in the upstream reservoir was designed to a specified level with air filling the rest of the grid.

3.1. *Calibration and verification*

The unsteady free-surface calculations required fine-enough grid sizes and small initial time steps. Therefore, various grid sizes and time steps were examined to check

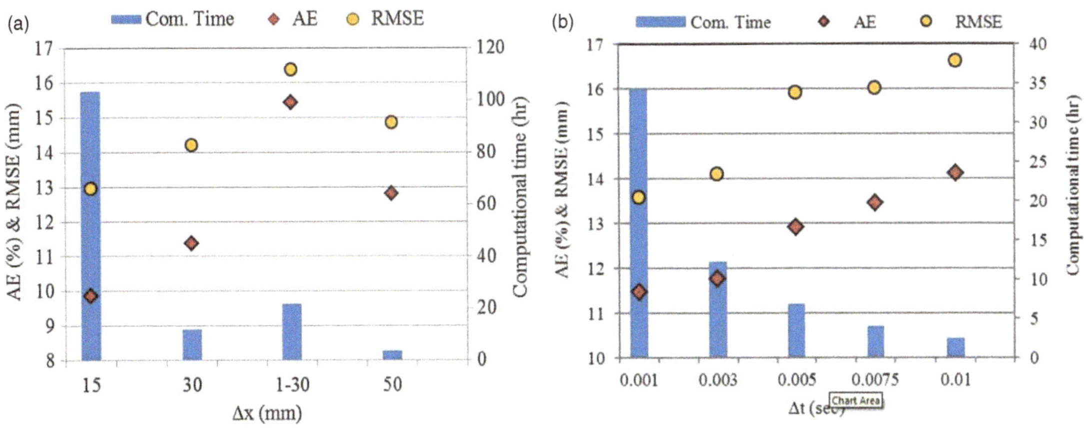

Figure 7. Sensitivity analysis of the different (a) mesh sizes, and (b) time steps.

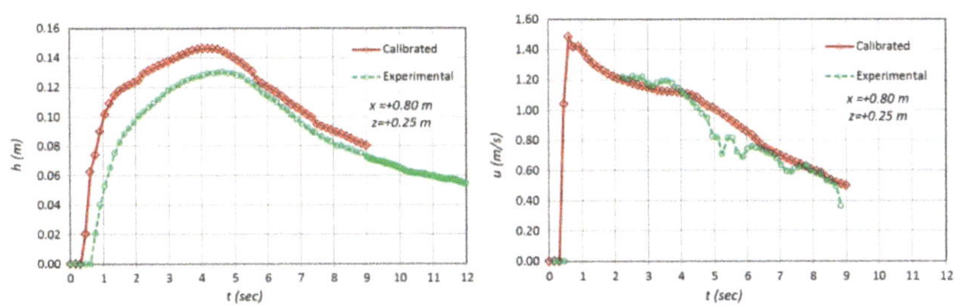

Figure 8. Verification of the numerical model for the 90° reservoir at point G6 for the temporal variation of (a) the water level and (b) the average velocity.

their effectiveness in the simulation. Four sets of grids were created to the dimensions of the experiments using *Gambit* (ver. 2.3.16), a mesh-generation package: uniform mesh in both the x and y directions with spacing of 15, 30, and 50 mm, and varied mesh sizes of 1–30 mm. Figure 7(a) illustrates the differences between the numerical and experimental water levels for different mesh sizes. Considering a 90° reservoir, two indices of (1) average error (AE) and (2) root mean square error (RMSE) are calculated and presented. In addition, the figure shows the computational time for each of the simulations. Regarding both the accuracy and the computational cost, the grid size of 30 mm was found to be appropriate for the purpose of this study. Again a sensitivity analysis was conducted on the value of the time step by considering $\Delta t = 0.001$, 0.003, 0.005, 0.0075, and 0.01 s (Figure 7(b)). Analysis of the results showed that $\Delta t = 0.003$ s is an acceptable value because it satisfies the accuracy and convergence criteria as well as the Courant-Friedrichs-Lewy condition. Considering $\Delta x = 30$ mm and $\Delta t = 0.003$ s, the sensitivity analyses on the different turbulence models, bed roughness values, and pressure-velocity coupling algorithms showed a negligible difference between the results.

Thus, the k-ϵ turbulence model was used to account for turbulence, the velocity pressure coupling was resolved via a PISO-type, and a nominal roughness height of 0.01 mm was assigned to the wall boundaries. In addition, the PRESTO! pressure discretization scheme and the first-order upwind momentum and turbulent kinetic energy discretization scheme were applied (Karki & Patankar, 1989). Convergence is reached when the scaled residual of each variable is in the order of 0.001. Moreover, the free surface is defined by a value of 0.5 for VOF (Fluent Manual, 2006). In *FLUENT*, a time step different from the one used for the rest of the transport equations is defined for the volume fraction calculation. This time step was refined based on the input for the maximum Courant number allowed near the free surface (Fluent Manual, 2006). In this study the Courant number of 0.25 was predefined. In order to verify the calibrated numerical model, the 90° reservoir was simulated in *FLUENT*. Figure 8 compares the numerical and experimental results at point G6. Table 3 shows the average difference between the numerical and experimental results at the points G5 and G6.

4. Analysis of the results

In Figure 9 the experimental results of the water level variations for the 90° reservoir and channel (points G4 and G6, respectively) are compared to the analytical solution of Ritter (1892). In the figure, two vertical dashed lines indicate the time limits in which the limited upstream

Table 3. Comparison between numerical and experimental results at points G5 and G6.

Non-dimensional parameter	G5 (X = 1.429) RMSE	G5 AE (%)	G6 (X = 2.286) RMSE	G6 AE (%)	Average at points G5 and G6 RMSE	Average AE (%)
Water level (H)	0.07	32.3	0.06	23.2	0.07	27.7
Velocity (U)	0.05	− 10.6	0.04	5.9	0.05	− 2.3
Flow discharge (Q)	0.01	− 8.1	0.03	18.5	0.02	5.2

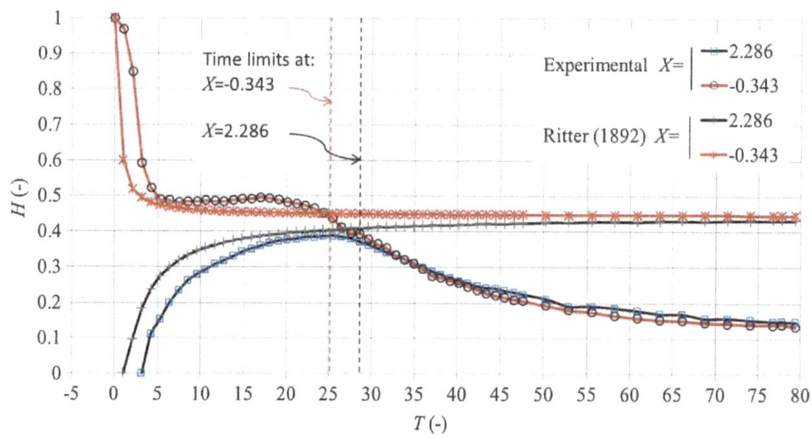

Figure 9. Experimental and analytic (Ritter, 1892) results of temporal water level variation at X = − 0.343 (point G4) and X = 2.286 (point G6).

reservoir length starts to influence the wave levels at the measuring points. The value of time limit is estimated as $T_{\text{limit}} = (2L + X)/C_{\text{NF}}$, where $L = l/h_u$ = dimensionless reservoir length and CNF = dimensionless negative wave-front celerity (Leal, Ferreira, & Cardoso, 2009). The figure indicates that the water level curve of the Ritter solution separates from the experimental curve from $T = T_{\text{limit}}$ such that for $T \geq T_{\text{limit}}$, the Ritter results get close to the constant value of $H = 4/9$, while the experimental data decrease gradually due to reservoir depletion. Figure 9 shows that the water level in the reservoir decreases more rapidly when the Ritter solution is used. Moreover, the wave-front in the Ritter solution reaches G6 sooner than the experimental results.

Figure 10 illustrates the attained experimental and numerical results at point G5. The vertical dashed lines show the T_{limit} values of the reservoirs and from left to right correspond to the 30°, 45°, 60°, and 90° reservoirs, respectively. According to Figure 10(a), the value of $H_{\max}$ ($= h_{\max}/h_u$) and its corresponding time depend on the side slope of the reservoirs such that the corresponding time of the maximum water level reduces as the side slope decreases. Thus, the rising limb of the curve corresponding to a 30° reservoir is sharper than the other ones. Note also that the 30° reservoir has the highest maximum water level but a faster attenuation of the peak points. Figure 10(b) compares the temporal velocities at G5. The figure shows that the side slope value highly affects the velocity of the dam-break flow downstream. Also, it can be seen that the experimental curves of the 45° and 60° reservoirs have some fluctuations, while the curve of the 90° reservoir is smooth. The fluctuation could relate to the side negative waves in the reservoir which are transported downstream. Note that Figure 10(b) does not include the experimental 30° side slope reservoir because of the high turbulent flow which affects the ADV performance. The results show that $U_{\max}$ [$= u_{\max}/(gh_u)^{0.5}$] of the 45° and 60° reservoirs are approximately 30% and 21% greater than that of the 90° reservoir, respectively; thus, the lower the side slope, the higher the maximum average velocity at downstream locations. Moreover, it can be observed that numerical modeling could not simulate these perturbations very well. Figure 10(c) shows the dam-break flow rates at point G5. According to this figure, the outflow hydrograph of the 30° reservoir is sharper than the others, i.e., the higher $Q_{\max}$ [$= q_{\max}/(gh_u^3)^{0.5}$] values occur at the earlier times, while the area under the curves is the same (due to the same capacity of the reservoirs). According to Figure 10(c) the peak outflow values of the 30°, 45°, and 60° reservoirs are almost 23%, 16%, and 11% more than that of the 90° reservoir, respectively. Note that the rapidly rising limb of the hydrographs is missed in the experiments. Also the $T_{Q_{\max}}$ value occurs faster when the reservoir has lower side slopes, such that this amount for 30°, 45°, 60°, and 90°

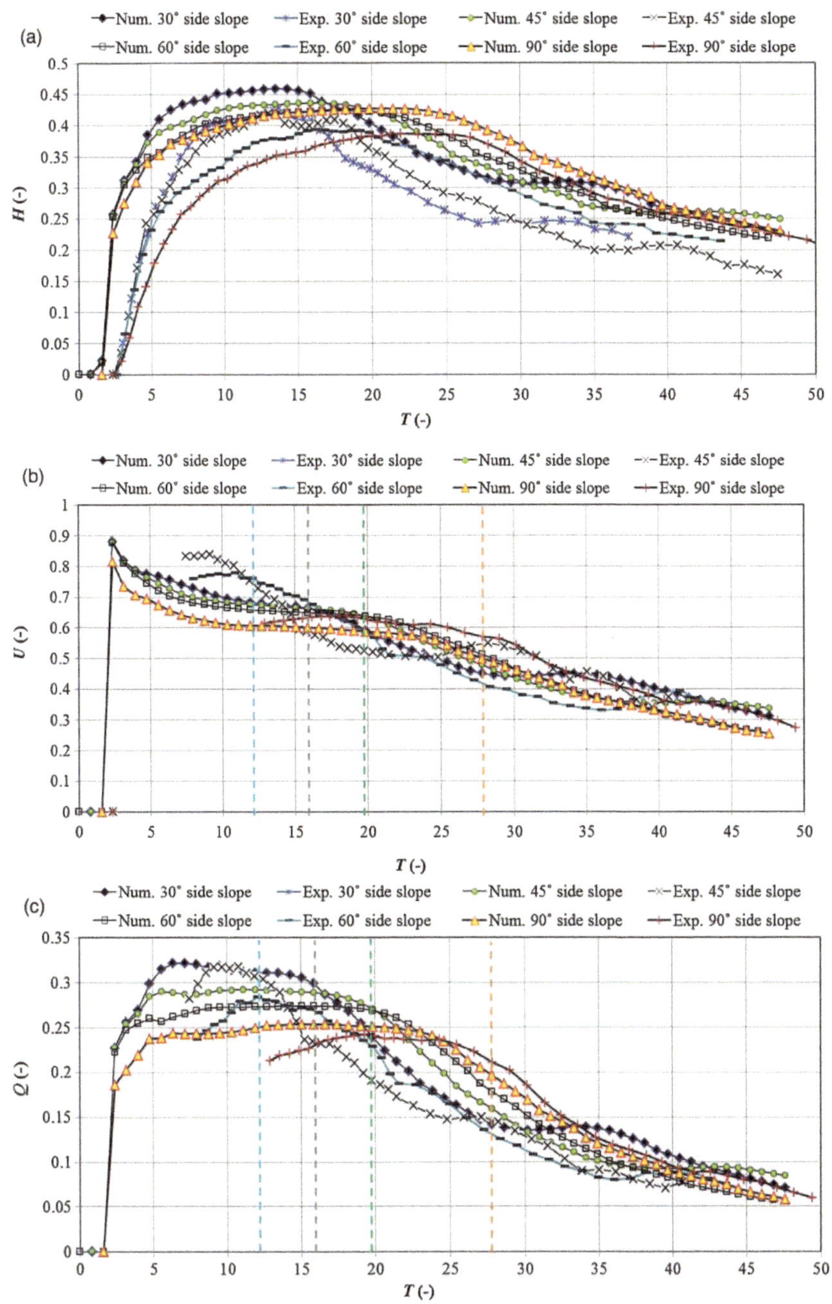

Figure 10. Experimental and numerical results at point G5 for the temporal variation of (a) the water level, (b) the velocity, and (c) the flow discharge.

reservoirs is calculated as 6.35, 11.12, 15.09, and 15.88, respectively (from the numerical results).

Figure 11 shows the numerical results in comparison with the data achieved from the experiments on the 45° reservoir. The figure also illustrates the analytical results based on the Ritter (1892) solution, which are time-independent at the gate location ($X = 0$). As it is shown, the Ritter solution underestimates the water level and overestimates the velocity; however, the results are somehow satisfactory for the flow rates. At G5, G10, and G13 it is observed that the numerical and experimental results are in good agreement for water level, although the fluctuations have not been simulated in the numerical model. Moreover, Figure 11(a) shows that in each location the numerical

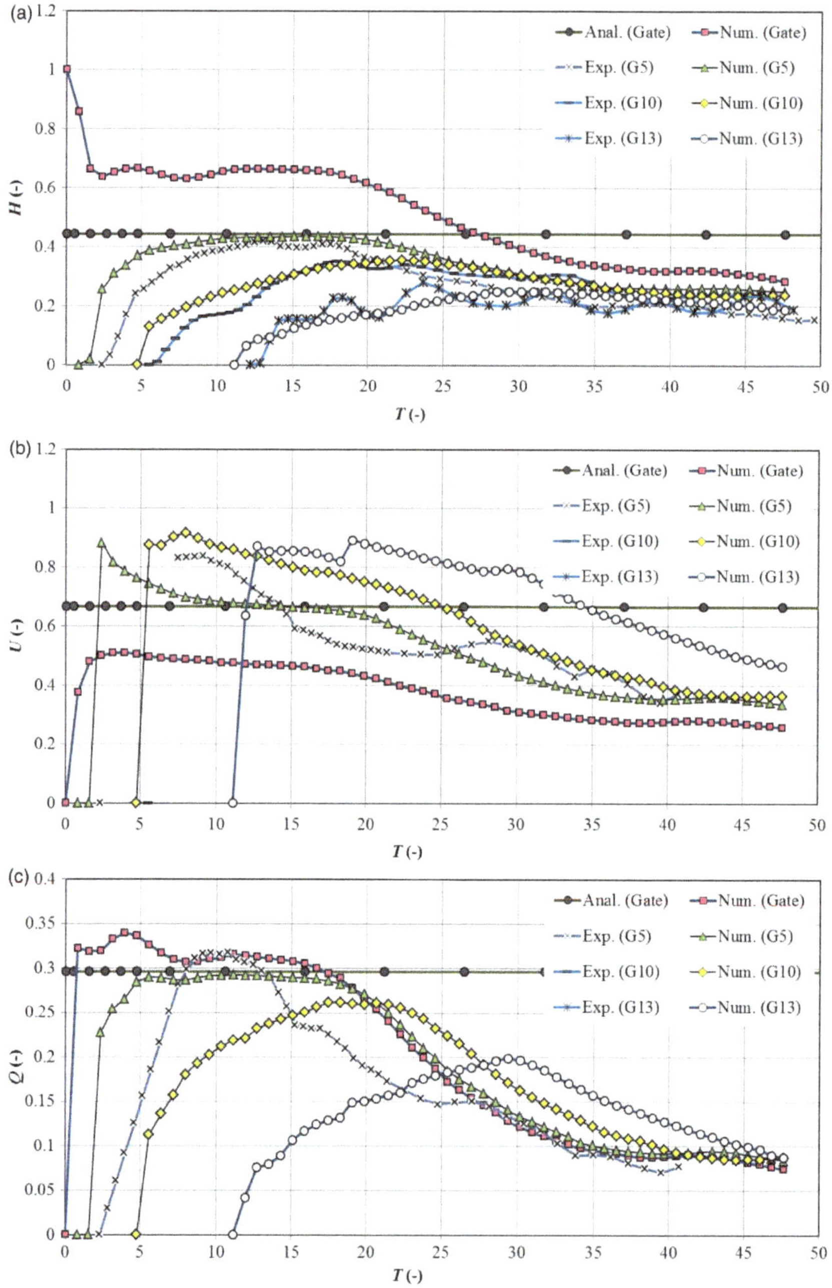

Figure 11. Comparison of the experimental and numerical results of the 45° reservoir at downstream locations.

wave is earlier than the experimental wave. According to Figure 11(b), in the case of the 45° reservoir the velocity measurement by ADV has not been possible at points G10 and G13, so the experimental hydrographs have not been calculated for these points.

In Table 4 the values of the maximum water level and the outflow discharge from the numerical analysis are calculated and summarized. In this table the values for another reservoir with 75° side slopes are also presented. These values are presented graphically in Figure 12. It can be seen that the influence of side slopes is much more considerable at the points close to the gate but negligible for $X \geq 5.714$. Moreover, the figure indicates that the water level is affected considerably by the reservoir side slope such that the maximum water level at the gate position for the case of the 30° reservoir is almost 58% greater than that

Table 4. Maximum value of dam-break outflow parameters at downstream locations.

		α (deg)				
X	Hydraulic parameter	90	75	60	45	30
0.0	H_{max}	0.733	0.666	0.596	0.534	0.465
	Q_{max}	0.371	0.339	0.312	0.291	0.257
	H_{max}	0.612	0.570	0.538	0.497	0.447
	$T_{H_{max}}$	7.145	11.910	16.670	19.850	19.850
0.571	Q_{max}	0.389	0.339	0.306	0.279	0.255
	$T_{Q_{max}}$	5.550	11.910	10.320	14.290	17.460
	H_{max}	0.458	0.436	0.423	0.420	0.428
	$T_{H_{max}}$	13.490	15.830	18.260	21.440	20.900
1.429	Q_{max}	0.321	0.292	0.274	0.261	0.253
	$T_{Q_{max}}$	6.350	11.110	14.290	17.460	15.880
	H_{max}	0.397	0.396	0.390	0.385	0.418
	$T_{H_{max}}$	15.850	18.260	20.640	23.030	22.230
2.286	Q_{max}	0.286	0.273	0.253	0.241	0.252
	$T_{Q_{max}}$	8.730	16.680	16.670	18.260	20.640
	H_{max}	0.366	0.355	0.330	0.343	0.340
	$T_{H_{max}}$	18.710	22.230	18.430	27.790	27.600
5.714	Q_{max}	0.271	0.262	0.247	0.234	0.235
	$T_{Q_{max}}$	16.670	17.470	19.850	23.810	23.810
	H_{max}	0.239	0.250	0.252	0.255	0.263
	$T_{H_{max}}$	27.790	30.170	32.550	35.730	35.730
15.714	Q_{max}	0.191	0.198	0.195	0.188	0.187
	$T_{Q_{max}}$	25.400	29.380	30.960	32.560	30.170

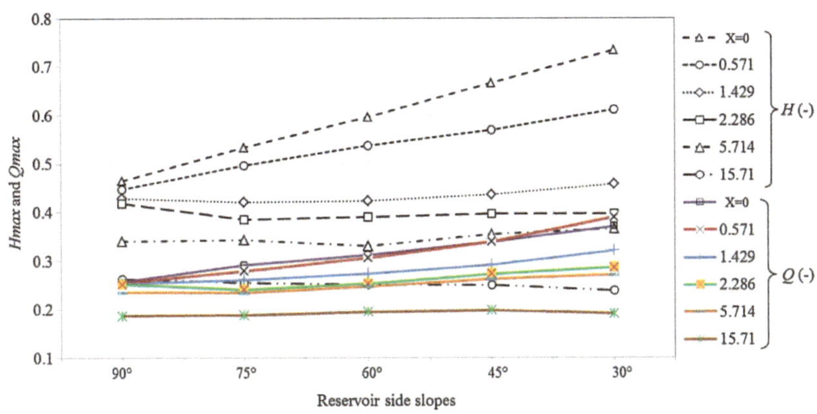

Figure 12. Effect of the side slope on the maximum water level and discharge.

of the 90° reservoir, while this percentile is about 44% for discharge values at the same position.

5. Conclusion

Several investigations have shown that reservoir geometry can affect dam-break outflow and consequently the flood damage that occurs downstream. The geometry of a natural reservoir is extremely complicated, so it is difficult to model such geometry in the lab – but it is possible to simplify its cross section as parabolic or trapezoidal. Assuming trapezoidal geometry for the reservoirs, in this paper experimental and numerical attempts were conducted on dam-break outflows from those reservoirs with different side slopes of 30°, 45°, 60°, and 90° to analyze the water level, velocities, and discharge values at different points. Such side slope values could represent the reservoirs located in the mountain valleys. To attain a robust comparison over the outflow hydrographs, the reservoir capacity was taken unchanged in the four models; thus, the reservoir length decreases as the degree of the side slope reduces. Measurement of the key parameters was not possible at every arbitrary point, so *FLUENT* CFD software was applied to attain the temporal and spatial variations of the parameters at any position. Comparison of the results showed that the reservoir side slope is a parameter that affects the emptying process, such that the outflow hydrograph from the 30° side slope reservoir had the highest peak-discharge and the earliest time to peak. According to the results, a shorter emptying time could be expected from reservoirs with lower side slopes. Results indicated that the side slope has a greater effect on the water level than the peak-discharges. Overall, in terms of the risk assessment, it can be concluded that reservoir geometry has a considerable impact on the dam-break flood wave. Therefore, because of the variety of shapes and sections of actual reservoirs, their complex behavior in the case of dam failure, and to the authors' knowledge the lack of studies in this area, it would be worthwhile to perform further investigations and study more deeply the effect of reservoir features on dam-break flow.

Disclosure statement

No potential conflict of interest was reported by the authors.

References

Chau, K. W., & Jiang, Y. W. (2001). 3D numerical model for Pearl river estuary. *Journal of Hydraulic Engineering, 127*(1), 72–82.

Chau, K. W., Jiang, Y. W. (2004). A three-dimensional pollutant transport model in orthogonal curvilinear and sigma coordinate system for Pearl river estuary. *International Journal of Environment and Pollution, 21*(2), 188–198.

Dargahi, B. (2006). Experimental study and 3D numerical simulations for a free-overflow spillway. *Journal of Hydraulic Engineering, 132*(9), 899–907.

Feizi Khankandi, A., Tahershamsi, A., & Soares-Frazão, S. (2012). Experimental investigation of reservoir geometry effect on dam-break flow. *Journal of Hydraulic Research, 50*(4), 376–387.

FLUENT. (2006). *Fluent 6.3 user manual.* Lebanon, NH: Fluent, Inc.

Fluent Inc. (2006). FLUENT [Computer software]. Centerra Resource Park, 10 Cavendish Court, Lebanon, NH 03766.

Fluent, Incorporated. (2006). Gambit [Computer software]. Centerra Resource Park, 10 Cavendish Court, Lebanon, NH 03766.

Gallegos, H. A., Schubert, J. E., & Sanders, B. F. (2009). Two-dimensional, high-resolution modeling of urban dam-break flooding: a case study of Baldwin Hills, California. *Advances in Water Resources, 32*(8), 1323–1335.

Hooshyaripor, F., Tahershamsi, A., & Golian, S. (2014). Application of copula method and neural networks for predicting peak outflow from breached embankments. *Journal of Hydro-Environment Research, 8*, 292–303.

Karki, K., Patankar, S. V. (1989). Pressure-based calculation procedure for viscous flows at all speeds in arbitrary configurations. *Engineering Applications of Computational Fluid Mechanics, 27*, 1167–1174.

LaRocque, L., Imran, J., & Chaudhry, M. (2013). Experimental and numerical investigations of two-dimensional dam-break flows. *Journal of Hydraulic Engineering, 139*(6), 569–579.

Lauber, G., Hager, W. H. (1998). Experiments to dam-break wave: Horizontal channel. *Journal of Hydraulic Research, 36*(3), 291–307.

Leal, J. G. A. B., Ferreira, R. M. L., & Cardoso, A. H. (2009). Maximum level and time to peak of dam-break waves on mobile horizontal bed. *Journal of Hydraulic Engineering, 135*(11), 995–999.

Liang, D. (2012). Shallow water equations: Magic and limitations. *Dam Engineering, XXIII*(1), 41–47.

Liang, D. F., Lin, B. L., & Falconer, R. A. (2007). A boundary-fitted numerical model for flood routing with shock-capturing capability. *Journal of Hydrology, 332*(3–4), 477–486.

Liu, T., & Yang, J. (2014). Three-dimensional computations of water-air flow in a bottom spillway during gate opening. *Engineering Applications of Computational Fluid Mechanics, 8*(1), 104–115.

Maghsoodi, R., Roozgar, M. S., & Sarkardeh, H. (2012). 3D simulation of dam break flows. *Dam Engineering, XXIII*(2), 53–69.

Ozmen-Cagatay, H., Kocaman, S., & Guzel, H. (2014). Investigation of dam-break flood waves in a dry channel with a hump. *Journal of Hydro-Environment Research, 8*, 304–315.

Pilotti, M., Tomirotti, M., Valerio, G., & Bacchi, B. (2010). Simplified method for the characterization of the hydrograph following a sudden partial dam break. *Journal of Hydraulic Engineering, 136*(10), 693–704.

Pu, J. H., Shao, S., Huang, Y., & Hussain, K. (2013). Evaluations of SWEs and SPH numerical modelling techniques for dam break flows. *Engineering Applications of Computational Fluid Mechanics, 7*(4), 544–563.

Ritter, A. (1892). Die Fortpflanzung de Wasserwellen. *Zeitschrift Verein Deutscher Ingenieure, 36*(33), 947–954.

Soares-Frazão, S., & Zech, Y. (2007). Experimental study of dambreak flow against an isolated obstacle. *Journal of Hydraulic Research, 45*(S1), 27–36.

Tahershamsi, A., Shetty, A. V., & Ponce, V. M. (2003). Embankment dam breaching: Geometry and peak outflow characteristics. *Dam Engineering, 14*(2), 73–87.

Vischer, D. L., & Hager, W. H. (1998). *Dam hydraulics*. Chichester, UK: Wiley.

7

Simulation of open channel bend characteristics using computational fluid dynamics and artificial neural networks

Azadeh Gholami, Hossein Bonakdari*, Amir Hossein Zaji and Ali Akbar Akhtari

Department of Civil Engineering, Razi University, Kermanshah, Iran

The ability to simulate flow characteristics is one of the most important issues in the design and application of open channel bends. Three-dimensional computational fluid dynamics (CFD) and multi-layer feed-forward artificial neural networks (MLFF-ANNs) are used and compared in modeling the flow depth and velocity field in sharp bends. CFD is modeled in two phases, water and air, using the volume of fluid method. The backpropagation algorithm is applied in the training process of the ANN model. An experimental study of a 90° curved channel is undertaken to verify and compare the efficiency of the CFD and ANN models. The results show that both CFD and ANN methods can be successfully applied to the modeling of open channel bend characteristics. However, the ANN method performs significantly better than CFD.

Keywords: artificial neural network; computational fluid dynamics; curved open channel; experimental modeling; velocity field

1. Introduction

Open channel bends are common hydraulic structures employed in natural and artificial channels. Because of the pressure gradient and centrifugal force, as well as the interaction between these two forces, strong secondary flows are created in open channel bends (Balen, Uijttewaal, & Blanckaert, 2010; Kimura, Uijttewaal, & Balen, 2008; Lien, Hsieh, Yang, & Yeh, 1999). Secondary flow causes full, three-dimensional complex fluid flow in bends (Ferguson, Parsons, Lane, & Hardy, 2003; Lu, Zhang, Cui, & Leung, 2004; Ottevanger, Blanckaert, & Uijttewaal, 2012).

Several experimental studies have been carried out on open channel bends to investigate flow characteristics. Rozovskii (1961) and Leschziner and Rodi (1979) studied sharp and mild bends, and developed some rules for specifying these. DeVriend and Geoldof (1983) conducted experimental studies on a 90° mild bend and reported that the change in the transverse profile of the water surface along the bend is linear. Experimental studies by Anwar (1986) demonstrated that the velocity pattern is not dependent on the radius at the bend's entrance but the flow pattern's bend-related effects are evident in the bend's outlet. Ye and McCorquodale (1998) conducted a detailed, vast experimental study on curved channels. They examined the flow patterns and secondary currents in the bends and showed that secondary currents and super-elevation begin in the upstream cross sections and gradually reach the inner cross sections of the bend. Blanckaert and Graf (2001) conducted experimental studies on turbulent flow in a movable bed with a 120° sharp bend. Zeng, Constantinescu, Blanckaert, and Weber (2008) experimentally examined the velocity patterns, vorticity, turbulent kinetic energy, and bathymetry of a 193° sharp bend. Ramamurthy, Han, and Biron (2013) conducted vast experimental and numerical research studies on the velocity profiles of sharp bends. Many numerical studies have been conducted to model the characteristics of open channel bends as well.

One of the major benefits of numerical simulations of open channel bends is the transfer of discrete laboratory measurements to the continuous velocity field. DeMarchis and Napoli (2006) numerically modeled the flow pattern of a 270° sharp bend and examined the velocity components and water surface level within the channel. Zhang and Shen (2008) generated a three-dimensional numerical model of the flow within curved channels. The authors stated that the secondary flow caused by the centrifugal force is a primary reason for changes that occur in velocity and water surface level. Huang, Jia, Chan, and Wang (2009) used three-dimensional modeling along with different turbulence models to study secondary flow cells, depth average velocity distribution, and water surface level in curved channels. Furthermore, their model was able to simulate the minor secondary flow's roll cell in the outer channel wall. Naji, Ghodsian, Vaghefi, and Panahpur (2010) used a numerical three-dimensional model to examine the flow pattern within a 90° bend and stated that

*Corresponding author. Email: bonakdari@yahoo.com

the secondary flow within this channel is the main reason why changes occur in longitudinal, transverse, and vertical velocity components. Ramamurthy et al. (2013) used two numerical models along with different turbulence models on a 90° sharp bend; they simulated the free surface using rigid lid, porosity, and volume of fluid methods. The authors stated that the numerical results of the RSM turbulence model and volume of fluid method are more consistent with the experimental results.

Soft computing methods have successfully been used in different areas of hydraulics (Bilhan, Emiroglu, & Kisi, 2010, 2011; Chau, Wu, & Li, 2005; Dursun, Kaya, & Firat, 2012; Ebtehaj & Bonakdari, 2013; Emiroglu, Bilhan, & Kisi, 2011; Jahangirzadeh et al., 2014; Kisi, Emiroglu, Bilhan, & Guven, 2012; Shamshirband et al., 2015; Wang, Chau, Cheng, & Qiu, 2009; Wang, Xu, Chau, & Lei, 2014; Wu & Chau, 2010; Wu, Chau, & Li, 2008; Yuhang and Wenxin, 2009). Sahu, Jana, Agarwal, and Khatua (2011) used artificial neural network (ANN) modeling to study and predict the velocity values within a meandering open channel. The high correlation coefficient between the results indicates the high accuracy of the presented network in predicting velocity within this channel. Bonakdari, Baghalian, Nazari, and Fazli (2011) predicted the velocity field values in a mild bend using ANN and the Genetic Algorithm. They used two-phase (air + water) three-dimensional modeling to verify the results where no experimental results were available. Their results indicated that there is an acceptable level of consistency between the results of the numerical and ANN models. Baghalian, Bonakdari, Nazari, and Fazli (2012) took a sediment tool into account and studied the flow in a 90° mild bend in a three-phase state (water + air + sediment). According to their results, the numerical and ANN models were more consistent with the experimental values compared with the analytical solution. Gholami, Akhtari, Minatour, Bonakdari, and Javadi (2014) experimentally and numerically investigated the flow pattern in a 90° sharp bend. They focused on the general characteristics of flow and evaluated the performance of these models. It was found that sharp bends have a full, three-dimensional complex flow pattern.

According to the studies mentioned above, only a few have been conducted to evaluate the flow characteristics of open channel sharp bends using numerical methods. Therefore in this study, the computational fluid dynamics (CFD) and ANN methods are applied in order to model open channel sharp bend flow characteristics. An extensive experimental study is undertaken by the authors to verify and train the models. The CFD models were run using FLUENT software with the $k - \varepsilon$ (RNG) turbulence method (Fluent Manual, 2005). Two different ANN models were developed to model the velocity field and flow depth. After comparing the performance of the CFD and ANN models, their performance was compared in case there are no experimental measurements available. The present study results regarding sharp bends are very efficient in the design and construction of artificial curved channels.

2. Experimental model

The ANN models were trained and tested and the CFD models validated using an experimental study conducted in the hydraulic laboratory at the Ferdowsi University of Iran (Figure 1). A square Plexiglas flume of 40.3 cm width and height was used. The set-up consisted of a straight channel with a 3.6 m entrance, a sharp open channel bend ($R_c/b = 1.5 < 3$) with a central angle of 90°, followed by a straight channel of 1.8 m length.

To measure the longitudinal velocity, the PROPLER one-dimensional velocity meter was used. The flow height was measured with a micrometer. During the experiments, flow discharge, downstream depth, Froude number, and Reynolds number were fixed at 25.3 lit/s, 15 cm, 0.34, and 44,705, respectively. Therefore, the flow was considered fully turbulent and subcritical. A more detailed explanation of the experimental study conducted is given in Gholami et al. (2014).

To perform this study, the velocity and depth at different points and sections in the channel were measured. Considering the effects of the channel walls and floor on the velocity index, 13 transverse points were selected in the channel width; at each transverse point, four depth points

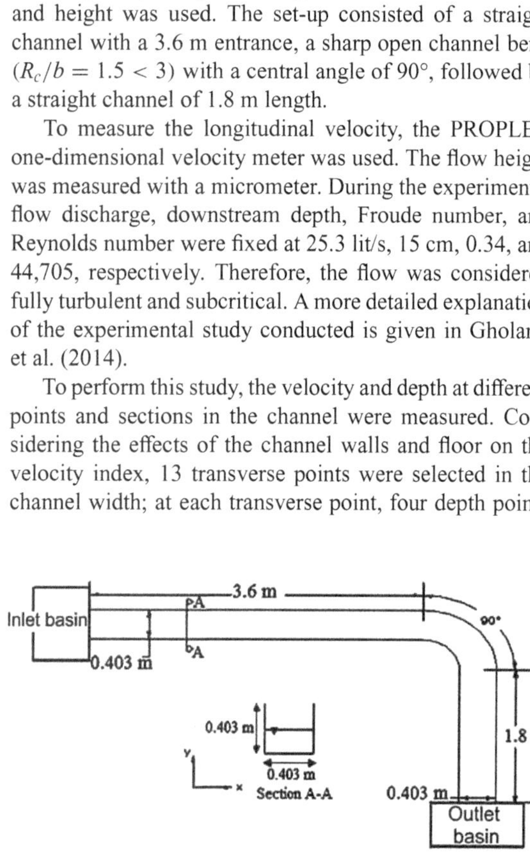

Figure 1. Schematic view of the 90° channel.

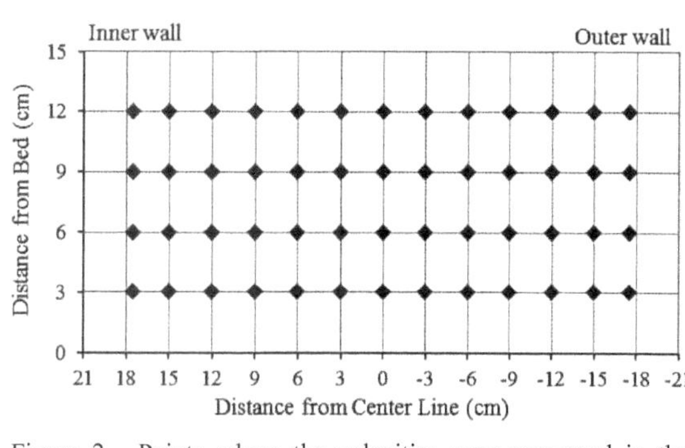

Figure 2. Points where the velocities were measured in the experimental model cross section.

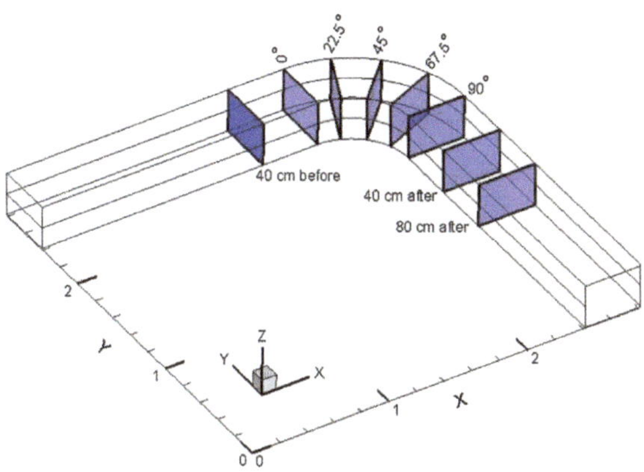

Figure 3. Situation of cross sections studied in the models.

were selected at 3, 6, 9 and 12 cm from the water surface. Figure 2 indicates the coordinates of the points used for measuring the velocity in each cross section. The positions of these sections are 40 cm before the bend, 40 and 80 cm after the bend, the beginning of the bend, the end of the bend, and the sections at the central angles 22.5°, 45°, and 67.5° in the bend (Figure 3).

3. Computational fluid dynamics (CFD) model

3.1. Governing equations

Governing equations for viscous and turbulent non-compressible flows include the continuity equation (Equation 1) and Reynolds-Averaged Navier-Stokes equation (Equation 2), written as follows:

$$\frac{\partial U_i}{\partial x_i} = 0, \quad \text{with} \quad i = 1, 2, 3, \tag{1}$$

$$\frac{\partial U_i}{\partial t} + U_j \frac{\partial U_i}{\partial x_i} = \frac{1}{\rho}\frac{\partial}{\partial x_i}(-P\delta_{ij} - \rho\overline{u_i.u_j}), \quad (i,j = 1,2,3), \tag{2}$$

where U_i is the average velocity component in the x_i direction, ρ is the density of the water, P is the pressure, δ_{ij} is the Kroncker delta, which is 1 for $i = j$ and 0 otherwise. and x_i and x_j are the general space dimensions. The Reynolds stress term is modeled by the $k - \varepsilon$ *(RNG)* turbulence model, first introduced by Yakhot and Orszag (1986). This is an isotropic two-equation turbulence model for computing the turbulent stresses using the Boussinesq approximation (Stoesser, Ruether, & Olsen, 2010):

$$\overline{-u_i u'_j} = \left[\nu_t\left(\frac{\partial U_i}{\partial x_j} + \frac{\partial U_j}{\partial x_i}\right)\right] - \left[\frac{2}{3}k\delta_{ij}\right]. \tag{3}$$

In the case of the $k - \varepsilon$ *(RNG)* model, the turbulent kinetic energy k and its dissipation rate ε for determining the eddy viscosity are obtained with the following equations (Rahimzadeh, Maghsoodi, Sarkardeh, & Tavakkol, 2012):

$$\nu_t = C_\mu \frac{k^2}{\varepsilon}, \tag{4}$$

$$\frac{\partial(k)}{\partial t} + \frac{\partial(kU_i)}{\partial x_i} = \frac{\partial}{\partial x_j}\left[\alpha_k \mu_{\text{eff}} \frac{\partial k}{\partial x_j}\right] + G_k + G_b - \varepsilon, \tag{5}$$

$$\frac{\partial(\varepsilon)}{\partial t} + \frac{\partial(\varepsilon\, U_j)}{\partial x_j} = \frac{\partial}{\partial x_j}\left[\alpha_\varepsilon \mu_{\text{eff}} \frac{\partial \varepsilon}{\partial x_j}\right] + C_{1\varepsilon}\frac{\varepsilon}{k}(G_k + C_{3\varepsilon}G_b) - C_{2\varepsilon}\frac{\varepsilon^2}{k} - \frac{C_\mu \eta^3(1 - \eta/\eta_0)}{1 + \beta\eta^3}\frac{\varepsilon^2}{k}, \tag{6}$$

where μ_{eff} is the effective viscosity, G_k and G_b are generation of turbulence kinetic energy due to the mean velocity gradients and generation of turbulence kinetic energy due to buoyancy, respectively, α_k and α_ε are the inverse effective Prandtl numbers for k and ε, respectively, and α_k and α_ε are computed using the following formula:

$$\left|\frac{\alpha - 1.3929}{\alpha_0 - 1.3929}\right|^{0.6321}\left|\frac{\alpha + 2.3929}{\alpha_0 + 2.3929}\right|^{0.3679} = \frac{\mu_{\text{mol}}}{\mu_{\text{eff}}} \tag{7}$$

where $\alpha_0 = 1.0$. In the high Reynolds number limit ($\mu_{\text{mol}}/\mu_{\text{eff}} << 1$), $\alpha_k = \alpha_\varepsilon \approx 1.393$.

In the present study, the parameters C_μ, $C_{1\varepsilon}$, and $C_{2\varepsilon}$ are constant and given the standard values of 0.0845, 1.42, and 1.68, respectively.

3.2. Gridding, CFD solving methods and boundary conditions

Gridding was constructed using GAMBIT, which is the preprocessor available in the commercial software FLUENT package. The execution time of computations from CFD modeling and the resulting accuracy and degree of consistency with the experimental results depend on the gridding size. A number of different grids of various sizes were examined in order to model this experimental study and to consider the initial results obtained. After carrying out sensitivity analysis on the meshing, a grid was eventually selected with a total of 67,500 nodes (25, 30, and 90 nodes in the length, width, and height, respectively) for a 90° bend. In order to regulate the solution field gridding near the bed, the smaller grid of the walls and water's free surface was determined suitable due to the steeper gradient. In addition, the grid was selected in such manner that it would be in line with the flow direction. As it approaches the bend the grid becomes smaller, while as it exits the bend the grid becomes larger with the aim to save computation time. Figure 4 illustrates a view of the solution field in the plan and cross section.

FLUENT software was used for CFD modeling in this research. This software uses the finite volume method

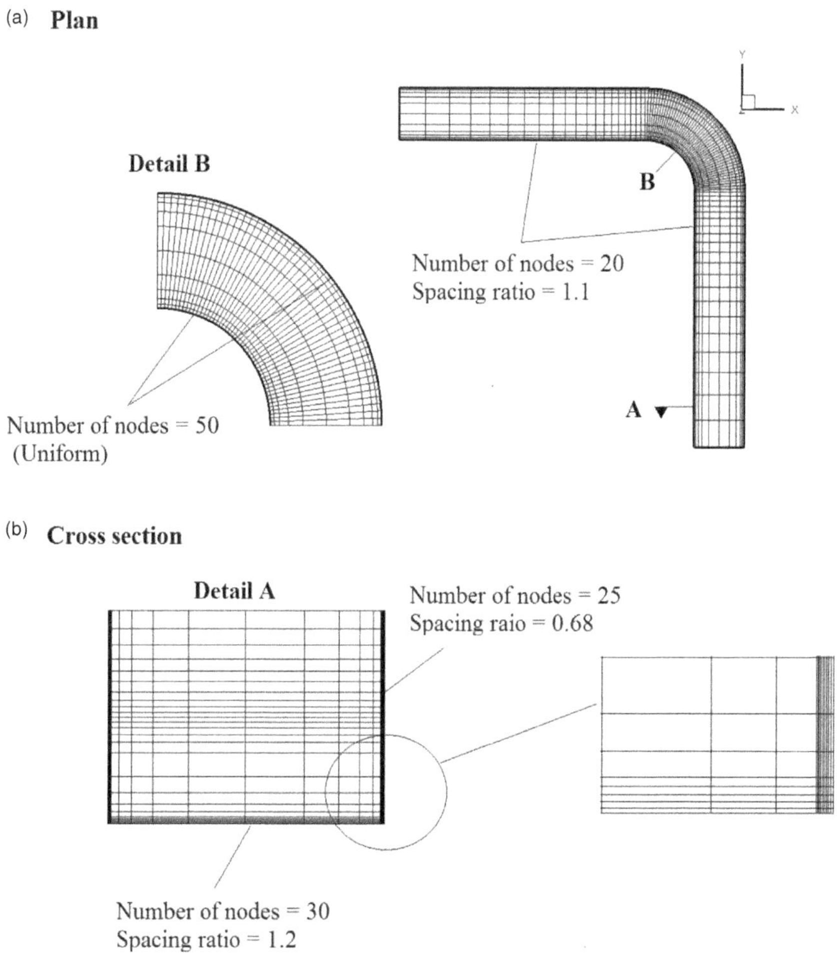

Figure 4. Gridding in CFD Model for (a) plan, and (b) cross section.

Table 1. Boundary conditions applied in the CFD model.

Boundary conditions	Type of boundary condition	Notes	Values
Inlet	Velocity inlet	The inlet was considered separately for air and water phase states.	V (water) = 0.419 m/s V (air) = 0.0001 m/s
Free surface	Pressure inlet	Two-phase flow.	Atmospheric pressure
Outlet	Pressure outlet	In *open channel* and *free surface level* states.	–
Bed and walls	Wall	Using the *standard wall function*.	–

(FVM) to solve the three-dimensional equations which govern flow. It considers the entire flow field as discrete control volumes and calculates the integral of the equations governing fluid flow on each control volume. The software uses different discretization schemes determined by the user, solves algebraic discrete equations and uses different algorithms to solve dependent equations to the point of reaching convergence.

In this study, the PISO pressure-velocity coupling algorithm was used because it is designed specifically for transient simulations. The PRESTO discretization scheme was used for pressure. The modified High Resolution Interface Capturing (HRIC) for volume fraction, QUICK, was applied for momentum and first-order upwind for turbulent kinetic energy discretization schemes (Fluent Manual, 2005). The flow was analyzed three-dimensionally and VOF served to model the two-phase flow of the free surface and solve the flow field until the remainder reached 10^{-4}. The boundary condition was also applied to the model in order to determine the solution shown in Table 1.

4. Application of artificial neural network (ANN) models

4.1. ANN structure

The ANN is an idea inspired by biological neural networks and is used to process data. The MATLAB R2011b software was utilized in order to prepare a proper ANN model (MATLAB, 2009). Two separate neural networks were used in this study: one to predict velocity (case 1) and the other to predict water depth (case 2). Both cases are multi-layer feed-forward (MLFF) neural networks. Figure 5 provides a general view of these networks, which are organized into three layers, namely input, hidden, and output (there may be one or more hidden layers). Each layer is made up of a number of neurons which are connected to the neurons of the adjacent layer through some weights. The network training process entails correcting the connective weights between neurons through the best training method.

In this study, 3, 40, and 1 neurons were used in the input layer, one hidden layer, and output layer, respectively, in case (1). The input for this case is the coordinate of points in the X, Y, and Z directions at various points in the channel and the output or target is the corresponding velocity of these points.

2, 10, 10, and 1 neurons were selected to be placed in the input layer, two hidden layers, and the output layer, respectively, in case (2). In this case, the input is the coordinate of the points in the X and Y directions at various channel points, while the output or target is the corresponding flow depth of each of these points.

Careful and actual analyses for the purpose of finding the number of neurons in the hidden layer are generally very complex. There is no particular law which specifies the number of neurons within hidden layers (Dawson & Wilby, 1998), but we can say that the number of hidden layer neurons is a function of the number of input axial elements and also the maximum number of input-space regions which can be linearly separated (Fausett, 1994; Lippman, 1987). Therefore, the number of neurons in the hidden layer is usually obtained experimentally. If the number of neurons is excessively low it reduces the analysis capability and consequently the numerical accuracy of prediction. However, if the number of hidden layer neurons is excessively high, the model will undergo overtraining and memorize the data instead of analyzing them. The number of neurons within the hidden layer(s) is commonly determined through trial and error (Cobaner, Unal, & Kisi, 2009; Kalteh, 2008; Kisi, 2008; Zaji & Bonakdari, 2015). Therefore, the trial and error method was utilized and different models were tested with various numbers of neurons considered within the hidden layer. Then the model that presented the best results was selected as the final ANN model. The training process continued for 1000 iterations, until both networks reached an acceptable level of error between the data obtained from the ANN model and the experimental data. Both networks were trained through the error backpropagation method in this research, and the tansig function served to activate the neurons in the hidden and output layers. This equation is shown below:

$$f(x) = \frac{2}{(1+\exp(-2x))} - 1. \tag{8}$$

A total of 416 experimental data were used in the present study. For case 1 the data were divided into two groups: 320 data for training the network and 96 for testing. 104 experimental data were used for case 2, with 80 data used to train the network and 24 to test it. To train both neural networks, the experimental data of flow velocity and depth of flow values at 13 different points of the channel width (see Figure 2) in the eight cross sections (40 cm before the bend, 0°, 22.5°, 45°, 67.5°, and 90°, and 40 and 80 cm after the bend) were used (see Figure 3). In case 2, however, the depth corresponding to these points in the channel bed was used and the velocity corresponding to these points at four different distances from the channel bed (Z = 3, 6, 9, and 12 cm) was applied in case 1.

4.2. The error backpropagation method

In order to train the network with the error backpropagation algorithm, the inputs are presented to the neural network

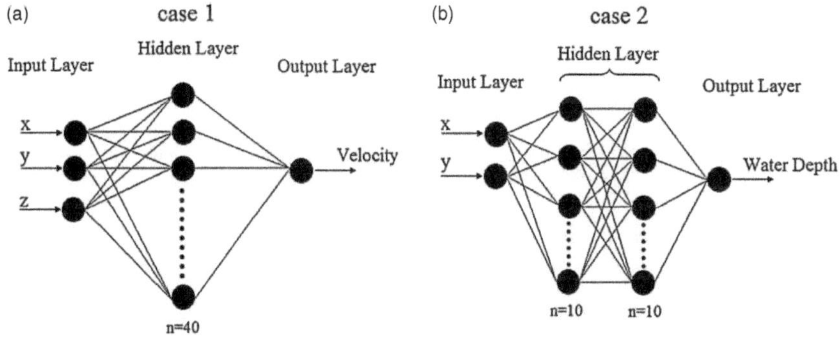

Figure 5. ANN structures for predicting (a) flow velocity (case 1), and (b) flow depth (case 2).

and the outputs obtained from the network are compared with the actual output values (target values) in order to compute the error. Then the computed error within the network is propagated backwards and the connected weights are corrected. This process continues until an acceptable level of convergence is achieved (Kalteh, 2008).

The weight changes in each W_{ji} specific weight, which connects the *jth* neuron to the *ith* neuron in the previous layer. It can be written as follows, where η is the training rate parameter and E is the error function (Rajurkar, Kothyari, & Chaube, 2004):

$$W_{ji} = -\eta \frac{\partial E}{\partial W_{ji}}. \quad (9)$$

The weight changes can be written as shown below using the chain rule in partial derivatives (Rajurkar et al., 2004):

$$\Delta W_{ji}^{n} = \eta^{n} \delta_{j}^{n} z_{j}^{n-1} \quad (10)$$

where z_j^{n-1} indicates the output value of the jth neuron in the $(n-1)th$ layer, $\delta_j^n = (T_i - z_j^n) g(v_j^n)$is for the output layer, $\delta_j^n = (\sum_{h=1} \delta_h^{n+1} W_{hj}^{n+1}) g(v_j^n)$ is for the hidden layers, $v_j = W_{hj}^n \cdot u_h^n$, and g() indicates the first derivative of the nonlinear activation function (Rajurkar et al., 2004).

4.3. *Data normalizing and model performance evaluation*

All input data are normalized as data in the range of 0 to 1 in the ANN model. Equation (11) is applied in the present study to normalize the data (Mohd Zahit, 2007):

$$X_n = \frac{X_i - X_{\min}}{X_{\max} - X_{\min}}, \quad (11)$$

where X_n is the normalized parameter, $X_{\max}$ is the maximum observational parameter, $X_{\min}$ is the minimum observational parameter and X_i is the observational parameter. It is crucial to determine the indexes and issues in order to analyze the results obtained from the models in the simulation results.

The performance of both the CFD model and the ANN model was examined through the Root Mean Square Error (RMSE), Mean Absolute Error (MAE) and coefficient of determination R^2 statistical parameters in the present research:

Equations (12) and (13) are for computing *RMSE* and *MAE*, respectively:

$$\mathrm{RMSE} = \sqrt{\frac{\sum (O_i - t_i)^2}{N}}, \quad (12)$$

$$\mathrm{MAE} = \frac{1}{N} \sum_{i=1}^{N} |O_i - t_i|. \quad (13)$$

Equation (14) is for calculating R^2:

$$R^2 = 1 - \frac{\sum (O_i - t_i)^2}{\sum (O_i - \bar{O}_i)^2}. \quad (14)$$

In the above-mentioned equations, O_i is the input observational parameter, t_i is the parameter predicted by the ANN model, $\bar{O}_i$ is the mean observational parameter and N is the number of parameters. R^2 is the linear regression line between the values predicted by the ANN model and the observational values to determine the network application.

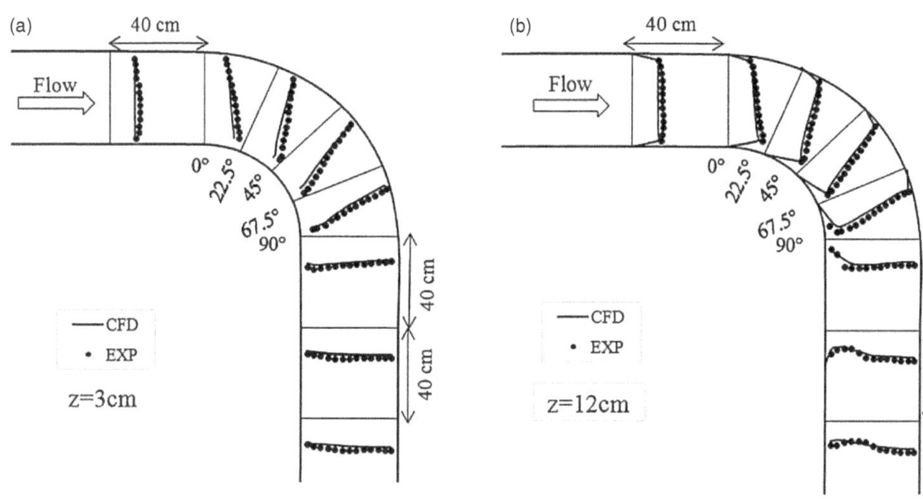

Figure 6. Distribution of longitudinal component of velocity at different sections and levels: (a) at water level $Z = 3$ cm (near the channel bed), and (b) at water level $Z = 12$ cm (near the water surface).

5. Results and discussion

5.1. *The CFD model*

5.1.1. *CFD model performance evaluation*

The experimental results were used to examine the accuracy of the CFD model. Figure 6 compares the longitudinal velocity values within different cross sections at two levels, $Z = 12$ cm (near the water surface) and $Z = 3$ cm (near the channel bed), which were obtained from the CFD model and the experimental values. The predicted transverse water surface profiles of the CFD model were compared with the experimental measurements at different cross sections in Figure 7. The CFD model performance evaluation for velocity ($Z = 3$ and 12 cm) and flow depth prediction is presented in Table 2.

Clearly, the consistency of the CFD and experimental results is generally acceptable, with an average RMSE and MAE of 2.05 and 1.8, respectively, for the level near the channel bed, and 2.07 and 1.52 for the levels near the water surface in velocity prediction. Regarding the velocity prediction values in Table 2, it can be stated that when $Z = 3$ cm (near the channel bed) the minimum error occurs in the cross section located 40 cm after the bend and the maximum error occurs near the middle cross section of the bend. In contrast, when $Z = 12$ cm (near the water surface) the minimum error value occurs in the middle cross section and the maximum error occurs in the cross section located 40 cm after the bend. This is because velocity decreases (due to secondary flows) beginning from the layers close to the water surface ($Z = 12$ cm) until it reaches the lower layers ($Z = 3$cm) at the end cross sections of the bend. Basically, when velocity decreases in the upper layers it does not decrease in the layers close to the bed, subsequently creating an insignificant difference between the error values obtained from the CFD model and the results of the experimental model in the cross sections located after the bend and in the layers close to the water surface. With regard to the depth prediction values in Table 2, CFD predicts flow depth with an RMSE and MAE of 0.13 and 0.11, respectively. In addition, it can be concluded that the amount of prediction error increases by moving from the beginning toward the end sections of the bend. This is due to the greater impact of the bend on fluid flow in the end sections.

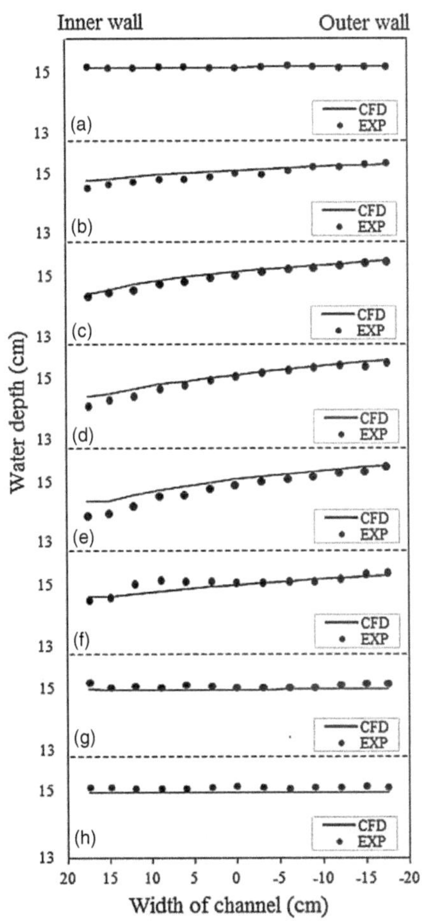

Figure 7. Comparing the transverse water surface profiles in the CFD model and the experimental model in the 90° channel at the following cross sections: (a) 40 cm before the bend, (b) 0°, (c) 22.5°, (d) 45°, (e) 67.5°, (f) 90°, (g) 40 cm after the bend, and (h) 80 cm after the bend.

As the flow enters the bend, the transverse pressure gradient between the inner and outer wall causes the velocity to reach a maximum in the inner wall and minimum in the outer wall. The velocity increase continues up to the 67.5° cross section and the maximum velocity continues in this area. The development of secondary flows in the final cross sections of the bend causes the maximum velocity to gradually move to the channel axis from the inner wall and then move towards the outer channel wall. Figure 8 shows the geometrical locations of the maximum velocity at different cross sections in the channel in the 3, 6, 9, and 12 cm layers from the channel bed. A careful study of this figure reveals that the maximum velocity occurs in the inner channel at half width and in different layers from the channel bed to the end of the bend becuase the bend is sharp. The maximum velocity is transferred to the outer wall in the cross sections after the bend. Molls and Chaudhry (1995) and Leschziner and Rodi (1979) found similar results from conducting studies on sharp bends. The velocity values in different channel layers will be explained thoroughly in the following sections as well as studying the longitudinal velocity-depth profile distribution.

5.1.2. *Secondary currents*

Secondary flows are obvious flow phenomena within curved channels. Figure 9 indicates the transverse velocities and secondary flow cells at 0°, 45°, and 90°, and 80 cm after the bend cross section. A one-way radial

Table 2. Comparison of the velocity distribution and depth of flow prediction in the CFD model and the experimental model at $Z = 3$ and $Z = 12$ cm from the bed at different cross sections.

	Velocity prediction				Depth prediction	
	RMSE (m)	MAE (m)	RMSE (m)	MAE (m)	RMSE (m)	MAE (m)
Cross section	$Z = 3$ cm		$Z = 12$ cm		$Z = 0$	
40 cm before the bend	2.58	2.3	1.66	1.53	0.048	0.046
0°	2.85	2.43	1.65	1.29	0.11	0.098
22.5°	3.03	2.7	1.62	1.45	0.096	0.087
45°	2.66	2.47	1.06	0.918	0.12	0.096
67.5°	2.18	1.91	2.45	1.33	0.22	0.201
90°	1.03	0.96	2.41	1.56	0.15	0.105
40 cm after the bend	0.94	0.77	1.93	1.34	0.13	0.12
80 cm after the bend	1.15	0.93	3.8	2.76	0.165	0.164
Averaged values	**2.05**	**1.8**	**2.07**	**1.52**	**0.13**	**0.11**

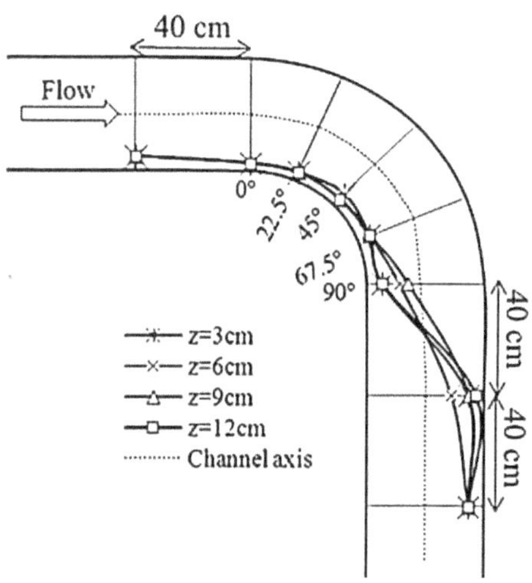

Figure 8. The geometrical location of the maximum velocity in different layers of the channel bed.

flow occurs in the initial cross section of the bend, to establish the continuity of the flow from the outer to the inner wall. In the following cross sections, the main secondary flow cell can be seen in all cross-sectional widths, and as it moves forward along the bend it can be seen moving towards the inner channel wall. The minor secondary flow cells are also clearly visible in the outer wall of the channel near the water surface. Blanckaert and Graf (2001) conducted experimental studies on 120° sharp bends while Huang et al. (2009) used a numerical model and also reported the existence of minor secondary flow cells. The minor roll cell indicates the $k - \varepsilon$ (RNG) turbulence model's capability to predict the cell, since in case of using the $k - \varepsilon$ (standard) isotropic (linear) turbulence model this cell is not predicted (Naji et al., 2010).

5.2. *The ANN models*

As previously mentioned, the defined neural networks are first trained using existing experimental data and the error backpropagation method, after which they are tested. The general specifications of both networks utilized are presented in Table 3. The correlation analysis between the data predicted by the ANN model and the experimental data is presented in Figure 10 for case 1 and Figure 11 for case 2. The R^2 values are shown for each figure, where R^2 is equal to 0.9834 for case 1 and 0.9801 for case 2 in the training state.

The performance of the ANN models' velocity and depth predictions is shown in Table 4, where the velocity fields are investigated at two heights, $Z = 3$ and 12 cm. According to this table, the ANN model that predicted velocity with an RMSE of 0.59 and 0.52 for $Z = 3$ and 12 cm, respectively, and flow depth with an RMSE of 0.037, performed fairly well in open channel bend characteristics prediction. Also according to Table 4, at $Z = 3$ cm, the velocity prediction performance was the highest in the cross sections located before and at the entrance of the bend, but at $Z = 12$ cm, the velocity prediction performance was highest at the cross section located after the bend. Moreover, the ANN model did not adequately predict the velocity decrease from the layers close to the water surface in the section located after the bend. This result is consistent with the result obtained in Section 5.1.1. Thus, it can be stated that the error value of both models (ANN and CFD) slightly increased when predicting the effects of secondary flows on the bend's flow patterns. In case 2 (water surface prediction model), the minimum error occurred at the entrance cross sections of the bend and the maximum error value at the end cross sections of the bend. It can be concluded that this model did not perform well in predicting flow depth, as it became constant in the cross sections located after the bend (returning to the primary normal depth of 15 cm) and so it slightly differed from the experimental values. Therefore, similar to the CFD model, the ANN was less accurate in the middle and end sections, and the section located after the bend where the impact of the bend is greater.

Considering these values and also Figures 10 and 11, it can be understood that both networks were trained well

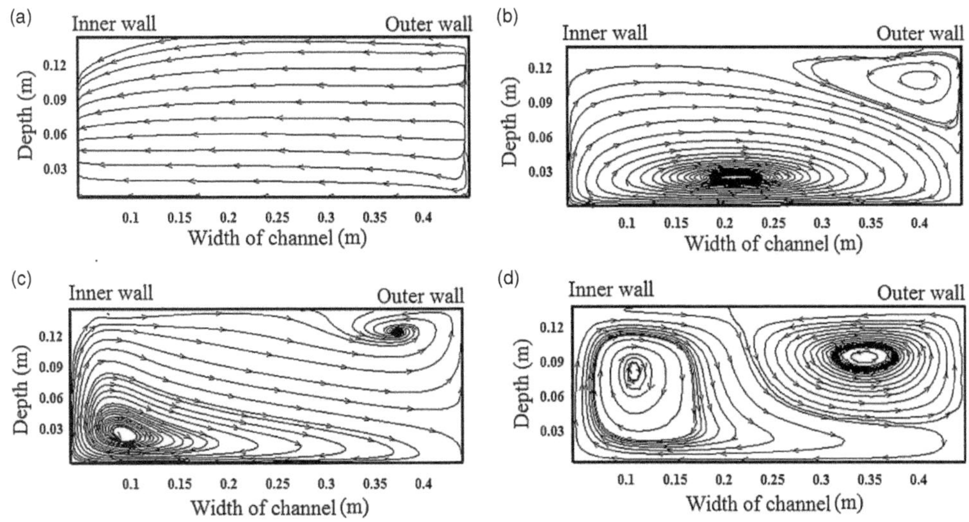

Figure 9. Secondary currents at (a) 0°, (b) 45°, (c) 90°, and (d) 80 cm after the bend.

Table 3. Overview of the two ANN models.

Properties	Case 1	Case 2
Number of data	416	**104**
Number of training data	320	**80**
Number of testing data	96	**24**
Input variables	*X*	***X***
	Y	***Y***
	Z	–
Output variables	Velocity	**Water depth**
Number of hidden layers	1	**2**
Input neurons	3	**2**
Hidden 1 neurons	40	**10**
Hidden 2 neurons	–	**10**
Output neurons	1	**1**
RMSE of training data	0.57	**0.039**
RMSE of testing data	0.55	**0.037**
MAE of training data	0.48	**0.05**
MAE of testing data	0.45	**0.03**
R^2 of training data	0.9834	**0.9801**
R^2 of testing data	0.9844	**0.9922**

and their performance in predicting velocity and flow depth parameters was satisfactory. Therefore, after ensuring performance accuracy of these networks, they can be used to predict the velocity and flow depth values at channel points where there are no available experimental data, as shown in the following section.

5.3. Comparison of the CFD and ANN model results

As mentioned earlier, the trained neural networks are used in this section to examine the water surface profile and velocity distribution at points in the channel where no experimental data are available. These values are finally compared with the values obtained from the CFD simulation. Experimental data were obtained at only eight cross sections in the present research, but ANN models can help to examine and study the water depth and velocity values at other cross sections.

5.3.1. Longitudinal velocity distribution in the channel width and depth

The transverse distributions of the depth-averaged longitudinal velocity at 20°, 40°, 50°, 60°, and 70°, and 30 cm after the bend cross sections as simulated by the CFD and ANN models (case 1) are compared in Figure 12. The RMSE and MAE values for the results of the CFD and ANN model results are shown in Table 5. In terms of the average values of the RMSE (2.33) and MAE (2.02), both methods were able to predict the depth-averaged velocity at different cross sections very well. There is also an acceptable level of consistency between the results of both models. The geometrical location of the maximum depth-averaged velocity at different cross sections also supports this in Figure 13. Considering this figure and as mentioned in previous sections, the maximum velocity commonly occurs in the inner section of the bend, the reason being the power of the longitudinal flow and the fact that it overpowers the secondary flows within these types of bends. This result is also consistent with the results obtained by Zhou, Wang, Shao, and Jia (2006). When examining the velocity pattern in the present 90° sharp bend and taking into account Figure 12, the crucial point is that the transverse distribution of the depth-averaged velocity is linear from the beginning to almost the end of the bend, where the secondary flow becomes stronger.

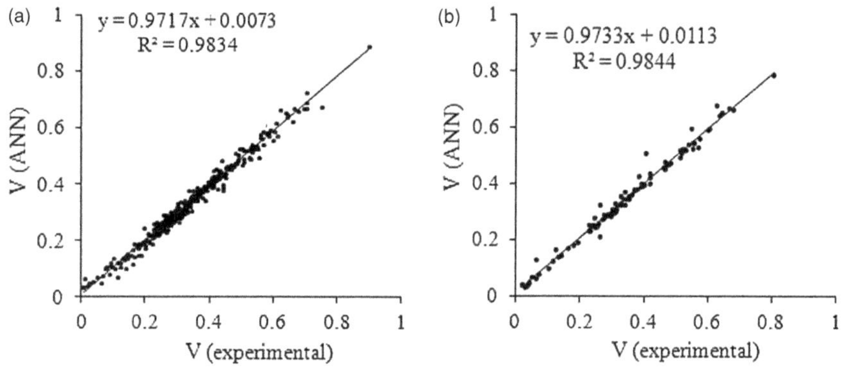

Figure 10. Comparison between ANN results (case 1) and experimental results used in: (a) Training, and (b) Testing.

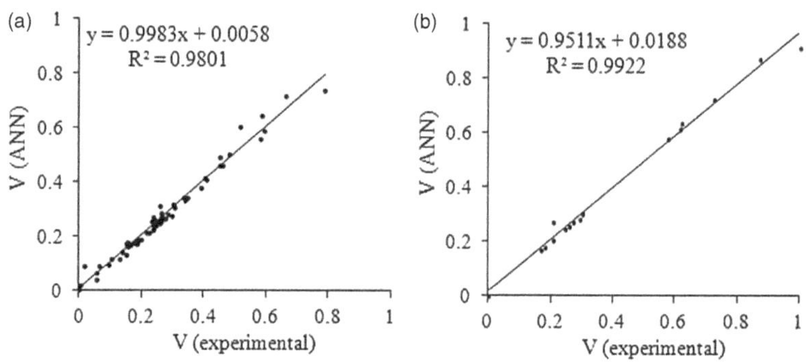

Figure 11. Comparison between ANN results (case 2) and experimental results used in: (a) Training, and (b) Testing.

In Figure 14, the depth distributions of the longitudinal velocity component at 20°, 50°, and 70°, and 30 cm after the bend cross sections are compared through the CFD and ANN models.

The acceptable level of consistency between the results of the two models can clearly be seen in these figures. In Figure 14, for the cross section located after the bend, the ANN model predicted lower velocity values between the water surface and channel mid-depth, which is slightly different from the CFD model results. In general, both the CFD model and the ANN model are capable of predicting the velocity pattern in different channel cross sections. The results obtained from both models with a mean RMSE of 2.5 are acceptably consistent with each other. The velocity

Table 4. Comparison of the velocity and depth of flow prediction in the ANN model and the experimental model at $Z = 3$ and $Z = 12$ cm from the bed at different cross sections.

	Velocity prediction				Depth prediction	
	RMSE (m)	MAE (m)	RMSE (m)	MAE (m)	RMSE (m)	MAE (m)
Cross section	$Z = 3$ cm		$Z = 12$ cm		$Z = 0$	
40 cm before the bend	0.79	0.57	0.47	0.39	0.021	0.02
0°	0.75	0.57	0.5	0.42	0.036	0.03
22.5°	0.6	0.54	0.48	0.41	0.0265	0.017
45°	0.54	0.44	0.47	0.4	0.023	0.02
67.5°	0.54	0.3	0.52	0.42	0.058	0.049
90°	0.53	0.49	0.53	0.49	0.039	0.03
40 cm after the bend	0.55	0.45	0.56	0.52	0.059	0.049
80 cm after the bend	0.42	0.38	0.65	0.57	0.038	0.028
Averaged values	**0.59**	**0.46**	**0.52**	**0.45**	**0.037**	**0.03**

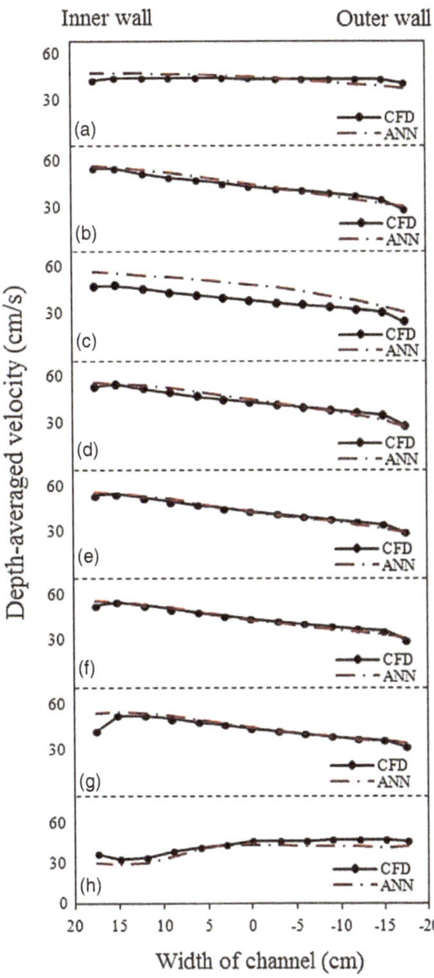

Figure 12. Comparing depth-averaged velocity distribution in the CFD model and the ANN model at the following cross sections: (a) 30 cm before the bend, (b) 20°, (c) 30°, (d) 40°, (e) 50°, (f) 60°, (g) 70°, and (h) 30 cm after the bend.

depth distribution is not logarithmic in any of the states. Ghamry and Steffler (2003) suggested that the depth velocity distribution is parabolic for the flow within channels. The results obtained from the CFD and ANN models are slightly different from each other in cross sections where the presence of the bend and its main characteristic (secondary flow) in the channel leads to changing flow patterns and velocity component distributions, as mentioned in prior sections.

5.3.2. *Water surface profile*

The longitudinal profile of the water surface predicted by the CFD model and the ANN model is illustrated in Figure 15 along the channel length, 30 cm before and 30 cm after the bend, as well as near the inner walls, outer

Table 5. The RMSE and MAE values for depth-averaged velocity distribution between the CFD and ANN models (case 1).

Cross section	RMSE (m)	MAE (m)
30 cm before the bend	2.86	**2.55**
20°	1.9	**1.81**
30°	3.65	**3.39**
40°	2.14	**1.82**
50°	1.3	**1.1**
60°	1.4	**1.18**
70°	2.36	**1.67**
30 cm after the bend	3.04	**2.67**
Averaged RMSE in the cross sections	**2.33**	2.02

walls, and the channel axis. The water surface rises in the cross section before the bend in both models, since the water needs to gain more power to enter the bend. As it exits the bend, the flow depth is restored to its original state (15 cm). Considering the figure and the mean RMSE of 1.3, both models were able to predict the water surface longitudinal profile well and both are acceptably consistent with each other except in the inner edges of the bend, as mentioned before. With respect to examining the flow pattern within sharp bend channels, Ye and McCorquodale (1998) also point out that water surface rises in the cross sections located before the bend.

Table 6 shows the water surface difference in the outer mid-width of the channel (subtraction of the water surface in the channel axis from the outer wall) and the inner mid-width (subtraction of the water surface in the inner wall from the channel axis) in the initial, end, and inner cross sections of the 90° bend with the CFD and ANN models. The values in this table signify that in all cross sections,

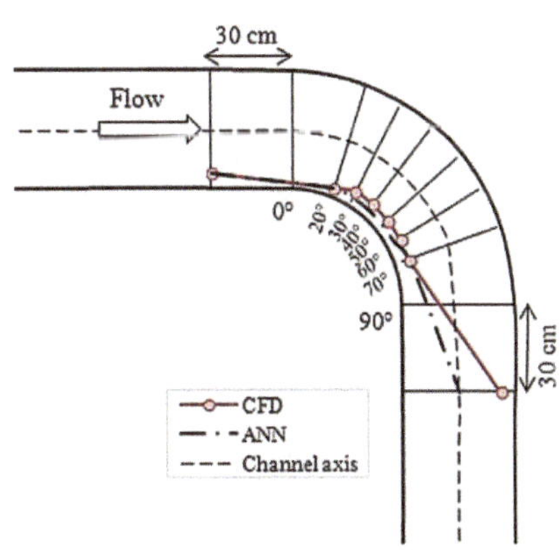

Figure 13. The geometrical location of the maximum depth-averaged velocity in different cross sections of the channel with the CFD and ANN models.

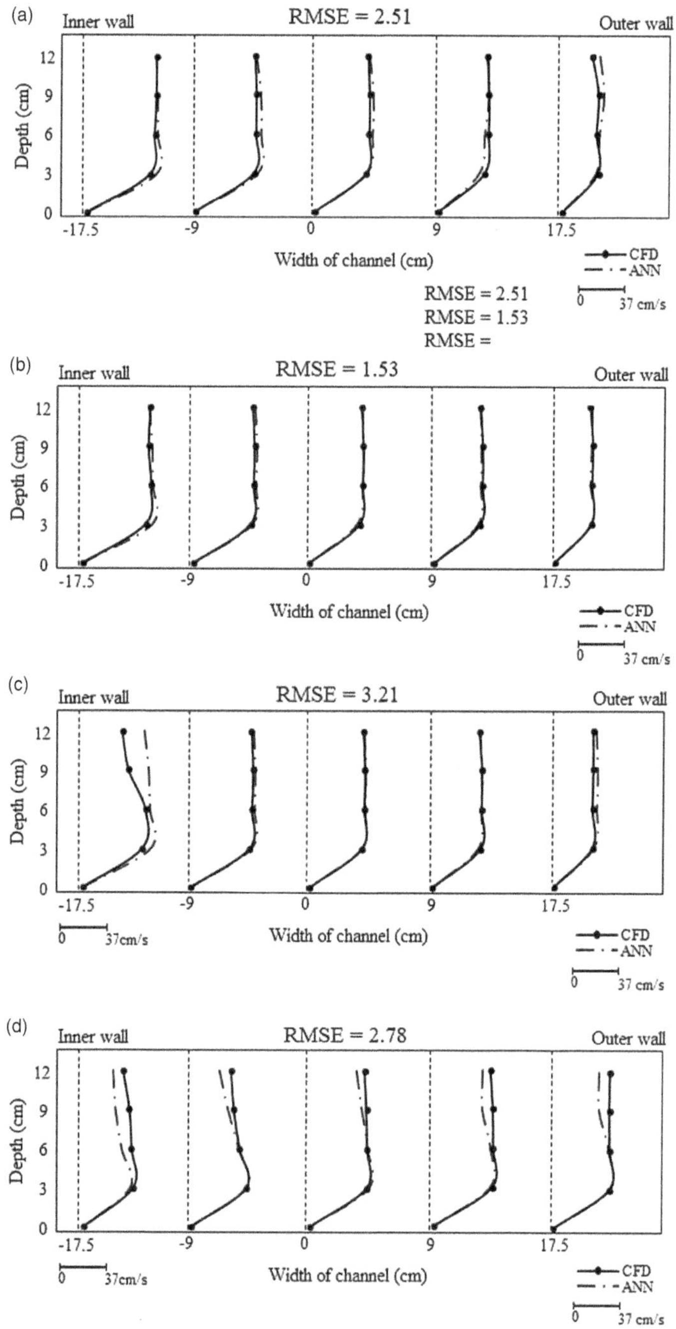

Figure 14. Comparison of longitudinal velocity distribution in the channel depth with the CFD and ANN models at the following cross sections: (a) 20°, (b) 50°, (c) 70°, and (d) 30 cm after the bend.

both models predicted significantly greater water surface change in the inner mid-width than the outer mid-width of the channel. Therefore, it can be stated that the transverse slope of the water surface is nonlinear in the 90° sharp bend. With regard to examining the flow pattern in sharp bends, Leschziner and Rodi (1979) concurred that the transverse slope of the water surface is nonlinear in sharp bends and the inner half has a greater slope than the outer half of the channel. In addition, Bodnar and Prihoda (2006) conducted a series of numerical studies on

Table 6. The water surface difference in the outer and inner mid-width of the channel cross sections in the CFD and ANN models.

	0° (entrance)		20°		40°		60°		80°		90° (end of the bend)	
Mid-width	Inner	Outer	Inner	Outer	Inner	Outer	Inner	Outer	Inner	Outer	Inner	Outer
CFD	0.31	0.16	0.7	0.33	0.59	0.46	0.63	0.45	0.58	0.44	0.22	0.19
ANN	0.42	0.21	0.83	0.26	0.64	0.14	0.89	0.37	0.87	0.48	0.35	0.28

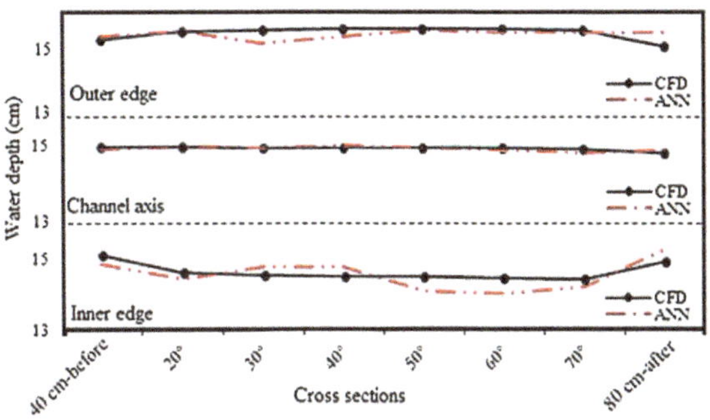

Figure 15. Comparison of longitudinal water surface profiles with the CFD model and the ANN model at the outer edge of the bend, channel axis, and inner edge of the bend.

a 90° sharp bend and also found that the water surface transverse slope is nonlinear. Meanwhile, DeVriend and Geoldof (1983) conducted an experimental study on a 90° mild bend and reported that the water surface transverse profile in the inner cross section of the bend is linear. Steffler, Rajartnam, and Peterson (1985) also pointed out that the water surface transverse profile is linear in a 270° mild bend.

In light of the above discussions, it can be concluded that both the CFD model and the ANN model are capable of predicting the general pattern governing flow, and the results obtained from both models are consistent. Therefore both methods can be used for the purpose of predicting flow pattern within bends. Nonetheless, each method has advantages and disadvantages.

The CFD model is more time-consuming than the ANN model. The ANN model helps achieve a desirable network in a shorter time for predicting values through network training. In addition, the ANN model can predict flow variable values at any point without requiring the CFD model or available experimental data.

6. Conclusion

The CFD and ANN methods were applied in this study to simulate flow characteristics in open channel bends. The CFD and ANN models were verified and trained through an experimental study. Two ANN models were developed, one for modeling the velocity field and another for modeling the flow depth. After that, the performance of the CFD and ANN models in velocity and flow depth simulation was compared. The results show that both models are reasonably accurate. However, the ANN models with an RMSE of 0.55 and 0.037 in velocity and depth prediction, respectively, performed significantly better than the CFD model with an RMSE of 2.06 and 0.13. In the next part of the study, the application of validated CFD and ANN models were investigated for cross sections where experimental data is not available. The results indicate that both the CFD and ANN models can predict flow characteristics without experimental data with an RMSE of 2.33.

Notations

R_c	Central radius of bend
b	Channel width (m)
U_i	Average velocity component (m/s)
ρ	Density of water (kg/m^3)
p	Pressure
δ_{ij}	Kroncker delta
x_i	The general space dimension in the x direction
x_j	The general space dimension in the y direction
$-\rho\overline{u'_i u'_j}$	Reynolds stresses
$\bar{u}_i$	Mean velocity component
u'_i	Fluctuating velocity component
μ	Viscosity of fluid
μ_t	Turbulent viscosity
μ_{eff}	Effective viscosity
μ_{mol}	Molecular viscosity
t	Time (s)
β	Constant number
ε	Turbulence dissipation rate (m^2/s^3)
k	Turbulence kinetic energy (m^2/s^3)
G_k	Generation of turbulence kinetic energy due to the mean velocity gradients
G_b	Generation of turbulence kinetic energy due to buoyancy
S	Modulus of the mean rate-of-strain tensor
α_k	Inverse effective Prandtl number for k
α_ε	Inverse effective Prandtl number for ε
C_μ	$k-\varepsilon$ Turbulence model constant
$C_{1\varepsilon}$	$k-\varepsilon$ Turbulence model constant
$C_{2\varepsilon}$	$k-\varepsilon$ Turbulence model constant
W_{ij}	Specific weight which connects the j th neuron to the ith neuron in the previous layer of the ANN
η	Learning rate parameter of the ANN
E	The error function of the ANN
W^n_{hj}	Weight between the hidden layer and the output layer of the ANN

u_i	Input of the ANN
u_h^n	Output of the hidden layer of the ANN
δ_j^n	The local gradient function in the nth output layer of the ANN
δ_h^n	The local gradient function in the nth hidden layer of the ANN
$g\ ()$	The first derivative of the nonlinear activation function of the ANN
T_i	Desired output of the ANN

Disclosure statement

No potential conflict of interest was reported by the authors.

References

Anwar, H. O. (1986). Turbulent structure in a river bed. *Journal of Hydraulic Engineering*, *112*(8), 657–669.

Baghalian, S., Bonakdari, H., Nazari, F., & Fazli, M. (2012). Closed-form solution for flow field in curved channels in comparison with experimental and numerical analyses and artificial neural network. *Engineering Applications of Computational Fluid Mechanics*, *6*(4), 514–526.

Balen, W. V., Uijttewaal, W. S. J., & Blanckaert, K. (2010). Large-eddy simulation of a curved open-channel flow over topography. *Physics of Fluids*, *22*(7), 1–18.

Bilhan, O., Emiroglu, M. E., & Kisi, O. (2010). Application of two different neural network techniques to lateral outflow over rectangular side weirs located on a straight channel. *Advances in Engineering Software*, *41*(6), 831–837.

Bilhan, O., Emiroglu, M. E., & Kisi, O. (2011). Use of artificial neural networks for prediction of discharge coefficient of triangular labyrinth side weir in curved channels. *Advances in Engineering Software*, *42*(4), 208–214.

Blanckaert, K., & Graf, W. H. (2001). Mean flow and turbulence in open channel bend. *Journal of Hydraulic Engineering*, *127*(10), 835–847.

Bodnar, T., & Prihoda, J. (2006). Numerical simulation of turbulent free-surface flow in curved channel. *Flow, Turbulence and Combustion*, *76*, 429–442.

Bonakdari, H., Baghalian, S., Nazari, F., & Fazli, M. (2011). Numerical analysis and prediction of the velocity field in curved open channel using artificial neural network and genetic algorithm. *Engineering Applications of Computational Fluid Mechanics*, *5*(3), 384–396.

Chau, K. W., Wu, C. L., & Li, Y. S. (2005). Comparison of several flood forecasting models in Yangtze river. *Journal of Hydrologic Engineering*, *10*(6), 485–491.

Cobaner, M., Unal, B., & Kisi, O. (2009). Suspended sediment concentration estimation by an adaptive neuro-fuzzy and neural network approaches using hydro-meteorological data. *Journal of hydrology*, *367*(1), 52–61.

Dawson, C. W., & Wilby, R. L. (1998). An artificial neural network approach to rainfall-runoff modeling. *Hydrological Sciences Journal*, *43*(1), 47–66.

DeMarchis, M., & Napoli, E. (2006). 3D numerical simulation of curved open channel flows. *Proceedings of the International Conference on Water Resources, Hydraulics & Hydrology*, May 11–13, Chalkida, Evia Island, Greece, 86–91.

DeVriend, H. J., & Geoldof, H. J. (1983). Main flow velocity in short river bends. *Journal of Hydraulic Engineering*, *109*(7), 991–1011.

Dursun, O. F., Kaya, N., & Firat, M. (2012). Estimating discharge coefficient of semi-elliptical side weir using ANFIS. *Journal of Hydrology*, *426*, 55–62.

Ebtehaj, I., & Bonakdari, H. (2013). Evaluation of sediment transport in sewer using artificial neural network. *Engineering Applications of Computational Fluid Mechanics*, *7*(3), 382–392.

Emiroglu, M. E., Bilhan, O., & Kisi, O. (2011). Neural network for estimation of discharge capacity of triangular labyrinth side-weir located on a straight channel. *Expert Systems with Applications*, *38*(1), 867–874.

Fausett, L. (1994). *Fundamental of neural networks: Architecture, algorithms and applications*. New Jersey, USA: Prentice Hall.

Ferguson, R. I., Parsons, D. R., Lane, S. N., & Hardy, R. J. (2003). Flow in meander bends with recirculation at the inner bank. *Journal of Water Resources Research*, *39*(11), 1–13.

Fluent Manual (2005). *Manual and user guide of fluent software*. New York, USA: Fluent Inc.

Ghamry, H. K., & Steffler, P. M. (2003). Effect of applying different distribution shapes for velocities and pressure on simulation of curved open channels. *Journal of Hydraulic Engineering*, *128*(11), 969–982.

Gholami, A., Akhtari, A. A., Minatour, Y., Bonakdari, H., & Javadi, A. A. (2014). Experimental and numerical study on velocity fields and water surface profile in a strongly-curved 90° open channel bend. *Engineering Applications of Computational Fluid Mechanics*, *8*(3), 447–461.

Huang, S. L., Jia, Y. F., Chan, H. C., & Wang, S. S. Y. (2009). Three dimensional numerical modeling of secondary flows in a wide curved channel. *Journal of Hydrodynamics*, *21*(6), 758–766.

Jahangirzadeh, A., Shamshirband, S., Petkovic, D., Basser, H., Sedaghat, A., Akib, S., & Karami, H. (2014). Adaptive neuro-fuzzy estimation of the influence of slot on local scour at bridge pier groups. *Journal of Coastal Conservation*. doi:10.1007/s11852-014-0357-5.

Kalteh, A. M. (2008). Rainfall-runoff modeling using artificial neural networks (ANNs): modeling and understanding. *Caspian Journal of Environmental Science*, *6*(1), 53–58.

Kimura, I., Uijttewaal, W. S. J., & Balen, W. (2008). Application of the non-linear $k - \varepsilon$ model for simulation curved open channel flows. *Proceedings of River Flow,* September 3–5, Cesme-izmir, Turkey, 1, 99–108.

Kisi, O. (2008). The potential of different ANN techniques in evapotranspiration modeling. *Hydrological Processes*, *22*(14), 2449–2460.

Kisi, O., Emiroglu, M. E., Bilhan, O., & Guven, A. (2012). Prediction of lateral outflow over triangular labyrinth side weirs under subcritical conditions using soft computing approaches. *Expert Systems with Applications*, *39*(3), 3454–3460.

Leschziner, M. A., & Rodi, W. (1979). Calculation of strongly curved open channel flow. *Journal of the Hydraulics Division*, *105*(10), 1297–1314.

Lien, H. C., Hsieh, T. Y., Yang, J. C., & Yeh, K. C. (1999). Bend flow simulation using 2D depth-averaged model. *Journal of Hydraulic Engineering*, *125*(10), 1097–1108.

Lippman, R. P. (1987). An introduction to computing with neural nets. *ASSP Magazine IEEE*, *4*(2), 4–22.

Lu, W. Z., Zhang, W. S., Cui, C. Z., & Leung, A. Y. T. (2004). A numerical analysis of free-surface flow in curved open channel with velocity-pressure-free-surface correction. *Computational Mechanics*, *33*, 215–224.

MATLAB (2009). User's Guide Version 7.8, R2009a. MathWorks Co., USA.

Mohd Zahit, A. (2007). The application of the artificial neural network model for river water quality classification with emphasis on the impact of land use activities: a case study from several catchments in Malaysia, *Ph.D Thesis*, University of Nottingham, Nottingham, UK.

Molls, T., & Chaudhry, M. H. (1995). Depth-averaged open-channel flow model. *Journal of Hydraulic Engineering, 121*(6), 453–465.

Naji, M. A., Ghodsian, M., Vaghefi, M., & Panahpur, N. (2010). Experimental and numerical simulation of flow in a 90° bend. *Flow Measurement and Instrumentation, 21*(3), 292–298.

Ottevanger, W., Blanckaert, K., & Uijttewaal, W. S. J. (2012). Processes governing the flow redistribution in sharp river bends. *Geomorphology, 163–164*, 45–55.

Rahimzadeh, H., Maghsoodi, R., Sarkardeh, H., & Tavakkol, S. (2012). Simulating flow over circular spillways by using different turbulence models. *Engineering Applications of Computational Fluid Mechanics, 6*(1), 100–109.

Ramamurthy, A., Han, S., & Biron, P. (2013). Three-dimensional simulation parameters for 90° open channel bend flows. *Journal of Computing in Civil Engineering, 27*(3), 282–291.

Rajurkar, M. P., Kothyari, U. C., & Chaube, U. C. (2004). Modeling of the daily rainfall-runoff relationship with artificial neural network. *Journal of Hydrology, 285*, 96–113.

Rozovskii, I. L. (1961). *Flow of water in bends of open channels.* Academy of Sciences of the Ukrainian SSR, Israel Program for Science Translation, Jerusalem, 1–233.

Sahu, M., Jana, S., Agarwal, S., & Khatua, K. K. (2011). Point form velocity prediction in meandering open channel using artificial neural network. *2nd International Conference on Environmental Science and Technology*, 6, IACSIT Press, Singapore, 209–212.

Shamshirband, S., Gocic, M., Petkovic, D., Javidnia, H., Hafizah, S., Hamid, A., ... Qasem, S. N. (2015). Clustering project management for drought regions determination: a case study in Serbia. *Agricultural and Forest Meteorology, 200*(15), 57–65.

Steffler, P. M., Rajartnam, N., & Peterson, A. W. (1985). Water surface change of channel curvature. *Journal of Hydraulic Engineering, 111*(5), 866–870.

Stoesser, T., Ruether, N., & Olsen, N. R. B. (2010). Calculation of primary and secondary flow and boundary shear stresses in a meandering channel. *Advances in Water Resources, 33*, 158–170.

Wang, W. C., Chau, K. W., Cheng, C. T., & Qiu, L. (2009). A comparison of performance of several artificial intelligence methods for forecasting monthly discharge time series. *Journal of Hydrology, 374*(3–4), 294–306.

Wang, W. C., Xu, D. M., Chau, K. W., & Lei, G. J. (2014). Assessment of river water quality based on theory of variable fuzzy sets and fuzzy binary comparison method. *Water Resources Management, 28*(12), 4183–4200.

Wu, C. L., & Chau, K. W. (2010). Data-driven models for monthly streamflow time series prediction. *Engineering Applications of Artificial Intelligence, 23*(8), 1350–1367.

Wu, C. L., Chau, K. W., & Li, Y. S. (2008). River stage prediction based on a distributed support vector regression. *Journal of Hydrology, 358*(1–2), 96–111.

Yakhot, V., & Orszag, S. A. (1986). Renormalization group analysis of turbulence I: basic theory. *Journal of Scientific Computing, 1*, 3–51.

Ye, J., & McCorquodale, J. A. (1998). Simulation of curved open channel flows by 3D hydrodynamic model. *Journal of Hydraulic Engineering, 124*(7), 687–698.

Yuhang, Z., & Wenxin, H. (2009). Application of artificial neural network to predict the friction factor of open channel flow. *Communications in Nonlinear Science and Numerical Simulation, 14*, 2373–2378.

Zaji, A. H., & Bonakdari, H. (2015). Application of artificial neural network and genetic programming models for estimating the longitudinal velocity field in open channel junctions. *Flow Measurement and Instrumentation, 41*, 81–89.

Zeng, J., Constantinescu, G., Blanckaert, K., & Weber, L. (2008). Flow and bathymetry in sharp open-channel bends: experiments and predictions. *Water resources research, 42*(1–2), 1–12.

Zhang, M. L., & Shen, Y. M. (2008). Three dimensional simulation of meandering river based on 3-D $k-\varepsilon$ (*RNG*) turbulence model. *Journal of Hydrodynamics, 20*(4), 448–455.

Zhou, G., Wang, H., Shao, X., & Jia, D. D. (2006). 2-D numerical simulation of flow in a curved open channel. *Advances in Water Resources and Hydraulic Engineering, 3*, 871–876.

8

Numerical investigation of vortex dynamics in an H-rotor vertical axis wind turbine

Yan Chen and Yongsheng Lian

Department of Mechanical Engineering, University of Louisville, Louisville, KY, 40292, USA

We study the vortex dynamics of a two-dimensional H-rotor wind turbine using a Navier-Stokes solver. The k-ε turbulence model with the wall function is used as the turbulence closure. A sliding mesh technique is employed to handle the blade rotation. The vortex-blade interaction is systematically investigated and its influence on the force generation is discussed. Our simulations show that the vortex-blade interaction largely depends on the solidity and tip speed ratio. We further study the impact of solidity on the turbine performance. Our simulations show that the peak torque per blade decreases with the solidity while the peak torque azimuthal angle increases with the solidity. Our simulations also show that the increase in the azimuthal angle is more significant at low tip speed ratios than at high tip speed ratios. The impact of blade thickness is studied. Our simulations show that a thicker airfoil has a higher torque coefficient than a thinner airfoil. However, because for the thinner airfoil its peak torque occurs at a high tip speed ratio, the thinner airfoil has an overall higher power coefficient than the thicker airfoil.

Keywords: H-rotor; CFD; vortex dynamics; dynamic stall

1. Introduction

Wind powered energy generation has experienced tremendous growth worldwide and it is considered as an environmentally friendly and economically competitive means of electric power generation. World Wind Energy reported that by the end of 2010 all installed wind turbines can generate 430 terawatthours per year, equaling 2.5% of the global electricity consumption (World Wind Energy Report, 2010). Wind energy is generated using energy conversion system to convert kinetic energy from the wind into electricity. Wind energy conversion systems generally can be categorized into two groups based on the orientation of their rotation axis: the horizontal axis wind turbine (HAWT), and the vertical axis wind turbine (VAWT). The VAWT mainly has three kinds: the Savonius, the Darrieus, and the H-rotor. Figure 1 shows one horizontal axis wind turbine and three vertical axis wind turbines.

The VAWT presents several advantages over the HAWT. First, the VAWT has low sound emission due to its low operating speed (Simão Ferreira, Van Zuijlen, Bijl, Van Bussel, & Van Kuik, 2010). Recent studies (Liu, Ouyang, Wu, Tian, & Du, 2014) are putting efforts on reducing noise of the rotor. Iida, Mizuno, and Fukudome (2004) predicts that the noise level of VAWTs is about 70 dB which is 10 dB lower than HAWTs. Second, the VAWT is insensitive to the wind direction (Eriksson, Bernhoff, & Leijon, 2008). Third, the VAWT has increased power output in skewed flows while the HAWT has decreased power output, making the VAWT more favorable on top of a roof. Mertens, Van Kuik, and Van Bussel (2003) numerically studied the performance of H-rotors in skewed flows. In their work computational fluid dynamics (CFD) was used to calculate the wind velocity over a rectangular model building. The calculated velocity was passed to a low fidelity model, the double multiple streamtube model (DMSM), to evaluate the turbine performance. They showed that the skewed flow increased the power coefficient of the H-rotor. This phenomenon was further confirmed by their wind tunnel experiments (Mertens et al., 2003) and hotwire measurement (Simão Ferreira, Van Bussel, & Van Kuik, 2006). Last, the VAWT can accept wind from any direction and can potentially be used in lieu of HAWT designs in urban and mountainous areas and gusty area (Riegler, 2003).

The H-rotor shown in Figure 1 is a promising wind energy conversion system suitable for the built environment. Its small dimension ensures that it can be installed in urban areas. It has excellent energy conversion efficiency and has low manufacturing cost due to its simple structure. Despite its simple geometry, flow phenomena involved in H-rotor are complex and simplification cannot be easily made. These challenges are summarized in recent studies (Simão Ferreira et al., 2010). First, the blade rotating motion implies a variation of angle of attack which causes variation of force on the blade. As a result, it requires

*Corresponding author. Email: yongsheng.lian@louisville.edu

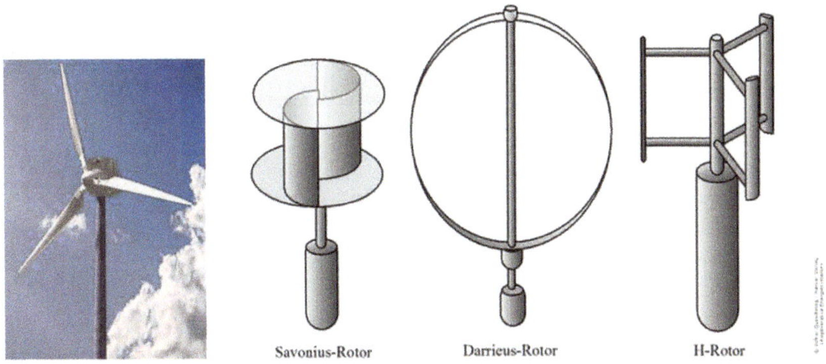

Figure 1. HAWT*, Savonius, Darrieus and H-rotor†.
*http://www.tfcenergy.com/images2/horizontal-axis-wind-turbine-design-225px.jpg
†http://envento-windenergie.com/images/Bild6-14.jpg

a time-accurate model to analyze the unsteady behavior. Second, wind turbines often operate beyond their individual static stall angle and experience dynamic stall. The dynamic stall is more severe at low tip speed ratio (TSR) and is a challenging topic by itself.

Both experimental and numerical studies have been reported in the study of H-rotors. In their wind tunnel experiment Oler, Strickland, Im, and Graham (1983) observed a large vortex on a single-bladed H-rotor. Simão Ferreira, Kuik, Bussel, and Scarano (2009) used the PIV method to visualize vortex shedding from a single-bladed H-rotor. Vortex shedding was observed at a low TSR of 2 but not at high TSRs of 3 and 4. Mohamed (2013) experimentally investigated the effect of solidity on the turbine performance to check the self-starting capability of Darrieus turbine.

Numerical methods with various fidelities have been proposed to study VAWTs. For example, Templin (1976) introduced the single streamtube model which is derived from blade element momentum theory without considering the solidity and TSP. Templin showed that the model can predict the power but the predicted value is higher than experimental results. Wilson and Lissaman (1974) proposed an improved model: the multiple streamtube model. In this improved model, the swept area of the turbine is divided into several parallel streamtubes and the blade element momentum theory is applied in each streamtube separately. The improved model predicted the instantaneous aerodynamic blade forces and the induced velocities better than the single streamtube model. Paraschivoiu (1988) presented a double multiple streamtube model to predict the performance of Darrieus wind turbine. The DMSM gave a better prediction than the original streamtube model. One drawback of most streamtube models is that they are limited to low TSRs and small solidities since the blade element momentum theory is not valid at high TSRs and solidities (Paraschivoiu, 2002).

Vortex models have also been applied to study VAWTs. Vortex models calculate the velocity field through the influence of vorticity in the wake of the blades (Islam, Ting, & Fartaj, 2008). Larsen proposed a vortex model to study a single bladed VAWT (1975). In his model the angle of attack is assumed to be small and the stall effect is neglected. Strickland, Webster, and Nguyen (1980) improved the vortex model to include the dynamic stall, pitching circulation and added mass effect. Cardona (1984) added a modified form of dynamic stall effects. His model showed improvement in the prediction of the instantaneous blade force.

With the advancement of computing power and numerical algorithms, computational fluid dynamics become a reliable tool in the study of H-rotor. Tchon and Paraschivoiu (1994) simulated two-dimensional H-rotors using a laminar Navier-Stokes solver. In their study the Reynolds number is 6700 which is one tenth of the experimental condition. Their results matched the experimental normal force but not the tangential force. Guerri, Sakout, and Bouhadef (2007) investigated the aerodynamic performance of a small rotating VAWT by using a Reynolds Averaged Navier-Stokes (RANS) solver and moving meshing technique. In their study the SST k-omega turbulence model was used. Their study showed that the RANS solver can capture the dynamic stall phenomenon and gave a better prediction of power coefficient than the DMSM model. Besides, the RANS model considers the airfoil section data while the DMSM model does not. Hamada etc. compared 2D and 3D results using the unsteady RNG k-ε turbulence model and claimed that an appropriate time step is required to obtain the accurate solution. Horiuchi, Ushiyama, and Seki (2005) examined the H-rotor performance using the detached eddy simulations (DES). Comparing the calculated and experimentally measured wind velocity, they concluded that the numerical simulation was applicable to the flow analysis for H-rotors. They also noticed that at the distance of 10 times of turbine radius from turbine center, flow could not recover its freestream condition, indicating a large spacing is needed when installing wind turbine in large quantity. Mohamed (2012) numerically investigated the influence of airfoil shape on H-rotor performance. He studied 20 different airfoils using the realizable $k-\varepsilon$ turbulence model. He was able to identify one airfoil that provides better performance than the conventional ones.

Among all the phenomena involved, dynamic stall plays a critical role in determining the H-rotor performance especially at low TSRs. The understanding of dynamic stall can help the design of more efficient and durable turbines. Ferreira et al. have done extensive research to understand the dynamic stall phenomenon on an H-rotor. They used PIV to visualize the shedding and convection of vortices (Simão Ferreira, 2009). They also numerically investigated dynamic stall using different turbulence models and further compared the influence of turbulence models on the accuracy of dynamic stall prediction. They found that the k-epsilon model and the Spalart Allmaras model give similar cycled averaged tangential and normal forces over the upwind semi-rotation. But the spatial distribution of vorticity is different among the two turbulence models (Simão Ferreira, Van Bussel, & Van Kuik, 2007). They also studied the same problem using large eddy simulations (LES) and detached eddy simulations. They found DES overall performances better than LES. The impact of dynamic stall on torque is also studied numerically by Hamada with the RNG k-epsilon turbulence model (Hamada, Smith, Durrani, Qin, & Howell, 2008).

To improve the efficiency, Hwang optimized a 2D H-rotor using a RANS solver and the k-epsilon turbulence model (Hwang, Lee, & Kim, 2009). Hwang showed that the performance can be improved by 25% with an optimized pitching angle motion. Paraschivoiu, Trifu, and Saeed (2009) computed the optimal variation of the blades' pitch angle to maximize the power coefficient at given conditions. The optimization was conducted by combining a genetic algorithm and a DMSM code. The optimized blade pitch was able to enhance the turbine performance at the operating wind speed. Takao et al. (2009) reported that an H-rotor can be improved by adding a directed guide vane row because of the guide vane row changes the airflow inlet angle to the rotor.

In this paper we study the aerodynamics of a 2D H-rotor using the RNG $k-\varepsilon$ turbulence model with a focus on the vortex dynamics and its impact on force histories. The influence of solidity is investigated. The blade thickness effect will also be studied. In the next we first introduce the aerodynamics of H-rotors, and then we present the numerical methods and computational set-up. Last we present the numerical results and discussion.

1.1. Aerodynamics of H-rotor VAWT

An H-rotor consists of straight blades with an airfoil cross section profile. The strut and supporting arms serve as the structural support. A typical H-rotor VAWT is shown in Figure 2. When wind blows toward the turbine, both lift and drag are generated on the blade. The force component in the tangential direction generates torque that spins the turbine.

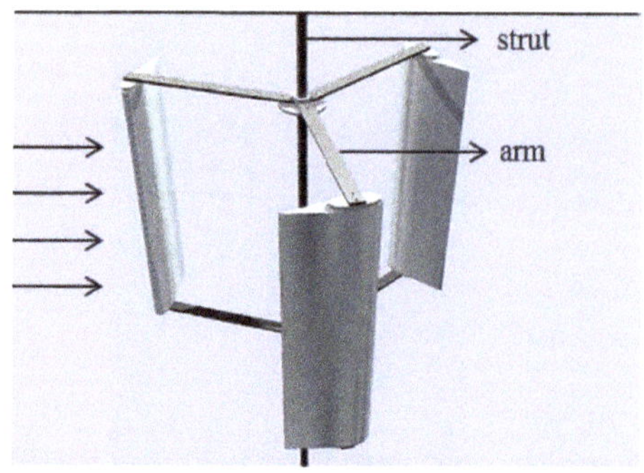

Figure 2. Prototype of a three-bladed H-rotor VAWT (Howell, Qin, Edwards, & Durrani, 2010).

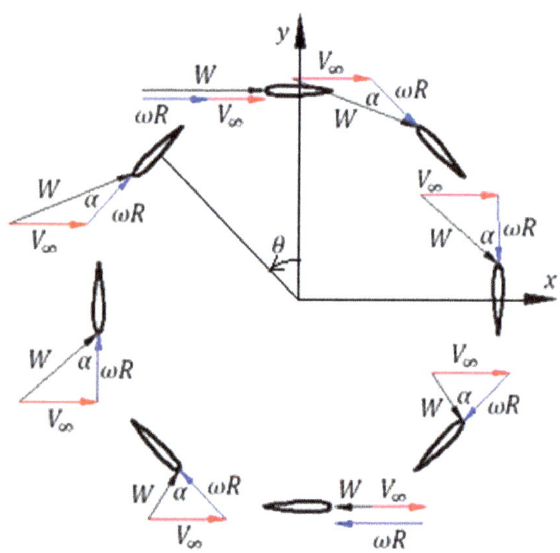

Figure 3. Effective AoA and effective wind velocity (W) at various azimuthal positions (θ is the azimuthal angle).

The blade rotation changes the velocity magnitude and direction perceived by the blade. Hence the angle of attack (AoA) varies with the azimuthal position of the blade. As shown in Figure 3 the AoA (α) is defined as the angle between the airfoil chord line and the effective wind velocity (W) perceived by the blade. The effective velocity is composed by the wind velocity (V_∞) and the blade rotating velocity (ωR). Figure 3 illustrates the variation of the effective velocity at different azimuthal positions of the blade. The maximum and minimum effective velocity is achieved when the blade is at the azimuthal angle of 180° and 0°, respectively. Figure 4 shows the effective wind velocity perceived by the blade at different azimuth angles. In the figure the effective wind velocity is normalized by the freestream wind speed.

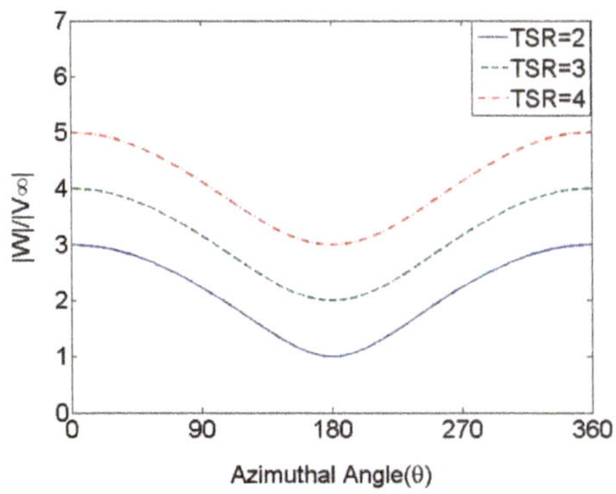

Figure 4. Variation of effective wind velocity perceived by the blade at different azimuth angles at different tip speed ratios.

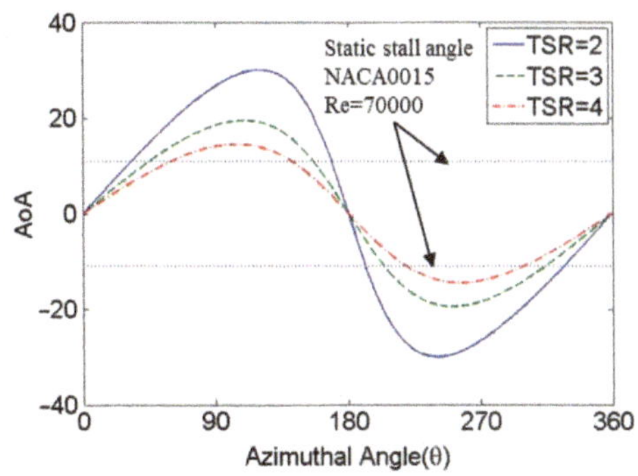

Figure 5. The effective angle of attack versus the azimuth angle at different TSRs.

For an H-rotor the effective AoA is not only a function of the azimuthal position but a function of the tip speed ratio which is defined as the ratio between the blade rotating speed (ωR) and the freestream speed (V_∞). The relationship between the effective AoA (α), the azimuthal angle, θ, and the tip speed ratio, λ, is defined as follows:

$$\alpha = \tan^{-1}\frac{\sin\theta}{\cos\theta + \lambda} \tag{1}$$

Figure 5 plots the effective AoA versus the azimuthal angle at different tip speed ratios. The static stall angle of NACA 0015 airfoil at the Reynolds number of 70,000 is also plotted. It can be seen that as the tip speed ratio decreases more portion of the rotation cycle exceeds the static stall angle.

Figure 6 shows the directions of the lift, drag and their normal and tangential components on a blade. Here the lift is the force that is perpendicular to the relative wind velocity (W) and the drag is the one parallel to W. In the study of H-rotor, tangential and normal forces are usually used. The tangential force is resultant force parallel to the rotational direction and the normal force is the one perpendicular to the rotational direction. The tangential force F_T and the

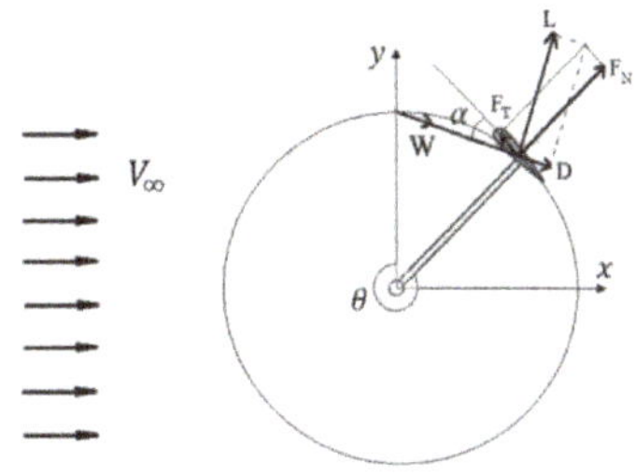

Figure 6. Force components on a rotating blade.

normal force F_N can be expressed as below:

$$F_T = L\sin\alpha - D\cos\alpha \tag{2}$$

$$F_N = L\cos\alpha - D\sin\alpha \tag{3}$$

where L and D are the lift and drag, respectively. The tangential and normal force coefficients can be expressed as:

$$C_T = \frac{F_T}{1/2\rho c V_\infty^2} \tag{4}$$

$$C_N = \frac{F_N}{1/2\rho c V_\infty^2} \tag{5}$$

The averaged tangential force (F_{Ta}) over one rotational cycle is

$$F_{Ta} = \frac{1}{2\pi}\int_0^{2\pi} F_T(\theta)\mathrm{d}\theta \tag{6}$$

The torque (Q) on the blade is generated by the tangential force and it is calculated as follows:

$$Q = NF_{Ta}R \tag{7}$$

where N is the number of blades and R is the turbine rotation radius as shown in Figure 3. And the torque coefficient can be expressed as:

$$C_Q = \frac{Q}{\frac{1}{2}\rho RAV_\infty^2} \tag{8}$$

where A is the cross-sectional area of the wind turbine. In 2D simulations, the height of the turbine blade is considered as unit length, therefore,

$$A = 2R \tag{9}$$

The total power (P) is

$$P = Q\cdot\omega \tag{10}$$

and the power coefficient is defined:

$$C_P = \frac{P}{\frac{1}{2}\rho AV_\infty^3} \tag{11}$$

2. Numerical methods

2.1. *Fluid solver*

We solve the unsteady incompressible Navier-Stokes equations to describe the flow fields. The least squares cell-based method is applied to evaluate all the gradients. The convection terms are discretized using a second-order upwind scheme. The diffusion terms are discretized using a second-order central difference scheme and the pressure term is by using second-order central difference scheme. Time marching is achieved using a second-order backward implicit method. The SIMPLE (Semi Implicit Method for Pressure Linked Equations) method is used to couple the velocity and pressure (Patankar, 1980). The following RNG k-ε turbulence model is used as the turbulence closure (Yakhot, Orszag, Thangam, Gatski, & Speziale, 1992):

$$\frac{\partial(\rho k)}{\partial t}+\frac{\partial}{\partial x_i}(\rho k u_i)=\frac{\partial}{\partial x_j}\left(\alpha_k \mu_{eff}\frac{\partial k}{\partial x_j}\right)+G_k-\rho\varepsilon \quad (12)$$

$$\frac{\partial}{\partial t}(\rho\varepsilon)+\frac{\partial}{\partial x_i}(\rho\varepsilon u_i)=\frac{\partial}{\partial x_j}\left(\alpha_\varepsilon \mu_{eff}\frac{\partial \varepsilon}{\partial x_j}\right)+C_{1\varepsilon}\frac{\varepsilon}{k}G_b-C^*_{2\varepsilon}\rho\frac{\varepsilon^2}{k}-R_\varepsilon \quad (13)$$

where k is turbulent kinetic energy, ε is turbulent dissipation rate, μ_{eff} is effective turbulent viscosity, G_k represents the generation of turbulent kinetic energy due to the mean velocity gradients. The quantities α_k and α_ε are the inverse effective Prandtl numbers for k and ε.The RNG k-ε turbulence model was proposed by Yakhot. In this model, the rigorous statistical technique is applied to the unsteady Navior-Stokes equations to capture the effect of smaller-scale motions (Pezzinga, 1994). Compared with the standard k-ε model, the RNG k-ε model uses an additional term (R_ε) in its dissipation equation to model the interaction between the turbulence dissipation and the mean shear. Tests showed that the additional term can significantly improve the model accuracy for problems with rapidly strained flows (Ansys, 2009). Furthermore, the effect of swirl on turbulence is included in μ_{eff}, enhancing the accuracy for swirling flow simulation.

2.2. *Sliding mesh*

To handle the blade rotation we used the sliding mesh technique to update the mesh around the rotating wind turbine blades. In this method, cell zones are surrounded by at least one interface zone where it meets its neighboring cell zone. The interface zones are associated with one another to form a mesh interface. And both zones will move relative to each other along the mesh interface. One advantage is that the sliding mesh technique does not require a point-to-point match at the interface which is often required for other moving grid methods (Lian, Steen, Trygg-Wilander, & Shyy, 2003). With the sliding mesh technique, the computational domain is divided into stationary areas and rotating areas. The rotating areas consist of the moving objects. As shown in Figure 7, the rotating area consists of the blades.

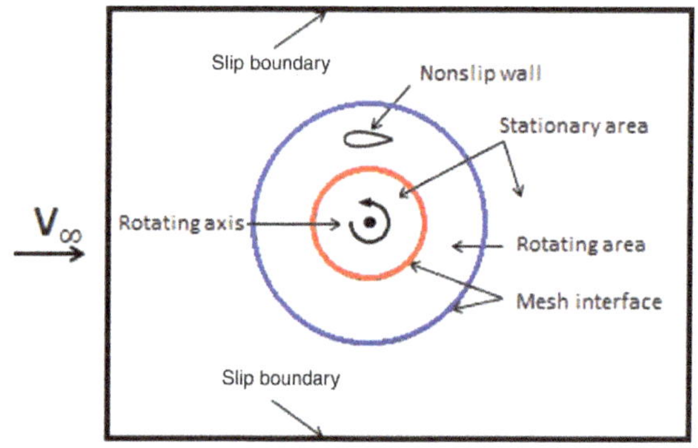

Figure 7. Computational setup of the problem.

In the rotating region, the governing equations need to be modified to incorporate the rotation effect. The modified continuity and momentum equations are as follows:

$$\nabla\cdot(\vec{u}-\vec{V})=0 \quad (14)$$

$$\frac{\partial\vec{u}}{\partial t}+\vec{u}\cdot\nabla(\vec{u}-\vec{V})=\frac{-\nabla p}{\rho}+\frac{1}{\rho}\nabla\cdot(\mu\nabla\vec{u}) \quad (15)$$

Here $\mathbf{\vec{u}}$ is the fluid velocity in the stationary frame and $\mathbf{\vec{V}}$ is the velocity due to mesh motion, p is the pressure and μ is the dynamic viscosity. At the sliding interface a conservative interpolation is used for both mass and momentum based on a set of fictitious control volumes.

3. Results and discussions

3.1. *Computational setup*

Figure 7 illustrates the computational setup. The annular region (including the blades) rotates with a prescribed speed defined by the TSR. In this rotating region the modified Navier-Stokes equations (Equations 13 and 14) are solved. The rest regions are stationary. In the simulation, the Dirichlet boundary condition is applied on the left boundary where the velocity is set as the freestream velocity; slip boundary condition is applied to the top and bottom surfaces; the pressure is prescribed on the right boundary. The airfoil surfaces are defined as nonslip walls. The inner boundary of the annulus (the red line) is defined as interface zones with the inner region boundary. The outer boundary of the annulus (the blue line) is defined as interface zones with the overlapping boundary of outer region.

3.2. *Grid sensitivity analysis*

A sensitivity analysis is carried out. The flow conditions and geometry are based on experiment of Oler et al. (1983). They tested a straight one-bladed rotor with NACA0015 profile in a water tow tank. The tow tank has a depth of 1.25 meters, a width of 5 meters, and a length of 10 meters. The solidity is 0.25 which is the ratio of the sum of chord lengths to the turbine rotation radius. In the experiment they maintained a constant angular velocity but varied the incoming flow velocity to get different TSRs. The case we simulate here has a TSR of 2.5. The Reynolds number is 67,000 based on the effective tip speed and the chord length.

The baseline grid has 400 points on the airfoil in the circumferential direction. The medium grid has 600 points and the fine grid has 800 points in the circumferential direction. In the radial direction the grid distribution is carefully adjusted to ensure the y^+ is in the range of 15 and 30 for turbulent flow simulations.

Figure 8 shows the tangential force coefficient (C_T) versus the azimuth angle with different grid densities. From 0° to 70° all the grids predict almost the same results. After 70° the coarse grid predicts a lower coefficient than the medium and fine grids. But there is almost no disparity between the medium and fine grids from 70° to 360°, indicating a grid-independent solution is obtained with the medium grid. At $\theta = 70°$ the effective angle of the airfoil is 20.6°, which is beyond its static stall angle shown in Figure 5. Figure 9 presents snapshots of vorticity contours with streamlines at $\theta = 85°$ with different grids. There is a slight difference in the recirculation zone within the separation bubble between the coarse and medium grids but there is almost no difference between the medium and fine grids. The medium grid is therefore used in the following study.

In Figure 10 we compare the simulated tangential and normal force coefficients with experimental data (Oler et al., 1983). The numerical results match the trend of the measured ones but the magnitudes are different. Simulations by various other groups show similar disparity in force coefficients (Deglaire, Engblom, Ågren, & Bernhoff, 2009).

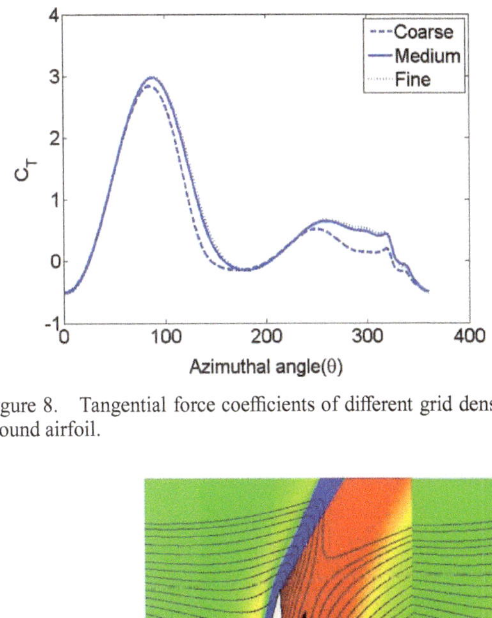

Figure 8. Tangential force coefficients of different grid density around airfoil.

3.3. *Tip speed ratio effect*

As shown in Figure 5, at the same azimuth angle the effective AoA encountered by the airfoil changes with the TSR. The higher the TSR, the less the airfoil exceeds its static stall angle. In Figure 11 we compare the tangential force coefficient histories at different TSRs. It is found the coefficient peak decreases with the increase of TSR. It is also observed that from $\theta = 0°$ to $\theta = 90°$ the coefficient increases with the TSR. At around 90°, the tangential force coefficient begins to drop at TSR = 2 but the coefficient does not reach its peak until at around 100° for TSR of 3 and 4.

Figure 12 compares the snapshots of vorticity contours at $\theta = 100°$ at different TSRs. It is clear that a high TSR

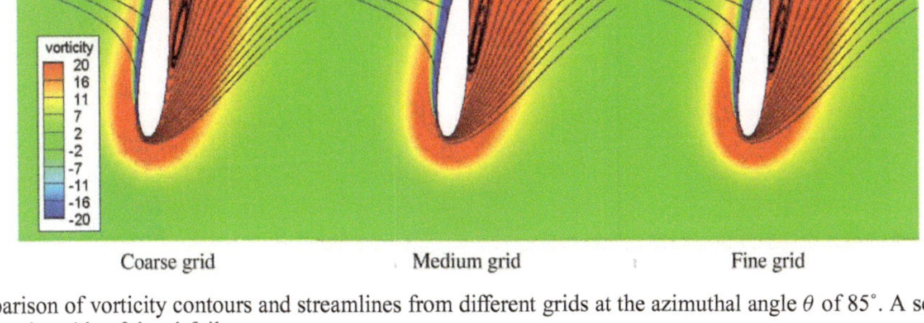

Figure 9. Comparison of vorticity contours and streamlines from different grids at the azimuthal angle θ of 85°. A separation bubble is observed on the suction side of the airfoil.

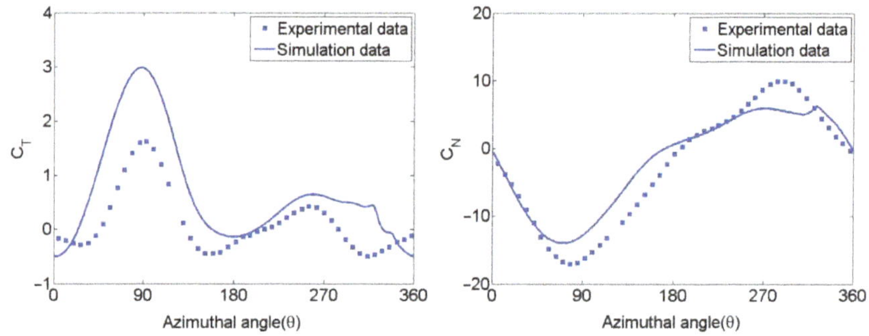

Figure 10. Comparison of instantaneous tangential and normal force coefficients between experiment (Oler et al., 1983) and simulation.

can defer the shed of the leading edge vortex. For TSR = 2 a large vortex has already shed into the wake; for TSR = 3 the vortex starts to shed but still attaches to the surface; for TSR = 4 the vortex attaches to the surface.

The torque coefficients of a three-bladed H-rotor at different TSRs are plotted in Figure 13. The peak torque coefficient is achieved at 2.5. Figure 14 plots the power coefficients. The peak occurs at TSR of 2.5.

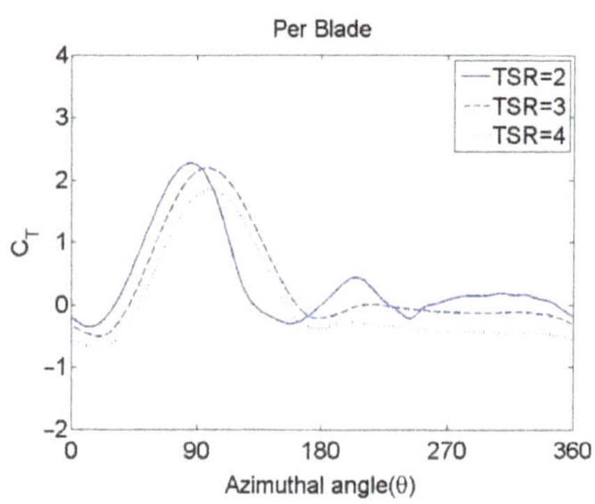

Figure 11. Comparison of instantaneous tangential force coefficient of three bladed H-rotor at different TSRs.

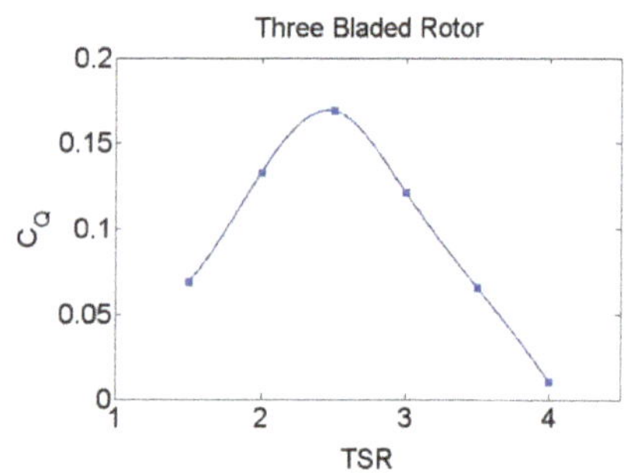

Figure 13. Torque coefficients of the three-bladed H-rotor at different TSRs.

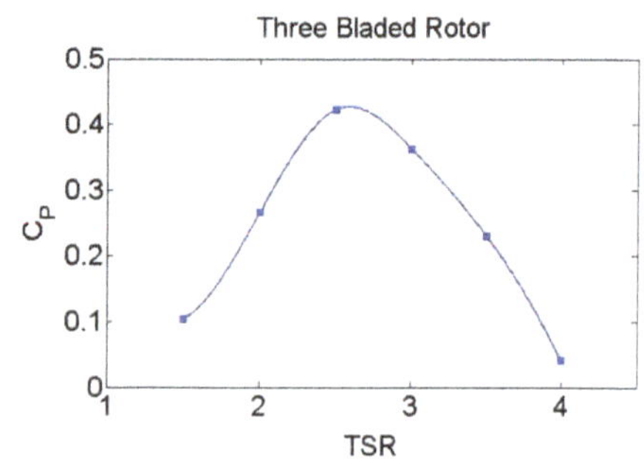

Figure 14. Power coefficients of the three-bladed H-rotor at different TSRs.

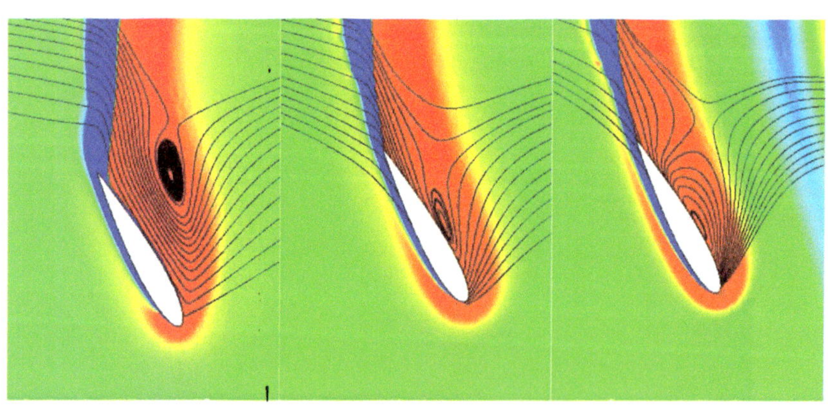

TSR=2 TSR=3 TSR=4

Figure 12. Comparison of vorticity contours at different TSRs at $\theta = 100^{\circ}$. A high TSR defers the shedding of leading edge vortex.

3.4. *Solidity effect*

Figure 15 compares the tangential force coefficient histories of the one-bladed, two-bladed, three-bladed, and four-bladed H-rotors at different TSRs. Figure 16 compares their cycle averaged per blade torque coefficients at different TSRs. It is clear that for all the tested TSRs increasing the number of blades reduces the tangential force coefficient magnitude but the reduction level highly depends on the TSRs: the higher the TSRs the more reduction the coefficient. This is also true for the cycle averaged torque coefficient (Figure 16). At TSR = 2 all the rotors have the nearly identical per blade torque coefficients but their coefficients are quite different at the TSR of 4. From Figure 15 we can also see that increasing the number of blades delays the appearance of tangential force peak, which is more obvious at low TSRs than at high TSRs.

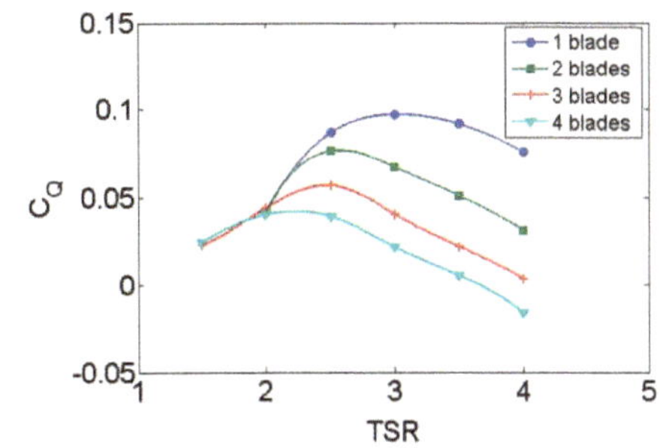

Figure 16. Cycle averaged per blade torque coefficient at different solidities and TSRs.

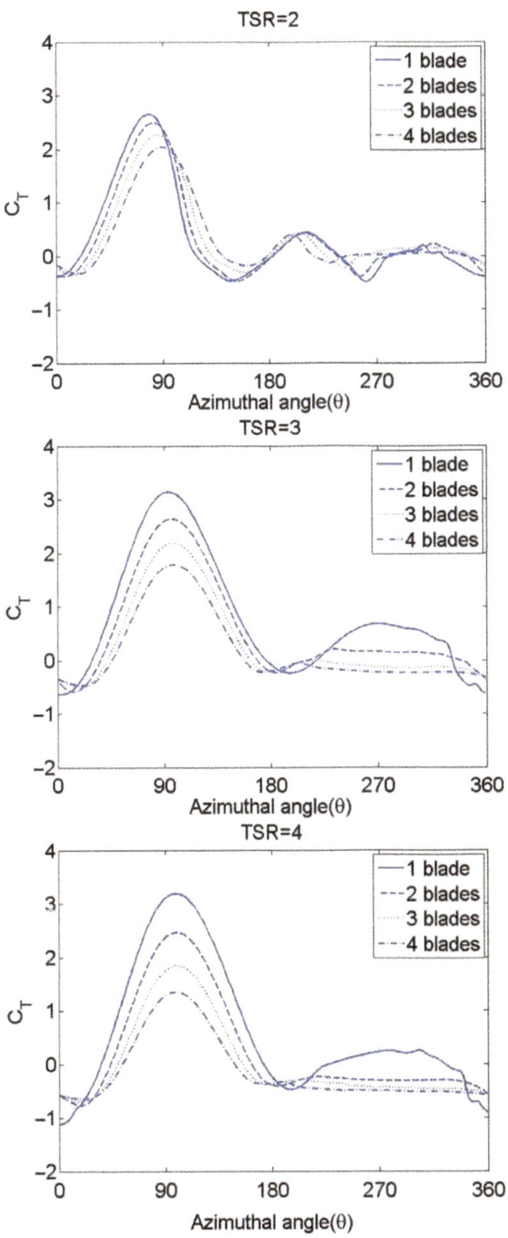

Figure 15. Comparison of tangential force coefficients at different solidities and TSRs.

We compare the vorticity contours at different solidities in Figure 17 and Figure 18 . In each case four representative instants are selected ($\theta = 57°$, 140°, 230° and 333°). In all cases strong vortex-blade interactions are observed but the interaction behavior varies with the azimuthal angle, solidity, and TSR.

Figure 17 is the case of TSR = 2. We compare the vorticity contours around the circled blades. At $\theta = 57°$ and 140°, the blade is in the upstream and no vortex-blade interaction is observed in all cases. At $\theta = 230°$, no interaction is observed in the one-bladed case; in the two-bladed case the blade is about to interact with the wake of the previous blade; in the three-bladed case vortex-blade interaction is observed. At $\theta = 333°$, the interaction is seen in both the two- and three-bladed cases. In the one-bladed case, the airfoil goes through a weak vortex shed from itself. In Figure 15, we can find the resulting force histories are very similar between the two- and three-bladed cases while they are different from the one-bladed case.

The vorticity contours at TSR = 3 are shown in Figure 18. From Figure 5 we can see that at TSR = 3 the effective AoA is lower than that at TSR = 2. This lower effective AoA leads to different votex dynamics. At $\theta = 140°$ a much smaller LEV is observed at TSR = 3 than at TSR = 2. At $\theta = 230°$ the blade already interacts with the wake in the case of TSR = 3 but not in the case of TSR = 2.

Figure 19 compares the cycle averaged power coefficients of turbines with different numbers of blades. Except for the one-bladed rotor, the power coefficient increases with the number of blade at low TSRs and decreases with the number of blades at high TSRs. Overall the

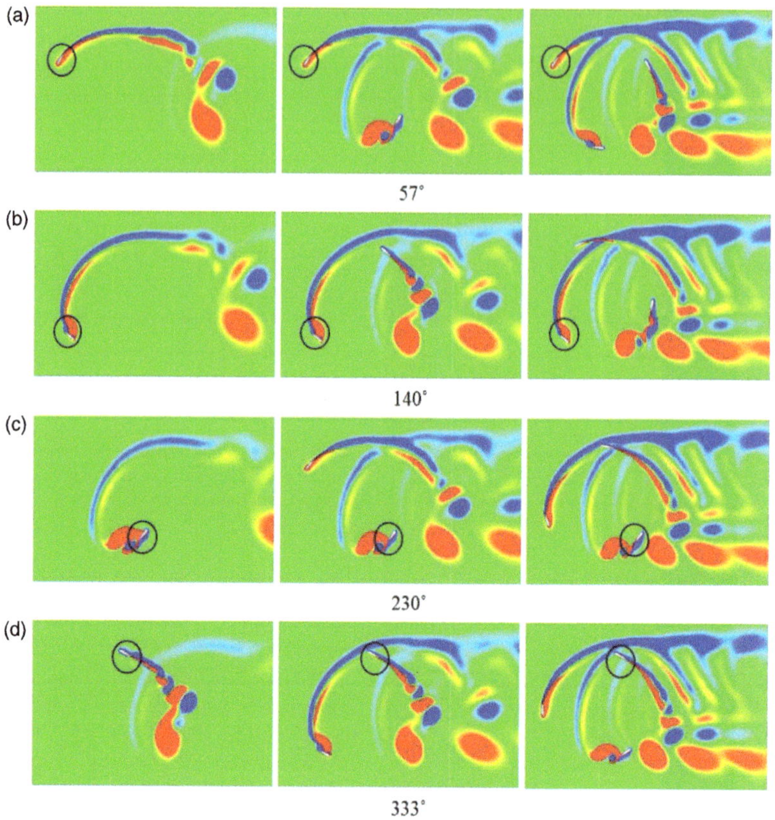

Figure 17. Comparison of vorticity contours at different solidities (TSR = 2).

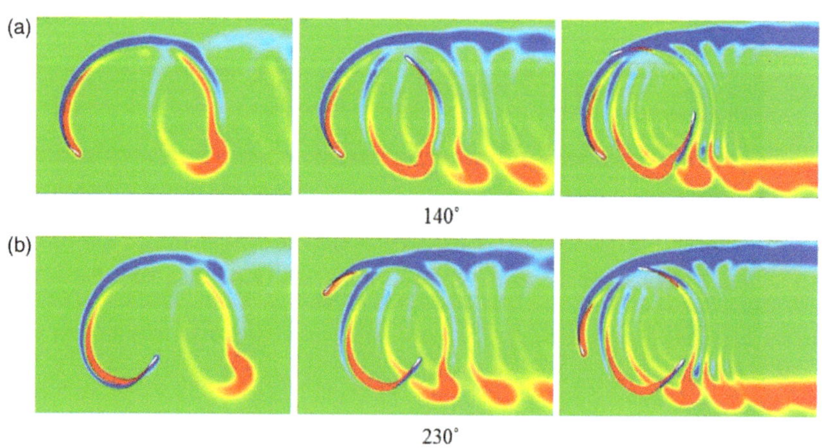

Figure 18. Comparison of vorticity contours at different solidities (TSR = 3).

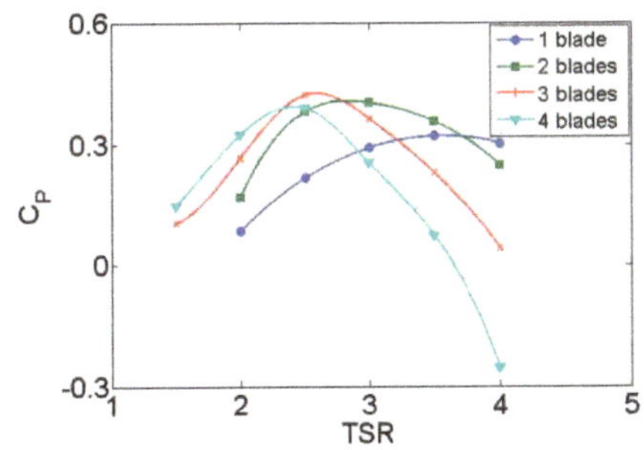

Figure 19. Cycle averaged power coefficient curves of rotors with different numbers of blades.

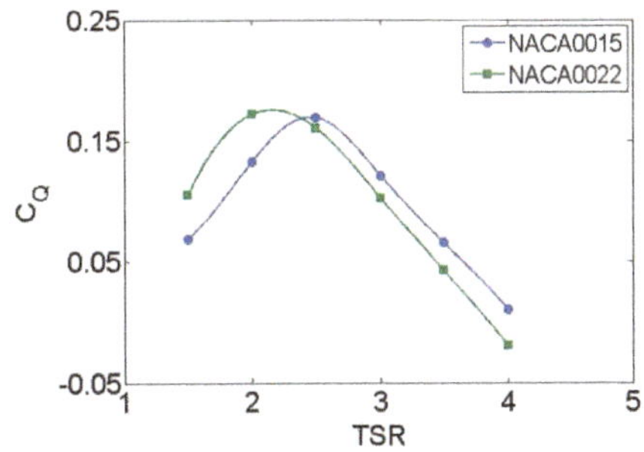

Figure 21. Comparison of torque coefficients of rotors of different cross sectional profiles.

three-bladed rotor has the highest coefficient at TSR of 2.5.

3.5. *Airfoil thickness effects*

The performance of two three-bladed rotors with different cross sectional profiles is further studied. One has the NACA0015 profile and the other one has the NACA0022 profile. Figure 20 compares the vorticity contours near their tangential force efficient peaks at different TSRs. Overall they look alike each other.

Figure 21 compares the cycle averaged torque coefficients at different TSRs. At low TSRs, the NACA0022 blade gives a better performance than the NACA0015 blade. The NACA0022 blade reaches its peak torque at a lower TSR than the NACA0015 blade. The peak torque of the NACA0022 blade is slightly higher than the NACA0015 blade. But the NACA0015

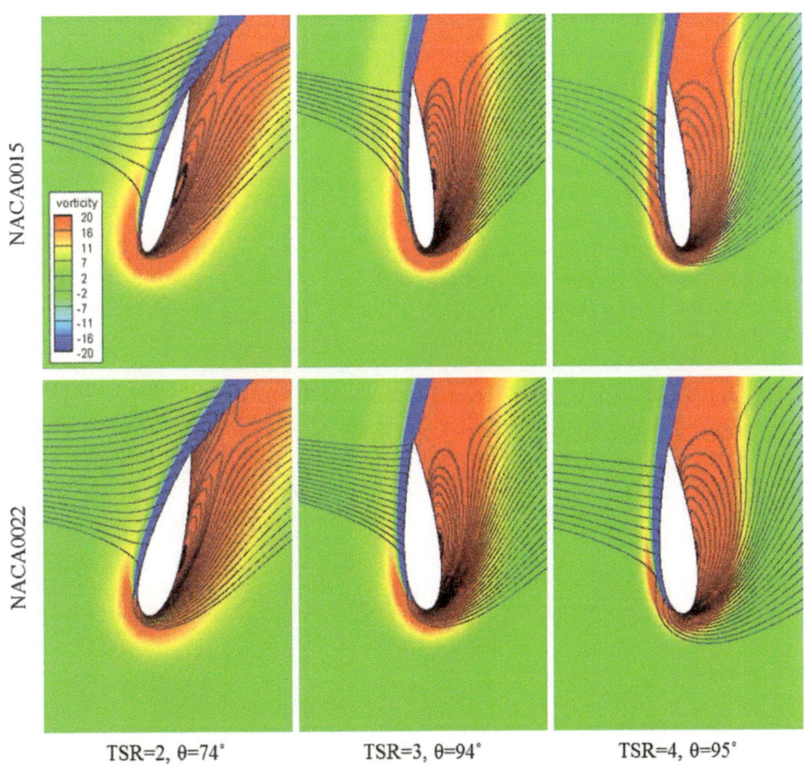

Figure 20. Vorticity contours & Streamlines around the peak of tangential force coefficient in a circle of NACA0015 and NACA 0022 at TSR = 2, 3 and 4.

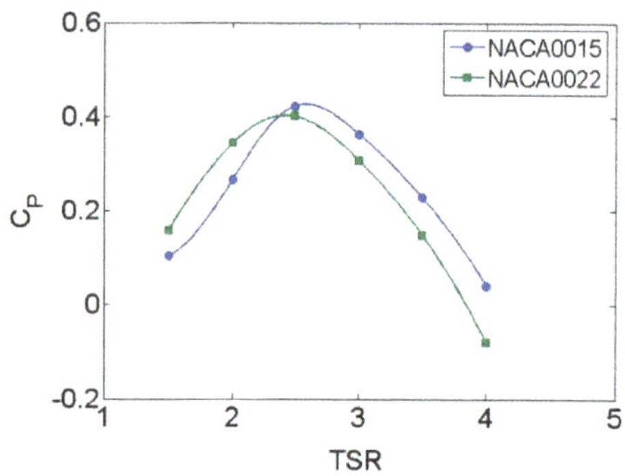

Figure 22. Comparison of power coefficients of rotors of different cross sectional profiles.

blade has an overall higher peak power coefficient (Figure 22).

4. Conclusion

We have studied the aerodynamic performance of an H-rotor using a Navier-Stokes solver. A sliding mesh method is used to handle the rotation motion of the blades. The transient RNG k-ε turbulence model is used as the turbulence closure. Detailed sensitivity analysis is presented. Due to the massive flow separation, we need a very fine grid to capture the flow phenomenon. Comparisons are made with available numerical and experimental data. Discrepancies exist between experimental data and simulation data. It is found that the power performance is increasing with TSR increasing at low TSRs. At high TSRs, Our study shows that vortex and blade interactions have an obvious impact on the blade performance. The power performance continues increasing until the number of blades equal to three. The four bladed H-rotor shows a lower power coefficient than three bladed H-rotor. The power performance is also significantly influenced by the airfoil thickness. Our results show that increasing the airfoil thickness improves the blade performance at the low tip speed ratio.

Since H-rotors are characterized with massive flow separation, vortical flows, and laminar-to-turbulent transition, high fidelity models such as detached eddy simulation or large eddy simulation is recommended for further work.

References

Ansys. (2009). *Users Guide Manual.* In Fluent 12.0/12.1 Documentation, Ansys.

Cardona, J. L. (1984). Flow curvature and dynamic stall simulated with an aerodynamic free-vortex model for VAWT. *Wind Engineering, 8*, 135–143.

Deglaire, P., Engblom, S., Ågren, O., & Bernhoff, H. (2009). Analytical solutions for a single blade in vertical axis turbine motion in two dimensions. *European Journal of Mechanics. B, Fluids, 28*, 506–520.

Eriksson, S., Bernhoff, H., & Leijon, M. (2008). Evaluation of different turbine concepts for wind power. *Renewable and Sustainable Energy Reviews, 12*(5), 1419–1434.

Guerri, O., Sakout, A., & Bouhadef, K. (2007). Simulations of the fluid flow around a rotating vertical axis wind turbine. *Wind Engineering, 31*(3), 149–163.

Hamada, K., Smith, T., Durrani, N., Qin, N., & Howell, R. (2008). *Unsteady flow simulation and dynamic stall around vertical axis wind turbine blades.* In 46th AIAA Annual Aerospace Sciences Meeting and Exhibit, pp. 1–11; AIAA Paper 2008-1319.

Horiuchi, K., Ushiyama, I., & Seki, K. (2005). Straight wing vertical axis wind turbines: A flow analysis. *Wind Engineering, 29*(3), 243–252.

Howell, R., Qin, N., Edwards, J., & Durrani, N. (2010). Wind tunnel and numerical study of a small vertical axis wind turbine. *Renewable Energy, 35*(2), 412–422.

Hwang, I. S., Lee, Y. H., & Kim, S. J. (2009). Optimization of cycloidal water turbine and the performance improvement by individual blade control. *Applied Energy, 86*(9), 1532–1540.

Iida, A., Mizuno, A., & Fukudome, K. (2004). *Numerical simulation of aerodynamic noise radiated form vertical axis wind turbines.* Proceedings of the 18 International Congress on Acoustics.

Islam, M., Ting, D. S. K., & Fartaj, A. (2008). Aerodynamic models for Darrieus-type straight-bladed vertical axis wind turbines. *Renewable and Sustainable Energy Reviews, 12*(4), 1087–1109.

Larsen, H. (1975). *Summary of a vortex theory for the cyclogiro.* In the second US national conferences on wind engineering research, Colorado State University.

Lian, Y., Steen, J., Trygg-Wilander, M., & Shyy, W. (2003). Low Reynolds number turbulent flows around a dynamically shaped airfoil. *Computers & Fluids, 32*(3), 287–303.

Liu, H., Ouyang, H., Wu, Y., Tian, J., & Du, Z. (2014). Investigation of unsteady flows and noise in rotor-stator interaction with adjustable lean vane. *Engineering Applications of Computational Fluid Mechanics, 8*(2), 299–377.

Mertens, S., Van Kuik, G., & Van Bussel, G. (2003). Performance of an H-Darrieus in the skewed flow on a roof. *Journal of Solar Energy Engineering, 125*(4), 433–440.

Mohamed, M. (2012). Performance investigation of H-rotor Darrieus turbine with new airfoil shapes. *Energy, 47*(1), 522–530.

Mohamed, M. (2013). Impacts of solidity and hybrid system in small wind turbines performance. *Energy, 57*, 495–504.

Oler, J., Strickland, J., Im, B., & Graham, G. (1983). *Dynamic stall regulation of the Darrieus turbine.* Albuquerque, NM: Sandia National Laboratories.

Paraschivoiu, I. (1988). Double-multiple streamtube model for studying vertical-axis wind turbines. *Journal of Propulsion and Power, 4*(4), 370–377.

Paraschivoiu, I. (2002). *Wind turbine design: With emphasis on Darrieus concept.* Canada: Polytechnic International Press.

Paraschivoiu, I., Trifu, O., & Saeed, F. (2009). H-Darrieus Wind turbine with blade pitch control. *International Journal of Rotating Machinery*, 1–7.

Patankar, S. V. (1980). *Numerical heat transfer and fluid flow.* Washington, DC; New York, NY: Hemisphere Pub. Corp; McGraw-Hill.

Pezzinga, G. (1994). Velocity distribution in compound channel flows by numerical modeling. *Journal of Hydraulic Engineering, 120*(10), 1176–1198.

Riegler, H. (2003). HAWT versus VAWT: Small VAWTs find a clear niche. *Refocus, 4*(4), 44–46.

Simão Ferreira, C., Kuik, G., Bussel, G., & Scarano, F. (2009). Visualization by PIV of dynamic stall on a vertical axis wind turbine. *Experiments in Fluids*, *46*(1), 97–108.

Simão Ferreira, C., Van Bussel, G., & Van Kuik, G. (2007). *2D CFD simulation of dynamic stall on a Vertical Axis Wind Turbine: verification and validation with PIV measurements*. 45th AIAA Aerospace Sciences Meeting and Exhibit/ASME Wind Energy Symposium, Reno, NV.

Simão Ferreira, C. J., Van Bussel, G. J. W., & Van Kuik, G. M. (2006). Wind tunnel hotwire measurements, flow visualization and thrust measurement of a VAWT in skew. *Journal of Solar Energy Engineering*, *128*(4), 487–497.

Simão Ferreira, C. J., Van Zuijlen, A., Bijl, H., Van Bussel, G., & Van Kuik, G. (2010). Simulating dynamic stall in a two-dimensional vertical-axis wind turbine: verification and validation with particle image velocimetry data. *Wind Energy*, *13*(1), 1–17.

Strickland, J. H., Webster, B. T., & Nguyen, T. (1980). *Vortex model of the Darrieus turbine: an analytical and experimental study*. NASA STI/Recon Technical Report 80-25887.

Takao, M., Kuma, H., Maeda, T., Kamada, Y., Oki, M., & Minoda, A. (2009). A straight-bladed vertical axis wind turbine with a directed guide vane row — effect of guide vane geometry on the performance. *Journal of Thermal Science*, *18*(1), 54–57.

Tchon, K. F., & Paraschivoiu, I. (1994). Navier-Stokes simulation of the flow around an airfoil in Darrieus motion. *Journal of Fluids Engineering*, *116*(4), 870–876.

Templin, R. J. (1976). *Aerodynamic performance theory for the NRC vertical-axis wind turbine*. NRC Lab. report, NASA STI/Recon Technical Report N, 76-16618.

Wilson, R. E., & Lissaman, P. B. S. (1974). *Applied aerodynamics of wind power machhines*. Oregon State University Corvallis.

World Wind Energy Report. (2010). Retrieved from http://www.wwindea.org/.

Yakhot, V., Orszag, S. A., Thangam, S., Gatski, T. B., & Speziale, C. G. (1992). Development of turbulence models for shear flows by a double expansion technique. *Physics Fluids A: Fluid Dynamics*, *4*(7), 1510–1520.

9

Simulation-based investigation of unsteady flow in near-hub region of a Kaplan Turbine with experimental comparison

B. G. Mulu[a], M. J. Cervantes[b,c*], C. Devals[d], T. C. Vu[e] and F. Guibault[d]

[a]Vattenfall Research and Development AB, Laboratorievägen, SE-814 26 Alvkarleby, Sweden; [b]Division of Fluid and Experimental Mechanics, Luleå University of Technology, SE-971 87 Luleå, Sweden; [c]Department of Energy and Process Engineering, Water Power Laboratory, Norwegian University of Science and Technology, Trondheim, Norway; [d]Computer and Software Engineering Department, Ecole Polytechnique de Montréal, Montréal, QC, Canada, H3C 3A7; [e]Andritz Hydro Ltd., 6100 TransCanada Hwy, Pointe-Claire, QC, Canada

This paper presents a detailed comparison of steady and unsteady turbulent flow simulation results in the U9 Kaplan turbine draft tube with experimental velocity and pressure measurements. The computational flow domain includes the guide vanes, the runner and the draft tube. A number of turbulence models were studied, including the standard $k - \varepsilon$, RNG $k - \varepsilon$, SST and SST-SAS models. Prediction of the flow behavior in the conical section of the draft tube directly below the runner cone is very sensitive to the prediction of the separation point on the runner cone. The results demonstrate a significant increase in precision of the flow modeling in the runner cone region by using unsteady flow simulations compare to stage simulation. The prediction of the flow in the runner cone region, however, remains delicate, and no turbulence model could accurately predict the complex phenomena observed experimentally.

Keywords: Kaplan turbine; hydropower; boundary condition; CFD; experimental validation

1. Introduction

Over the last decade, the constant increase in demand for energy has pushed current electric production facilities to their operational limits. In this context, the rehabilitation of aging hydraulic installations often appears as a way to increase the production capacity in a way that is both economically advantageous and socially acceptable. From an engineering design perspective, however, rehabilitation projects carry enormous challenges. Designers must either modify existing components or develop new components that are both highly specific to site operating conditions, and well adapted to existing components, for which replacement cannot be economically justified. An accurate analysis of existing installations, therefore, constitutes a central competitive tool both during feasibility studies, to determine which components should be replaced, as well as during the engineering design phases, to ensure that contractual guarantees can be met. Short design cycle times may impose severe constraints on the acceptable duration of those analyses, and may prevent analysts from resorting to laboratory tests. Even when available, model tests are very expensive and mainly offer integral values such as efficiency and parameters at a limited number of measurement points. Such information is crucial for design validation but is restrictive in its ability to support design optimization. Numerical simulations, on the other hand, provide detailed values everywhere in the computational domain.

The fidelity of flow simulations currently varies from the most inexpensive potential flow models to the highest fidelity LES and DNS simulations. In this spectrum, high fidelity unsteady, turbulent viscous flow models constitute a powerful alternative to understand existing turbine characteristics, predict the global system behavior and optimize specific flow regions, offering detailed information over the entire flow field. The current performance of large computer clusters allows for consideration of unsteady Reynolds averaged Navier-Stokes (URANS) simulations in multiple turbine components as the highest level of model fidelity currently available to analysts in production environments. These simulations, however, still fall short in their attempts to capture the full complexity of hydraulic turbine flow. This deficiency, is especially true in axial machines, such as propellers and Kaplan turbines, and is even more obvious when only some components of the turbine are analyzed. There is, consequently, a need to validate the RANS and URANS simulations in axial turbines against precise model measurements carried out in the laboratory.

*Corresponding author. Email: michel.cervantes@ltu.se

2. State of the art

Comparisons between numerical simulations and experiments in draft tubes have been carried out for several combinations of components and machine types. The most advanced simulations have been carried out in Francis turbines (Ciocan, Iliescu, Vu, Nennemann, & Avellan, 2007; Vu, Devals, Zhang, Nennemann, & Guibault, 2011a), for instance, as part of the FLINDT project, which took place at Ècole Polytechnique Fédérale de Lausanne (EPFL) and involved several turbine manufacturers. High fidelity RANS simulations, both steady and unsteady, of the runner and draft tube have shown very good agreement between experimental measurements and simulation results. A number of validation cases have also been carried out using Large-Eddy Simulations (LES) for a draft tube downstream a Francis runner (Duprat et al., 2009), and more recently for a whole Francis turbine, including distributor, runner and draft-tube in Su, Li, Li, Wei, and Zhao (2012).

Similar work is currently also underway for propeller machines as part of the Axial-T project conducted at Laval University. Comparisons between model measurements and RANS and URANS simulations in the draft tube cone region of this propeller turbine have highlighted some of the difficulties involved in predicting accurately the flow below the runner in axial machines (Gagnon et al., 2012; Gagnon, Flemming, Qian, Desch Aanes, & Coulson, 2010; Henau et al., 2010; Vu, Koller, Gauthier, & Deschêne, 2011b).

There are, however, few experimental results currently available that allow comparisons in Kaplan turbines, with the notable exception of the Turbine-99 test case (Cervantes, Engström, & Gustavsson, 2005), where the draft tube geometry together with the experimental results have been published describing the flow and the geometry. These experimental results include the velocity inlet boundary conditions immediately after the runner and the velocity profiles in the draft tube cone and diffuser. These results, along with several comparisons using numerical simulations, indicate that the boundary conditions at the draft tube inlet have a large influence on the flow prediction accuracy. Swirling flows that are subject to a strong adverse pressure gradient are difficult to simulate accurately because they are very sensitive. It is important to feed the draft tube with the correct swirl to be able to capture the correct draft tube flow. In addition to the problems associated with the inlet boundary conditions, the flow computation in a Kaplan draft tube is also challenging due to the different flow phenomena appearing simultaneously. The region near the runner cone is of fundamental importance. In this area, a highly unsteady turbulent swirling flow is affected by a strong adverse pressure gradient. Slight variations in the numerical prediction of the flow separation point along the runner cone produces important variations in the flow prediction in the cone region, which may affect the overall performance estimation of the turbine. This problem was investigated numerically by Nilsson (2006), where runner simulations without the draft tube were performed in the Turbine-99 test case. The trailing blade-hub clearance was found to be important in keeping the flow attached to the runner cone because the flow was feeding momentum to the boundary layer. In recent work, Nilsson and Cervantes (2012) have presented unsteady simulations in the Turbine-99 draft-tube using the experimental phase average velocity proposed at the workshop as inlet boundary conditions. From this study, it was concluded that draft-tube alone simulations were very sensitive to the inlet boundary flow resolution, in particular in the region below the hub.

The problems related to this complex flow physics are compounded by the fact that turbulence needs either to be modeled or resolved accurately, and, in the latter case, at an extremely high computational cost. Recently, Galvan, Reggio, and Guibault (2011) investigated the sensitivity of steady-state flow simulations in the Turbine-99 Kaplan draft tube, by comparing several two-equation turbulence models. They concluded that the RNG $k-\varepsilon$ model provided the best agreement with experimental measurements, at the expense, however, of good convergence behavior of the simulations. Similar conclusions were also drawn by Krzemianowski, Banaszek, and Tesch (2011), once again for steady flow simulations. In terms of unsteady flow simulations in Kaplan turbines, most studies have been devoted to the prediction of pressure pulsation at off-design conditions. For instance, Kurosawa and Satou (2006) carried out unsteady flow simulations in a complete Kaplan turbine, but presented no comparison between experimental and simulated velocities. Similarly, Liu, Shao, Wu, and Wu (2008) focused on the comparison of fluctuating pressure signals at the draft-tube wall.

In the current paper a Kaplan turbine model known as Porjus U9 is investigated numerically. Recently, Mulu, Cervantes, Vu, Devals, and Guibault (2011) have investigated the influence of the boundary conditions at the draft tube inlet on the flow prediction accuracy downstream of the Porjus U9 runner using a detailed comparison of the steady RANS simulations and the experimental results at the best efficiency point. The influence of the mesh resolution and turbulence model were investigated, as well as the impact of the flow quantity averaging process at the draft tube inlet. Two types of computations were carried out: the first involved the draft tube alone, and the second computation used stage averaging between the runner and the draft tube. The RANS two-equation eddy-viscosity turbulence models ($k-\varepsilon$, RNG $k-\varepsilon$ and SST) and the Reynolds stress models (SSG and BSL) were used in the study. The computations of the draft tube with the experimental inlet boundary conditions demonstrated the difficulty of the two-equation models to capture the draft tube cone central vortex that is produced by the runner cone. Two-equation turbulence models predicted an early separation on the runner cone. The Reynolds stress models (RSM) performed satisfactorily and demonstrated the anisotropy of the flow. As for the stage calculations results, they gave the general characteristics of the draft tube flow and were found to be surprisingly insensitive to the turbulence model used. All turbulence models predicted an early separation on the runner cone and over-predicted the size of the draft tube cone central vortex below the runner cone. Furthermore, the effect of the blade wake and the hub-blade clearance were not accurately predicted. Further downstream, the numerical results better matched the experimental results because of numerical diffusion. These results and other investigations using the Porjus U9 draft tube carried out by Petit, Mulu, Nilsson, and Cervantes (2010), Petit, Nilsson, Vu, Manole, and Leonsson (2008), have allowed to determine that unsteady analysis of the turbine, including the guide vanes, the runner and the draft tube, are needed to better capture the complex flow behavior below the runner cone. To the best of the authors' knowledge, up to now, no work has been devoted to the comparison of fully coupled unsteady flow simulations in a Kaplan runner and draft-tube with experiments and the analysis of the near hub velocity field in the presence of jets generated by near-hub blade gap.

This paper presents the URANS simulation results and a detailed comparison with the experimental results of the U9 model turbine at the best efficiency operating point. The computational flow domain includes the guide vanes, the runner and the draft tube. This paper is structured as follows: the next section describes the turbine model, followed by a section describing the numerical models and the evaluation method used. Section 7 presents the results and comparisons, and the paper ends with the discussions and conclusions.

3. Test case

The experimental investigation was performed in a Kaplan turbine model known as Porjus U9. The measurements were made in a draft tube that was built at a 1:3.1 scale model of the prototype turbine. The experiment was conducted under a constant runner blade angle and at the three following loads: the best efficiency point (BEP) and two off-design operating points (part load and high load). The runner angular velocity was 696.3 rpm. The velocities were measured with a two-component LDA technique (Mulu & Cervantes, 2009).

Four Plexiglas windows were used to perform the LDA measurements. The windows were placed at an angular positions a, b, c, and d, with a 90° spacing around the cone circumference. At each angular position three locations (sections I to III) along the vertical direction were investigated. At angular positions a and d, two additional profiles were measured perpendicular to the cone axis (BC in Figure 1), which were used as inlet boundary conditions for the numerical simulations. A detailed description of the test rig and the experimental results can be found in Mulu, Jonsson, and Cervantes (2012) and Jonsson, Mulu, and Cervantes (2012).

The pressure measurements were performed using membrane-type pressure transducers with an accuracy of 0.1%. Several pressure sensors were installed along the draft tube. Twenty taps were mounted in the conical diffuser at four angular positions circumferentially, i.e., at each angular position equally spaced five taps were installed in the vertical direction (Figure 1). The angular positions were the same as the LDA measurement sections (a–d). A total of 13 pressure taps were mounted on the draft tube elbow; seven taps along the inner radius and six taps along the outer radius. A reference pressure tap was located near the outlet of the straight diffuser. L^* is

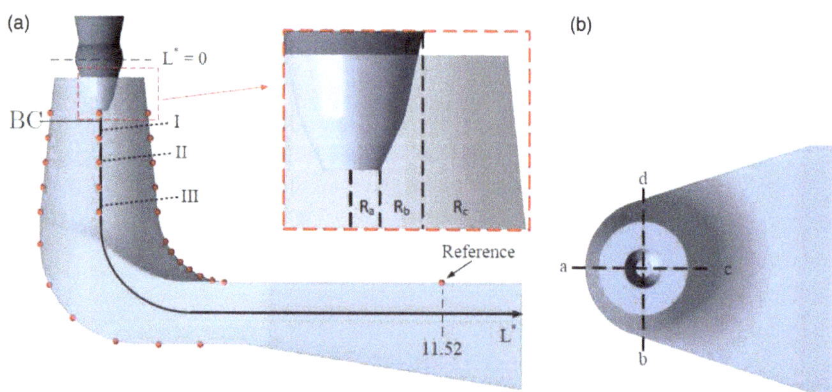

Figure 1. U9 draft tube with measurement locations. (a) The positions for the pressure taps are marked by red dots. In this paper, the pressure taps are numbered p1–p5 in the cone (starting from the top) and p6–p11 and p6–p12 along the outer and inner radius of the elbow, respectively. Sections I to III correspond to the LDA sections. (b) Circumferential positions for the pressure and LDA measurements in the draft tube cone.

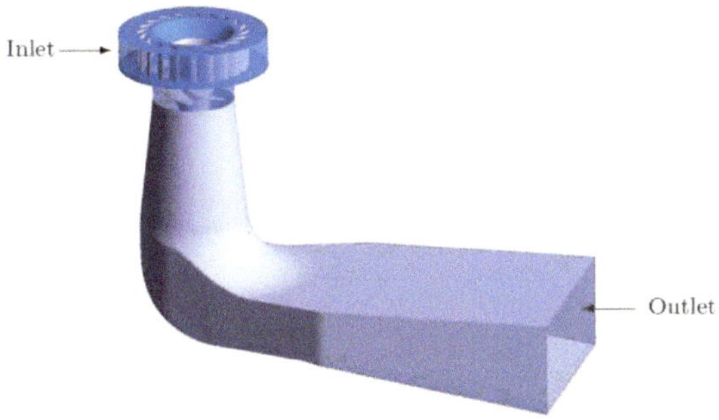

Figure 2. The computational domains, i.e., the guide vanes, the runner blades and the draft tube, and the boundary conditions.

the normalized center line used to express the positions of the measurements (Figure 1). The pressure measurements were performed by Jonsson et al. (2012).

4. Numerics

In the current study, a transient rotor–stator simulation is performed with the entire turbine assembly (Figure 2). Such a configuration allows for capturing the effect of unsteadiness in the simulations and should be more accurate than previous stage simulations, which neglect the effect of unsteadiness.

4.1. *Flow solver*

The commercial software ANSYS-CFX-13.0 was used in this study. All computations were performed using the high-resolution discretization scheme for the continuity equation and the advection term in the momentum equation. The high-resolution scheme has a blend factor that varies between second and first-order accuracy as a function of the variable gradients in the flow regions to avoid non-physical oscillations. First-order upwind scheme was used for the advection terms in the turbulence equations. Because, for some of the turbulence models (RNG $k-\varepsilon$ and SST-SAS) used, the solution did not converge properly with high resolution. Since the current work compares different turbulence models it was decided to use the same setup for all cases. The $k-\varepsilon$ model simulations were performed with the high resolution (bland factor $\beta = 0.75$) and first-order upwind schemes for the advection terms in the turbulence equations. However, any significant difference in the results was not observed. The transient term in the Navier-Stokes equations was solved using a second-order backward Euler scheme. This scheme is implicit and conservative in time and, therefore, does not require a Courant number stability limit. However, to achieve accurate resolution of the time behavior of the flow, and for numerical stability reasons as well, the time step needs to be small enough to resolve the relevant details. The time step was set to 0.0002 s, which corresponds to approximately 1° of revolution of the runner, for a total simulation time of 10 s.

4.2. *Boundary conditions*

In low-head hydraulic turbines, there is a strong interaction between the runner and the draft tube. It is necessary to introduce this interaction into the simulations to achieve accurate results. During the Turbine-99 workshop, Cervantes and Engström reported the importance of the runner cone region for the flow simulation in the draft tube (Cervantes et al., 2005). They also noted the importance of the runner cone in estimating the pressure recovery in the Kaplan elbow draft tube.

Generally, the entire turbine assembly should be used for the CFD simulation of the flow in water turbines. However, using the entire turbine assembly might require excessive computer resources and simulation time. Additionally, it is nearly impossible to generate one complete good-quality mesh for the entire turbine. The flow domain is usually divided into several flow domains that are connected through interface connections. As the flow in the distributor is accelerating and is well defined in a 345° wrap angle spiral casing, the guide vanes region represents a satisfactory inlet region. The flow domains used in the current investigation are presented in Figure 2.

There are three standard techniques to simulate rotor–stator: the frozen rotor, the stage interface and the rotor–stator interface. In this work, unsteady simulations are performed and a transient rotor–stator interface is used. The frozen rotor interface calculation is used to initialize the simulations performed with the rotor–stator interface because it is numerically robust and computationally inexpensive. Previously, Mulu et al. (2011) investigated the Porjus U9 draft tube with a stage interface. Because the circumferential effect of the quantities are filtered through the averaging process using a stage approach, the use of a single guide vane and runner blade flow channel was possible. This reduced the computational domain substantially. The authors noted that the stage interface approach gives a general understanding of the draft tube flow; however, this approach fails to accurately predict the effect of the blade wakes and the hub-blade clearance. Furthermore, the turbulence models predicted an early separation on the runner cone (Figure 9). The flow near the runner cone is equivalent to a bluff body flow. RANS prediction of such flow is limited as RANS describes flows in a statistical manner, i.e., quasi-periodic large scale and chaotic small scales features cannot be distinguished (Breuer, Jovicic, & Mazaev, 2003). Nevertheless, the different models used predict differently separation. Figure 3 presents the shroud-blade and hub-blade clearances for the chosen blade tilt angle. The shroud-blade clearance at the leading and trailing edge is 2.25 and 1.25 mm, respectively; the minimum is 0.3 mm at the middle of the blade. For the hub-blade clearance the minimum is 0 mm and the maximum is 3 mm.

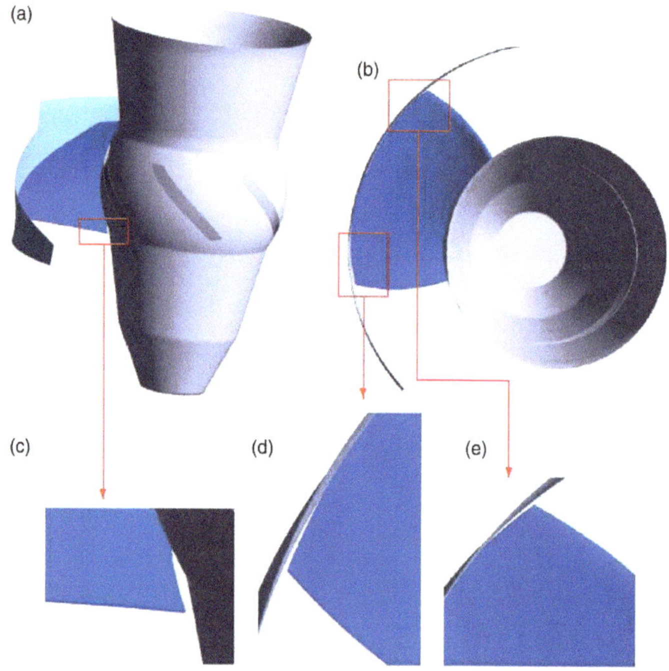

Figure 3. Shroud-blade and hub-blade clearances used in the simulation.

Fully conforming multiblock hexahedral meshes where generated for all components, including gaps in the runner part. A grid independence study was performed with steady state simulations to evaluate the effect of grid size on the solution. The pressure recovery

$$Cp_{cfd} = \frac{\frac{1}{A_{out}} \iint\limits_{A_{out}} PdA - \frac{1}{A_{in}} \iint\limits_{A_{in}} PdA}{\frac{1}{2}\rho\left(\frac{Q}{A_{in}}\right)^2} \tag{1}$$

And loss factor

$$\varsigma = \frac{\iint\limits_{A_{in}} \left(P + \rho\frac{|U|^2}{2}\right) U.\hat{n}dA - \iint\limits_{A_{out}} \left(P + \rho\frac{|U|^2}{2}\right) U.\hat{n}dA}{\left|\iint\limits_{A_{in}} \rho\left(\frac{|U|^2}{2}\right) U.\hat{n}dA\right|} \tag{2}$$

were the two quantities investigated because they are directly linked to the pressure and velocity variations in the flow domain. The performance of the draft tube is usually quantified with these two parameters as well when using CFD. The size of the three selected meshes and the value of the engineering quantities investigated are presented in Table1. In addition, the mean velocity profiles, at the location where the experimental measurements were performed, were compared for the three meshes and showed no significant differences. For the unsteady simulations the fine mesh was used.

The resolution of the stage interface was studied with the parameter "GGI stage groups max", which determines the smallest size for any given span range. Three different setups were used to study the sensitivity of the radial discretization: first without defining the "GGI stage groups max" parameter and then two values, 50 and 100, were used. When the parameter is set to a large number, the stage bands will be fine near the boundary layer regions where the mesh is very fine and denser where the mesh is coarser. However, in the current study varying this values shows no significant difference, therefore "GGI stage groups max" = 100 was used.

In the current study, the complete geometry of the guide vanes and the runner is used for the simulation together with the elbow draft tube. The three domains are connected to each other with a general grid interface. The mesh density used for the entire assembly is composed of approximately 17 million hexahedral cells. The draft tube mesh was generated with ICEM and the guide vane and the runner meshes were generated with ANDRITZ hydro/Poly

Table 1. Mean pressure recovery (Cpcfd) and loss factor (ς) obtained from the three different mesh sizes.

Grid size [millins]	17	8.9	3.9
Cp_{cfd}	0.90784	0.9078	0.9077
ς	0.1403	0.1416	0.1433

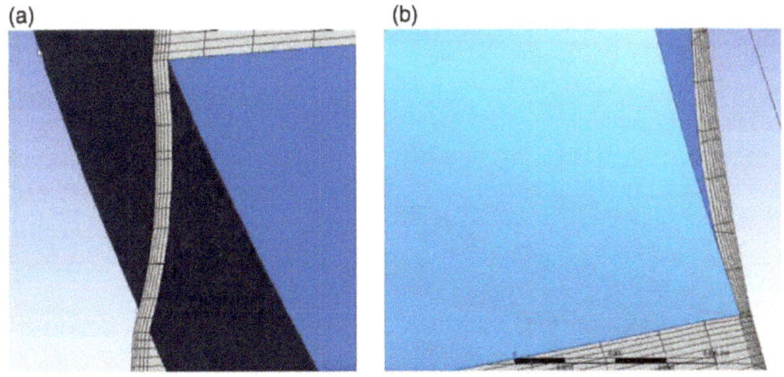

Figure 4. Mesh density at (a) the shroud and (b) the hub clearance, the number of cells in the clearance varies between 5 and 10.

Table 2. Mesh properties used in the simulations.

	Cell	Face angle (min) [deg.]	Aspect ration	y+ (min/avg/max)
Guid vane	7485100	14	483	1.2/83.2/421.8
Runner	5926290	9	138	3.0/98.8/578.3
Draft tube	3700000	31	372	4.4/41.3/93.2

in-house code. In both methods an o-grid technique is used. Figure 4 presents the mesh resolution around the runner blades, hub and shroud. In the current study the number of cells used in the clearance at hub and shroud varied between 5 and 10. The mesh density and properties, such as the minimum angle, the aspect ratio and the y^+ values, are presented in Table 2. At the inlet, the mass flow rate with a flow direction is imposed. Previously, Mulu et al. (2011) performed a sensitivity study on the turbulence intensity for the draft tube and the stage simulations with different intensity values (1%, 5% and 10%). They found that the turbulence intensity did not play a significant role. Therefore, a turbulence intensity of 5% at the inlet was used in the current study. All walls were assumed to be smooth, and a no-slip wall condition was imposed for all solid surfaces. The wall velocity for the runner hub in the draft tube is set to 696.3 rpm. An angular velocity was imposed to the complete runner assuming no-slip boundary at the runner-fluid interface. At the outlet, an average static pressure over the entire outlet was specified to be 0 Pa. The transient solution at a given time step was considered converged when the value of the root mean square (RMS) residuals for each equation solved was less than 10^{-5}.

4.3. *Turbulence models*

The governing equations are solved using the Reynolds average method and assuming an incompressible and isothermal fluid. This method is based on a statistical approach where time-averaging of variables is carried out to separate the mean from the fluctuations. The unsteady Reynolds averaged Navier-Stokes continuity and momentum equations are given as:

$$\frac{\partial \bar{u}_i}{\partial x_i} = 0 \tag{3}$$

$$\frac{\partial \bar{u}_i}{\partial x_i} + \bar{u}_j \frac{\partial \bar{u}_i}{\partial x_j} = -\frac{1}{\rho}\frac{\partial \bar{p}}{\partial x_i} + \nu \frac{\partial^2 \bar{u}_i}{\partial x_j^2} - \frac{\partial (\overline{u_i' u_j'})}{\partial x_j} \tag{4}$$

where $\overline{u_i}$ are the time-averaged velocity components, $\overline{p}$ is the time-averaged pressure, ρ is the fluid density, ν is the fluid kinematic viscosity, t is the time and u are the fluctuating velocity components.

As a result of the Reynolds averaging, in Equation (4), there is an additional stress term acting on the mean flow due to the fluctuating velocity, which is known as the Reynolds stress ($\tau_{ij} = -\overline{u_i' u_j'}$). This term represents the effect of the turbulence on the mean flow. The Reynolds stress term introduces six unknown variables in the governing equations, which must be modeled. Generally, the RANS turbulence models can be broadly divided into two types: eddy-viscosity models and Reynolds stress models. The eddy-viscosity models are based on the Boussinesq hypothesis, which relates the Reynolds stresses to the mean velocity gradients via the eddy viscosity:

$$\tau_{ij} = -\overline{u_i' u_j'} = \nu_t \left(\frac{\partial \bar{u}_i}{\partial x_j} + \frac{\partial \bar{u}_j}{\partial x_i} \right) - \frac{2}{3}\left(k + \nu_t \frac{\partial \bar{u}_k}{\partial x_k} \right) \delta_{ij} \tag{5}$$

where $k = \frac{1}{2}(\overline{u_i' u_j'})$ is the turbulent kinetic energy, δ_{ij} is the Kronecker delta, and ν_t is the turbulent viscosity (also known as the eddy viscosity) and has to be modeled. In the two-equation eddy-viscosity models, the turbulent properties of the flow are represented with the inclusion of two extra transport equations: one equation for the turbulent kinetic energy, k, and one equation for the turbulent dissipation rate, ε, or the specific dissipation rate, ω. This means that both the velocity and turbulent length scale are solved using separate transport equations.

The flow prediction in the runner and draft tube is very sensitive to the boundary conditions applied and the turbulence model selected (Cervantes, 2003; Menter, 2003).

In this situation, the flow exhibits characteristics (e.g., streamline curvature) that are outside the scope of the experiments for which the turbulence models have been calibrated (Menter, 2003). Different two-equation RANS models were used in the present work to evaluate their capability to reproduce the flow in a Kaplan turbine at the BEP ($k - \varepsilon$, RNG $k - \varepsilon$ and SST) and the scale adapted simulation (SAS) model.

In the standard $k - \varepsilon$ model, the eddy viscosity is related to the turbulent kinetic energy and the dissipation rate using the following assumption: $\nu_t = C_\mu \frac{k^2}{\varepsilon}$, where C_μ is a model constant (Durbin & Reif, 2001). The standard $k - \varepsilon$ model has proven to be robust and relatively accurate for a large number of engineering flows and is widely used in industry because it can be solved relatively fast compared with other turbulence models. However, this model performs poorly for complex flows involving a large adverse pressure gradient, separation, rotation, and a strong streamline curvature. Another major drawback associated with this model is the over prediction of the eddy viscosity near the wall; thus, this model requires special treatment at the wall (Durbin & Reif, 2001; Pope, 2000; Wilcox, 1993). In the present study, scalable wall functions (see Vieser, Esch, & Menter, 2002) were used with the $k - \varepsilon$ model. Wall functions allow not resolving the flow near the wall. It assumes that mesh points near the wall lie in the logarithmic region of the boundary layer, thereby allowing modeling the wall shear-stress using a logarithmic profile assumption. In cases where points are very close to the wall, the logarithmic assumption does not hold, and a limit on the minimum acceptable y_+ is automatically imposed in the computation of the wall shear-stress, making the wall function scalable.

The RNG $k - \varepsilon$ model is based on the renormalization group analysis of the Navier-Stokes equations and attempts to overcome some of the limitations of the standard $k - \varepsilon$ model. The transport equation for the turbulent kinetic energy is the same as the equation in the standard $k - \varepsilon$ model. The major difference between these models is that the model constants of the transport equations for the dissipation rate vary. Additionally, the constant coefficient related to the turbulence production term due to the viscous forces is replaced by a function in the case of the RNG model. This function includes the effect of rapid strain and the streamline curvatures. The RNG model also includes the effect of the swirl on the turbulence, which enhances the accuracy for swirling flows (Wilcox, 1993). Furthermore, the RNG model takes into account the different turbulent length scales (Yakhot, Orszag, Thangam, Gatski, & Speziale, 1992). Scalable wall functions were also used with this turbulence model.

In the near-wall boundary layer, the Wilcox model ($k - \omega$) is more accurate than $k - \varepsilon$. However, the Wilcox model has the disadvantage of a strong sensitivity to the value of ω in the free stream outside the boundary layer (Menter, 1992; Wilcox, 1993). This problem can be overcome by combining the $k - \omega$ in the near-wall region with the free stream independence of the $k - \varepsilon$ model far from the wall, which is the base for the zonal shear stress transport (SST) model. In the SST model, blending functions are used to ensure a proper selection of $k - \varepsilon$ and $k - \omega$ zones. The main advantage of the SST model is its ability to accurately predict the onset and the amount of flow separation under high adverse pressure gradients. This is because the model accounts for the transport effects of the turbulent shear stress (Menter, 1994). Bardina, Huang, and Coakley (1997) have shown the ability of the SST model to predict separation in an adverse pressure gradient. The results obtained by Marjavaara et al. (2005) also confirm the ability of the SST model to perform well in a Kaplan draft tube.

The SST-SAS turbulence model is used for transient cases. The SST-SAS model differs from the URANS SST model by adding an additional scale adapted source term (SAS) in the ω transport equation. The SST-SAS model uses the von Karman length scale for the turbulence scale equation to detect the unsteadiness. For flows with transient behavior, the SAS model dynamically adjusts to resolve the turbulent structures based on the information provided by the von Karman length scale and results in a LES-like behavior in unsteady regions of the flow field. The model provides the standard RANS model in stable flow regions. More details about the SST-SAS model can be found in Egorov, Menter, Lechner, and Cokljat (2010).

The Reynolds stress models solve the transport equations for each Reynolds stress and for the ε or ω that provide the length and time scales of the turbulence (Pope, 2000). The isotropic eddy-viscosity hypothesis is avoided. However, some terms must still be modeled, such as the pressure-strain, the dissipation rate, and the turbulent diffusion. The RSM models have the potential to accurately predict complex flows because they account for the effects of streamline curvature, swirl, rapid changes in the strain rate and anisotropy of the turbulent stress. A drawback of the model is that it demands more computer resources because more equations are solved, and the model is less robust than the linear eddy-viscosity models. Mulu et al. (2011) show that the RSM model performs well in a draft tube simulation without including the upstream geometry. However, in the current study, it was very difficult to achieve convergence using the RSM model. A very small time step was necessary to avoid divergence, and made the simulation convergence impossible in a reasonable period of time, i.e., less than one week using 100 cores.

5. Engineering quantities

There is significant quantity of kinetic energy leaving the runner in reaction turbines, especially in low head turbines. The amount of kinetic energy that is converted to pressure energy at the runner exit defines the performance of the draft tube. The pressure recovery factor, Cp, is a common engineering quantity used to quantify the amount of kinetic energy recovered along the draft tube. Cp is defined as the ratio of the pressure difference at the outlet and inlet of the draft tube to the dynamic pressure of the mean axial velocity at the inlet. In the current study, the performance of the draft tube was quantified by calculating the pressure recovery factor based on the wall pressure (Equation 6) (Gubin, 1973), because this value is experimentally available from Mulu et al. (2012).

$$Cp_w = \frac{P_{wall} - P_{wall\_inlet}}{\frac{\rho}{2}\left(\frac{Q}{A_{inlet}}\right)^2} \quad (6)$$

where P_{wall_inlet} and A_{inlet} are the static wall pressure and the cross-sectional area at position p1 in the cone, P_{wall} is the static wall pressure along the draft tube and Q is the flow rate (Figure 1).

The wall pressure recovery is an estimation of draft tube efficiency because it only considers the wall pressure rather than the average pressure across the whole cross section, thus it doesn't take into account the pressure variation in the radial direction. Cp_w is also based on the average kinetic energy at the inlet, which means that the swirling motion and non-uniform velocity profiles effects are neglected. Experimentally, it is difficult to obtain the average pressure over the cross-section along the radius, therefore, Cp_w is a common parameter used to characterize draft tubes.

At the runner exit there is a residual swirl, depending on the operating conditions and the runner type. Some swirl may be advantageous in the conical diffuser. Swirl can also cause instability that leads to vibration, noise, fatigue failures, power swings, vertical movements of the turbine, and reversed flow at the centre of the draft tube. Hydraulic turbines are usually designed to have an optimal swirl at the exit of the runner when operating at the BEP, which allows high pressure recovery and no flow separation at the draft tube cone wall (Gubin, 1973). The intensity of the swirl is characterized by the swirl number, which is determined by the ratio of the axial flux of the angular momentum to the axial momentum flux times the equivalent radius, as shown in Equation 7 (Senoo & Nagata, 1972).

$$S_w = \frac{\int_0^R UVr^2 dr}{R\int_0^R U^2 r dr} \quad (7)$$

where U is the axial velocity component, V is the tangential velocity component and R is the equivalent exit radius.

6. Data reduction

The experimental velocity measurements were performed using a laser Doppler anemometry (LDA) technique. The data was collected at several points along the radius because LDA is a point-wise measurement method. The time-averaged profile was obtained after several rotations of the runner at a fixed radial position. The phase-resolved velocities were obtained by decomposing the instantaneous velocity with a Reynolds triple decomposition method,

$$u(x,t) = \bar{u}(x) + \tilde{u}(x,t) + u'(x,t) \quad (8)$$

where $u(x,t) = \langle u(x,t)\rangle + u'(x,t)$ is the instantaneous quantity, $\bar{u}(x,t)$is the time-averaged quantity, $\tilde{u}(x,t)$ is the periodic fluctuation quantity and $u'(x,t)$ is the random fluctuation quantity.$\langle u(x,t)\rangle$ is the phase-averaged quantity over the cycles. The velocity measurements were related to the runner angular position using an encoder signal. A circumferential resolution of 3° was used to calculate the time-averaged velocity data. A detailed description of the method can be found in Mulu et al. (2012).

In the case of the CFD, the velocity and pressure information were obtained at every time step for a complete rotation of the runner. The simulation time step corresponds to approximately 1° of runner rotation. The time-averaged velocities in the simulation were obtained in a similar fashion as in the experiments, i.e., the velocity information at a fixed radial position was used to calculate the averaged and angular velocities. Because the experimental values are obtained with a circumferential resolution of 3°, the simulation results are also phase-resolved with the same circumferential resolution. The experimental axial velocity obtained at sections I-III is slightly different from the true axial velocity because the LDA probe was orthogonal to the draft tube cone wall. The same component was extracted from the simulations and is presented in various figures.

For the CFD calculations, the wall pressures are extracted over a period at the same point where the experimental measurements were performed (Figure 1).

7. Results and discussions

The simulated time-averaged and phase-resolved pressures and velocities are presented and compared to the experimental results at different locations in the draft tube. The time-averaged and phase-resolved velocities are non-dimensionalized using the bulk velocity obtained from the flow rate and the area at section I, unless otherwise mentioned. The measured radii are non-dimensionalized with respect to the runner radius. The positive directions for the axial velocity and tangential velocity are defined as vertically down through the draft tube cone and clockwise from the top of the runner, respectively.

7.1. *Engineering quantities*

Figure 5 presents the wall pressure recovery of the conical diffuser obtained experimentally and numerically using four different turbulence models at four angular positions (Figure 1). The wall pressure recovery is defined

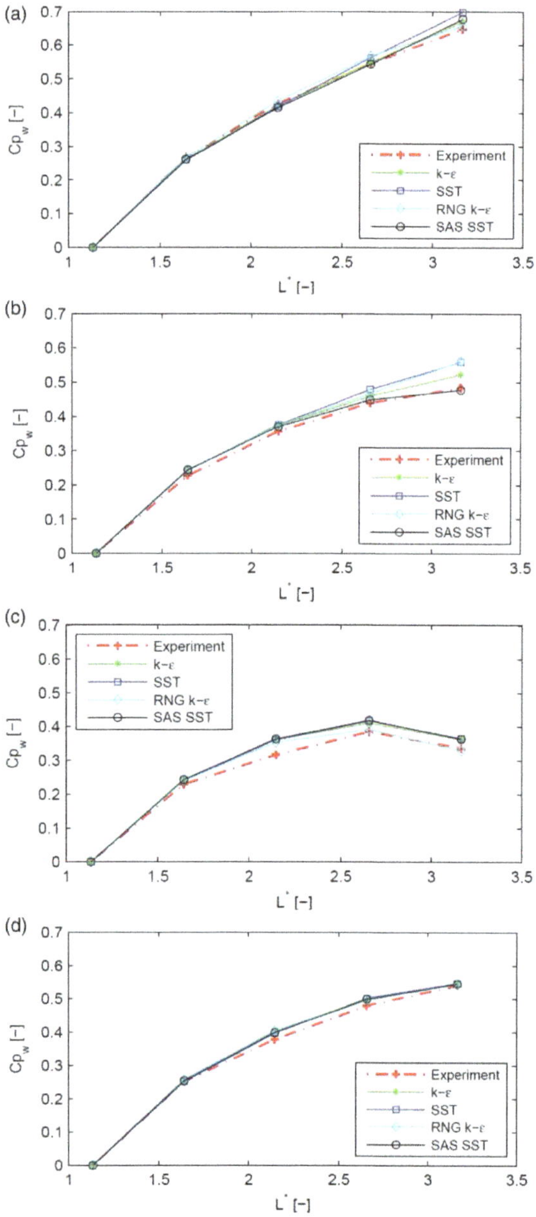

Figure 5. Wall pressure recovery, Cp_W, of the draft tube cone obtained from experimental measurements and CFD calculations (a) Angular position a. (b) Angular position b. (c) Angular position c. (d) Angular position d (Figure 1). L* is the normalized centerline of the draft tube.

by Equation 6. The pressure recovery factor indicates the degree of conversion of the kinetic energy into static pressure, and a higher value indicates a higher efficiency of the region considered. Both the experimental and the numerical pressure recovery Cp_W increase in the downstream direction, which indicate the correct operation of the draft tube cone. The numerical results show similar values of Cp_W at all angular positions compared with the experimental results. From both the experiments and the simulations, the highest pressure recovery is obtained at section a, specifically at the end of the conical diffuser (Figure 5a). This is expected because the flow decelerates towards the outer radius of the elbow, and thus, the static pressure increases. At section c, the recovery is low due to the acceleration of the flow toward the inner radius of the elbow and, as a result, the static pressure decreases (Figure 5c).

The wall pressure recovery is well predicted with all turbulence models in this part of the draft tube. The largest discrepancies between the experimental and the numerical results occur at section b (Figure 5b). At approximately L* = 3.2, the wall pressure recovery is overestimated in the simulations, except when using the SAS model, compared with the experiment.

Figure 6 shows the wall pressure recovery Cp_W along the upper and lower paths in the draft tube for the four turbulence models together with the experimental results. The results from both the experiments and the simulations show that Cp_W along the upper path is low at the beginning of the inner radius of the elbow because the flow accelerates in that area. Then, the pressure recovery increases along the radius from L* = 4 to L* = 7 (Figure 6b). Conversely, Cp_W along the lower path continues increasing up to L* = 5 and starts to decrease along the elbow radius. In the elbow of the draft tube, all turbulence models predict the same behavior as the experiments with slight differences between each model. These differences may be attributed to how each turbulence model handles the strong streamline curvatures.

Figures 5 and 6 indicate that most of the pressure recovery occurs in the conical diffuser in both the experiments and the simulations. From the results presented, all the investigated turbulence models did not experience any difficulty predicting the pressure distribution along the draft tube.

Figure 7 presents the flux of angular momentum ($FAM = \int_0^R UVr^2 dr$) obtained from the experiments and the RNG $k - \varepsilon$ turbulence model. The results from the other turbulence models are similar to that of the RNG $k - \varepsilon$ model; therefore, the results are not presented here. The flux of angular momentum increases from the centre to the draft tube cone wall in both the experimental and the simulation results. The flux of the angular momentum is conserved, i.e., the tangential velocity is nearly unaltered by the diffuser, cf. Figure 8. The experimental and the numerical results are similar from the centre until r' = 0.75. However, there is a difference close to the draft tube cone wall.

Swirl plays a vital role in the transformation of kinetic energy into pressure energy in the draft tube cone, because it can prevent flow separation at the cone wall. The swirl numbers calculated, using Equation 7, from the experiments and the simulations are presented in Table 3. The swirl numbers obtained from each turbulence model at

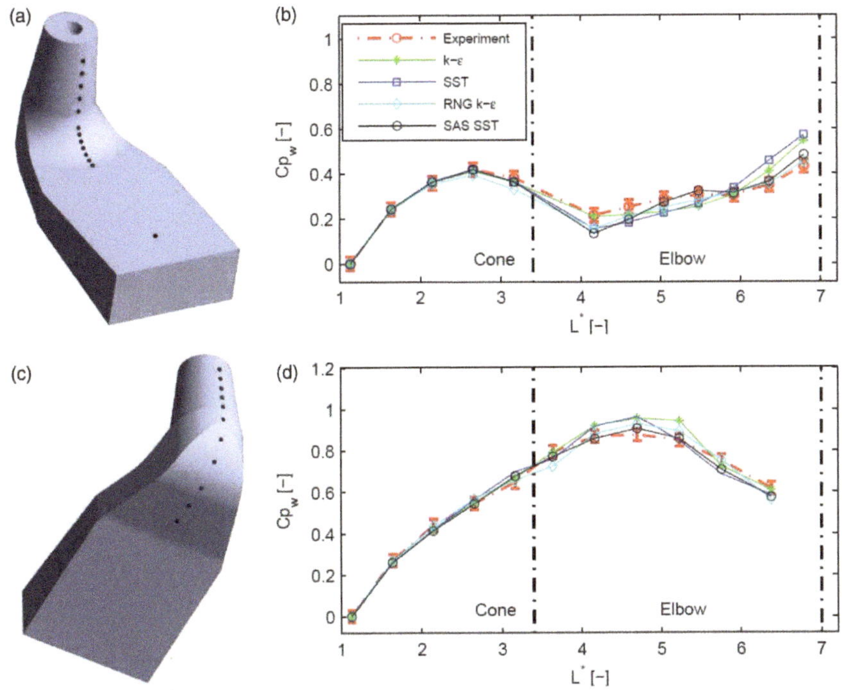

Figure 6. (a) Positions along the upper sensor path in the draft tube cone and elbow. (b) Wall pressure recovery along the upper path. (c) Positions along the lower sensor path in the draft tube cone and elbow. (d) Wall pressure recovery along the lower path. L* is the normalized centerline of the draft tube.

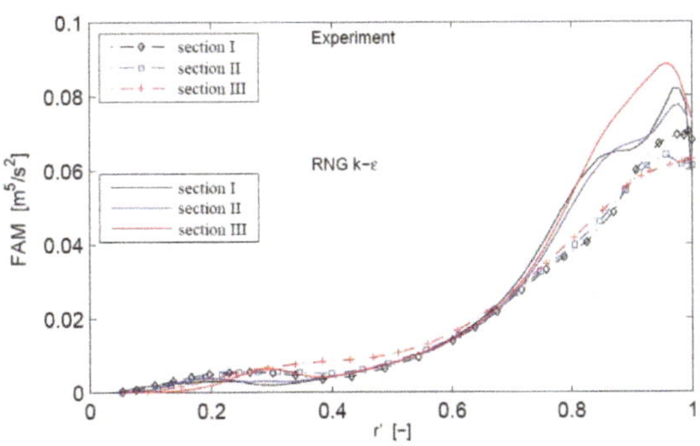

Figure 7. Experimental and numerical (RNG $k-\varepsilon$ model) angular momentum flux at sections Id, IId, and IIId. The radius is normalized by the radius at each section.

each measurement section are approximately equal to the experimental values. However, there are small differences between the results of the different turbulence models. The $k-\varepsilon$ model seems to predict the swirl number best.

7.2. *Time averaged quantities*

The time-averaged velocity profiles obtained using the unsteady state computations are compared to the experimental data at three different sections (I, II, and III) in the draft tube cone below the runner hub.

As shown in Figure 8, the results from the four turbulence models ($k-\varepsilon$, RNG $k-\varepsilon$, SST and SAS-SST) are compared with the experimental data. The CFD results fit well with the measurement in the region R_c for all turbulence models used. This region corresponds to the blade wake (hub-shroud) where the effects of the viscosity are negligible. Some difference appears between the experimental and the numerical results of the axial velocity near the wall at section I, especially for the RNG and $k-\varepsilon$ models. Flow at the draft tube wall is sensitive to the tip vortex developing at the blade-shroud clearance (Breuer et al., 2003). The bump at $r^* = 0.9$ is characteristic. The axial velocity is somehow overestimated in measurement section III. The major differences appear in the regions R_a and R_b for the axial and tangential velocity at section I, II and III. In measurement sections I and II, the $k-\varepsilon$ and RNG $k-\varepsilon$ models overestimate the axial and tangential velocity components in region R_a, whereas the SST and SAS-SST models underestimate these velocities. In section III, all turbulence models underestimate the axial and tangential velocities.

For this application, the unsteady computation yields a significant improvement over the steady state stage computation, especially in section I, as shown in Figure 9 (Mulu et al., 2011). In the previous stage simulations, the computational flow domain comprised a single guide vane and runner blade flow channel together with the draft tube.

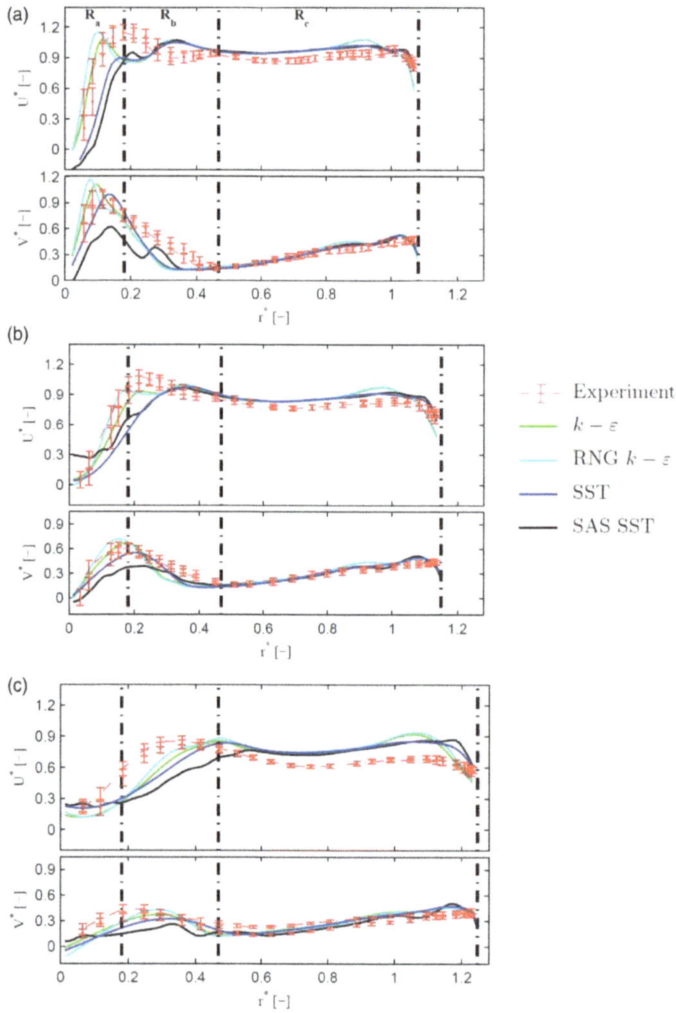

Figure 8. Experimental and simulated mean axial (U*) and tangential (V*) velocities (a) Section I, (b) Section II and (c) Section III. The experimental results are presented with bars representing the velocity fluctuations at angular position d.

Table 3. Experimental and numerical swirl numbers.

Sw	Id	IId	IIId
Experiment	0.255	0.277	0.332
$k - \varepsilon$	0.255	0.277	0.316
RNG $k - \varepsilon$	0.254	0.272	0.294
SST	0.253	0.273	0.231
SAS SST	0.250	0.271	0.300

However, in the present transient rotor–stator simulation, the entire turbine flow domain was used. The simulation results obtained with the $k - \varepsilon$ model are similar to that of the RNG $k - \varepsilon$ model; therefore, only the $k - \varepsilon$ results are presented.

In the stage computation, the flow recirculation occurs early in the hub region, which explains why the axial and tangential components are underestimated in the R_a and R_b regions for sections I, II, and III. In both Figures 8 and 9 there is a difference between the experimental and numerical results between R_a and R_b. For the stage simulation due to the early separation the maximum velocity is observed in R_b. During the transient rotor–stator simulation, there is no early separation. The velocity between R_a and R_b decreases and then increases in R_a. This behavior is attributed to the wakes created by the jets and blade channels interaction where the velocity decreases within the wakes.

The draft tube entrance is a double diffuser; the area increases outwards and inwards. In order to ensure an attached flow to the draft tube cone wall, some swirl must still be present in the flow leaving the runner. To keep the flow attached to the runner cone, the required conditions are different. The runner cone rotates faster than the fluid and thus increases the angular momentum of the fluid leaving the runner. The condition is not appropriate

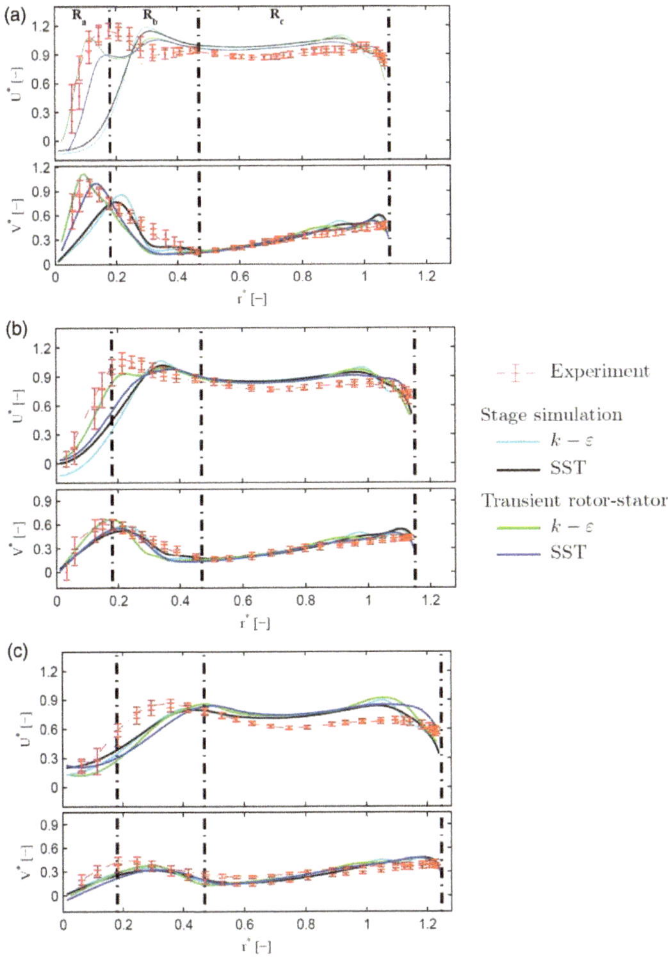

Figure 9. Experimental and simulated mean axial (U*) and tangential (V*) velocities (a) Section I, (b) Section II and (c) Section III. The experimental results are presented with bars representing the velocity fluctuations at angular position d. The stage simulations were performed with one guide vane flow channel, one runner blade flow channel and the draft tube. The transient rotor-stator simulations were performed with the entire turbine assembly, i.e., twenty guide vanes, six runner blades and the draft tube. A detailed description of the stage simulation can be found in Mulu et al. (2011).

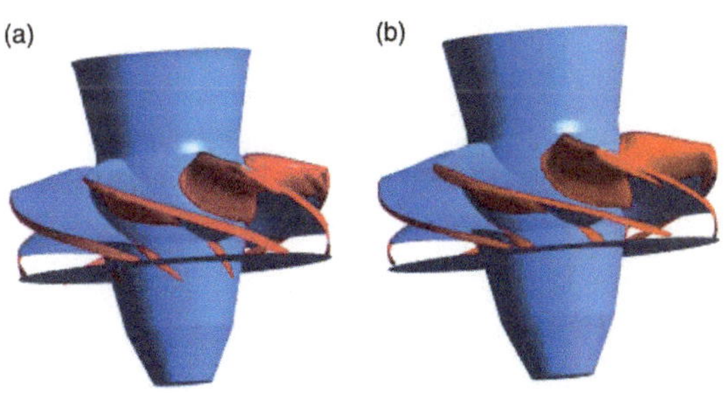

Figure 10. Isosurface of the axial velocity in the stationary frame (U = 6.5 m/s) in the runner and draft tube regions. The disk below the runner blades represents the interface between the regions. (a) unsteady results of the complete rotor–stator simulations and (b) results of the stage simulation.

to keep the flow attached to the runner cone. To avoid separation, the boundary layer needs to be energized; this is done with a special clearance on the blade allowing the formation of a jet in the runner cone boundary layer, see Figure 10a.

The stage calculation averages the flow velocity upstream the interface. The jet thus disappears from the calculation at the runner–runner cone junction, see Figure 10b. The jet absence promotes early separation on the runner cone explaining the main difference between the transient and stage simulations.

Figure 11 shows a comparison of the computed and measured mean flow angle at section Id, IId, and IIId. The experimental flow angle is compared with those obtained with the $k - \varepsilon$, the RNG $k - \varepsilon$ and the SST model. All turbulence models demonstrate similar results except the SST model in region R_a at section IIId. The numerical simulations are unable to capture the correct angle towards the end of the draft tube cone. The main reason is an under-estimation of the axial velocity in the vortex core leading to an over-estimation of the axial velocity in region R_c.

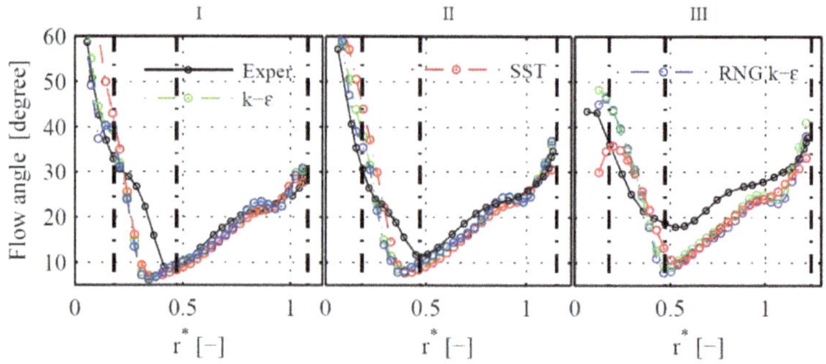

Figure 11. Flow angle, i.e., arctan (*V/U*), at sections Id – IIId, from the experiment and the simulations.

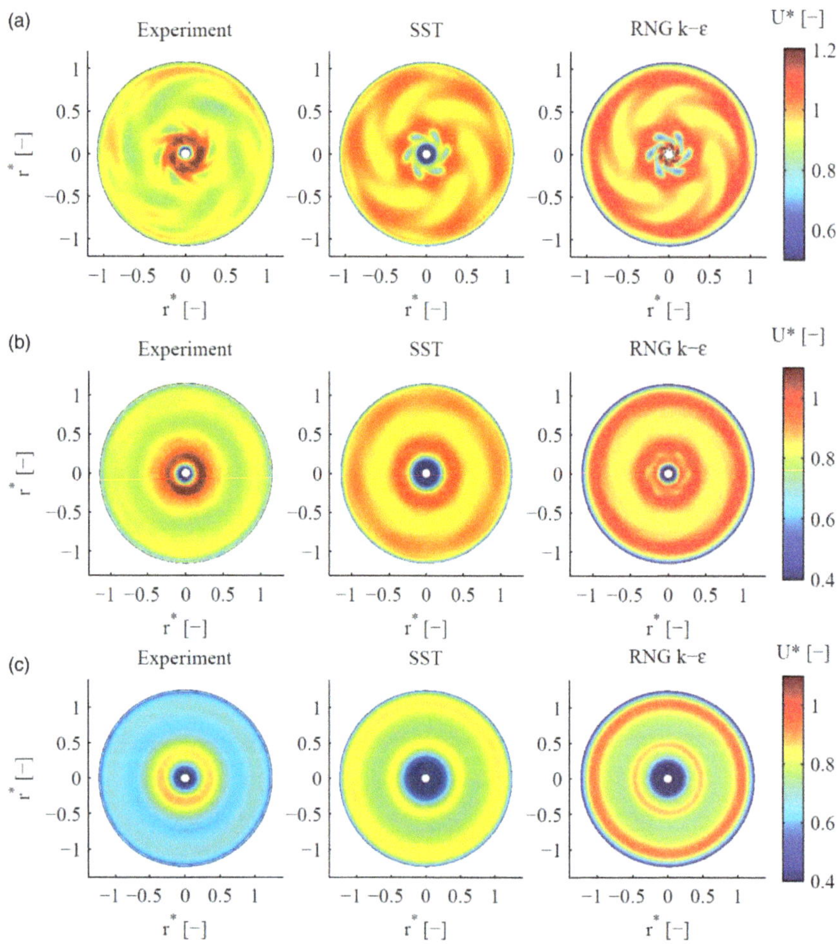

Figure 12. Phase-resolved axial velocity for one runner revolution. (a) Section Id (b) Section IId and (c) Section IIId.

7.3. Phase-averaged quantities

The experimental and numerical results of the phase-resolved axial and tangential velocities are presented in Figures 12 and 13, respectively. The results of two turbulence models, SST and RNG $k-\varepsilon$, are presented since they are representative of the results obtained using the other turbulence models.

In Figure 12, the blade wakes, which are low axial velocity regions, are clearly discernible at section I for both the experiments and the simulations. The blade wakes signatures are, however, more pronounced in the simulations than in the experiments, where they are smeared out and join each other. The velocity is slightly lower below the blades in the experiments compared with the simulations,

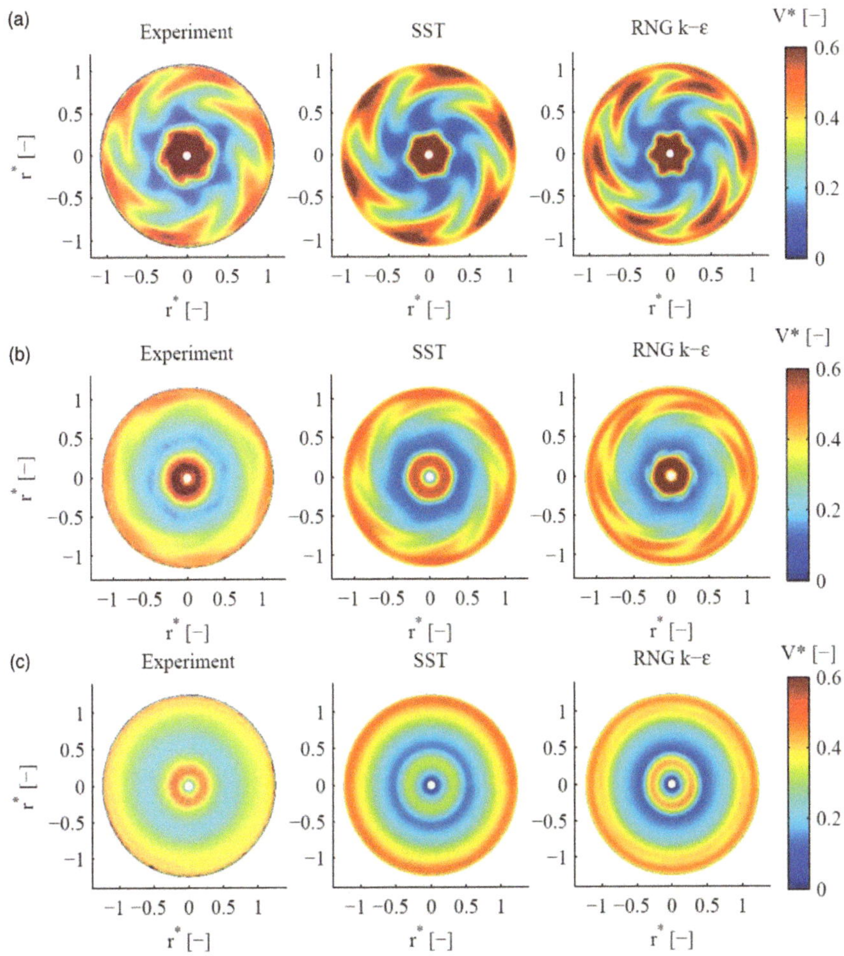

Figure 13. Phase-resolved tangential velocity for one runner revolution. (a) Section Id (b) Section IId and (c) Section IIId.

because higher velocities are obtained below the runner cone. The main difference between the SST and the RNG $k-\varepsilon$ turbulence models is found below the runner cone.

The small wakes that were counter rotating below the runner cone due to the blade-hub clearances, i.e., due to the flow jet coming out of the blade-hub clearances with high velocity, are also captured in the simulations, see Figure 14. However, the large axial velocity beneath the runner cone is not captured by any turbulence model. The reason is attributed to the inability of the turbulence models to predict flows with a strong streamline curvature (Menter, 2003). Figure 14 presents the swirling strength of the vortex core region which represents the strength of the local swirling motion. The main flow domain rotates clockwise except for the portion of the flow leaving through the blade-hub clearance, where the swirling strength shows counter-clockwise vortices.

At section II and III, the blade wakes have disappeared in both the experiments and the simulations. The results are now similar to the mean results presented in Figure 8. The Figures point out an overestimation of the velocity below the blades and an underestimation below the runner cone

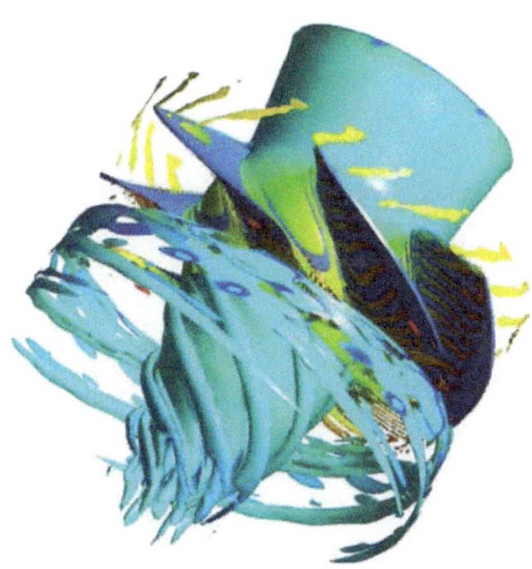

Figure 14. Swirling strength of the vortex core region.

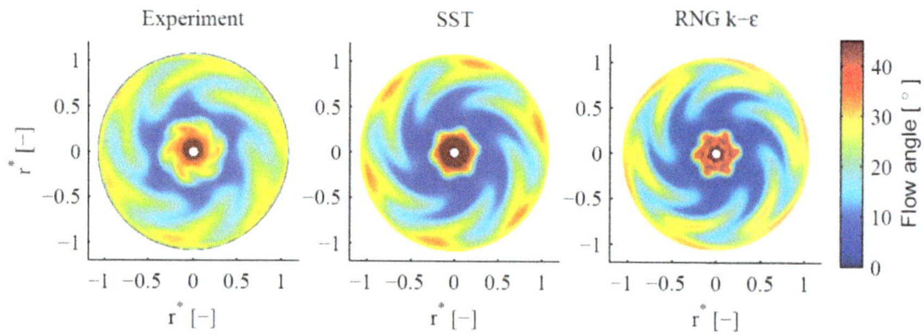

Figure 15. Phase-resolved flow angle at section Id for one runner revolution.

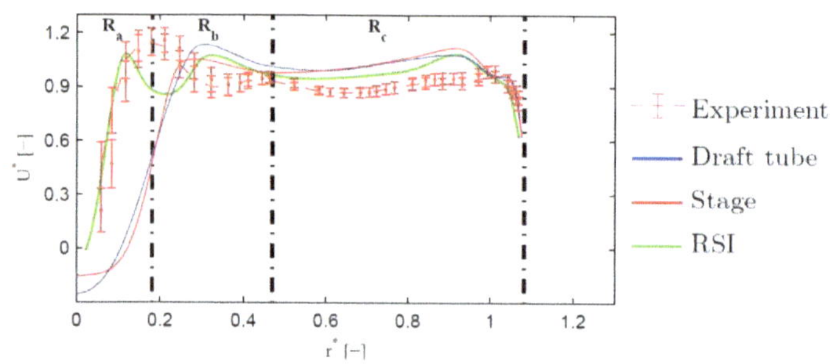

Figure 16. Axial velocity profile at section Ia; experimental and numerical results with $k - \varepsilon$.

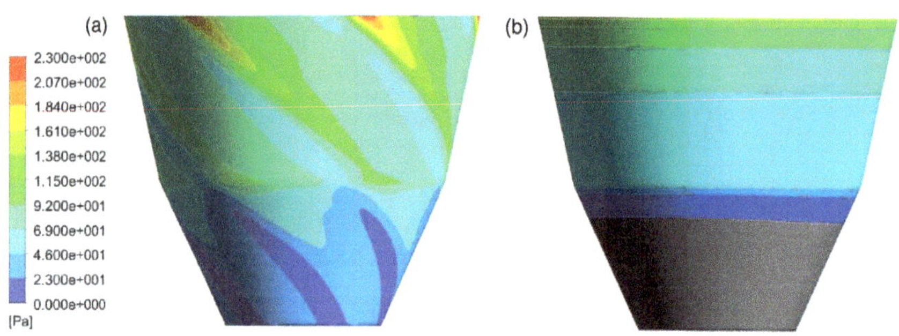

Figure 17. Axial wall-shear stress on the runner cone for the (a) RSI and (b) draft tube simulations with $k - \varepsilon$. The grey color indicates a negative wall shear stress, i.e., separation on the runner cone.

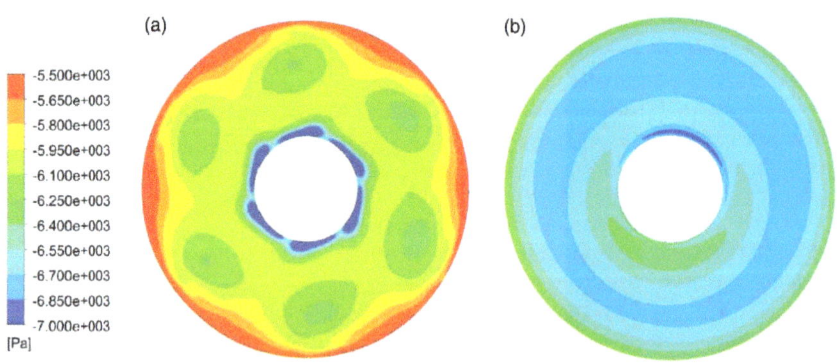

Figure 18. Pressure on a plane normal to the runner cone for the (a) RSI and (b) draft tube simulations. The plane is situated just below the circle delimiting the runner cone in 2 regions, see Figure 17.

compare to the experimental results. The results are similar for both turbulence models.

The phase-resolved tangential velocity results obtained from the simulations are in better agreement with the experiments than for the axial velocity (Figure 13). The blade wakes are represented by high velocity, dark red regions. The amplitudes are similar, independent of the turbulence model used with few noticeable differences. The blade velocity deficit, which is the dark blue regions, near the hub is larger in the simulations. Furthermore, the higher tangential velocity spots are the blade trailing edges near the draft tube cone, which are difficult to discern in the experiments. This difference may be attributed to an insufficient number of samples in each bin in the experiment, which may affect the phase value.

At section II, the blade wakes are still present in the simulations and are function of the turbulence model used. Furthermore, the amplitude of the tangential velocity at section II is higher in the simulations, similarly to section III, where all wakes have disappeared.

The phase-resolved flow angle at section I is presented in Figure 15. The results at the other two sections are not presented because they are similar to the graph presented in Figure 11. The lower flow angles, blue region, are found between the blade wakes where the flow is nearly axial. In the blade wake regions, high shear induces large tangential velocity and flow angles. These three figures are found to be similar with a smaller angle in the inter-blade wakes for the simulation.

In order to quantify the effect of the jets, the velocity and turbulent quantities were extracted from the transient simulations (RSI) at the runner-draft tube interface. Numerical simulation of the draft tube alone was performed with the circumferentially averaged quantities from the transient simulations as inlet boundary conditions. The axial velocity profiles at section Ia for the RSI, stage and draft tube simulations are presented in Figure 16. The stage and draft tube simulation results similarly point out the specific effect of the jets. An early separation appears on the runner cone without the jets. The circumferential distribution of the energy affects the flow near the runner cone.

The axial wall-shear stress on the runner cone is presented in Figure 17 for the RSI and the draft tube. The effect of the jets is significant on the runner cone. Higher wall-shear stress is obtained locally. The flow is completely attached until the end of the runner cone for the RSI case. The gray color on the runner cone of the draft tube simulations signifies a negative wall-shear stress, i.e., separation.

The pressure on a plane normal to the runner-cone is presented in Figure 18 to illustrate better the phenomenon. The high-velocity jets near the runner cone create a lower pressure region for the RSI simulation, see the dark blue region. The lower pressure allows bending the streamlines toward the runner-cone, avoiding separation on the runner cone despite the centrifugal force imposed to the fluid. The pressure contour plot for the draft tube simulation is different. Lower gradient is captured in the pressure variation due to the circumferential averaging of the inlet quantities. Furthermore, a high pressure region near the runner cone is captured leading to an earlier separation.

8. Conclusion

URANS simulations of a Kaplan model were performed at the BEP. The numerical domain comprised the guide vanes, the runner and the draft tube and was discretized with a high-density mesh. The simulations were performed for four different turbulence models: $k-\varepsilon$, RNG $k-\varepsilon$, SST and SAS SST.

A comparison of the engineering quantities, the mean velocity profiles and the phase-resolved velocities obtained from the simulations with the experimental results show the ability of such simulations to capture the main draft tube flow characteristics. A good agreement was globally obtained independently of the turbulence model used. The methodology used, transient rotor–stator simulation, is a step forward compared to previously presented stage simulations.

The numerical wall pressure results presented the best agreement with the experiments. Some discrepancies appeared at the draft tube elbow inlet.

The numerical mean velocity profiles are similar in the blade wake region. The numerical results slightly overestimate the axial velocity, while the tangential velocity is well predicted and matched the experimental results. The main discrepancies appear below the runner cone. The experimental bump in the axial velocity is not captured by any turbulence model. The $k-\varepsilon$ and RNG $k-\varepsilon$ models best fit the results, while the SST and SAS SST models clearly underestimate the axial velocity. Further down the draft tube, the discrepancy between the experiments and the numerical results is amplified. The discrepancy may be attributed to the difficulty in predicting flows with strong streamline curvature for the turbulence models used.

The blade wake signatures are clearly visible in the numerical phase-resolved axial velocity at the inlet of the conical diffuser. These wakes are mixed and dissipated further downstream, as in the experimental results. The blade wake signatures are more pronounced for the simulations than for the experiments. Beneath the runner cone, the main effect of the blade-hub clearance is captured by the simulations. However, all turbulence models have difficulties predicting the high axial velocity region produced by the blade-hub clearance to ensure an attached flow to the runner cone. Independently of the turbulence model used, the phase-resolved tangential velocity results are in better agreement with the experiments when compared with the axial velocity. The hub clearance plays an important role in Kaplan draft tube flow. Accurate simulation of this region is necessary to obtain physically representative results. A mesh sensitivity analysis with various turbulence models is necessary to quantify the exact contribution of the hub clearance modeling to the final numerical results.

The turbulence models are able to predict the behaviour of the free and forced vortexes with the exception of the SAS SST model. The $k - \varepsilon$ and RNG $k - \varepsilon$ models underestimate the effects of blade tip clearances, while the SST model captures this effect much better. The reason for this is attributed to its ability to perform better under a strong adverse pressure gradient. Even if the RNG $k - \varepsilon$ model accounts for the effect of the rapid strain, the strong streamline curvature and the swirling motion (vorticity), the numerical results demonstrated no significant gain over the $k - \varepsilon$ model. The SAS SST model performed poorly in predicting the phase-resolved velocity compared with the other models. The blade-shroud clearance seems sensitive to the turbulence model used. As draft tube flows are sensitive to initial boundary layer, this issue should be further analyzed.

Acknowledgements

The authors are grateful to the Swedish Waterpower Centre (SVC) for the financial support. The authors also want to thank the staff at VRD for the help during the measurements.

References

Bardina, J. E., Huang, P. G., & Coakley, T. J. (1997). *Turbulence modelling validation testing and development* (Technical memorandum). NASA.

Breuer, M., Jovicic, N., & Mazaev, K. (2003). Comparison of DES, RANS and LES for the separated flow around a flat at plate at high incidence. *International Journal for Numerical Methods in Fluids, 41*, 357–388.

Cervantes, M. J. (2003). *Effects of boundary conditions and unsteadiness on draft tube flow* (PhD thesis). Sweden: Lulea University of Technology.

Cervantes, M. J., Engström, T. F., & Gustavsson, L. H. (Eds.). (2005). *Proceedings of Turbine-99 III: 3rd IAHR/ERCOFTAC Workshop on Draft Tube Flow*. Luleå, Sweden: Luleå University of Technology.

Ciocan, G. D., Iliescu, M. S., Vu, T. C., Nennemann, B., & Avellan, F. (2007). Experimental study and numerical simulation of the FLINDT draft tube rotating vortex. *Journal of Fluids Engineering, 129*(2), 146–158.

Duprat, C., Balarac, G., Metais, O., Tridon, S., Barre, S., Ciocan, G. D., Laverne, T., & Tomas, L. (2009). *Large-eddy simulation of draft tube flow and validation from experimental measurements*. 3rd IAHR International Meeting of the Workgroup on Cavitation and Dynamic Problems in Hydraulic Machinery and Systems.

Durbin, P. A., & Reif, B. A. P. (2001). *Statistical theory and modeling for turbulent flows*. Chichester: John Wiley.

Egorov, Y., Menter, F. R., Lechner, R., & Cokljat, D. (2010). The scale-adaptive simulation method for unsteady turbulent flow predictions. part 2: application to complex flows. *Flow Turbulence and Combustion, 85*, 139–165.

Gagnon, J. M., Aeschlimann, V., Houde, S. B., Flemming, F., Coulson, S., & Deschenes, C. (2012). Experimental investigation of draft tube inlet velocity field of a propeller turbine. *Journal of Fluids Engineering, 134*(10), 101102.

Gagnon, J. M., F. Flemming, R. Qian, Desch, Aanes C, & Coulson, S. (2010). *Experimental and numerical investigations of inlet boundary conditions for a propeller turbine draft tube*. ASME, editor, 3rd Joint US-European Fluids Engineering Summer Meeting and 8th International Conference on Nanochannels, Microchannels, and Minichannels, Montreal.

Galvan, S., Reggio, M., & Guibault, F. (2011). Assessment study of k-e turbulence models and near-wall modeling for steady state swirling flow analysis in draft tube using fluent. *Engineering Applications of Computational Fluid Mechanics, 5*(4), 459–478.

Gubin, M. F. (1973). *Draft tubes of hydroelectric stations*. New Delhi, India: Amerind Publishing.

Henau, V. D., Payette, F. A., Sabourin, M., Desch, Aanes C, Gagnon, J. M., & Gouin, P. (2010). Computational study of a low head draft tube and validation with experimental data. *IOP Conference Series: Earth and Environmental Science, 12*(1), 012084.

Jonsson, P. P., Mulu, B. G., & Cervantes, M. J. (2012). Experimental investigation of a Kaplan draft tube -Part II: Off-design conditions. *Applied Energy, 94*, 71–83.

Krzemianowski, Z., Banaszek, M., & Tesch, K. (2011). Experimental validation of numerical model within a flow configuration of the model Kaplan turbine. *Mechanics and Mechanical Engineering, 15*, 297–307.

Kurosawa, S., & Satou, S. (2006). *Turbulent flow simulation for the draft tube of a Kaplan turbine*. 23rd IAHR Symposium on hydraulic machinery and systems. Yokohama, Japan: IAHR.

Liu, S., Shao, J., Wu, S., & Wu, Y. (2008). Numerical simulation of pressure fluctuation in Kaplan turbine. *Science in China Series E: Technological Sciences, 51*(8), 1137–1148.

Marjavaara, D. B., Lundström, S. T., Wright, J., Kamakoti, R., Thakur, S., & Shyy, W. (2005). *Steady and unsteady simulations of the turbine-99 draft tube using cfx-5 and stream*. Proceedings of Turbine-99 III Sweden: Luleå University of Technology, number 2005:20.

Menter, F. R. (1992). Influence of freestream values on $k - \omega$ turbulence model predictions. *AIAA-Journal, 30*, 6.

Menter, F. R. (1994). Two-equation eddy-viscosity turbulence models for engineering applications. *AIAA Journal, 32*, 1598–1605.

Menter, F. R. (2003). Turbulence modelling for turbomachinery. *QNET-CFD Network Newsletter, 2*, 10–13.

Mulu, B. G., & Cervantes, M. J. (2009). *Experimental investigation of a Kaplan model with LDA*. Proceedings of the 33th Congress of IAHR, Vancover, Canada.

Mulu, B. G., Cervantes, M. J., Vu, T. C., Devals, C., & Guibault, F. (2011). *Effects of inlet boundary conditions on Kaplan draft tube simulation accuracy*. Proceedings of the Hydro 2011 Conference.

Mulu, B. G., Jonsson, P. P., & Cervantes, M. J. (2012). Experimental investigation of a Kaplan draft tube -Part I: Best efficiency point. *Applied Energy, 94*, 71–83.

Nilsson, H. (2006, October). *Evaluation of OpenFOAM for CFD of turbulent flow in water turbines*. 23rd IAHR Symposium-Yokohama, Japan.

Nilsson, H., & Cervantes, M. J. (2012). Effects of inlet boundary conditions, on the computed flow in the turbine-99 draft tube, using OpenFOAM and CFX. *IOP Conference Series: Earth and Environmental Science, 15*(3), 032002.

Petit, O., Mulu, B. G., Nilsson, H., & Cervantes, M. J. (2010). Comparison of numerical and experimental results of the flow in the U9 Kaplan turbine model. *IOP Conference Series: Earth and Environmental Science, 12*(1), 012024.

Petit, O., Nilsson, H., Vu, T. C., Manole, O., & Leonsson, S. (2008). *The flow in the U9 Kaplan turbine -Preliminary and planned simulations using CFX and OpenFOAM*. Proceedings of the 24th Symposium on Hydraulic Machinery and Systems, IAHR.

Pope, S. B. (2000). *Turbulent flows*. Cambridge: Cambridge University Press.

Senoo, Y., & Nagata, T. (1972). Swirl flow in long pipes with different roughness. *Bulletin of the JSME, 15*(90), 1514–1521.

Su, W. T., Li, F. C., Li, X. B., Wei, X. Z., & Zhao, Y. (2012). Assessment of LES performance in simulating complex 3D flows in turbo-machines. *Engineering Applications of Computational Fluid Mechanics*, *6*(3), 356–365.

Vieser, W., Esch, T., & Menter, F. (2002). *Heat transfer predictions using advanced two-equation turbulence models* (Technical report). ANSYS Corp.

Vu, T. C., Devals, C., Zhang, Y., Nennemann, B., & Guibault, F. (2011a). Steady and unsteady flow computation in an elbow draft tube with experimental validation. *International Journal of Fluid Machinery and Systems, 4*(1), 87–96.

Vu, T. C., Koller, M., Gauthier, M., & Deschêne, C. (2011b). Flow simulation and efficiency hill chart prediction for a propeller turbine. *International Journal of Fluid Machinery and Systems*, *4*(2), 243–254.

Wilcox, D. C. (1993). *Turbulence modeling for CFD*. La Canada, CA: DCW Industries.

Yakhot, V., Orszag, S. A., Thangam, S., Gatski, T. B., & Speziale, C. G. (1992). Development of turbulence models for shear flows by a double expansion technique. *Physics of Fluids A, 4*(7), 1510–1520.

10

Experimental and computational investigations of flapping wings for Nano-air-vehicles

Weixing Yuan*, Richard Lee and Luc Levasseur

Aerospace Portfolio, National Research Council (NRC) Canada, Ottawa, Ontario, K1A 0R6, Canada

This paper presents the finalized results of a recent project which investigated the aeromechanical aspects of aerodynamic force generation by making use of flapping wings. Flapping-wing experiments using small wings have some unique challenges posed by the low force level (~ 1 N) and the cyclic wing motion. A tailored experimental water tunnel facility was developed for flapping wings operating at high reduced frequency with a complex two-dimensional and a three-dimensional motion profile. The experimental capability is demonstrated by the test cases of two-dimensional and three-dimensional flapping wings, designed according to a proposed notional nano-air-vehicle at a hovering condition. The features of the water tunnel, the geometric and kinematic parameters of the airfoils/wings, and the setups of the motion rigs for each test case are described. Measured forces and particle image velocimetry data are analyzed and cross-checked with the numerical results obtained from a code developed in-house. The comparisons of the experimental and numerical results show that the established experimental approach obtained a quantitatively reliable solution for the development of flapping wings and can serve for numerical validation of engineering tool developments. The investigation reveals that the kinematics of a rigid airfoil or wing is the dominant influence in the generation of aerodynamic forces, while the cross-section profile plays a secondary role. An asymmetric-wake-in-time is found behind the single airfoils and wings, which contributes to an asymmetry behavior of the resulting aerodynamic forces. In addition to the findings of single airfoils and wings, further analyses of the numerical and experimental results confirm that wing-wing interaction through the clap-fling mechanism can intensify the generation of the thrust force while accompanied by a small reduction in the overall propulsion efficiency.

Keywords: flapping wing; low Reynolds number flow; micro air vehicle; water tunnel; computational fluid dynamics

Nomenclature

b	= length of the wing span
$C_D(C_d)$	= drag coefficient based on reference area and reference velocity, $C_D = F_x \Big/ \frac{1}{2}\rho U_{ref}^2 s$
$C_L(C_l)$	= lift coefficient based on reference area and reference velocity, $C_L = F_y \Big/ \frac{1}{2}\rho U_{ref}^2 s$
C_p	= pressure coefficient, $C_p = (p - p_\infty) \Big/ \frac{1}{2}\rho U_{ref}^2$
c	= airfoil chord length
F_x, F_y	= forces in the x- and y-directions
f	= frequency of the flapping motion
H	= plunging amplitude (peak value) of the point about which the airfoil pitches
$\bar{H}$	= normalized plunging amplitude, $\bar{H} = H/c$
h	= plunging displacement of the point about which airfoil pitches
k_c	= reduced frequency, $k_c = \pi f c / U_\infty$
Re	= Reynolds number based on airfoil chord, $\mathrm{Re} = U_{ref}\, c/(\mu/\rho)$
s	= projection area of the wing body
T	= period of the flapping motion
t	= time
U_{ref}	= reference velocity, $U_{ref} = \lVert U_\infty + V_{induced} \rVert$
U_∞	= freestream velocity
u, v, w	= velocity components
$V_{induced}$	= peak velocity of the plunging/flapping motion, $V_{induced} = H\omega$ or $V_{induced} = 0.5b\Gamma\omega$
x, y, z	= Cartesian coordinates
x_0	= pivot location
α	= effective angle of attack, positive nose-up
δ	= phase difference between pitching and plunging (90 degs for this study)
Γ	= amplitude (peak value) of the root flapping angle
γ	= flapping angle of the leading edge about which the airfoil pitches
μ	= fluid dynamic viscosity
Θ	= pitching amplitude (peak value)
θ	= pitching displacement (angle) of the airfoil, positive nose-down (tail-up)
θ_0	= mean pitch position (zero for this study)
ρ	= fluid density
ω	= angular frequency of the flapping motion, $\omega = 2\pi f$
ω_z	= spanwise vorticity, $\omega_z = \frac{\partial v}{\partial x} - \frac{\partial u}{\partial y}$

*Corresponding author. Email: weixing.yuan@nrc-cnrc.gc.ca

1. Introduction

The concept of a new class of air vehicle is emerging amid the progress in technological areas such as aerodynamics, micro-electronics and sensors, micro-electromechanical systems, and micro-manufacturing methods. Referred to as a "nano-air-vehicle" (NAV), its mission will typically be to penetrate and reconnoitre confined spaces without being detected, most likely in an urban environment. To complete this mission successfully, the vehicle will be required to hover steadily and quietly, perch, maneuver with ease and swiftness, and be responsive to sudden disturbances such as gusts, which are inevitable in the urban setting. Such vehicles are envisioned to be small in size, being less than 7.5 cm in length typically, with a mass of less than 10 grams. Satisfying the demanding mission requirements with a conventional fixed-wing/forward-thrust platform, or a rotary-wing approach, seems unlikely because the aerodynamics of these vehicles provides less propulsive efficiencies at a small scale (Young, 2005). Insect flight, on the other hand, offers an attractive alternative solution. Insects and tiny birds, such as hummingbirds, are small airborne bodies which rely on the unsteady aerodynamics of flapping wings to produce lift and thrust. The unsteadiness of the aerodynamics arises from the rapid complex motion of wings – which flap and rotate through large angles, and sweep fore and aft at the same time – combined with rapid adjustments of the shape of the wing. Developing an understanding of the kinematics of insect wings and the aerodynamics arising from the motion are essential toward the development of an engineering tool capable of rapidly optimizing the design and motion of a wing set for a nano-air-vehicle.

A great deal of progress has been made in the past decade in understanding the flapping-wing aerodynamics. Mueller (2001), Shyy, Lian, Tang, Viieru, and Liu (2008), Platzer, Jones, Young, and Lai (2008) and Ol (2010) provided broader collection and detailed review of previous research work in flapping-wing aerodynamics at very low Reynolds numbers. It was first recognized by Knoller (1909) and Betz (1912) that a flapping airfoil generates thrust. Thrust generation is crucial for insect-like and NAV flights, in particular at hovering conditions – a primary topic of the present study. Of fundamental and applied interest, Lai and Platzer (2001) visualized that a finite thrust was generated due to the vortex shedding from the trailing edge of a NACA 0012 airfoil in pure plunging motion at zero freestream velocity. Back to 1990, Freymuth (1990) visualized the flow past an airfoil in combined harmonic plunging and pitching motion in still air and revealed strong vortices shed from both leading and trailing edges that organize efficiently into a jet structure and hence produce a large thrust. Sunada, Kawachi, Matsumoto, and Sakaguchi (2001) used a load cell with strain gages to measure the instantaneous forces acting on a two-dimensional (2D) wing undergoing a combined pitch–plunge motion and determined the combinations of these forces for maximum time-averaged thrust and for maximum efficiency. Although 2D models showed advantages in elucidating conical vortices forming on the leading and trailing edges, three-dimensional (3D) models are needed to cover the spanwise flow features thus understand thoroughly the physics of force generation by hovering insects. Since complexities of flow visualizations in 3D experiments, computational fluid dynamics (CFD) tools showed complementary effectiveness. Back in the 1990s, Liu, Ellington, Kawachi, van Den Berg, and Willmott (1998), Liu & Kawachi (1998) had initiated CFD simulations of hawkmoth hovering. In these simulations, the artificial compressibility method following (Chorin, 1968) was used by adding a pseudo-time derivative of pressure to the continuity equation. Yuan, Khalid and Huang (2008) attempted 3D flow simulations past an insect-like wing which was similar to the ones used by Dickinson, Lehmann, and Sane (1999) and Sane and Dickinson (2001). Young, Lai, and Germain (2008) investigated the effect of flapping amplitude, flapping frequency, and wing rotation timing and duration on the aerodynamic performance of the dragonfly hind wing using a 3D Navier–Stokes solver. It was found that the mean vertical force and power are essentially independent of frequencies exceeding 5 Hz and the hovering efficiency peaks at a flapping amplitude of about 35 deg, very close to the observed amplitude of the dragonfly hind wing in hover. More recently, Fairuz, Abdullah, Yusoff, and Abdullah (2014) investigated the fluid-structure interaction of unsteady aerodynamics of a flapping wing at a low Reynolds number.

Although there exists a large body of experimental work on flapping-wing aerodynamics and they provide good qualitative or even quantitative descriptions, because of the difficulties associated with doing experiments on low force level (~ 1 N) and the complex flapping motion, the reported quantitative results are often incomplete or hard to be used for engineering applications. A collaboration involving Defence Research and Development Canada in Valcartier (DRDC-V), Advanced Subsonics Inc. (ASI), and the Aerodynamics Laboratory of the National Research Council Canada (NRC-Aerospace), has been investigating flapping-wing aerodynamics for NAV applications. The immediate goal of the collaboration was to develop the means to optimize rapidly the wing design and motion o f a notional NAV (Figure 1) and, in turn, develop an optimized configuration; in the long term, the collaboration envisioned building a functional prototype in stages. The work of the collaboration was introduced by Lesage et al. (2008). The progress of NRC-Aerospace concerning its CFD contribution has been reported subsequently by Yuan and Khalid (2008) and Yuan, Lee, Hoogkamp, and Khalid (2010). The detailed experimental setup and preliminary experimental results have been elaborated and presented by Lee, Yuan, Levasseur, and Hoogkamp (2011). This paper will continue to chronicle the collaboration, concentrating on the further analyses of experimental and

Length (nose to tail)	50 mm
Span	75 mm
Semi-span	37.5 mm
Chord	19 mm
Disk area	4180 mm^2
Mass	10 g
Weight (= Thrust)	~ 0.1 N
Frequency	~ 80 Hz
Flap angle amplitude	38 deg
Pitch angle amplitude	50 deg
Freestream velocity	3 m/s (hovering)

Figure 1. The notional nano-air-vehicle using the "clap-fling" mechanism. The wings flap about the central axis of the body and twist about the leading edge (Lesage et al., 2008; Zdunich, 2008).

numerical results, focusing on thrust generation and its related physics in hovering conditions.

Insects generate thrust and lift by controlling wing kinematics that includes flapping motion during stroke cycles. To facilitate the investigation, two standard test cases, based on a notional NAV, are defined in this study: a plunging and pitching two-dimensional airfoil (2D test case), and a flapping and pitching three-dimensional wing (3D test case). These test cases, including the described kinematics, are easy to follow numerically or even experimentally. The flapping-wing experiments using small wings as in the present study have unique challenges posed by the low force level (~ 1 N) and the complex wing motion. The low force levels required careful design and calibration of the load cells and motion control system. After detailed description of the experimental procedure, combined experimental and computational analysis is further performed. The results generated by the CFD codes are validated by experimental results produced from water tunnel tests of the standard test cases at NRC-Aerospace; in turn, CFD results provide detailed insights in flow physics for cross-check of the experimental data. In addition to the further analyses of the finalized results of the single flapping airfoils and wings, the paper will present the contributions on multi-wing interaction known as the clap-fling effect (Ellington, 1984; Weis-Fogh, 1973). These results could help calibrations of panel codes as engineering tool to optimize wing shape and motion with rigid or flexible wings for NAVs (Yuan et al., 2010; Zdunich, 2010).

2. Standard test cases for the notional NAV

A notional NAV was developed by the collaboration team to provide a context for vehicle size, motion characteristics, and performance parameters, and to serve as a starting point for the optimization process (Lesage et al., 2008; Zdunich, 2008). The definition of the vehicle was based on the mission requirements, anticipated progress in complementary technologies, and the experience gained from the development of the Mentor micro-air-vehicle (MAV), which flew successfully in March 2002 (Zdunich et al., 2007). Similar to the Mentor, the wings of the notional NAV flap in a three-dimensional manner (Figure 1). The wing is finite and the wing tip sweeps farther and faster than the wing root.

Table 1. Summary of scaling parameters and model kinematics and geometry for the water tunnel test cases.

		2D	3D
Scaling parameter	Reduced frequency	1.59	1.59
	Amplitude-to-chord ratio	0.75	1.0
	Induced Reynolds number	9,500	14,300
Model kinematics	Frequency	0.5 Hz	0.46 Hz
	Plunge/flap amplitude	48 mm	15 degs
	Pitch amplitude	30 degs	40 degs
	Freestream velocity	64 mm/s	64 mm/s
Model geometry	Reference chord	64 mm	70 mm
	Airfoil/wing thickness	3.18 mm	3.18 mm
	Semi-span		350 mm

The flapping mechanism is a single degree-of-freedom (1-DOF) actuation mechanism, which is light and robust because the mechanism is not overly complicated. The notional NAV possesses a four-wing "double hummingbird" configuration and takes advantage of the clap-fling mechanism to obtain an improved thrust-to-power ratio and a high thrust for a limited disk area. In addition, the notional vehicle also possesses other features including thin flat wings, passive aeroelastic-tailoring of wing camber and span-dependent twist.

Having developed a concept for a full-scale NAV, the similitude parameters can be developed subsequently for the standard test cases suitable for the NRC water tunnel. The standard test cases consider a hovering condition. Although the ground velocity of the vehicle is zero, the freestream velocity (3.0 m/s) is estimated according to the induced slipstream behind the flapping wings based on the actuator disk theory. The similarity parameters for dynamic, kinematic, and geometric scaling are, respectively, Reynolds number, reduced frequency, and flap-amplitude to chord ratio (Table 1). Of the three scaling parameters, only the matching of Reynolds number is relaxed.

The flap-amplitude to chord ratio (H/c) is an important geometrical similarity parameter for flapping wing flight. The value of this ratio is determined at approximately 77% of the wing semi-span, measured from the flap axis. This location is arbitrarily chosen as representative of the main "working" portion of the wing. For the standard 3D test case, H/c for the notional NAV is approximately 1.0. For the 2D test case, however, H/c was reduced to 0.75 because the motion rig was unable to produce sufficient plunge amplitude to render $H/c = 1.0$. $H/c = 0.75$ corresponds to that at 58% of the semi-span of the notional NAV.

2.1. *Standard 2D test case*

To facilitate the understanding of the flow physics in the early going, the 2D case imposes two-dimensional flow over an airfoil and simplifies the flapping motion of the NAV by transforming its rotational motion to linear motion, i.e., plunge. The 2D test case was developed well suited for the water tunnel and its motion rig.

The instantaneous position of the airfoil is expressed as a linear superposition of motion due to plunging (translation) and that due to pitching (rotation) (Figure 2). The motion in both pitch and plunge is assumed to be harmonic. The airfoil pitches about its leading edge, which is consistent with a passively twisted wing of the type envisioned for the notional nano-air-vehicle. The position due to plunging is given by:

$$h(t) = H\sin(\omega t + \delta), \tag{1}$$

where H is the plunging amplitude and $\omega = 2\pi f$ is the circular frequency. The pitching motion of the airfoil is described by:

$$\theta(t) = \theta_0 + \Theta\sin(\omega t), \tag{2}$$

with a positive angle when pitching counter-clockwise as illustrated in Figure 2. In this study, $\theta_0 = 0$ degs. and $\Theta = 30$ degs. A phase angle of $\delta = 90$ degs was selected because it was found to produce good results during the Mentor program (Zdunich et al., 2007), as well as by researchers of larger-scale flapping-wing vehicles (DeLaurier, 1993a, 1993b).

The counter-clockwise direction for the pitch angle was arbitrarily defined although it provided convenience in coding and thus was adopted by some researchers, e.g. Guglielmini & Blondeaux (2003) in polar coordinate system and Zdunich (2008) for a panel code. However, it did not alter the conventional definition of the aerodynamic angle of attack that is positive in clockwise direction. The instantaneous angle of attack can be expressed as:

$$\alpha(x,t) = -\theta(t) - \tan^{-1}\left[\frac{\dot{h}(t) + \dot{\theta}(t)(x - x_0)\cos(\theta)}{U_\infty + \dot{\theta}(t)(x - x_0)\sin(\theta)}\right], \tag{3}$$

where $\dot{h}$ and $\dot{\theta}$ are the first time-derivatives of Equations 1 and 2, respectively. The terms in the brackets in Equation 3 represent the contributions induced by the plunging motion and the pitching oscillation, respectively, which causes an equivalent cambered wing effect (Yuan & Khalid, 2008). In this work, the pivot point is the leading edge; thus $x_0 = 0$. The instantaneous angle of attack at the leading edge, as described by Equation 3, is plotted in Figure 3. To help readers compare with the predictions of thin airfoil theory or panel methods, the effective angle of attack as defined by the three-quarter chord is also plotted in the figure.

Given a reference length and a reference velocity, one can define the Reynolds number. In forward flight, the freestream velocity is often used as the reference velocity. For the 2D case, the small freestream velocity results in a Reynolds number $\mathrm{Re}_\infty = \rho U_\infty c/\mu = 4 \times 10^3$ and the reduced frequency is $k_c = \omega c/2U_\infty = 1.59$. In hovering, the wing tip velocity may be used as the reference velocity (Shyy et al., 2008). Identifying the peak velocity of the plunging motion as $V_{induced} = H\omega$, the induced Reynolds number and the reduced frequency could be recovered as $\mathrm{Re}_f = \rho V_{induced} c/\mu = 9.5 \times 10^3$ and $k_{induced} = \omega c/2V_{induced} = 0.67$, respectively. Since the speed of the airfoil/wing due to flapping may be very large compared to the freestream velocity, especially near the hover condition as in this study, a generalized Reynolds number based on a combination of these two velocities is used:

$$\mathrm{Re} = \frac{\rho \| U_\infty + V_{induced} \| c}{\mu}. \tag{4}$$

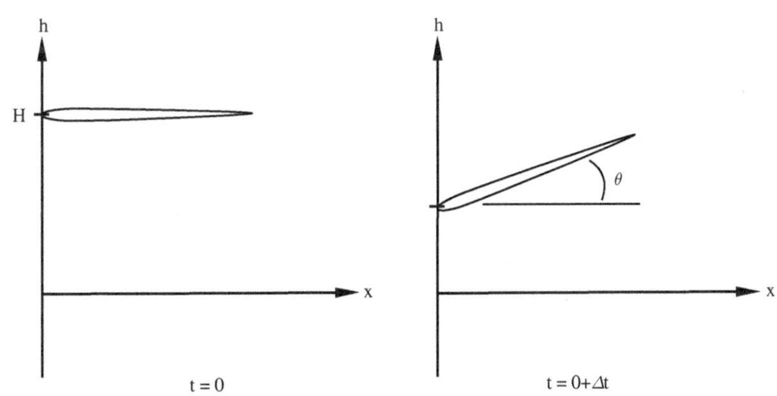

Figure 2. Schematic of the airfoil in plunging and pitching motion.

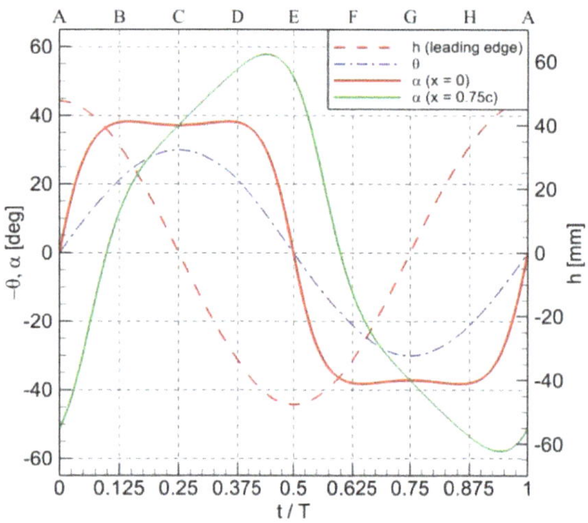

Figure 3. Time history of the effective angle of attack in accordance with the airfoil plunging and pitching motion.

As a result, the generalized Reynolds number of the 2D test case is $Re = 10.5 \times 10^3$. This practice had been employed by some researchers in the flapping-wing community. Zdunich (2008) used the magnitude of the vector sum of the freestream velocity and the velocity due to airfoil plunging as reference velocity to nondimensionalize the forces and moments. In fact, this reference velocity was termed as the relative inflow by Azuma, Okamoto, and Yasuda (2001) and relative velocity of the airfoil through the fluid by Hall, Pigott, and Hall (1997) and Hall and Hall (2001). Since all resulted Reynolds numbers using the three different reference velocities are listed here, one can easily rescale the nondimentional parameters for individual preference.

2.2. *Standard 3D test case*

The test case parameters are listed in Table 1. The flapping amplitude is limited to 15 degs because of the constraint by the width of the test section. A schematic of the wing model is shown in Figure 4. The model is tapered both in planform and thickness. The radius of flapping rotation is geometrically representative of the NAV. In order to avoid free surface effects, some clearance between the root of the wing and the surface of the water was deemed necessary.

Similar to the 2D case, the position of the wing at any time is expressed as a linear superposition of motion due to flapping and pitching, which are both rotational motions and assumed to be harmonic. The wing also pitches about its leading edge. The position due to flapping is described by:

$$\gamma(t) = \Gamma \sin(\omega t + \delta). \tag{5}$$

As in the 2D test case, the pitching motion lags the flapping motion by a phase angle of 90 degs. The description of the pitching motion of the wing is identical to Equation 2.

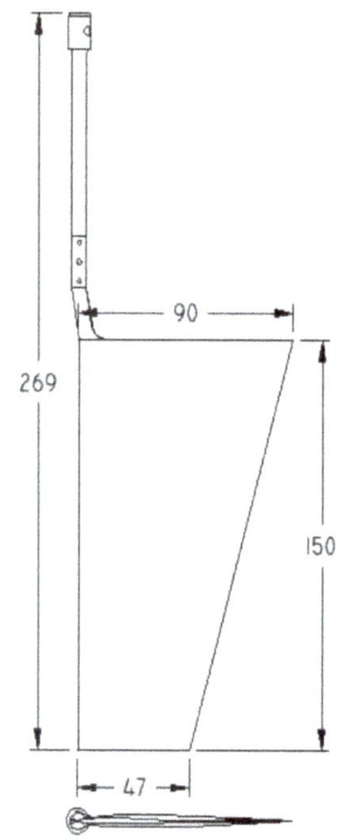

Figure 4. A schematic of the 3D wing model featuring a tapered NACA 0005 cross-section. The CFD simulation did not model the supporting arm. The dimensions are expressed in millimeters.

The setup results in a generalized Reynolds number based on the reference chord: $Re = 19 \times 10^3$.

3. Description of the experimental setup

The standard test cases were carried out experimentally in the Eidetics Flow Visualization Water Tunnel at NRC-Aerospace (Figure 5). The water tunnel features an open-surface test section with a return circuit. The nominal dimensions of the test section are 38 cm wide, 51 cm high, and 163 cm long. The maximum freestream velocity achievable in the test section is 305 cm/s; 20% of this range was required for the standard test cases. The freestream turbulence intensity in the test section is rated at less than 1%. Even so, dye flow visualization revealed that fluctuations in the test section increased after approximately 240 seconds of continuous operation of the facility. The increased fluctuations were attributed to remnants of the turbulent wake, which was developed by the large amplitude plunge motion of the 2D airfoil model but not attenuated adequately by the return circuit or the flow conditioner elements in the delivery plenum. For this reason, the operation of the motion rigs and the measurement of load data were limited to 180 seconds for both the 2D and 3D test cases.

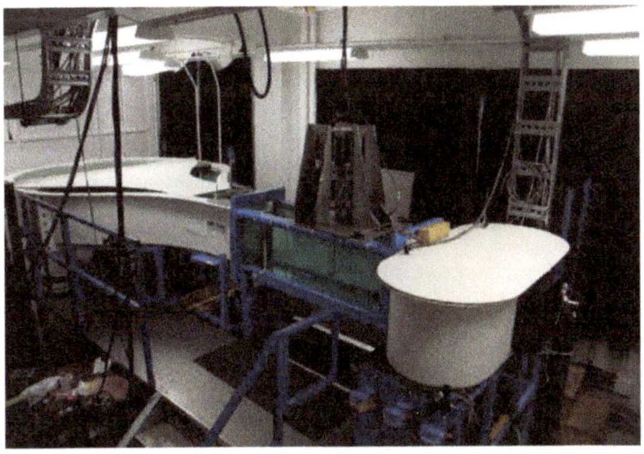

Figure 5. The water tunnel at NRC-Aerospace with the two-axis motion system installed.

3.1. *Standard 2D test case*

The experimental setup for the standard 2D test case is illustrated in Figure 6. As mentioned earlier, Lai and Platzer (2001) visualized that a finite thrust was generated due to the vortex shedding from the trailing edge of a NACA 0012 airfoil in pure plunging motion at zero freestream velocity. However, Guglielmini and Blondeaux (2003) showed that an elliptic airfoil generated zero time-averaged thrust due to the symmetry of the resulting flow. To confirm possible airfoil effects in this study, two rigid 2D airfoil models were designed and manufactured: one with a NACA 0005 cross-section; the other, a flat plate with a constant thickness equal the maximum thickness of the NACA 0005 airfoil. Both models featured a span of 25.4 cm and were fabricated from stainless steel to protect against corrosion and for stiffness, especially in the case of the NACA 0005 model which has a very thin trailing edge. Tall glass endplates were placed carefully at both tips of a model to mitigate tip effects and to encourage two-dimensional flow conditions over the airfoil; a nominal clearance of 1 mm was maintained between an endplate and the model, throughout its motion profile. The open surface above the airfoil was removed with the installation of a large plate just below the waterline; centered over the model, the plate was approximately five chords long and its width covered the test section. Reflection planes – appearing above and below the 2D airfoil model in the figure – are available for simulating the clap-fling phenomena mimicking multiwing interaction in a simplified way.

The motion of the airfoil model is controlled by a two degree-of-freedom (2-DOF) system. The model is connected, at its spanwise center, to a pair of vertical plunge rods. Each plunge rod is driven by a linear shaft motor integrated with a slider system; a linear encoder conveys the position of the motor to the motion controller, closing the feedback loop. The linear motion of the plunge rods is programmed to provide the sinusoidal plunge and pitch motions at the frequencies and amplitudes prescribed by the test case. The deviation of the actual motion of the airfoil model from the prescribed kinematics was around 0.5% of the maximum amplitude, see Lee, Yuan, Levasseur, and Hoogkamp (2011).

Two uniaxial load cells – each with a rating of 44.5 N (10 lbs) – sensed the lift force developed by the airfoil

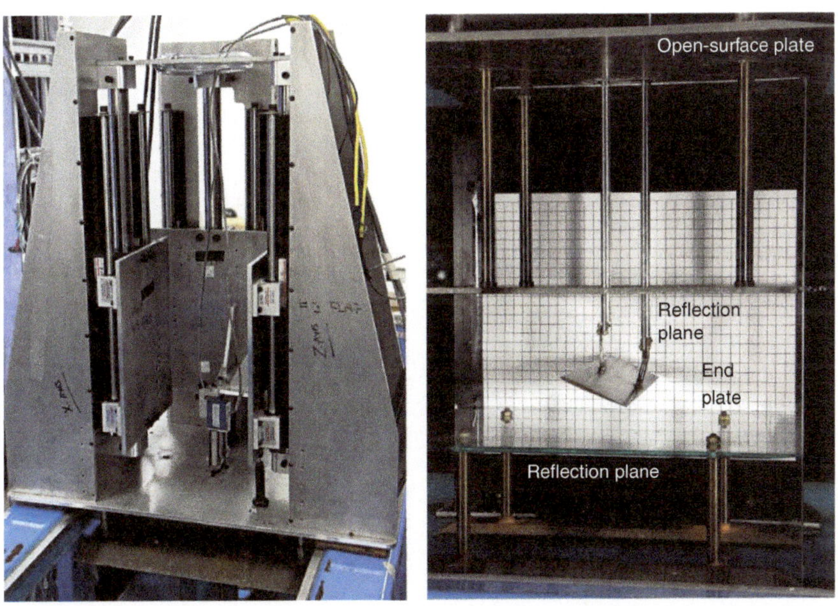

Figure 6. 2D model two-axis motion system. Left: the 2-DOF motion rig; Right: the two-dimensional airfoil model is shown between two reflection planes. The presence of the large plate seen at the top of the photograph removes the free surface condition above the airfoil model, when the reflection planes are not present.

and acting on the plunge rods. Each load cell was mounted inline between a plunge rod and its drive mechanism. The estimated uncertainty in the total lift force measurement is 0.084 N (8.6 g). The natural frequency of a plunge rod and load cell combination is 147 Hz. Drag force was not measured because the desired approach to acquire this quantity was not facilitated.

The output signals of the load cells and position encoders were sampled at rate of 1 kHz and acquired for 81 cycles of motion. Data were collected for both "wind" and "tare" conditions. The output analog signals of the load cells were filtered before conversion to a digital signal, and then digitally filtered once more with a low-pass cutoff of 10 Hz. The lift force signal was corrected for weight, inertial loading, and buoyancy. After the first six motion cycles were discarded – to account for transient flow conditions at the start of a run – the lift force data were parsed to formulate a three-cycle ensemble-average.

As mentioned, the tare scheme accounted for weight, inertial loading, and buoyancy. The lift force was adjusted by each tare component on a sample-by-sample basis. Unlike the practice of Isaac, Rolwes, and Colozza (2008) who used a numerical approach for calculating the inertia and gravity forces, the weight and inertial tares were acquired together experimentally by running the 2D model through the motion profile in a test section emptied of water. Care was taken to ensure that this tare signal was synchronized with its corresponding wind signal throughout the motion profile. A weight and inertial tare signal was acquired for each wind signal acquired. The buoyancy tare, on the other hand, was an analytical model which comprised the mean buoyancy force, estimated with the airfoil model placed at the zero-displacement plunge position; and a small fluctuating component, which increased or decreased the buoyancy tare as changes of the wetted volume of the plunge rods occurred during the motion profile.

Phase-locked particle imaging velocimetry (PIV) measurements of two components of velocity – streamwise and vertical – were performed for the 2D test case to identify features in the flow over the surface of the airfoil and in its wake. A LaVision PIV system was configured to receive a trigger signal from the motion control system; once triggered, images were acquired at intervals of 1/16 cycle to correspond with CFD results. The streamwise measurement plane was aligned with the quarter-span of the airfoil model; the laser illuminated the plane from beneath the test section. The airfoil model was coated with a Rhodamine-based paint to mitigate the reflection of laser light. The PIV recording parameters are listed in Lee et al. (2011).

3.2. *Standard 3D test case*

The experimental setup for the standard 3D test case is illustrated in Figure 7. Similar to the 2D airfoil models, two rigid 3D wing models were manufactured with a NACA 0005 and a flat plate cross-section. The models were fabricated from aluminium for lightness and shared an interchangeable supporting arm, which connected the wing model to the balance. The center of the cross-section of the supporting arm coincides with the leading edge of the wing model so that the pitching motion rotates the wing about the leading edge.

The sinusoidal motion prescribed for the wing model was controlled by a three degree-of-freedom (3-DOF) motion system that is capable of providing flapping, pitching, and sweeping motions directly. The sweeping motion was not activated for this work. Each motion was driven by a brushless DC servomotor equipped with a gear reduction box. An angular position encoder, integrated with the servomotor, closed the feedback loop for the motion control system. The difference between the prescribed kinematics of the wing model and the motion sensed by the encoders was less than 1% of the maximum amplitude (Lee et al., 2011).

A bi-fold five-component balance was mounted between the wing and the motion system, inline with the leading edge of the wing (Lee et al., 2011). The balance measured the normal and axial forces, and the pitching, rolling, and yawing moments in the body axes of the balance. The limit to which each component was calibrated, as well as the estimated measurement uncertainty and natural frequency are listed in Lee et al. (2011).

The acquisition and reduction of data – especially the adjustments for weight, inertial loading, and buoyancy in the force and moment data – followed the same procedures as for the standard 2D test case, described earlier. A difference between the procedures of the two test cases is the frequency cutoff of the low-pass digital filter; for the 3D test case, the cutoff was set to 2.3 Hz because 95% of the fluctuating component of normal force was captured below this frequency. All loading due to the supporting arm by itself was deemed negligible and was not considered in the data reduction process. The normal and axial forces were transformed into lift and drag forces in the test-section axes. The rolling and yawing moments were transferred from the resolving center of the balance to, respectively, the flapping axis and the intersection of the flapping and pitching axes in the 3-DOF motion rig; the transfer of the pitching moment was not necessary.

No PIV measurements were acquired for the 3D test cases.

4. Description of the in-house CFD solver

The computational-fluid-dynamics code, INSflow, was developed in-house at NRC-Aerospace for solving three-dimensional unsteady incompressible flows. The code has been used for a number of applications including large-eddy and unsteady Reynolds-averaged Navier–Stokes

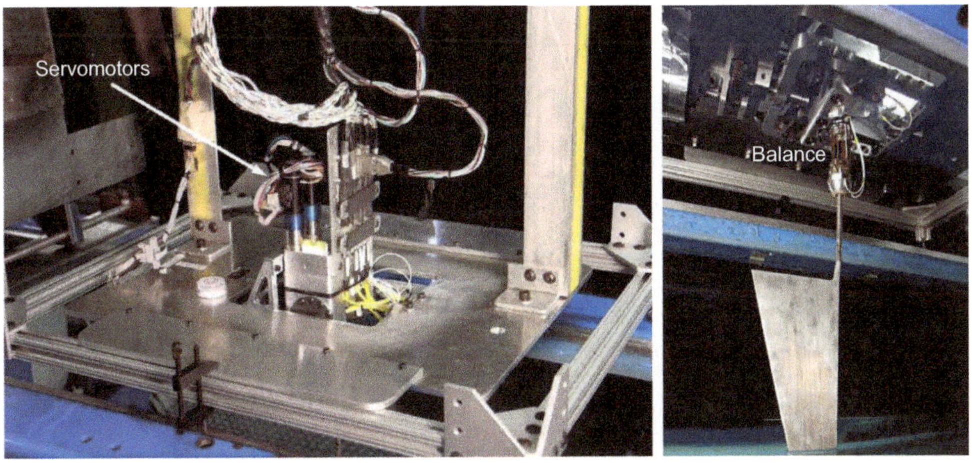

Figure 7. The 3-DOF motion rig. The three-dimensional wing model is installed in the balance.

simulations (Yuan, Poirel, Wang, & Benaissa, 2014), low-Reynolds flows (Yuan, Khalid, Windte, Scholz, & Radespiel, 2007), laminar-separation flutter (Yuan, Poirel, & Wang, 2013), and flapping-wing aerodynamics (Yuan et al., 2008, 2010).

INSflow applies the integral form of the conservation laws of mass and momentum. A fully implicit second-order temporal differencing scheme was implemented in the discretization, which made the algorithm stable for large timesteps. The discretization of the convective and diffusive fluxes was carried out in a co-located variable arrangement using a finite-volume approach that was second-order accurate in space. A first-order upwind scheme is available. The coupling of the pressure and velocity was handled using a modified SIMPLE algorithm (Ferziger & Perić, 1996; Patankar, 1980). The calculations were performed on moving grids; the velocity of the grid movement was included in the governing equations in an inertial frame of reference. In order to avoid artificial mass sources generated by the grid velocity, a space conservation law was introduced to ensure a fully conservative property in the computations, as applied by Demirdžić and Perić (1990).

4.1. *Standard 2D test case*

The 2D test case was considered a laminar flow. The 2D calculations were performed on O-type meshes. The farfield boundaries were located about 25 chords away from the surface of the airfoil. The nominal mesh had 481 × 129 grid points. This mesh was designed with a grid density increasing toward the wall at a rate of 6% for the near-wall region and 8.7% for the outer region. The grid lines in the normal-to-the-wall direction were simple straight lines. Improved grid orthogonality would accelerate the calculation convergence and increase the simulation accuracy, but the modification was not investigated in this study. The calculations were started from a stationary fluid. In general, the third cycle produced comparable results to the second cycle for the cases reported in this study.

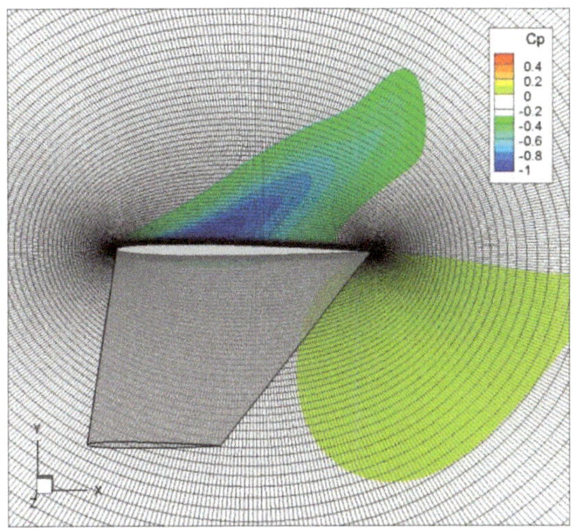

Figure 8. Computational mesh of the 3D NACA 0005 wing configuration in flapping motion.

To select a suitable grid for the investigations, computations for the standard 2D test case were conducted using second-order spatial discretization on three O-meshes: 241 × 97, 481 × 129, and 961 × 161, with corresponding timesteps for one plunging cycle of 2888, 3840, and 5760, respectively. Eight flapping cycles were conducted for the coarse and medium grids. Only six cycles were completed for the fine mesh. The discrepancies of the results of the aerodynamic forces obtained on the three grids were minimal. The aerodynamic coefficients matched well starting from the second cycle (Yuan et al., 2010). Based on this observation, the results obtained from the medium grid will be discussed in this paper.

4.2. *Standard 3D test case*

Same as the 2D test case, the 3D test case was considered a laminar flow. A grid with 481 × 129 × 33 nodes was used for the flow domain around the wing body (see Figure 8), with a farfield boundary located at about 25 reference chord lengths. This grid size is comparable to the medium grid used in the earlier 2D calculations, but with 33 stations located along the wing span. The computational domain contained two extensions. The first one stretched from the wing root to the flapping axis, while the second extended beyond the wing tip for 150 mm, which was equal to the spanwise dimension of the wing body. These two extended domains were discretized using grids with 481 × 133 × 33 nodes, thus constituting a three-block structure. Compared with the grid around the wing body, there were four extra grid points in the direction normal to the flapping arm, to mesh the thickness of the wing root and tip surfaces.

In the calculations, the farfield and the spanwise boundaries were subject to a freestream condition. The calculations were initiated from a stationary state. The computations were found to be time consuming and no time convergence study was performed. Since the Reynolds number was close to the one in the 2D test case, as many as 384 timesteps per flapping cycle were used to discretize the governing equations in time, which was confirmed to be sufficient for obtaining major features of the flow physics in the timestep refinement for the 2D test case reported in Yuan et al. (2010). An upwind scheme was used for the discretization in space as the second-order scheme encountered numerical instabilities.

5. Experimental and computational results

Measured and computed forces and spanwise vorticity are presented for the standard 2D and 3D test cases. For the 3D test case, only forces are available because PIV measurements were not undertaken. The experimental results are reflective of a mean flow field, whereas the computed results are instantaneous. Most results are expressed in engineering units since we wish to illustrate, specifically, the small magnitude of the loading encountered.

5.1. *Standard 2D test case*

5.1.1. *Lift force*

Two cycles of lift measurements and lift predictions, produced by INSflow, are superimposed in Figure 9. The experimental results are taken from an ensemble-average of 24 three-cycle time-series; the CFD results are the last two instantaneous cycles from a series of eight computed cycles. Compared with the previous publications (Lee et al., 2011; Yuan et al., 2010), a slight discrepancy is seen in the experimental data, which rectifies an inaccuracy in the three-cycle ensemble-averaging. The motion profiles synchronized the presentation of the two sets of results.

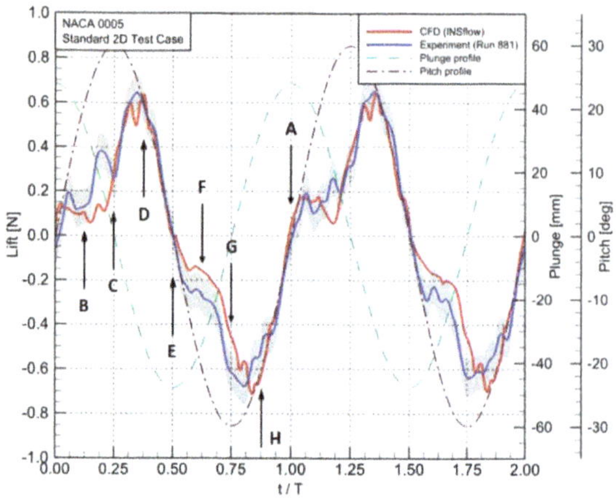

Figure 9. Measured and computed lift force for the pitching–plunging NACA 0005 airfoil. Standard 2D test case: Re = 10.5 × 10^3, U_∞ = 0.0635 m/s, f = 0.5 Hz. The shaded band represents uncertainty in the lift measurement.

In general there is good agreement between the results; for instance, there is repeatable behavior corresponding with the pitching-plunging motion schedule; the magnitudes of peak lift match well; and there are no indications of obvious or significant phase shift. For the most part, the computed lift force falls within the uncertainty band of the measured lift force. The CFD simulation reproduces successfully the sluggish development of lift force at the start of both the downstroke and the upstroke. Interestingly, the magnitude of the lift forces at the upstroke and the downstroke differ slightly; the difference could be indicative of an interaction between the airfoil and its asymmetric wake, which will be discussed later.

The most noticeable discrepancy between the experimental and computed results begins to appear following the start of the upward plunge at t/T = 0.5. Here, the magnitude of measured lift is higher than the computed results up to t/T = 0.8. This discrepancy is attributed to the attachment of the plunge rods to the upper surface of the airfoil (Figure 6); the attachment is suspected of interfering with the convection of a spanwise vortex from the leading edge to trailing edge of the airfoil. The convection of this spanwise vortex is illustrated in the CFD results depicted in Frames E, F, and G (t/T = 0.5, 0.625, and 0.75, respectively) in Figure 10. The interference disrupts the surface pressures at mid-span, giving rise to a higher local pressure than normal on the upper surface which, in turn, contributes to a small increment in the magnitude of the lift force from t/T = 0.5 to 0.8.

5.1.2. *Flowfield over the NACA 0005 airfoil*

Figure 10 illustrates the flowfield over the NACA 0005 airfoil in terms of spanwise vorticity. Particle imaging

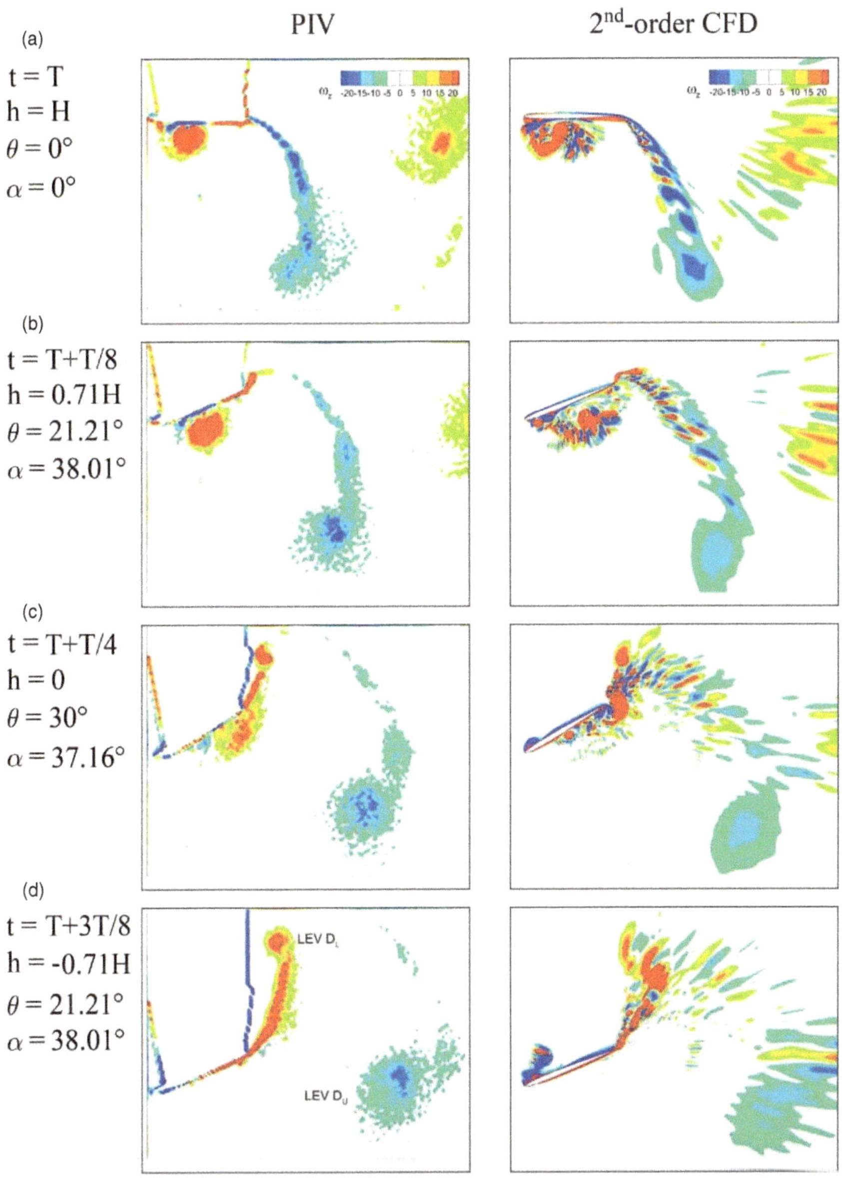

Figure 10. Downstroke: Experimental (at left) and computed spanwise vorticity (ω_z) distribution over the pitching-plunging NACA 0005 airfoil. Standard 2D test case: Re $= 10.5 \times 10^3$, $U_\infty = 0.0635$ m/s, $f = 0.5$ Hz. The temporal occurrence of each frame is identified in Figure 3. The abbreviation "LEV" represents "leading-edge vortex"; the subscripts "L" and "U" represent lower and upper surface, respectively.

velocimetry measurements and the results of CFD simulation are compared against each other at eight equally spaced points of the motion schedule as marked in Figure 9. The PIV results are an ensembled-average of 24 motion cycles; the CFD results are instantaneous and are the product of second-order calculations. Unlike the CFD simulation, PIV results for the upper surface of the airfoil are not available because the surface lay in the shadow of the laser-light sheet.

Frame A corresponds to the start of the downstroke at t/T $=$ 0 as well as the end of the upstroke after Frame H. At this instant of the motion schedule, the effective angle of attack, α, is zero because the plunging velocity is zero. A vortex-pair is attached to the airfoil. The first counter-clockwise vortex (red) of this pair stands on the lower surface of the airfoil at the leading edge; the development of this vortex has its origin in the later stages of the proceeding upstroke (Frames G and H). Originating from

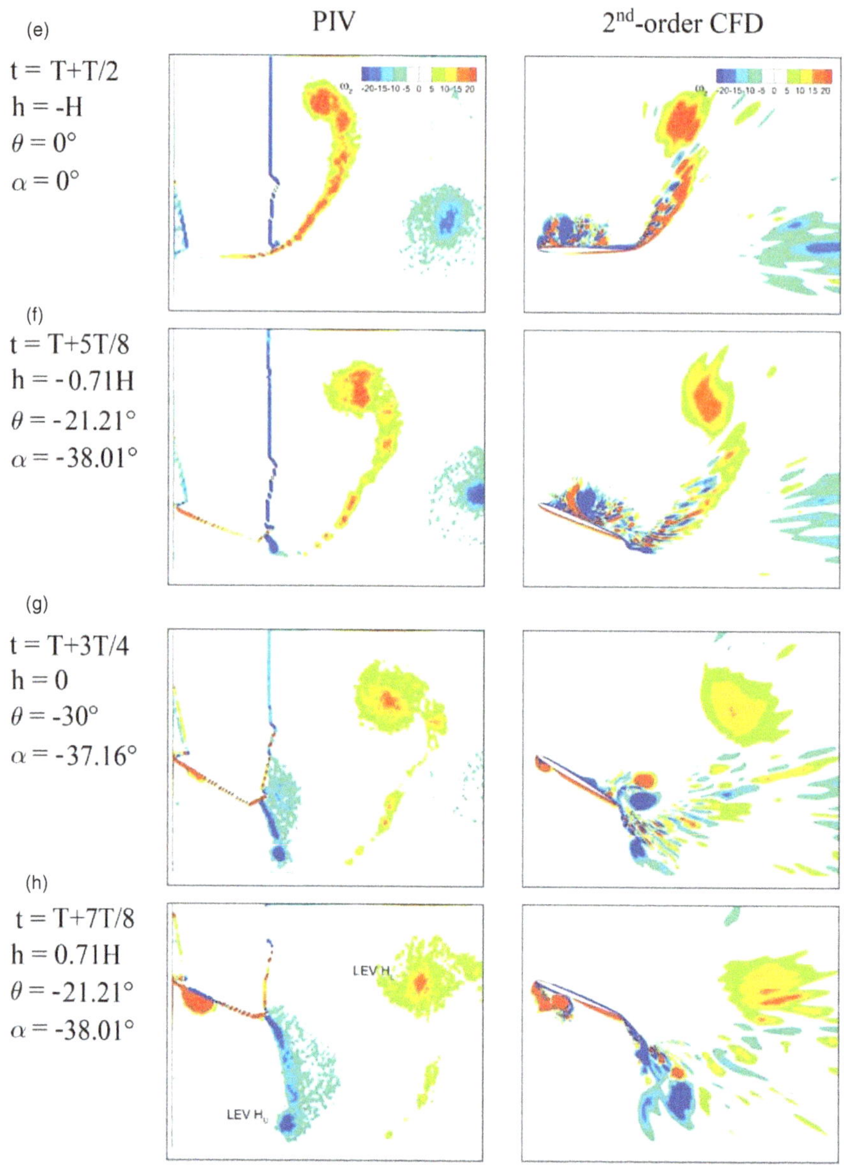

Figure 10a. Upstroke: Continued from previous page.

the upper surface of the airfoil, the second vortex (blue) of the pair is counter-rotating and has been shed from the trailing edge; however, it is considered attached because it is connected with the airfoil trailing edge through the counter-rotating vortex street and continues to receive vorticity from the airfoil. The PIV results detect a third, less intense vortex (red-yellow) downstream in the wake of the airfoil. This vortex is "free" – it does not receive vorticity from the airfoil – and is a product of the preceding cycle of motion, corresponding to the current vortex standing at the leading edge of the airfoil. In Frame B, α has passed its first peak (Figure 3). No new vortex has formed on the upper surface owing to a phase delay. The first vortex of the pair convects along the lower surface of the airfoil; the second vortex becomes free. The free vortex downstream convects further away in the wake of the airfoil. At the instant shown in Frame C, α is midway on its approach to its second peak. Of the first pair, the first vortex has shed from the trailing edge but continues to receive vorticity from the airfoil; the second vortex continues shedding. At the leading edge of the upper surface, the first vortex (blue) of a second vortex-pair begins to emerge. By Frame

D, the airfoil is approaching the end of its downstroke. The leading-edge vortex on the upper surface continues to develop in situ. The first vortex of the first vortex-pair has shed further from the trailing edge, but still receives vorticity from the airfoil; the second vortex of the pair is convecting as a free vortex.

After examining Frames A, B, C, and D in connection with the measured and computed lift forces in Figure 9, it becomes evident that the lift force is correlated with the leading-edge vortex, but the realization of lift is phase delayed. The counter-clockwise vortex (red) on the lower surface travelling from the leading edge to the trailing edge causes the delay of lift increase, owing to the downward aerodynamic force it induces (Frames A and B). As soon as this vortex sheds from the trailing edge (Frame C), the lift begins increasing significantly. At the same time, the generation of the leading-edge vortex on the upper surface augments the lift increase. The lift reaches its maximum when this vortex takes form and begins to convect downstream.

The upstroke is represented by Frames E to H inclusive; essentially, the development of the flowfield over the airfoil is the reverse of Frames A to D, with the leading-edge vortex developing on the lower surface instead. As we have noted, the mid-span interference of the plunge rod attachment points with the convection of the upper surface vortex occurs during this portion of the motion schedule.

As demonstrated above, vortices are shed alternately from the upper and lower surfaces. The counter-clockwise rotating vortices are shed from the lower surface while the clockwise rotating vortices are shed from the upper surface; however, the shedding starts at the central (neutral) position as seen in Frames C and G for the current pitching-plunging motion. In pure plunging motion, on the other hand, according to Lai and Platzer (1999), shedding takes place when the airfoil has reached its top and bottom positions in the motion profile. This vortex shedding process forms an upper row of counter-clockwise rotating vortices and a lower row of clockwise rotating vortices in the wake, leading to a wake with velocity excess (rather than deficit) or a jet-like vortex pattern corresponding to a thrust-generating body. The thrust generation can be readily verified by the mean velocity profile in the wake (Koochesfahani, 1989). The instantaneous thrust generation will be confirmed later in Section 5.4.

It should be noted that Lai and Platzer (1999) limited the sketch of vortex shedding process at the trailing edge only. Definitive visualization of the leading-edge vortex was lacking. We clearly visualized the vortex forming on the leading edge, convecting downstream, eventually shedding from the airfoil and contributing to the jet-like vortex pattern.

In general, the flowfield around the airfoil has been simulated reasonably well by the current computations. The numerical simulation has captured successfully the alternating development, convection, and shedding of a leading-edge vortex over both the upper and lower surfaces of the airfoil. However, there are deficiencies: first, the free vortex convecting downstream in the wake of the airfoil was not detected well; and second, in the latter stages of the downstroke (Frames C and D) and the upstroke (Frames G and H), the vorticity sheet feeding the vortex shed from the trailing edge of the airfoil was also not reproduced well. These are attributed mainly to an insufficient grid density in the second-order computations, in particular, in the region extending one chord length from the trailing edge of the airfoil into its wake. The coarse density of the grid generates numerical dissipation which tends to smear the vortex-shedding process. On the other hand, the numerical dissipation introduced by the first-order numerical scheme smeared intensively the free vortices in the wake. Nevertheless, these deficiencies did not affect the prediction of the flowfield on the airfoil surfaces and the resulting aerodynamic performance.

Earlier, we introduced the notion of an asymmetric wake. This phenomenon is illustrated in the PIV and CFD results presented in Figure 10. Consider the leading-edge vortices shed from the airfoil near the end of the downstroke ("LEV D_L" in Frame D) and near the end of the upstroke ("LEV H_U" in Frame H). From the PIV results, a cursory sampling of the vorticity magnitudes at the core of these vortices has revealed that the vorticity within LEV D_L is greater than the vorticity within LEV H_U by a factor of approximately two. Farther downstream in the wake, vortices LEV H_L and LEV D_U have dissipated somewhat, having comparable vorticities. Although the motion of the airfoil is symmetric, the wake it creates seemingly is not, which results in an asymmetric lift curve in time. At the end of the downstroke, vortex LEV D_L acts like a counter-clockwise circulation which induces an additional downward force on the upper surface of the airfoil. Understandably, this force is larger than the upward force induced by the vortex LEV H_U – at the end of the upstroke – causing the aforementioned asymmetry of lift in time.

Although our PIV window did not cover a wide range of the downstream wake, it is believed that the asymmetric wake has some similarities with the observations by Koochesfahani (1989) for pure pitching airfoils and by Lai and Platzer (1999) for pure plunging airfoils. They observed vortex streets in the wake region which are inclined at an angle pointing upward or downward relative to the neutral position of the airfoil. As mentioned by Lai and Platzer (1999), by applying numerical simulations, Jones, Dohring, and Platzer (1998) found that the mode of the deflected vortex street is determined by the starting condition of the plunging oscillation. Similarly, further computations in this study, starting from a mirrored or another initial flowfield solution, confirmed that the asymmetric force history is dependent on the initial solution of the flowfield.

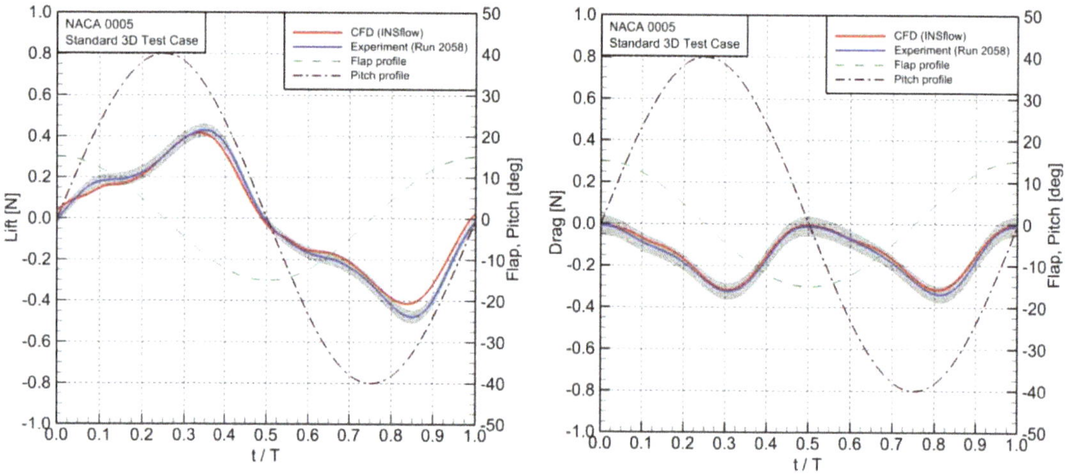

Figure 11. Measured and computed lift and drag forces for the flapping NACA 0005 wing. Standard 3D test case: Re $= 19 \times 10^3$, $U_\infty = 0.0635$ m/s, $f = 0.46$ Hz. The shaded bands represent the uncertainty in the lift and drag measurements.

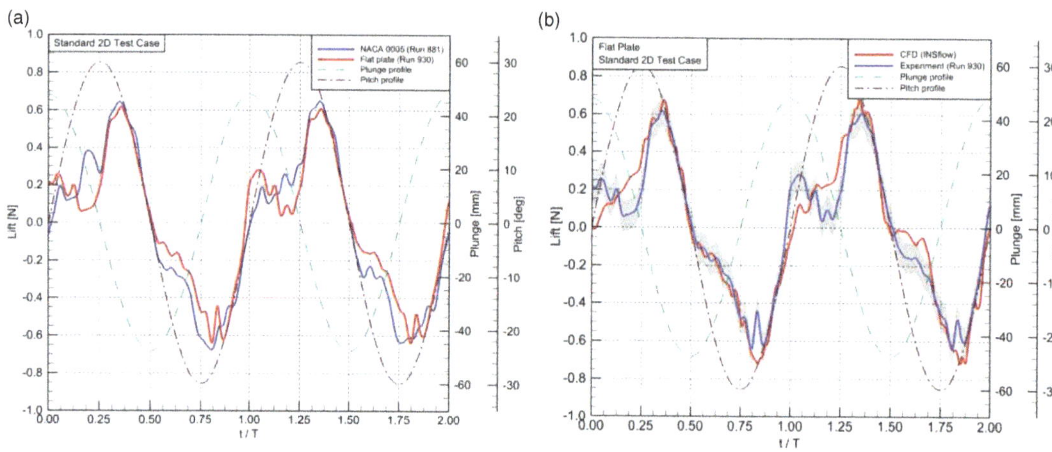

Experiments: NACA 0005 vs. flat plate   Flat plate: numerical vs. experimental

Figure 12. Lift forces for the pitching-plunging NACA 0005 airfoil. Standard 2D test case: Re $= 10.5 \times 10^3$, $U_\infty = 0.0635$ m/s, $f = 0.5$ Hz.

5.2. Standard 3D test case

One cycle of measured and computed lift and drag forces are superimposed in Figure 11. Again, the experimental results are a subset of an ensemble-average of 24 three-cycle time series; the CFD results are from the fourth cycle of a four-cycle computation. The motion profiles, again, synchronize the presentation of the two data sets. The experimental results presented here reflect a correction that was not detected prior to publication of the results elsewhere (Malhan, Lakshminarayan, Baeder, & Chopra, 2011; Yuan et al., 2010) – the correction rectifies an error in the buoyancy compensation algorithm of the data reduction process.

Similar to the results of the 2D test case, there is good agreement between the measured and computed results. There is repeatable behavior with the flapping schedule and a phase shift between the two sets of results is not noticeable. For the most part, the computed lift force falls within the uncertainty band of lift measurement, while the computed drag force agrees very well with the drag measurement and its associated uncertainty. The peaks in the measured lift and drag forces exhibit an asymmetry in time which is attributed to the interaction of the wing model with its asymmetric wake. The computed lift and drag forces, on the other hand, are symmetric, indicating that the simulation has not captured the interaction of the wing model and an asymmetric wake, which we suspect exists based on the findings from the 2D test case. The simulation may have missed the interaction because the computed results are the product of

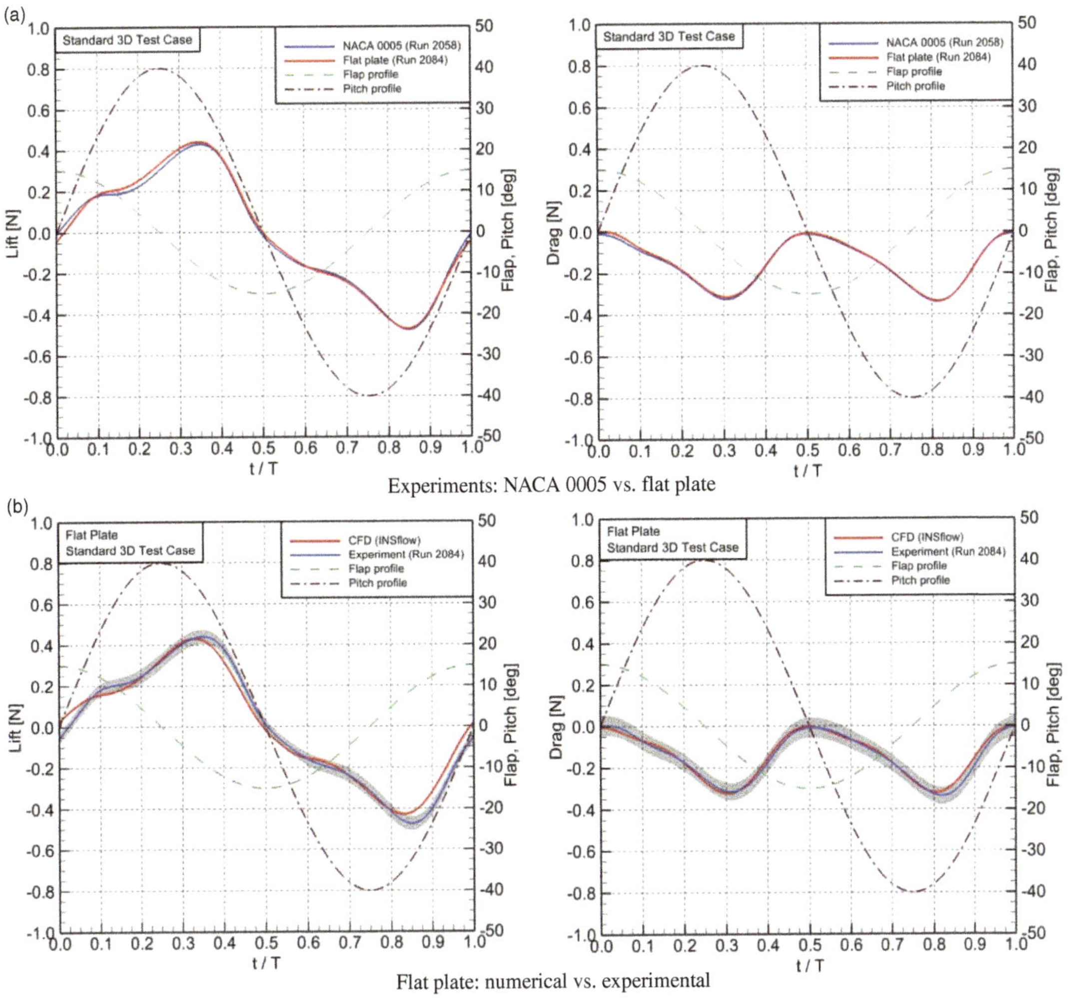

Figure 13. Lift and drag forces for the flapping NACA 0005 and flat-plate wings. Standard 3D test case: Re $= 19 \times 10^3$, $U_\infty = 0.0635$ m/s, $f = 0.46$ Hz.

first-order calculations which tended to be too dissipative.

5.3. *Effect of cross-section profile*

5.3.1. *Standard 2D test case*

The measurements of lift force acting on an airfoil with a NACA 0005 cross-sectional profile and a flat plate of constant thickness (i.e., 5% of chord) are compared in Figure 12. The two sets of results agree reasonably well, although the peak lift developed by the flat plate is slightly less than that developed by the NACA 0005 airfoil. The discrepancies between the numerical and the experimental results are also marginal. The good agreement suggests the development of aerodynamic loading is driven mainly by the kinematics of the airfoil, not its cross-sectional profile. The suggestion is founded on the general view that, in unsteady aerodynamics, lift is composed of noncirculatory and circulatory components (Nguyen, 2008). Although the components cannot be separated from each other in this study, the current results suggest that the noncirculatory component played a dominant role in the test cases. The noncirculatory component is due to the apparent mass and inertia effects which are generated when the wing undergoes a nonzero acceleration; specifically, it is a reaction to the inertial force which develops when the fluid experiences this nonzero acceleration. The circulatory component is more important for wing sections because of the vortical intensity of circulation which generates lift (Nguyen, 2008).

5.3.2. *Standard 3D test case*

The measurements of lift and drag forces on wing models with a NACA 0005 cross-sectional profile and a profile of constant thickness (flat plate) are compared in Figure 13.

Similar to the 2D test case (Figure 12), the two sets of data agree very well, but the peak lift force developed by the flat-plate wing is not slightly higher. The numerical results compare well with the experimental measurements. The good agreement reinforces the earlier suggestion that the development of aerodynamic loading is influenced largely by the kinematics of the airfoil/wing.

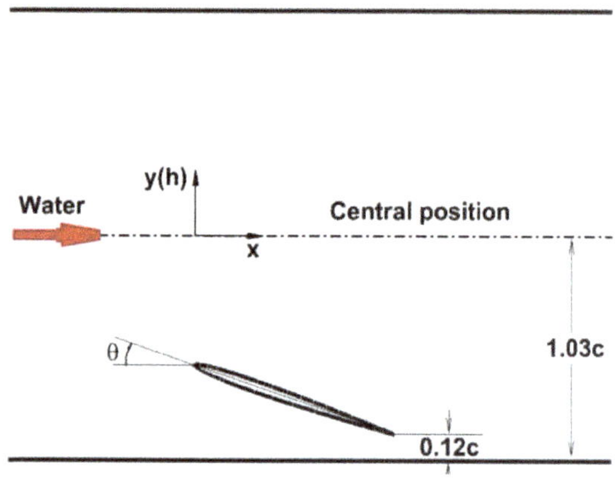

Figure 14. The setup of the reflection planes simulates clap-fling wing-wing interaction.

5.4. *Clap-fling effect of the wing-wing interaction*

Reflection planes were used to simulate experimentally and numerically the clap-fling effect associated with the wing-to-wing interaction. The flow condition was the same as the standard 2D test case. Originally, the closest position of the reflection planes was proposed to be located at 1.0 chord from the central position of the airfoil. Due to mechanical

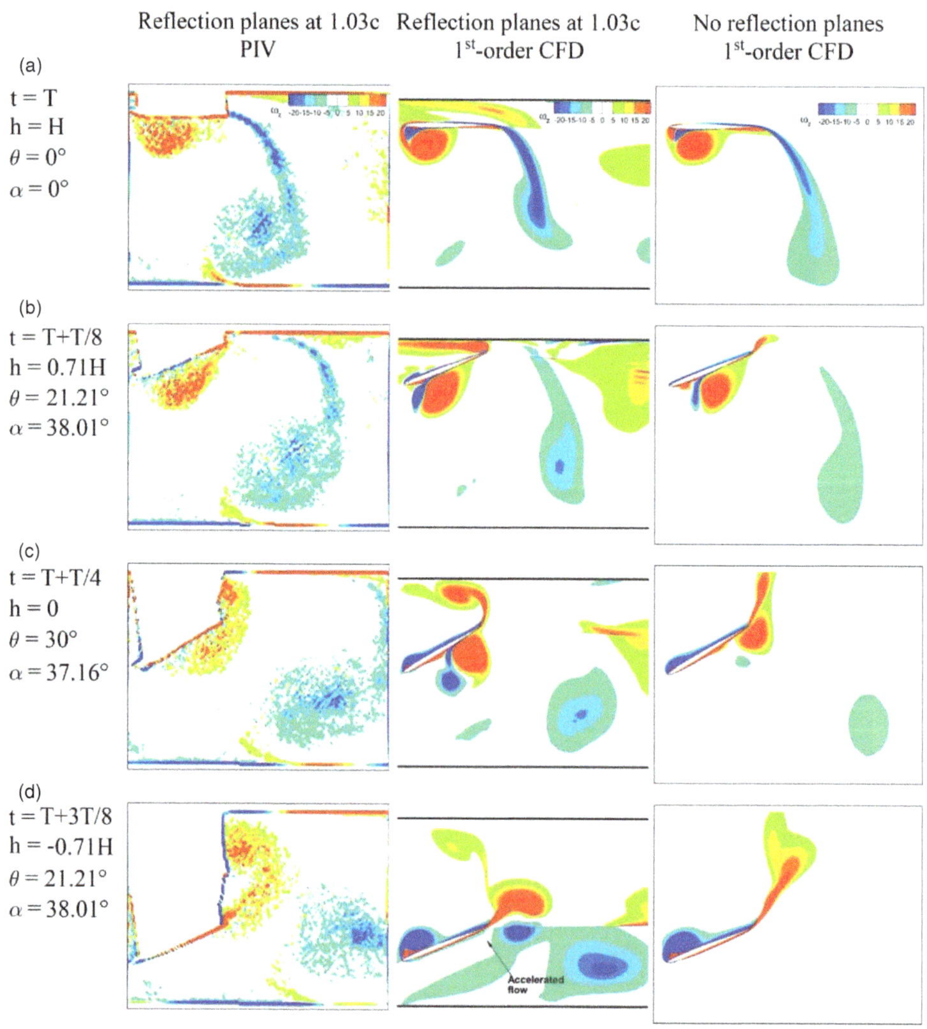

Figure 15. Downstroke: Spanwise vorticity (ω_z) distributions over the pitching-plunging NACA 0005 airfoil at Re = 10.5×10^3.

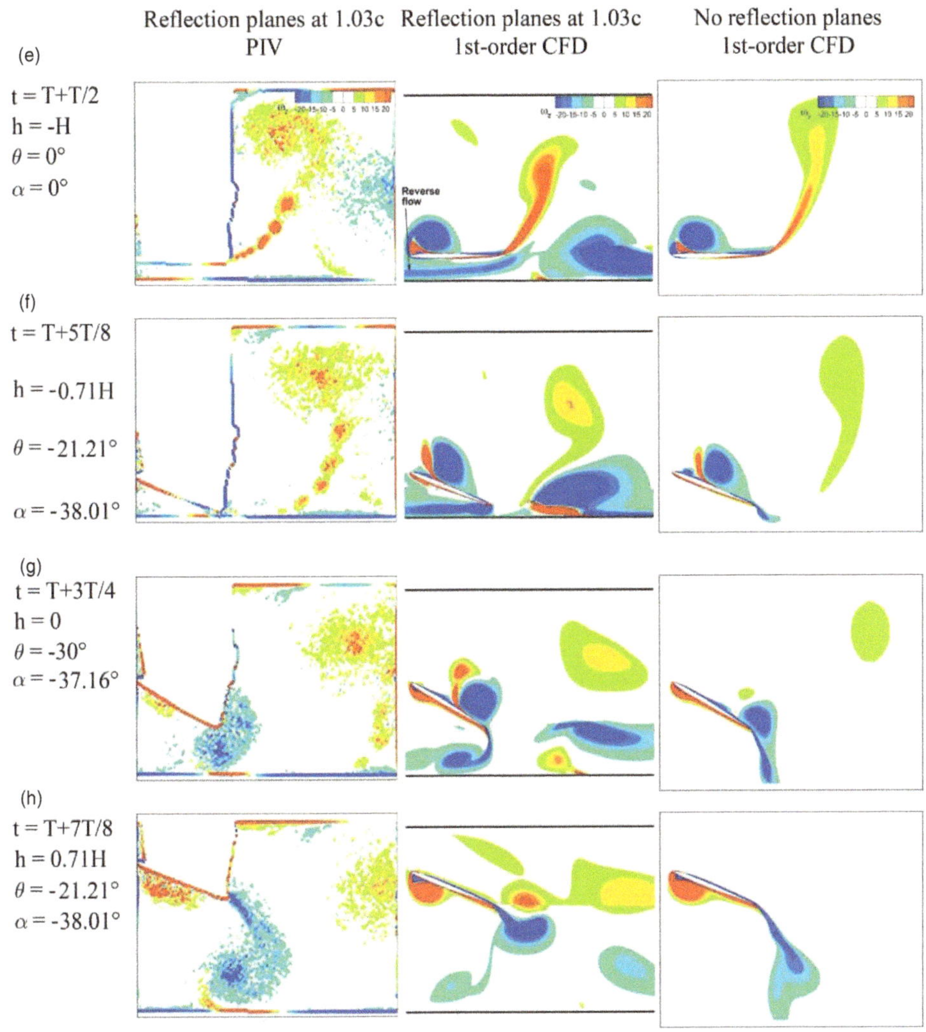

Figure 15a. Upstroke: Continued from previous page.

restrictions of the 2D motion rig and the presence of the reflection planes in the test section, the final position of the reflection planes was chosen to be 1.03 chords, as shown in Figure 14. Accordingly, the closest distance between the airfoil and the reflection planes was $0.12c$ rather than originally planned $0.088c$. The longitudinal length of the reflection planes was 6 chords (38 cm) in the experiments while it was 21 chords in the CFD simulations. In the numerical simulations, the reflection planes were assumed to be symmetry planes, which indicate fully symmetric flowfields about the reflection planes or, in reality, between interacting wings. The computational domain was split into two blocks (meshes) as the upper and lower parts. Two-dimensional H-type dynamic meshes were used, which allowed for mesh deformation during the computations. Each mesh had 497×129 grid points. Laminar flow was assumed. Since numerical instabilities were experienced in the second-order computations, results from first-order simulations are reported only.

The PIV-measured and CFD-computed flowfields of one pitching-plunging cycle are depicted in Figure 10. The PIV results are the product of an ensemble-average over 23 cycles. The CFD results presented are for an instantaneous solution. For the sake of intuitive comparisons, the first-order CFD results of the standard no-reflection-plane case are included in the right column. In the CFD simulations for the no-reflection-plane case, the third cycle produced results comparable to the second cycle; however, for the computations with the reflection planes, more than 20 cycles were needed to converge to a steady state of the periodic solution. Although there are discrepancies between the experimental and numerical results, again,

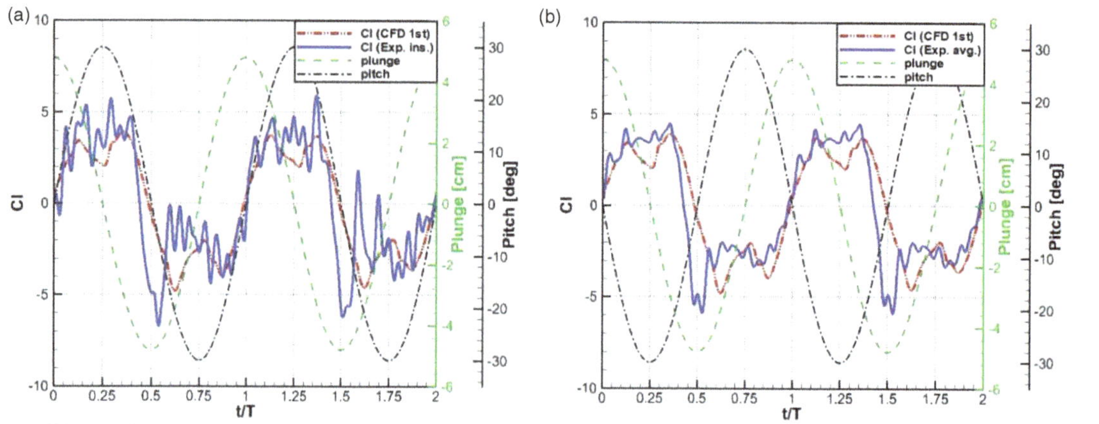

Measured instantaneous (cycles 9 & 10) vs. computed

Ensemble-averaged experimental (23 cycles) vs. Computed

Figure 16. Computed lift coefficients compared with experimental results for the case with reflection planes located at 1.03c from the central position.

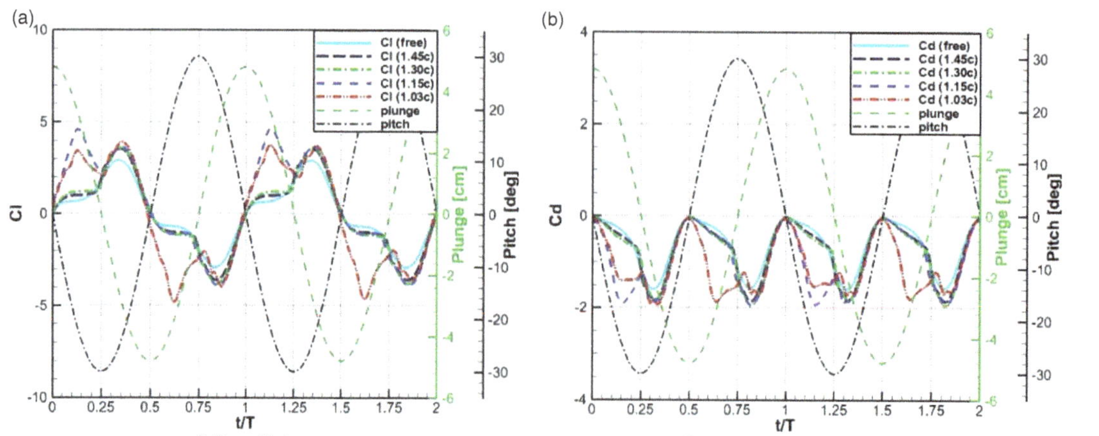

Lift coefficients

Drag (thrust) coefficients

Figure 17. Effect of the proximity of the reflection plane on the lift and drag coefficients. First-order CFD results.

the first-order CFD solution captured the major physical features of vortex generation and shedding.

As shown in Figure 15, the flowfields showed some similarities to those of the test case without reflection planes. However, noticeable differences were observed:

a) The sizes of the vortices are larger in the wing-wing interaction as compared with those in the no-reflection-plane case, suggesting that the reflection planes, or wing-wing interactions, intensify generation of the vortices.
b) During the downstroke period of the cycle (Frame B), the clockwise-rotating leading-edge vortex (in blue) on the upper surface of the airfoil occurred earlier, compared with the no-reflection-plane condition. Furthermore, an additional counter-clockwise trailing-edge vortex (in red) was also generated on the upper surface. These features are believed to intensify the generation of thrust, as discussed later.
c) At the same time, it appears that suction at the upper side affects vortex shedding on the lower surface. As shown in Frames B and C, the leading-edge vortices (i.e., the counter-clockwise vortex in red on the lower surface) generated during the previous half cycle, shed slower than those in the no-reflection-plane case. In Frame C, the counter-clockwise vortex has just reached the trailing edge; at this instance in the cycle, it had already started shedding from the trailing edge at the no-reflection-plane case.
d) When the airfoil approached the bottom reflection plane (Frame D), an accelerated flow was observed on the lower surface. The flow speed reached five times higher than the freestream speed, while it was about only three times the freestream speed

in the no-reflection-plane case. In Frame E, a flow reversal was observed between the airfoil and the reflection plane, as a result of the downwards push from the airfoil trailing edge.

Figure 16 compares the computed lift coefficient with that obtained in the experiment. While both instantaneous and ensemble-averaged experimental data are presented here, only instantaneous CFD results are depicted since the first-order CFD simulations produced a nearly periodic solution as soon as the computations converged. Although apparent discrepancies between the numerical and experimental results are noticeable, the agreement is suitable for the validation of an engineering tool.

Figure 17 further compares the effects of the reflection plane location on the aerodynamic coefficients. In general, the reflection planes intensified the generation of the aerodynamic forces at the test conditions. The closer the reflection planes are, the stronger the observed effects. Table 2 summarizes the reflection plane effects on the thrust generation and propulsion efficiency. Negative drag coefficients

Table 2. Summary of the clap-fling effect in wing-wing interaction (1st-order CFD results).

Reflection plane location	Averaged drag coefficient Cd	Propulsion efficiency η_p
No planes (Freestream)	– 0.64	25.4%
1.45c	– 0.74	24.7%
1.30c	– 0.78	24.7%
1.15c	– 0.99	24.5%
1.03c	– 1.08	24.6%

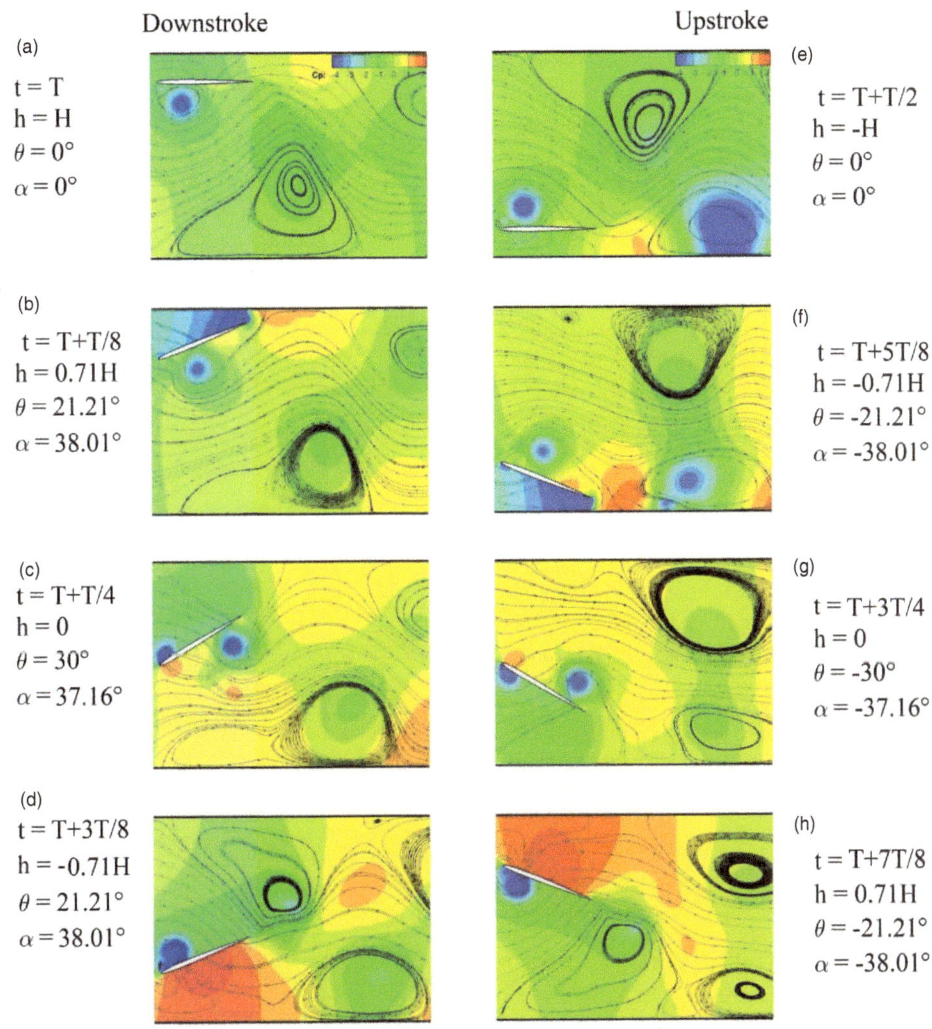

Figure 18. Computed pressure distribution and streamlines over the pitching-plunging NACA 0005 airfoil with reflection planes located at 1.03c, Re = 10.5×10^3.

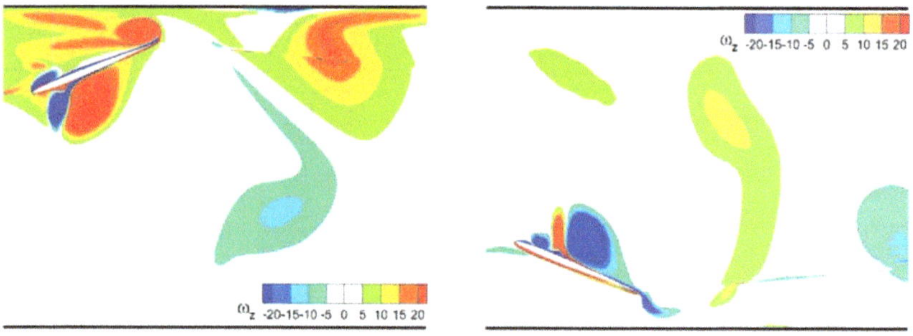

Figure 19. Computed spanwise vorticity (ω_z) distribution over the pitching-plunging NACA 0005 airfoil at Re $= 10.5 \times 10^3$, with reflection planes at 1.15c.

indicate thrust generation. As expected, the clap-fling mechanism improved the thrust generation. At the closest reflection plane placement (1.03c), the thrust coefficient is greater by 170%, compared with the no-reflection-plane case. Surprisingly, the propulsion efficiency deteriorated only slightly. For these results, the propulsion efficiency was defined as

$$\eta_p = \frac{\bar{C}_{Power,out}}{\bar{C}_{Power,in}} = \frac{-(1/TU_{ref})\int_0^T C_d U_\infty dt}{(1/TU_{ref})\int_0^T (C_l \dot{h} + C_m \dot{\theta} c)dt}. \quad (6)$$

This definition has countered for both the translation power input (for force acting on the airfoil) and the rotational (pitching) power input (for torque acting on the airfoil) in the denominator and is commonly used in the flapping-wing community (e.g. by Fitzgerald, Valdez, Vanella, Balaras, & Balachandran, 2011; Guglielmini & Blondeaux, 2003; Windte & Radespiel, 2008; Young & Lai, 2007). In the case when the power required for pitching motion is much smaller than that for plunging motion, the second term in the denominator can be ignored (Sunada et al., 2001). In the present study, the contribution of the pitching moment was included in the definition for completeness. However, the power input for the pitching motion was approximately 1% of the total input power.

After examining flowfield (Figure 15) in connection with the computed aerodynamic coefficients in Figure 17, it becomes evident that the augmentation of the thrust force is correlated with the vortex formation. Compared with the no-reflection-plane case, the thrust coefficient at the condition with the reflection planes located at 1.03c showed two extra peaks at t/T $= 0.125$ and 0.625. At these time instants, the airfoil moved apart from the reflection planes (plots B and F in Figure 15) and additional vortices were generated at the trailing edge. These additional trailing-edge vortices resulted in a large area of low pressure near the trailing edge on the forward side of the airfoil (cf. plots B and F in Figure 18). This low pressure contributed significantly to the augmentation of the thrust generation. On the other hand, when the airfoil approached the reflection planes (plots D and H in Figure 15), the accelerated flow between the airfoil and the near planes augmented also slightly to the thrust generation as shown in the thrust coefficients at t/T $= 0.25 - 0.375$ and $0.75 - 0.875$.

It is interesting that the test case with reflection planes located at 1.15c showed a completely different asymmetry with time. In Figure 17, a peak in the lift coefficient was observed clearly at t/T $= 0.125$, but not correspondingly at t/T $= 0.625$, which is different from the 1.03c case. In order to help understand the relevant flow physics or correlations, the flowfield of the selected test cases at these time instances in the cycle are illustrated in Figure 19. At t/T $= 0.125$, as the airfoil plunged downward from the upper reflection plane, leading-edge and trailing-edge vortices were observed on the upper surface of the airfoil, which is similar to that of the 1.03c case (Frame B in Figure 15). A flow reversal was also observed between the airfoil and the upper reflection plane. However, this was not the case for t/T $= 0.625$; neither the leading-edge vortex, nor the reverse flow was observed on the lower surface of the airfoil. This complex phenomenon suggests that the wings should operate in proximity close enough to another at extremes of stroke.

6. Conclusions

We have successfully performed, in a water tunnel, the standard test cases for a pitching-plunging 2D airfoil and a flapping 3D wing, as part of a collaborative partnership with DRDC Valcartier and Advanced Subsonics Inc. Force and particle imaging velocimetry measurements were carried out to acquire sets of data that are considered acceptable for the validation of corresponding CFD simulations. The experimental data have been presented specifically alongside the corresponding computational results produced by INSflow, an NRC in-house flow solver. In general there is good agreement between the experimental and numerical results. The analyses and comparisons of the results have confirmed that our experimental approach has produced a realistic simulation. Reliable experimental data can be acquired for code validation purposes, and reliable CFD data can be generated for the development of flapping wings.

The investigation has visualized the process of vortex forming on the airfoil leading edge, convecting downstream, and eventually shedding to the wake. The visualization confirmed the major contribution of the leading-edge vortex to the reverse Karman street – a wake with velocity excess or a jet-like vortex pattern. As an ancillary finding of this research work, a perfunctory investigation of the PIV and CFD results revealed that the immediate wake of the pitching-plunging 2D airfoil is not symmetric, i.e., the leading-edge vortices of the upper and lower surfaces of the airfoil possess significantly different magnitudes of vorticity after shedding from the trailing edge. This finding has prompted the notion that these vortices, in turn, have an asymmetric influence on the loading of the airfoil. Further reflection of this notion, however, is necessary. In addition, a cross-check with the experiments has shown that further refinement of the grid in the wake region is needed, if a more accurate wake solution – in terms of resolving shed vortices — is required from the numerical simulations.

On the other hand, the kinematics of a rigid airfoil or wing – not the cross-sectional profile – is the dominant influence in the development of the lift and thrust forces, based on the current investigation. Also, the formation of the leading-edge vortex on the suction surface of the airfoil/wing augments the generation of aerodynamic forces (lift and thrust); however, the influence of vortices, convecting downstream on the pressure side of the airfoil, causes a delay in the increase of these forces.

Finally, the investigation has confirmed that the use of wing–wing interaction, known as the clap-fling effect (Ellington, 1984), intensifies the generation of the aerodynamic forces. The augmentation of the thrust generation is mainly due to the formation of the trailing-edge vortices in addition to the leading-edge ones, when they peel apart. To realize the clap-fling effect, flapping wings should operate in close proximity to another at extremes of stroke.

Facilitating the desired approach to measure the drag force for 2D test cases and PIV techniques for 3D test cases would further help improving the data quality and elucidating the insights of flow physics. Further studies on the wake flow, including signal spectrum analysis, would help quantify the contribution from both leading-edge and trailing-edge vortices to the reverse Karma vortex street.

Acknowledgments

The authors wish to thank Francois Lesage of DRDC Valcartier – for his leadership and management of the collaboration – and the Technology Investment Fund (TIF) from Department of National Defence Canada. We are also thankful for the meaningful discussions with other members of the team, namely Patrick Zdunich (formerly of Advanced Subsonics, currently of the Flight Research Laboratory at the NRC) and Nicolas Hamel (DRDC Valcartier). Patrick's role as the diligent custodian of the standard test cases was greatly appreciated. Special thanks are also given to our colleagues at the NRC: Dr. Stuart McIlwain, who has played a supervision role in the early stage of the experimental preparation; Pierre Gravelle, who assembled, wired and calibrated the bi-fold five-component balance; and Alain Bilodeau, for the excellent craftsmanship he invested in the fabrication of the two-dimensional airfoil and three-dimensional wing models.

References

Azuma, A., Okamoto, M., & Yasuda, K. (2001). Aerodynamic characteristics of wings at low Reynolds number. In T. J. Muell (Ed.), *Fixed and flapping wing aerodynamics for micro air vehicle applications* (pp. 341–398). Reston, VA: American Institute of Aeronautics and Astronautics.

Betz, A. (1912). Ein Beitrag zur Erklaerung des Segelfluges. *Zeitschrift fuer Flugtechnik und Motorluftschiffahrt, 3*(21), 269–272.

Chorin, A. J. (1968). Numerical solution of the Navier-Stokes equations. *Mathematics of Computation, 22*: 745–762.

DeLaurier, J. D. (1993a). An aerodynamic model for flapping-wing flight. *The Aeronautical Journal, 97*(964), 125–130.

DeLaurier, J. D. (1993b). The development of an efficient ornithopter wing. *The Aeronautical Journal, 97*(965), 153–162.

Demirdžić, I., & Perić, M. (1990). Finite volume method for prediction of fluid flow in arbitrarily shaped domains with moving boundaries. *International Journal for Numerical Methods in Fluids, 10*, 771–790.

Dickinson, M. H., Lehmann, F. O., & Sane, S. P. (1999). Wing rotation and the aerodynamic basis of insect flight. *Science. 284*: 1954–1960.

Ellington, C. P. (1984). The aerodynamics of hovering insect flight. *Philosophical Transactions of the Royal Society of London, Ser. B Biological Sciences, 305*(1122), 79–113.

Fairuz, Z. M., Abdullah, M. Z., Yusoff, H., & Abdullah, M. K. (2014). Fluid structure interaction of unsteady aerodynamics of flapping wing at low Reynolds number. *Engineering Applications of Computational Fluid Mechanics, 7*(1), 144–158.

Ferziger, J. H., & Perić, M. (1996). *Computational methods for fluid dynamics*. Berlin: Springer-Verlag.

Fitzgerald, T., Valdez, M., Vanella, M., Balaras, E., & Balachandran, B. (2011). Flexible flapping system: computational investigations into fluid-structure interactions. *The Aeronautical Journal, 115*(1172), 593–603.

Freymuth, P. (1990). Thrust generation by an airfoil in hover modes. *Experiments in Fluids, 9*(1), 17–24.

Guglielmini, L., & Blondeaux, P. (2003). Propulsive efficiency of oscillating foils. *European Journal of Machanics B/Fluids, 23*, 255–278.

Hall, K., Pigott, S., & Hall, S. (1997, January 6–9). *Power requirements for large-amplitude flapping flight*. Proceedings of 35th AIAA Aerospace Sciences Meeting & Exhibit, Reno, NV, AIAA paper 1997–0827.

Hall, K., & Hall, S. (2001). A rational engineering analysis of the efficiency of flapping flight. In T. J. Mueller (Ed.), *Fixed and flapping wing aerodynamics for micro air vehicle applications* (pp. 249–274). Reston, VA: American Institute of Aeronautics and Astronautics.

Isaac, K. M., Rolwes, J., & Colozza, A. (2008). Aerodynamics of flapping and pitching wing using simulations and experiments. *AIAA Journal, 46*(6), 1505–1515.

Jones, K. D., Dohring, C. M., & Platzer, M. P. (1998). Experimental and computational investigation of the Knoller–Betz effect. *AIAA Journal, 36*(7), 1240–1246.

Knoller, R. (1909). Die Gesetze des Luftwiderstandes. *Flug- und Motortechnik (Wien), 3*(21), 1–7.

Koochesfahani, M. (1989). Vortical patterns in the wake of an oscillating airfoil. *AIAA Journal, 27*(9), 1200–1205.

Lai, J. C. S., & Platzer, M. F. (1999). Jet characteristics of a plunging airfoil. *AIAA Journal, 37*(12), 1529–1537.

Lai, J. C. S., & Platzer, M. F. (2001). Characteristics of a plunging airfoil at zero freestream velocity. *AIAA Journal, 39*(3), 531–534.

Lee, R., Yuan, W., Levasseur, L., & Hoogkamp, E. (2011, June 27–30). *Experimental simulation of flapping wings for nano-air-vehicles*. Proceedings of 29th AIAA Applied Aerodynamics Conference, Honolulu, Hawaii, AIAA paper 2011–3790.

Lesage, F., Hamel, N., Yuan, W., Khalid, M., Huang, X., & Zdunich, P. (2008, August 18–21). *Aerodynamic study of a flapping-wing NAV using a combination of numerical and experimental methods*. Proceedings of 26th AIAA Applied Aerodynamics Conference, Hawaii, AIAA 2008–6396.

Liu, H., Ellington, C. P., Kawachi, K., van Den Berg, C., & Willmott, A. P. (1998). A computational fluid dynamic study of hawkmoth hovering. *The Journal of Experimental Biology, 201*, 461–477.

Liu, H., & Kawachi, K. (1998). A numerical study of insect flight. *Journal of Computational Physics, 146*, 124–156.

Malhan, R., Lakshminarayan, V. K., Baeder, J., & Chopra, I. (2011, January 25–27). *Investigation of aerodynamics of rigid flapping wings for MAV applications: CFD validation*. Proceedings of AHS Specialists Conference of the American Helicopter Society, Tempe, Arizona.

Mueller, T. J. (2001). *Fixed and flapping wing aerodynamics for micro air vehicle applications*, Reston, VA: American Institute of Aeronautics and Astronautics.

Nguyen, N. (2008, January 7–10). *Integrated flight dynamic modeling of flexible aircraft with inertial force-propulsion-aeroelastic coupling*. Proceedings of 46th AIAA Aerospace Sciences Meeting and Exhibit, Reno, NV, AIAA paper 2008–194.

Ol, M. (2010). *AVT-149 task group. Unsteady aerodynamics for micro air vehicles* (Final Report of Task Group AVT-149, RTO Technical Report). North Atlantic Treaty Organization.

Patankar, S. V. (1980). *Numerical heat transfer and fluid flow*, Washington: Hemisphere Publishing Corporation.

Platzer, M., Jones, K., Young, J., & Lai, J. (2008). Flapping-wing aerodynamics: progress and challenges. *AIAA Journal, 46*(9), 2136–2149.

Sane, S. P., & Dickinson, M. H. (2001). The control of flight force by a flapping wing: lift and drag production. *The Journal of Experimental Biology, 204*, 2607–2626.

Shyy, W., Lian, Y., Tang, J., Viieru, D., & Liu, H. (2008). *Aerodynamics of low Reynolds number flyers*, New York: Cambridge University Press.

Sunada, S., Kawachi, K., Matsumoto, A., & Sakaguchi, A. (2001). Unsteady forces on a two-dimensional wing in plunging and pitching motions. *AIAA Journal, 39*(7), 1230–1239.

Weis-Fogh, T. (1973). Quick estimates of flight fitness in hovering animals, including novel mechanisms for lift production. *Journal of Experimental Biology, 59*, 169–230.

Windte, J., & Radespiel, R. (2008). Propulsive efficiency of a moving airfoil at transitional low Reynolds numbers. *AIAA Journal, 40*(9), 2165–2177.

Young, J. (2005). *Numerical simulation of the unsteady aerodynamics of flapping airfoils* (PhD thesis, School of Aerospace, Civil and Mechanical Engineering). New South Wales: The University of New South Wales, Australian Defence Force Academy.

Young, J., & Lai, J. (2007). Mechnisms influncing the efficiency of oscillating airfoil propulsion. *AIAA Journal, 45*(7), 1695–1702.

Young, J., Lai, J. C. S., & Germain, C. (2008). Numerical simulation and parameter variation of insect wing motion based on dragonfly hovering. *AIAA Journal, 46*(4), 918–924.

Yuan, W., Khalid, M., Windte, J., Scholz, U., & Radespiel, R. (2007). Computational and experimental investigations of low-Reynolds-number flows past an airfoil. *The Aeronautical Journal, 111*(1115), 17–29.

Yuan, W., & Khalid, M. (2008). Preliminary CFD studies of flapping-wing aerodynamics. *Canadian Aeronautics and Space Journal, 54*(3/4), 51–63.

Yuan, W., Khalid, M., & Huang, X. (2008, August 18–21). *Computations of flows past an insect-like flapping wing*, Proceedings of 26th AIAA Applied Aerodynamics Conference, Hawaii, AIAA 2008–6395.

Yuan, W., Lee, R., Hoogkamp, E., & Khalid, M. (2010). Numerical and experimental simulations of flapping wings. *International Journal of Micro Air Vehicles, 2*(3), 181–209.

Yuan, W., Poirel, D., & Wang, B. (2013). Simulations of pitch-heave limit-cycle oscillations at a transitional Reynolds number. *AIAA Journal, 51*(7), 1716–1732.

Yuan, W., Poirel, D., Wang, B., & Benaissa, A. (2014). Effect of freestream turbulence on airfoil limit-cycle oscillations at transitional Reynolds numbers. *Journal of Aircraft.* Advanced online publication. doi:10.2514/1.C032807.

Zdunich, P., Bilyk, D., MacMaster, M., Loewen, D., DeLaurier, J., Kornbluh, R., Low, T., Stanford, S., & Holeman, D. (2007). Development and testing of the Mentor flapping-wing micro air vehicle. *Journal of Aircraft, 44*(5), 1708–1711.

Zdunich, P. (2008, August 18–21). *Separated-flow discrete vortex model for nano-scale hovering flapping wings*. Proceedings of 26th AIAA Applied Aerodynamics Conference, Hawaii, AIAA 2008–6245.

Zdunich, P. (2010). *Development and application of a 3D vortex panel model with leading edge separation suitable for hovering flapping-wing flight* (Technical Report). Montreal: Advanced Subsonics.

11

Assessment of experimental optical techniques for characterizing heat transfer using numerical simulations

G.L. Juste[a*] and P. Fajardo[b]

[a]*Aerospace Propulsion and Fluid Mechanics, Universidad Politécnica de Madrid, Plaza Cardenal Cisneros 3, 28040 Madrid, Spain;* [b]*Aerospace Engineering Group. Universidad Carlos III de Madrid, 28911 Leganés, Spain*

This manuscript addresses the application of numerical simulations for assessing the error in the measurement of the bulk temperature along the laser beam of a 3D flow using a 2D Moiré deflectometry analysis. To analyze the effect of different flow parameters on the error, a 3D computational model of an experimental system was developed. The simulated domain represents the well-known solution of the backward facing step in a rectangular channel but includes a hot-plate at the bottom of the step to enhance the heat transfer effects. The geometry resembles that found in a general heat exchanger. The difference between the computed bulk temperature of the flow and the average temperature obtained via the 2D Moiré is analytically evaluated for various assumed general temperature profiles; the numerically computed profiles of temperature indicates that the error decreases with the channel aspect ratio.

The use of CFD enables the determination of the flow topology and thus an evaluation of the 3D flow behavior that will cause the measurement error. A parametric study was performed for different flow conditions, namely, the aspect ratio of the channel, the inflow conditions (flow velocity or Reynolds number), and the temperature of the hot wall. The results indicate that the Moiré technique is suitable for evaluating the bulk temperature in typical heat exchange devices and flow conditions.

Keywords: Moiré deflectometry; error quantification; backward facing step; heat transfer

1. Introduction

The analysis of forced convection heat transfer in closed channels is a very interesting topic for many researchers because it is encountered in many applications, such as gas-turbine blade cooling, as can been seen in the works of Han (2004) and Tanda (2004), or the widely used plate-fin compact heat exchangers (Jang & Kim, 2005).

Moreover, the design of a high effectiveness heat exchanger requires high values of heat transfer area per unit volume, which is achieved by the use of narrow channels as a basic component of such systems (Sehgal, Murugesan, & Mohapatra, 2012). To maintain the pressure drops at a reasonable value, the flow velocity must be lowered, and this, together with the downsizing of passages, results in operation under laminar flow conditions (Chung, Tucker, & Roychowdhury, 2003).

In laminar flows, the effect of turbulence is negligible, and hence, the convective heat transfer coefficient is drastically reduced. To overcome this limit, pin fins inside the flow channel, with abrupt changes of the flow area, can promote turbulence to enhance the convective heat transfer rate (Togun, Kazi, & Badarudin, 2011). As consequence of this increment of area in the flow direction, flow separation and posterior reattachment might appear, and an adequate knowledge of the temperature field inside these closed channels is required for further optimization.

The availability of sophisticated numerical methods has not diminished the need for measurements, and experimental quantitative and qualitative observations are still required for developing a comprehensive phenomenological understanding of the flow-field phenomena and for validation of numerical models.

Different optical techniques have been developed and are commonly used in experimental heat transfer studies. A review of those techniques can be found in the literature in works of Merzkirch (1987) and Newport, Sobhan, and Garvey (2008). Due to the nonintrusivity of such optical techniques, which allows for obtaining local heat transfer data without influencing the investigated process, these techniques are well-suited for the verification of numerical results.

Among the existing optical techniques, those methods relying on the changes of the refractive index of the fluid, such as interferometry, holographic interferometry, or Moiré deflectometry, can be used in flows with uniform pressure to determine simultaneously the temperature field at every point of the selected spatial region. A comprehensive review of the advantages and disadvantages of different experimental techniques for the temperature measurement in flows has been presented by Shakher and Nirala (1999), Naylor (2003), Childs, Greenwood, and Long (2000) and Jordan, Tauscher, and Mayinger (1997).

*Corresponding author. Email: gregorio.lopez@upm.es

These optical techniques are line-of sight methods, and therefore, they are influenced by the path of transmitted light, requiring the integration of the flow variables along the path. Consequently, the measurement of the refractive index gradient, or density gradient, distribution in the flow is straightforward only in the two-dimensional case and in the case of axisymmetric flow, by means of the Abel transform.

For a general three-dimensional field, a tomographic technique is applied to the two-dimensional fringe data obtained from multiple views to reconstruct the complete temperature field, as reported in Muralidhar (2012), Hirano, Kawaguchi, Satoh, and Saito (2006), and Michael and Yang (1992). However, there are configurations in which, due to the geometry, it is not possible to obtain several interferograms with different viewing angles, making the reconstruction impossible.

In applications in which the flow movement occurs mainly in one direction, such as the flow in rectangular closed channels, an alternative could be to obtain the two-dimensional bulk temperature in a plane parallel to the channel wall and the main direction of the flow. Obviously, this bulk temperature will depend on the velocity and temperature profiles in the plane perpendicular to the mainstream. These three-dimensional phenomena introduce errors on the bulk temperature measurement using two-dimensional techniques. The objective of this paper is to assess the effect of the flow variables on these errors. The amount of error is determined using CFD simulations to evaluate the difference between the beam-averaged and the bulk average temperature of the flow.

Several works in the literature report interferometric beam-averaged temperature and heat transfer data in a three-dimensional temperature field. For example, Naylor and Machin (2001) performed an investigation of beam-averaged temperature measurements in a three-dimensional field by the use of a two-dimensional analysis, but the measured average temperature did not agree with the defined bulk temperature. Juste and Benavides (2011) studied the end-wall errors in temperature measurement with Moiré deflectometry, but the technique was applied to external convective flow and mainly considered the edge effects only.

In the present work, the error produced when using two-dimensional Moiré deflectometry analysis to evaluate the bulk temperature of a three-dimensional internal flow is assessed. The analysis is performed in a closed rectangular channel that is straight and has a sudden expansion of flow area. This geometry is representative of many heat exchange applications and other engineering fields in which flow detachment and reattachment occur (Dufresne, Dewals, Erpicum, Archambeau, & Pirotton, 2011). The analysis is performed using a 3D-CFD numerical simulation of the laminar flow in the widely known backward facing step (BFS) channel configuration to provide a complete description of the expected flow topology, i.e., the velocity and temperature fields. The BFS configuration is commonly used as a benchmark for numerical analyses, and therefore, its solution is well known (Rajabi & Kavianpour, 2012). The heat transfer mechanism is modeled by a heated plate at the foot of the step. The effect of the parameters of the aspect ratio of the channel, the Reynolds number, the maximum temperature difference and the temperature distribution along the beam are analyzed.

2. Fundamentals of the laser Moiré deflectometry

To understand the error committed in the measurements due to the 3D flow phenomena, the basis of the experimental method should be reviewed first (as presented in Juste and Benavides, 2014). A drawing of a typical Moiré deflectometry configuration for temperature measurements in a backward-facing step is shown in Figure 1.

Moiré deflectometry obtains the ray deflections caused by the change of refractive index through phase objects. A parallel light beam passes, first, through the flow field, and then, through two identical gratings G1 and G2 separated by a pitch P and tilted an angle θ. Finally, the light reaches a screen, where an overlapped Moiré pattern is observed or registered with a camera.

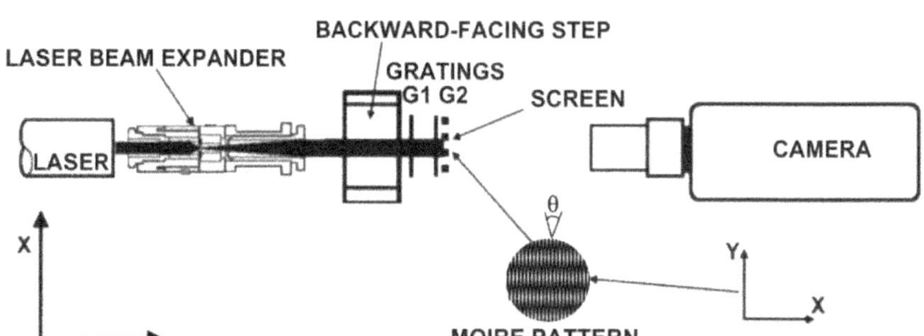

Figure 1. Diagram of the experimental procedure for Moiré deflectometry obtained from Juste and Benavides (2014).

If the flow field is uniform, a straight and unperturbed Moiré pattern with spatial period, P', of Moiré fringes is obtained. This unperturbed solution is given by:

$$P' = \frac{P}{2\sin(\theta/2)} \approx \frac{P}{\theta} \tag{1}$$

Disturbances in the flow induce a shift s at each point of the fringe and a deflection angle ϕ, which are linked with the refractive index n by means of the following system of equations:

$$\phi = \frac{1}{n_\infty}\int_o^L \frac{\partial n}{\partial y}dz \tag{2}$$

$$\phi = \frac{2s\,\sin(\theta/2)}{D} \tag{3}$$

The phase change φ of the Moiré pattern is also related to the displacement of the fringes and thus to the refractive index gradient. These dependencies are given by the following equations:

$$\varphi = \frac{2\pi s}{P'} \tag{4}$$

$$\varphi = \frac{2\pi D}{P}\frac{1}{n_\infty}\int_0^L \frac{\partial n}{\partial y}dz \tag{5}$$

If the flow field is two-dimensional, then the refractive index gradient in the fringe direction can be obtained as

$$\frac{\partial n}{\partial y} = \frac{P n_\infty}{2\pi D L}\varphi \tag{6}$$

Applying the Gladstone–Dale law, the density can be obtained from the molar refractivity of the fluid in the following form:

$$n - 1 = K\rho \tag{7}$$

Considering the flow pressure to be constant, the temperature field can be calculated from the density field through the use of the ideal gas law, if the pressure and gas composition are known ($R_g$being the Gas Constant):

$$T = \frac{p}{R_g \rho} \tag{8}$$

An excellent theoretical review of refractive index mapping by Moiré deflectometry is presented in Kafri and Glatt (1990).

3. Evaluation of two-dimensional bulk temperature

In general, the movement of a fluid through a pipe or duct can be approximated to be *one-dimensional.* This means that the flow properties can be assumed to vary in the direction of flow only. As a result, all properties are assumed to be uniform at any cross-section normal to the flow direction, and the properties are assumed to have a *bulk average value* over the entire cross-section (Lienhard & Lienhard, 2000). The bulk average temperature is computed as:

$$T_b = \frac{\int_A \rho u c_P T dA}{\int_A \rho u c_P dA} \tag{9}$$

In this way, the energy flowing through a given cross-section is equal to $\dot{m}\bar{c}_p T_b$,where $\bar{c}_p$is an average value. If all of the thermodynamic coefficients, e.g., c_P, are considered

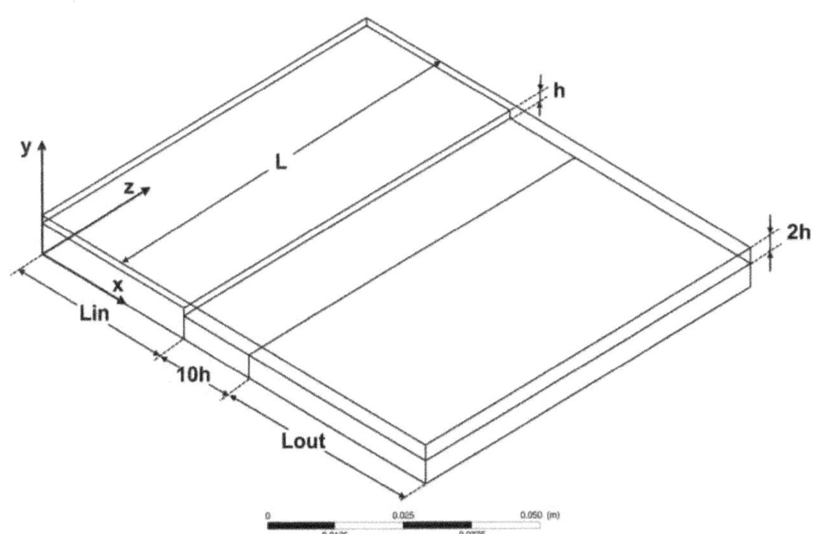

Figure 2. Sketch of the computational domain used in the simulations. The surface below the step (length 10 h) is heated.

to be constant, Equation (9) is reduced to

$$T_b = \frac{\int_A uTdA}{\int_A udA} \tag{10}$$

In a rectangular channel, as the one shown in Figure 2, the concept can be extended approximating the real flow to be two-dimensional, where the properties can be assumed to vary in the direction of flow (x axis) and in a perpendicular direction to the flow (y axis). The properties are then assumed to have *bulk average values* over the entire z-axis.

In this case, the bulk-averaged temperature along the z-axis is calculated as follows

$$T_b(x,y) = \frac{\int_{z_1}^{z_2} \rho(x,y,z)u(x,y,z)T(x,y,z)dz}{\int_{z_1}^{z_2} \rho(x,y,z)u(x,y,z)dz} \tag{11}$$

Considering the laser beam to be parallel to the z axis, the refractive index gradient averaged through the beam can be obtained as:

$$\left(\frac{\partial n}{\partial y}\right)_{2D} = \frac{1}{L}\int_{z_1}^{z_2} \frac{\partial n}{\partial y} dz \tag{12}$$

The refractive index is evaluated as:

$$(n)_{2D} = \overline{n_{ref}} + \int_{y_{ref}}^{y_2} \left(\frac{\partial n}{\partial y}\right)_{2D} dy \tag{13}$$

where $\overline{n_{ref}}$ is the average refractive index of refraction at the reference location y_{ref} and the 2D subscript in the previous expressions indicates that the magnitude is obtained from the Moiré analysis. The refractive index at any z-position along y-axis is related to the refractive index gradient by the following equation:

$$n = n_{ref} + \int_{y_{ref}}^{y_2} \frac{\partial n}{\partial y} dy \tag{14}$$

The average refractive index along z-axis can be expressed as:

$$\bar{n} = \frac{1}{L}\int_{z_1}^{z_2} n\, dz \tag{15}$$

Substituting Equation (14) into the integral in Equation (15), and then, applying Fubini's theorem and using Equation (12), gives

$$\bar{n} = \frac{1}{L}\int_{z_1}^{z_2} n_{ref}\, dz + \int_{y_{ref}}^{y_2} \left(\frac{\partial n}{\partial y}\right)_{2D} dy \tag{16}$$

The first term in Equation (16) is the average reference refractive index; therefore, the refractive index evaluated with 2D Moiré deflectometry analysis is the same as the average refractive index given by Equation (15).

Using the Gladstone–Dale relationship and Equation (15), the average temperature along the laser beam, obtained from 2D analysis of Moiré deflectometry and assuming constant pressure in the flow, is related to the temperature field, T(z), as follows

$$\bar{T} = \frac{L}{\int_{z_1}^{z_2} \frac{dz}{T(x,y,z)}} \tag{17}$$

The bulk average temperature computed in Equation (11) under the assumption of uniform pressure and ideal gas behavior, Equation (8), becomes

$$T_b(x,y) = \frac{\int_{z_1}^{z_2} u(x,y,z)dz}{\int_{z_1}^{z_2} \frac{u(x,y,z)}{T(x,y,z)} dz} \tag{18}$$

and for the hypothetical case of uniform velocity along the z-axis, Equation (17) becomes

$$T_b(x,y) = \frac{L}{\int_{z_1}^{z_2} \frac{dz}{T(x,y,z)}} \tag{19}$$

which is the same as the average temperature calculated from the 2D analysis of Moiré pattern, as stated in Equation (17). This means that under the previous hypothesis, the results will be the same. However, to achieve this result, some important assumptions have been made. In general, the average fluid temperature measured using 2D Moiré analysis is not equal to the bulk temperature along the light beam. The remaining part of this investigation is dedicated to assessing the magnitude of the difference between both temperatures and analyzing the main effects promoting this difference.

4. Numerical simulation

The problem tested numerically is a 3D steady laminar convective flow in a rectangular closed channel with a backward-facing step downstream. To solve the numerical problem, a commercial CFD package (ANSYS-Fluent) was used. The code is capable of solving the full Navier-Stokes equations in two-dimensional or three-dimensional geometries. For this purpose, the governing equations are discretized on a non-uniform Cartesian grid using a finite volumes procedure (Ansys, 2011). A second-order upwind scheme was used for the discretization of all of the equations to achieve higher-order accuracy (Moharana, Singh, & Khandekar, 2011; Singh, Paul, & Ranjan, 2011). A pressure-based solver was used. Pressure-velocity coupling was achieved using a SIMPLE (semi-implicit method for pressure-linked equations) algorithm, and the convergence was assured by using the default under-relaxation factors, as used in Singh et al. (2011) and Kim, Yadav, and Kim (2014). The convergence criterion used in the simulations requires that the residuals decrease by more than four orders-of-magnitude for all of the equations, with the exception of the energy equation, for which the criterion is to decreased by seven orders-of-magnitude.

Figure 2 shows the geometry of the rectangular duct and the backward-facing step used in this work along with the proper the nomenclature and dimensions, similar to the one used by Juste and Benavides (2014) for an aspect ratio 50. The stream-wise length upstream of the step used in the presented study, L_{in} (L_{in}/h = 21.76), for Re $\leq$ 400 is larger or equal than the length obtained with the correlation provided by Durst, Ray, Ünsal, and Bayoumi (2005), who conducted a detailed numerical study and proposed a nonlinear correlation for laminar flow in two-dimensional channels given by:

$$L_{dl}/h = [(0.619)^{1.6} + (0.0567 \cdot \mathrm{Re})^{1.6}]^{1/1.6} \quad (20)$$

Furthermore, the velocity profiles along the y-axis obtained numerically at the step for all Reynolds numbers analyzed in this work are equivalent. This extra length ensures the full development of the flow along the y-axis in the duct before reaching the step.

The length of the channel behind the step, L_{out}, is selected to ensure that the outlet boundary condition does not affect the flow near the step. This selection is performed following the results of the work by Demuren and Wilson (1994), who demonstrated numerically that the flow field is not affected by increasing L_{out}/h beyond a value of 7.

To account for variations of temperature similar to the actual experimental configurations (Juste & Benavides, 2011), the heated surface behind the step with a length 10 h is an aluminum block (thermal conductivity = 180 W K^{-1} m^{-1}). The heater is a simple sheet resistor held to constant temperature under the aluminum block. The facing step and the bottom wall upstream of the step and downstream of the aluminum block are taken non-adiabatic (with a thermal conductivity equal to k = 0.19 W K^{-1} m^{-1}). Therefore, computation of thermal conduction through a solid material, coupled with heat transfer in fluid, was performed. This procedure, known as Conjugate Heat Transfer (CHT), is described in Ansys (2011). The top and sidewalls are assumed to be adiabatic.

The inlet flow condition consists in an air stream with uniform temperature, T_0, and velocity, u_0, profiles. A pressure condition was applied as the boundary condition at the exit plane (zero gauge pressure), which is located far downstream from the step to reduce the influence of the outflow conditions on the results. It was confirmed that the use of a longer computational domain did not change the flow behavior in the step region. On all solid walls, the no-slip boundary condition was applied.

Numerical simulations were performed for different values of free stream velocity u_0, resistor temperature (T_{hot}), and aspect ratio ($AR = L/h$), holding the height of the upstream duct and the expansion ratio (ER) of the step at a constant value. The Reynolds number characteristic of the flow behavior is based on the inlet velocity and the characteristic length D_h (hydraulic diameter = 2 h), which is in agreement with Armaly, Dursts, and Pereira (1983):

$$\mathrm{Re} = \frac{u_0 2h}{\nu}, \quad (21)$$

where the kinematic viscosity, ν, is evaluated for the air at the inlet temperature. A parametric study was performed to evaluate the effect of different variables. The main parameters used in the study are summarized in Table 1. As shown in Equation (20), both the inlet velocity and Reynolds number can be used to characterize the flow operation because they are equivalent parameters in this problem.

4.1. *Grid independence*

A mesh-independence analysis was performed to ensure the independence of the results; therefore, several quadrilateral meshes were considered. The velocity profiles in

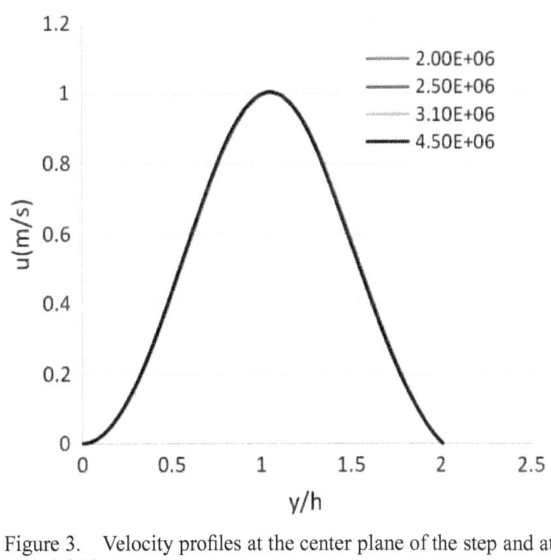

Figure 3. Velocity profiles at the center plane of the step and at x = 0.046 m.

Table 1. Parameters used in the simulations.

Geometrical parameters	Step height	h = 1.7 mm
	Expansion ratio	ER = 2
	Aspect ratio	$AR = \frac{L}{h} = \{50; 25; 12.5; 2.94; 1.47\}$
Operational conditions	Static pressure	P_0 = 101325 Pa
	Inlet temperature	T_0 = 293 K
	Inlet velocity	$u_0 = \{0.25\,\mathrm{m/s}; 0.5\,\mathrm{m/s}; 1\,\mathrm{m/s}; 2\,\mathrm{m/s}; 4\,\mathrm{m/s}\}$
	Corresponding Reynolds number	$\mathrm{Re} = \{47.8; 95.7; 191.3; 382.7; 765.3\}$
	Resistor temperature	$T_{hot} = \{345\,\mathrm{K}; 400\,\mathrm{K}\}$

Table 2. Results of the mesh independence analysis.

Mesh number	Number of cells	E
1	$2.0 \cdot 10^6$	0.00975
2	$2.5 \cdot 10^6$	0.00614
3	$3.1 \cdot 10^6$	0.00452
4	$4.5 \cdot 10^6$	–

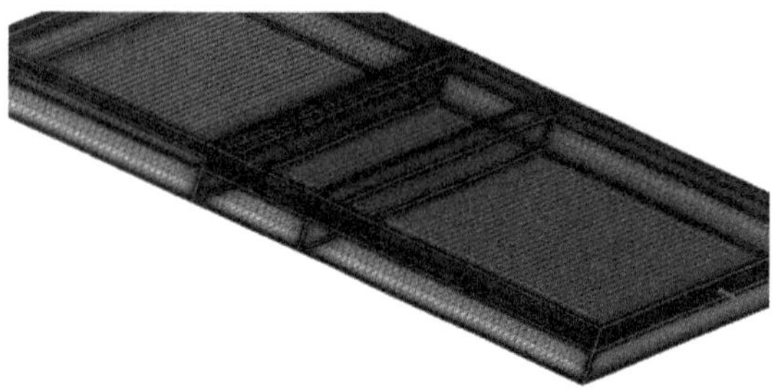

Figure 4. Computational mesh used in this work.

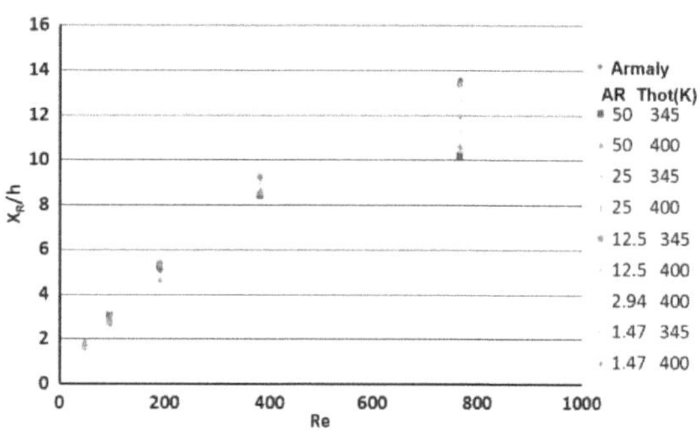

Figure 5. Reattachment length obtained in the simulations compared with the results presented in Armaly et al. (1983). The results correlate well up to a Reynolds number of 400.

the central plane of the step and at different x-positions for the case of Re = 226 were analyzed. As evidence, the results obtained at x = 0.046 m are shown below. Figure 3 presents the velocity profiles for at the center plane of the step and x = 0.046 m. All of the profiles appear to overlap into a single curve.

To provide a quantitative evaluation of the difference between the profiles, a similar procedure to the one described by Wang, Zhong, Zhang, and Kang (2014) is performed. The procedure is based on evaluating the root-mean-square error in the velocity, that is:

$$\varepsilon = \sqrt{\frac{1}{N}\sum_{i=1}^{N}\left(\frac{V_{i,n-1}-V_{i,n}}{V_{ave}}\right)^{2}} \qquad (22)$$

where $V_{i,n-1}$ is the velocity in the former grid, $V_{i,n}$ is the velocity in the current grid, subscript I denotes the ith sampling point, V_{ave} is the mean velocity of the profile, and N is the number of sampled points. The results are presented in Table 2.

The procedure was stopped at mesh number 3 because the root-mean-square error in the velocity was less than 0.5%.

Finally, a mesh with 3.1×10^6 grid cells was used in the computations in the case of the highest aspect ratio considered and a number of cells proportional to its volume in the rest of them. The region corresponding to the step is shown in Figure 4.

4.2. *Validation*

The accuracy of the numerical procedure was validated against the benchmarked experimental results of Armaly et al. (1983) by comparing the reattachment length in the mid-plane of z-axis. This parameter is the distance along the x-axis from the step corner to the point characterized by a zero longitudinal velocity in the extreme vicinity of the bottom wall.

The results obtained with the code ANSYS-Fluent for the different cases listed in Table 1 and for experimental results presented by Armaly are presented in Figure 5 and Table 3.

The agreement with the experimental results is good for Reynolds number values lower than 400 and for all of the aspect ratios analyzed. Moreover, one should remember that the results reported by Armaly et al. (1983) do not present an analysis of uncertainty evaluation. The disagreement found for higher Reynolds number values is

Table 3. Reattachment length.

		x_R/s									
	Armaly	AR	50		25		12.5		2.94	1.47	
Re		T_{hot} (K)	345	400	345	400	345	400	400	345	400
47.83	1.75		1.82	1.84	1.83	1.86	1.8	1.71	1.87	1.61	1.56
95.66	2.98867806		3.07	3.09	3.08	3.12	2.74	3.08	3.05	2.63	2.65
191.32	5.05126073		5.22	5.29	5.24	5.3	5.37	5.39	5.26	4.64	4.65
382.65	8.41018554		8.37	8.65	8.51	9.02	9.21	9.29	8.62	8.45	8.46
765.31	13.588147		10.13	10.62	10.2	10.5	13.3	11.3	12.7	11.2	12

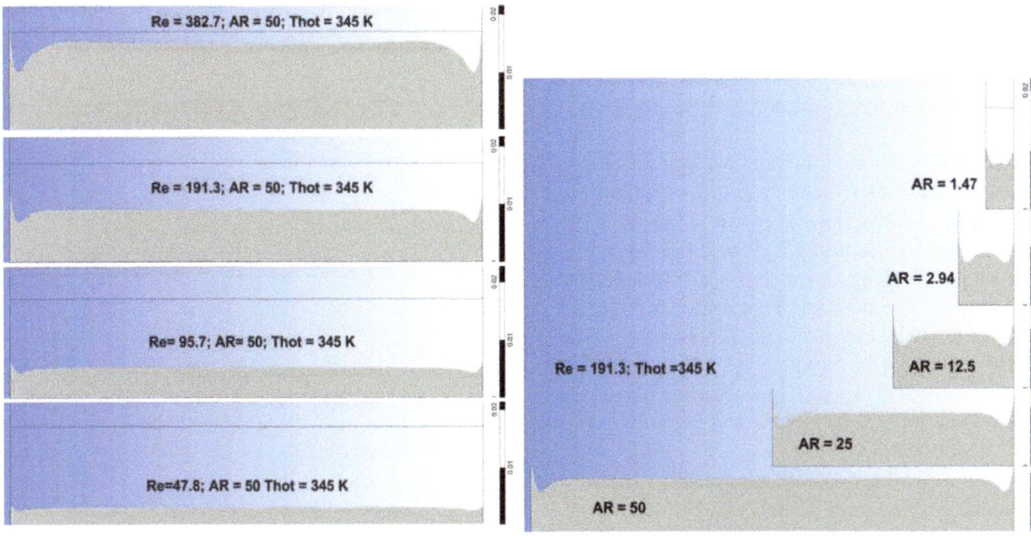

Figure 6. Variation of the reattachment length in the span-wise direction for different Re (left) and aspect ratios, AR, of the channel (right).

explained by the fact that the numerical simulation was performed assuming laminar flow and steady state conditions, but it is known that as the Reynolds number increases, the flow becomes turbulent. In the experiments performed by Armaly et al. (1983), the regime of laminar to turbulent transition was found in the range of Reynolds number of 1200–6600, but the actual representative range can be lower, as was shown in the works of Mouza, Pantzali, Paras, and Tihon (2005) and Tihon, Pěnkavová, Havlica, & Šimčík (2011). Other works, such as Kaiktsis, Karniadakis, and Orszag (1991) and Beaudoin, Cadot, Aider, and Eduardo Wesfreid (2004), studied the turbulent transition and the appearance of turbulent instabilities in the BFS geometry. Those results demonstrated that if an unstable turbulent region embedded in the surrounding laminar flow survives in the stream-wise direction, then the outflow plane cuts through this structure, and the accuracy of predicted profiles is strongly influenced by the outflow boundary condition.

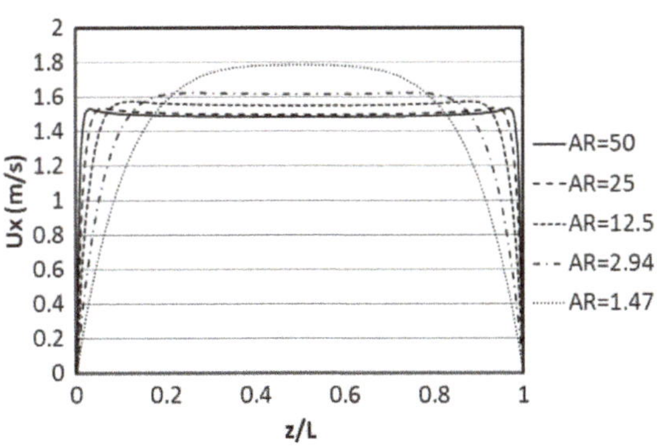

Figure 7. Variation of the span-wise velocity profiles at the mid-plane y of the channel with the aspect ratio (AR).

The discrepancies with the experimental data of Armaly et al. (1983) are attributed to the onset of three-dimensionality due to the turbulence for Reynolds

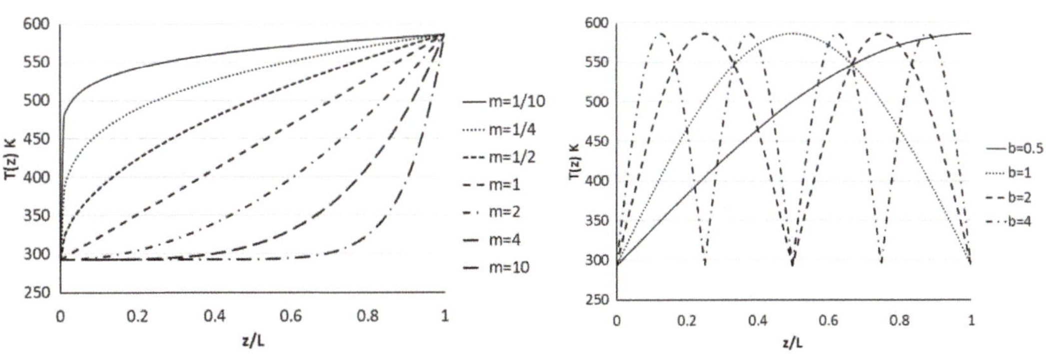

Figure 8. Different temperature profile shapes considered in this work. The panel figure represents different temperature profiles assuming a power-law shape (Equation (23)), while the right panel illustrates the temperature profile as a periodic function.

number values exceeding 600. Kaiktsis et al. (1991) identified a bifurcation of two-dimensional laminar flow to three-dimensional flow at Re = 700 as the primary source of mismatch between the simulation and the experimental data.

Moreover, the span-wise flow that develops adjacent to the side wall in the region close to the step modifies the reattachment length of 3D flow, as shown in Figure 6. The numerical simulation was performed for the full width in the z-axis instead of using a symmetry plane to represent a real experiment. The boundary represented in Figure 6 consists of the points in the span-wise direction, where the longitudinal velocity in the proximity of the bottom wall is zero, i.e., the onset of separation. Figure 6 (left) shows the variations with the Reynolds number and Figure 6 (right) shows the variation with aspect ratio.

The results show a similar behavior to those obtained in the work of Barbosa et al. (2005) and Beaudoin et al. (2004). The minimum x_R value of the reattachment line is located near the side walls, and the maximum is exactly on the side walls, while the central section of the curve is approximately constant (except for small aspect ratios). The three-dimensional effects increase as the Reynolds number increases and represent a larger extent of the span-wise width when decreasing the aspect ratio.

5. Discussion

5.1. Results in the channel upstream the step

Before analyzing the flow downstream of the step, we first studied the flow in the rectangular channel upstream of it at axial location $x/h = 14$. Span-wise direction velocity profiles for different aspect ratios are presented in Figure 7.

Although the distance from the inlet section was selected to ensure the full development of the flow in the y-direction, this does not occur in the span-wise direction. For large aspect ratios, the span-wise velocity distribution is close to a uniform velocity profile. When decreasing the aspect ratio, the velocity profile approaches the fully developed flow.

5.2. Importance of the temperature profile

To evaluate the importance of the temperature profile shape on the difference between the average Moiré temperature and the two-dimensional bulk temperature along the laser beam in closed rectangular channels, different hypothetical shapes of temperature profiles were considered, i.e., power functions with varying exponents:

$$T = T\left[1 + \left(\frac{z}{L}\right)^{m}\right] \tag{23}$$

such as those shown in Figure 8 (left) or periodic functions with varying frequencies:

$$T = T_{ref}\left\{1 + \left|\sin\left(\pi b \frac{z}{L}\right)\right|\right\} \tag{24}$$

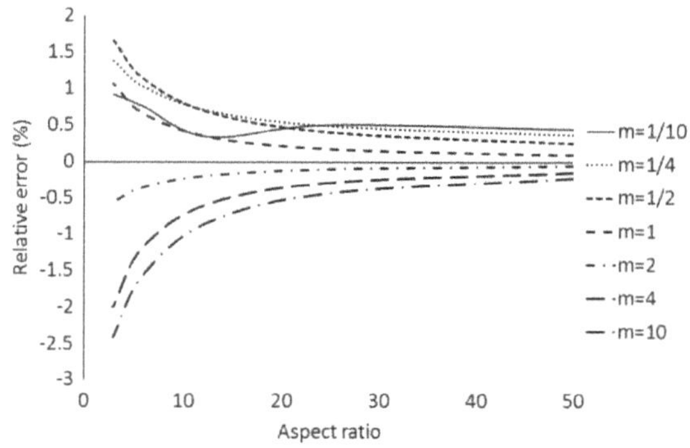

Figure 9. Relative error between the bulk and two-dimensional Moiré temperature as a function of the channel aspect ratio and for different exponents in the assumed temperature distribution (power-law Equation (23)).

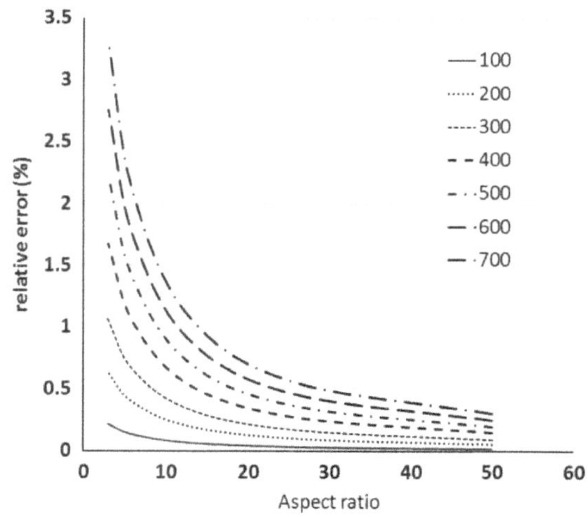

Figure 10. Relative error between the bulk and two-dimensional Moiré temperature as a function of the channel aspect ratio and for different ΔT (assuming a linear temperature distribution).

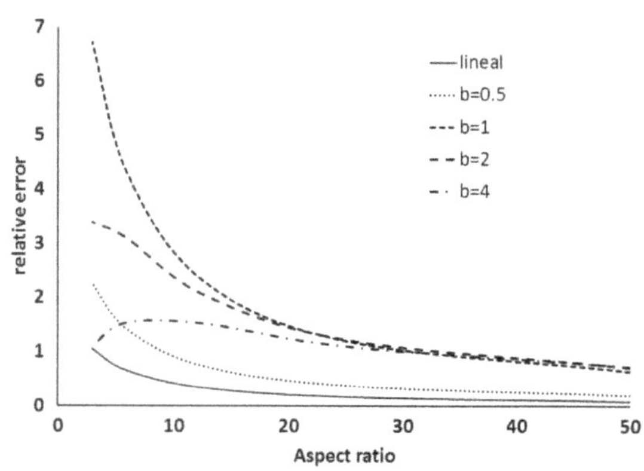

Figure 11. Relative error in the determination of the bulk temperature using a 2D Moiré analysis as a function of the channel aspect ratio and for different frequencies in the assumed temperature distribution (periodic function, Equation (24)).

as presented in Figure 8 (right). The former group represents temperature profiles close to a hot wall surface, while the latter group can simulate a heat exchanger with periodic fins. Figure 9 shows the relative error between bulk temperature and two-dimensional Moiré temperature with respect to the channel aspect ratio, assuming a power-law distribution of flow temperatures with different exponents and $\Delta T = (T_{max}-T_{min}) = 293$ K.

From the results, it is clear that, in addition to the small errors due to numerical integration, the temperature error is always small, that is, lower than 0.5, for aspect ratios larger than 20, which is obviously due to a nearly uniform velocity profile in the duct (see Figure 7). As the aspect ratio decreases, the velocity profile approaches a fully developed profile, the three dimensional effects arise, and the temperature error increases. This increment

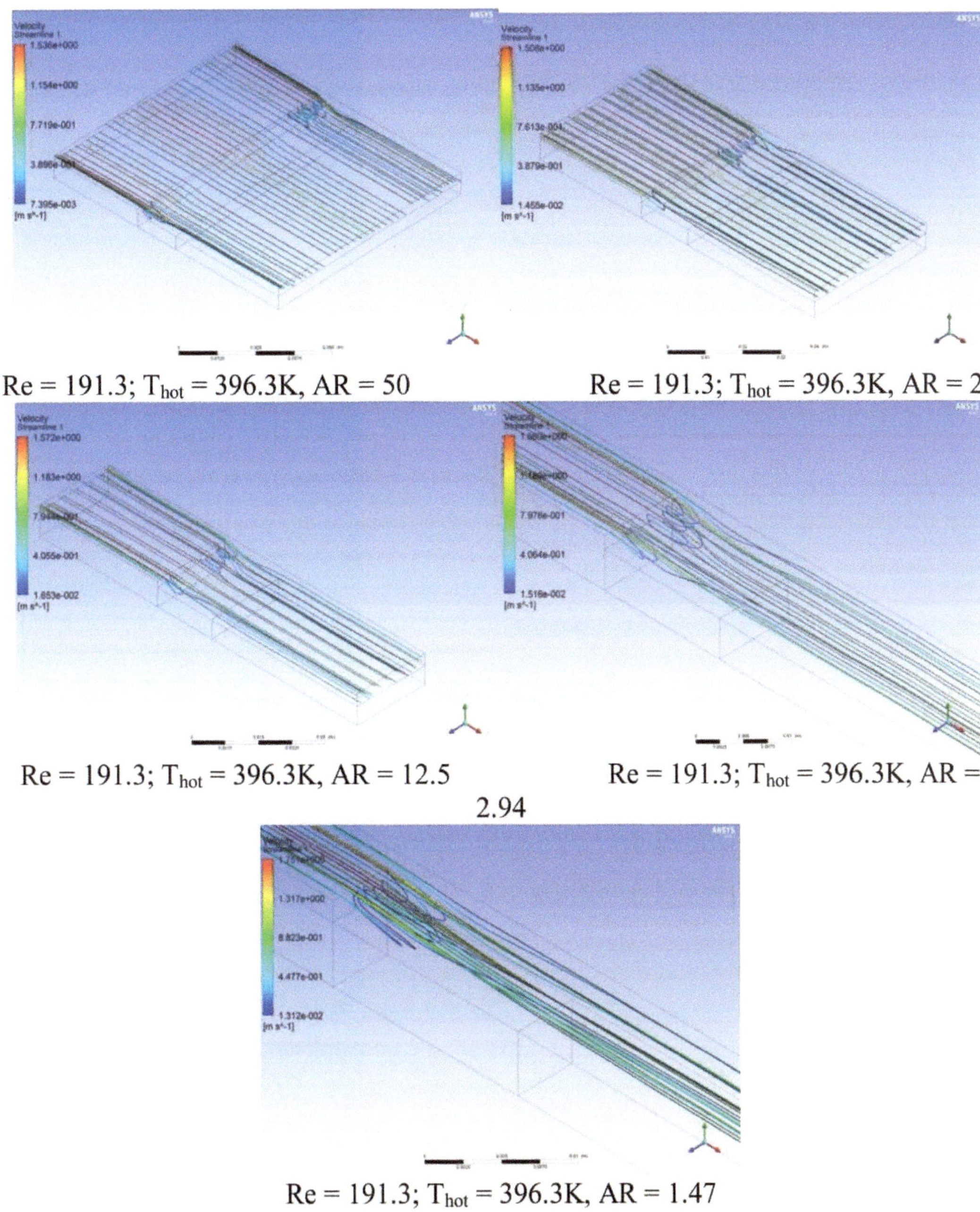

Figure 12. Streamlines after the backward-facing step for different channel aspects ratios.

in the relative error is higher as the exponent of the assumed power-law distribution deviates from the linear behavior. In any case, the error is relatively small for any value of the exponent of the power distribution of temperature.

The flow maximum temperature difference selected for computations, ΔT, is an overall limit for most heat exchanger applications. Figure 10 shows the variation of error with the aspect ratio for several values of ΔT and for the exponent, m = 1. The error in the estimation of bulk

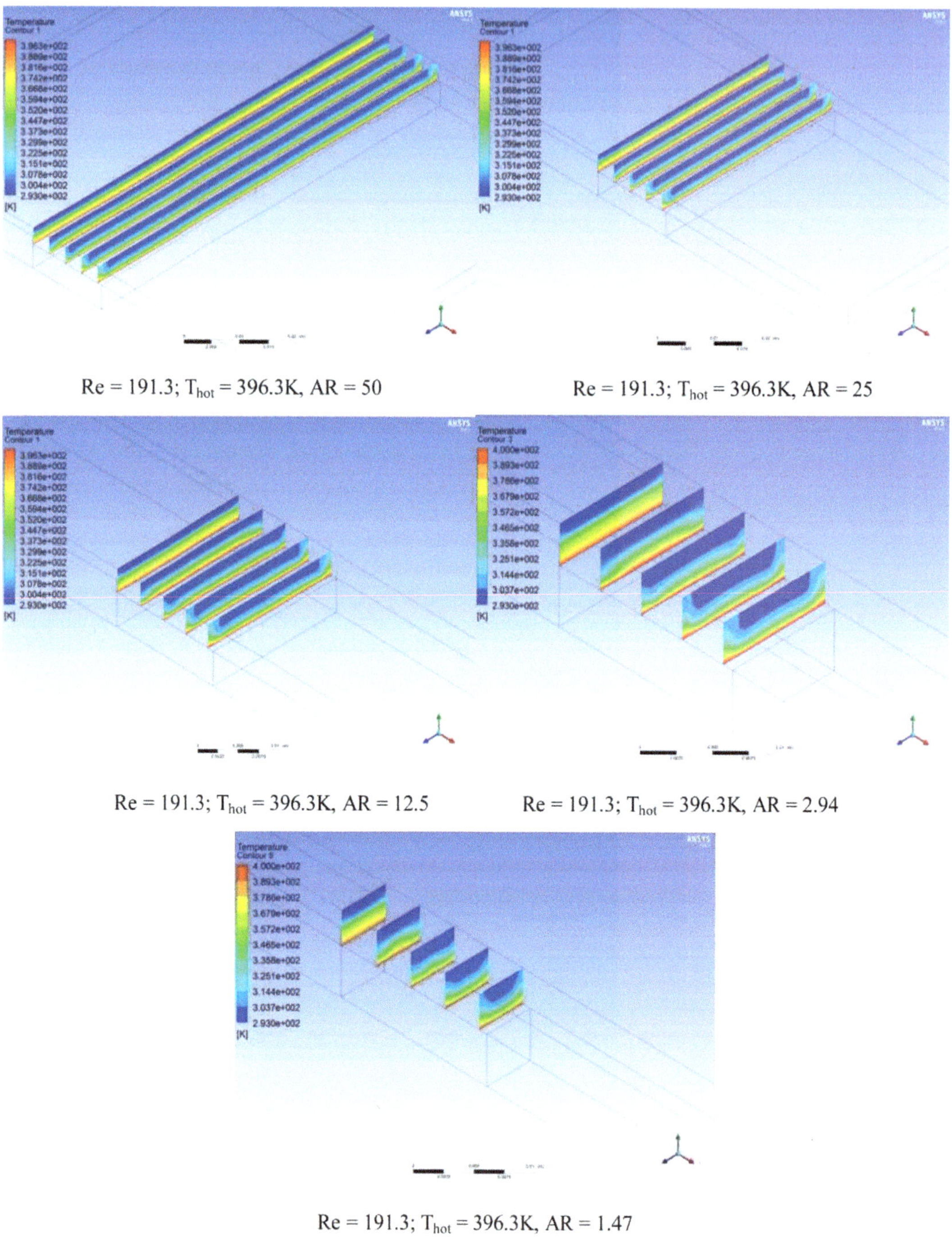

Figure 13. Temperature field in several y–z planes after the step.

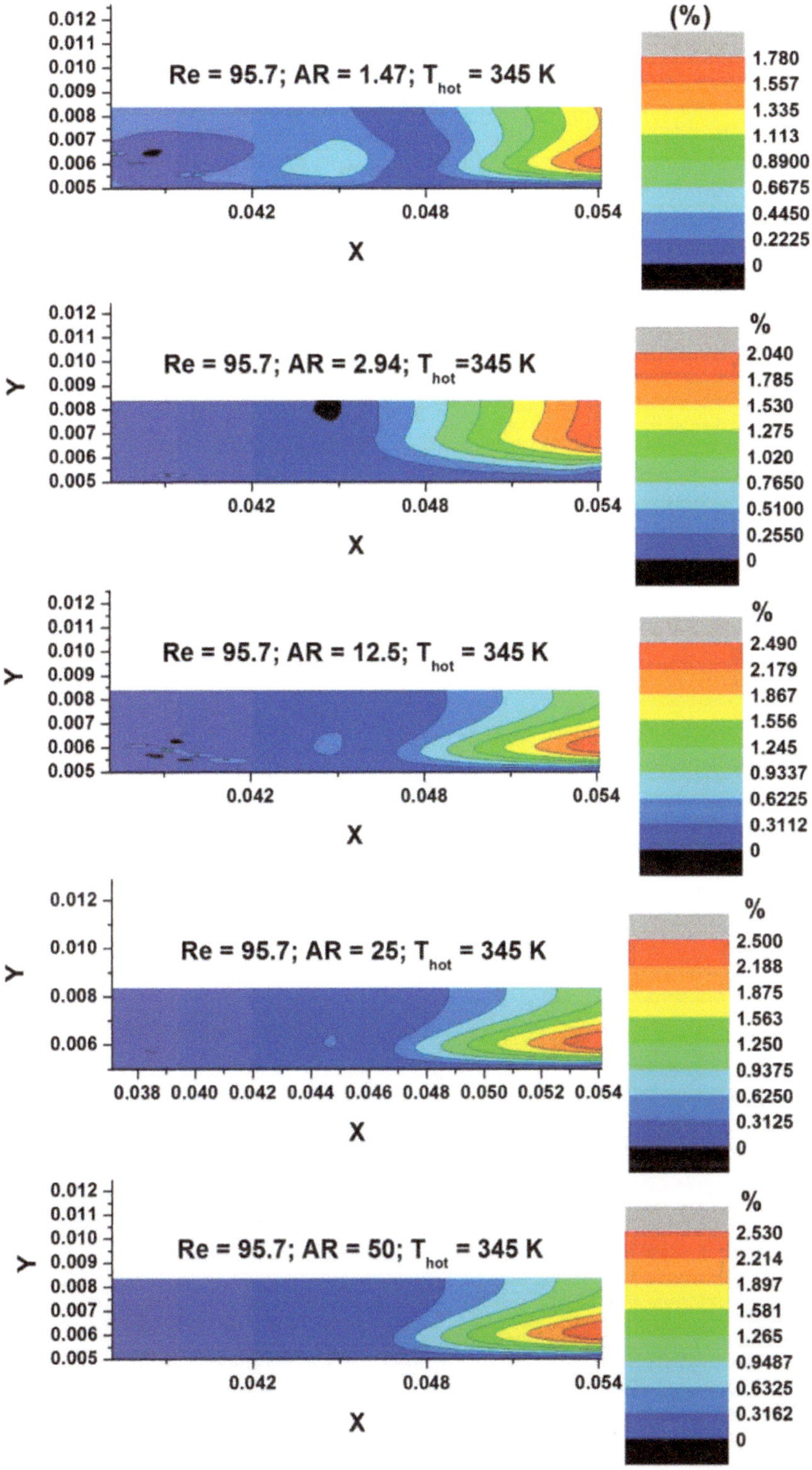

Figure 14. Error in the evaluation of the bulk temperature: effect of the aspect ratio (AR).

temperature is observed to increase with the temperature difference; this error can be of importance (more than a 1%) for aspect ratios lower than 10, i.e., with a velocity profile fully developed.

The results for the case of periodic temperature distribution are illustrated in Figure 11. Just as the case of the power-law distribution, the error in temperature estimation is small, below 1%, for aspect ratios larger than 20, i.e., for velocity profiles approximately uniform. When increasing the frequency, the error decreases. This behavior occurs because as the temperature profile approaches a uniform distribution, as derived from Equation (16) and Equation (18), and when the temperature distribution is uniform, the bulk temperature and the two-dimensional Moiré temperature are the same, as was previously demonstrated.

From these theoretical cases, the importance of 3D effects can be understood. It is possible to conclude that the relative error is reduced as the aspect ratio decreases and that the error is as small as a 3% for aspects ratios higher than 10 for all of the temperature profiles considered.

5.3. *Results after the step*

Rectangular ducts are basic components of cooling systems, but the main focus of the current work is on flow configurations with real three-dimensional features. The backward facing step allows for the study of the three-dimensional aspects of flows with separation and subsequent reattachment caused by a sudden change in the channel geometry. Next to a backward facing step, the swirling flow appearing in the span-wise direction, developed adjacent to the stepped wall, modifies the reattachment region of this separated-reattached three-dimensional flow, and consequently, the heat transfer downstream of the step, as shown in Rani, Sheu, and Tsai (2007).

Features of the three-dimensional laminar flow adjacent to the backward-facing step for changing aspect ratios are shown in Figure 12. The figure shows how the streamlines near the sidewall take different paths as they flow downstream from the step, leading to swirling flow in the spanwise direction inside the recirculation region. A similar result was obtained in the works of Nie and Armaly (2002, 2003). As the aspect ratio decreases, the portion of the flow governed by the vortex increases, until it represents the full width of backward facing step ($AR \approx 1.5$).

The temperature distribution in span-wise direction is also influenced by the three-dimensional behavior of flow. Figure 13 shows the temperature distribution in several yz-planes after the step for different aspect ratios.

The three-dimensional behavior of the temperature field increases as the flow moves downstream, and it also represents a larger extent of the flow in the span-wise direction, mainly far from the bottom wall, as the aspect ratio decreases.

The following figures are representative samples of the error in the evaluation of bulk temperature using 2D Moiré analysis in the cases listed in Table 1. As a rule of thumb, it can be established that the errors are small (less than 5%) for all of the studied cases in the region

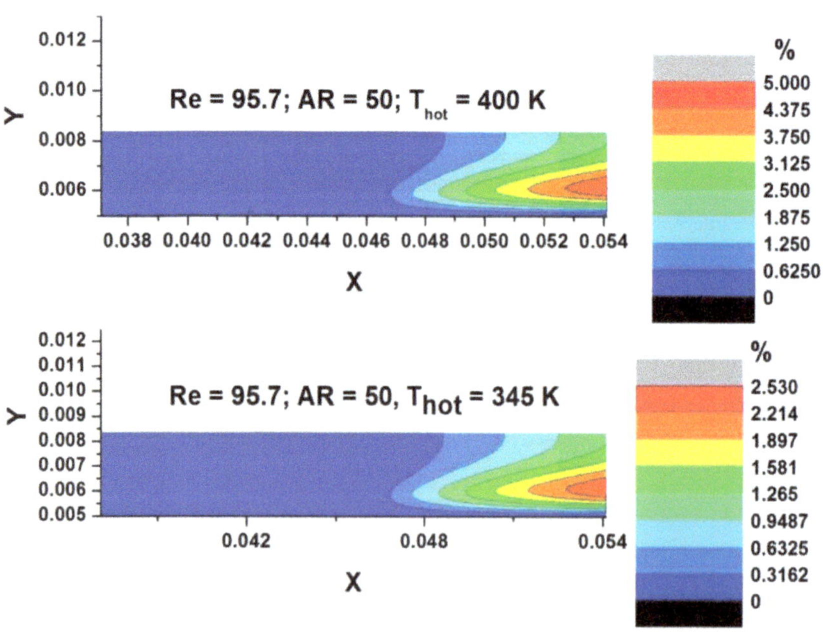

Figure 15. Error in the evaluation of the bulk temperature: effect of the resistor temperature (T_{hot}).

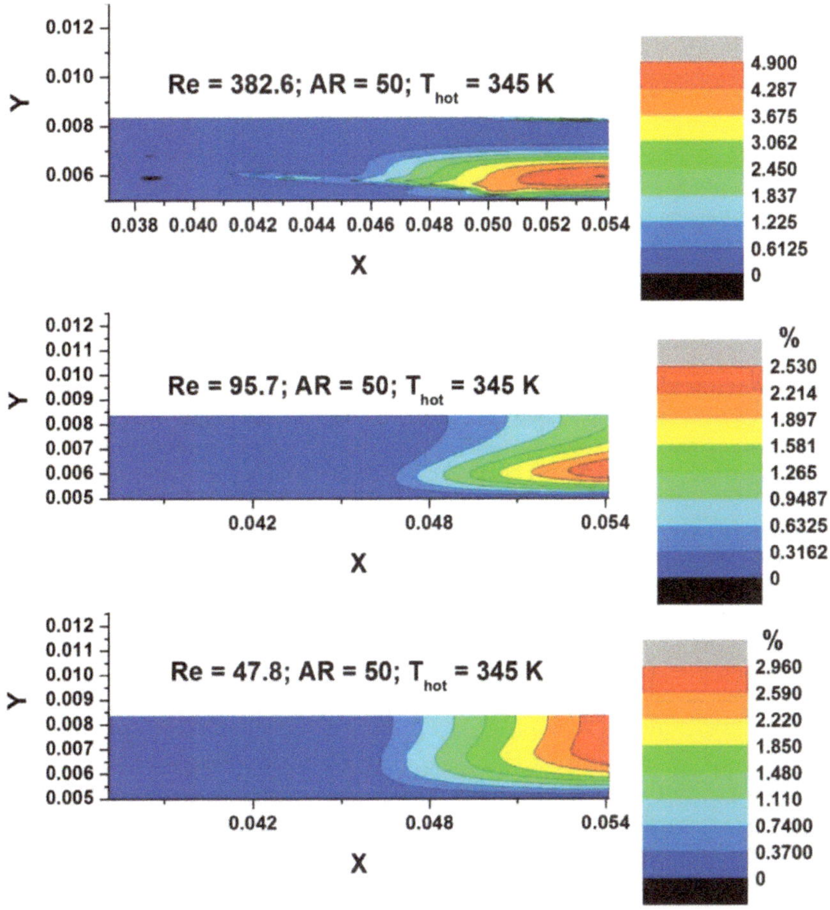

Figure 16. Error in the evaluation of the bulk temperature: effect of the Reynolds number (Re).

close to the step. The figures represent the error field in the region immediately downstream of the step covering the heated length (10 h). The y axis covers a range from 5 to 8.4 mm (2 h), and the magnitudes are established in a manner that reproduces the geometry used in the work of Juste and Benavides (2014). The maximum is found at the end of the heated plate and approximately at the central point.

Separating the effect of the different parameters, the influence of aspect ratios is analyzed in Figure 14. For a Re of 95.7 and with the resistor temperature at the smallest value considered in this analysis, $T_{hot} = 345$ K, the maximum error obtained is approximately 2.5%. As the aspect ratio decreases, the discrepancy between the 2D Moiré and the actual bulk temperature increases, probably due to the vortex flow in span-wise direction.

The effect of the temperature difference between the heated plate (bottom wall) and the flow is presented in Figure 15. The shape of the error fields are quite similar, but the magnitude of the error increases with temperature, mainly in the region downstream, where the three-dimensional effects in temperature are more relevant. This result agrees with the results of a previous study performed in the rectangular channel upstream of the step.

Finally, the effect of the Reynolds number, which is equivalent to the stream-wise inlet velocity, on the evaluation of bulk temperature agrees with that shown in Equation (17). Increasing the Reynolds number increases the inlet velocity, and therefore, the error in the evaluation of bulk temperature is also increased. These results are presented in Figure 16.

6. Conclusions

The main interest in using optical techniques for flow measurements is their non-intrusivity, which allows for obtaining the main characteristics of the flow without influencing the investigated process. This work focused on one of these techniques, namely, Moiré deflectometry, which is a line-of-sight technique in which a collimated laser beam is used to measure the optical properties of the inspected flow domain. The technique is particularly suited for the heat transfer measurements, such as the one required in heat exchange applications.

In this work, an analysis of the error introduced when measuring the bulk flow temperature along the laser beam using a 2D Moiré deflectometry analysis was presented. First, the process of evaluating the flow temperature using this technique was described in the paper, and an analytical estimation of the error for different temperature profiles along the laser beam was performed, indicating that the error decreases with the aspect ratio for the different profiles.

The analysis was performed by means of numerical simulations on the flow in the well-known geometry of the backward facing step in a rectangular channel, in which the bottom wall is heated to introduce the heat transfer effects. This geometry represents a simplified geometry of a general heat exchanger. A parametric analysis of the error was performed for different aspect ratios of the channel, different inlet Reynolds numbers (or inlet velocities) and different temperatures of the hot plate. The values were chosen as characteristic values for general heat exchanger applications. The simulations were previously validated against the benchmarked experimental results from the literature, finding good agreement for the different operating conditions. The difference between the experimental results and the simulations increases with the increase in the Reynolds number. This effect can be attributed to the onset of turbulent transition on the flow. The use of CFD allows for the recovery of the full flow field description. From the velocity and temperature fields, it is clear that the 3D effects increase as the aspect ratio decreases.

In conclusion, from the estimated 2D error field obtained via numerical simulation, the bulk temperature can be closely approximated using a two-dimensional Moiré analysis because the error is less than 5% (but increases with T_{hot}). The maximum error increases also with the increasing Reynolds number (or inlet velocity). However, the variation with the aspect ratio is less effective. Therefore, the Moiré technique is suitable for evaluating the bulk temperature in typical devices and flow conditions encountered in heat exchangers, with strong three-dimensional effects, but with a driving flow direction.

Nomenclature

A	cross section area
AR	aspect ratio (L/h)
c_P	specific heat at constant pressure
D	distance between gratings
D_h	hydraulic diameter
ER	expansion ratio (H/h)
G1, G2	gratings
g	gravity constant
h	height of the step
H	height of the channel
K	Gladstone-Dale constant
L	width of the backward facing step
L_{in}	channel length before the step
L_{dl}	development length
L_{out}	channel length after the step
n	refractive index
N	number of samples
P	distance between two grating lines
P'	distance between two fringes
p	pressure
p_0	inlet fluid pressure
Re	Reynolds number
R_g	gas constant
T	temperature
T_b	bulk temperature
T_{hot}	hot wall temperature
T0	inlet fluid temperature
s	fringe shift
u	fluid velocity in the x direction
u_0	free stream velocity
V	Velocity
x,y,z	Cartesian coordinates
x_R	reattachment length
ϵ	mean root-square velocity error
ϕ	deflection angle of the rays lights
φ	phase change of Moiré pattern
ν	kinematic viscosity
θ	angle between gratings
ρ	density

References

Ansys. Ansys Fluent User's Guide Fluent Inc. (2011, November). *Release 14.0*. Ansys, Southpointe, Canonsburg, PA.

Armaly, B. F., Dursts, F., & Pereira, J. C. F. (1983). Experimental and theoretical investigation of backward-facing step flow. *Journal of Fluid Mechanics*, *127*, 473–496.

Beaudoin, J. F., Cadot, O., Aider, J. L., & Eduardo Wesfreid J. (2004). Three-dimensional stationary flow over a backward-facing step. *European Journal of Mechanics-B/Fluids*, *23*(1), 147–155.

Childs, P. R. N., Greenwood, J. R., & Long, C. A. (2000). Review of temperature measurement. *Review of Scientific Instruments*, *71*(8), 2959–2978.

Chung, Y. M., Tucker, P. G., & Roychowdhury, D. G. (2003). Unsteady laminar flow and convective heat transfer in a sharp 180° bend. *International Journal of Heat and Fluid Flow*, *24*(1), 67–76.

Demuren, A., & Wilson, R. (1994). Estimating uncertainty in computations of two-dimensional separated flows. *Journal of Fluids Engineering*, *116*, 216–220.

Dufresne, M., Dewals, B., Erpicum, S., Archambeau, P., & Pirotton, M. (2011) Numerical investigation of flow patterns in rectangular shallow reservoirs. *Engineering Applications of Computational Fluid Mechanics*, *5*(2), 247–258.

Durst, F., Ray, S., Ünsal, B., & Bayoumi, O. A. (2005). The development lengths of laminar pipe and channel flows. *Journal of Fluids Engineering*, *127*(6), 1154–1160.

Han, J. C. (2004). Recent studies in turbine blade cooling. *International Journal of Rotating Machinery*, *10*(6), 443–457.

Hirano, T., Kawaguchi, T., Satoh, I., & Saito, T. (2006, June 1–10). *Non-intrusive temperature measurement of curved surface using laser interferometer and computer tomography*. Proceedings of 13th International Symposium on Applications of Laser Techniques to Fluid Mechanics.

Jang, S. P., & Kim, S. J. (2005). Fluid flow and thermal characteristics of a microchannel heat sink subject to an impinging air jet. *Journal of Heat Transfer*, *127*(7), 770–779.

Jordan, M., Tauscher, R., & Mayinger, F. (1997). New challenges in thermo-fluid dynamic research by advanced optical techniques. *International Journal of Heat and Technology*, *15*(1), 43–54.

Juste, G. L., & Benavides, E. M. (2011). End-wall errors in temperature measurement of external convective heat transfer with Moire deflectometry. *Optics and Lasers in Engineering*, *49*, 8–12.

Juste, G. L., & Benavides, E. M. (2014). Moiré-Fourier deflectometry for local heat transfer measurement over a backward-facing step. *International Journal of Thermal Sciences*, *77*, 244–251.

Kaiktsis, L., Karniadakis, G. E., & Orszag, S. A. (1991). Onset of three-dimensionality, equilibria, and early transition in flow over a backward-facing step. *Journal of Fluid Mechanics*, *231*, 501–528.

Kafri, O., & Glatt, I. (1990). *The physics of Moiré metrology*. New York: John Wiley.

Kim, J., Yadav, M., & Kim, S. (2014). Characteristics of Secondary Flow Induced by 90-Degree Elbow in Turbulent Pipe Flow. *Engineering Applications of Computational Fluid Mechanics*, *8*(2), 229–239.

Lienhard, J. H., & Lienhard, J. (2000). *A heat transfer textbook*. Cambridge: Phlogiston Press.

Merzkirch, W. (1987). *Flow visualization* (2nd ed.). San Diego: Academic Press.

Michael, Y. C., & Yang, K. T. (1992). Three-dimensional Mach-Zehnder interferometric tomography of the Rayleigh-Bénard problem. *Journal of Heat Transfer*, *114*(3), 622–629.

Mouza, A. A., Pantzali, M. N., Paras, S. V., & Tihon, J. (2005). *Experimental and numerical study of backward-facing step flow*. 5th National Chemical Engineering Conference. Thessaloniki, Greece.

Moharana, M. K., Singh, P. K., & Khandekar, S. (2011, January). *Axial heat conduction in the context of developing flows in microchannels*. ASME 2011 9th International Conference on Nanochannels, Microchannels, and Minichannels.

Muralidhar, K. (2012). Temperature field measurement in buoyancy-driven flows using interferometric tomography. *Annual Review of Heat Transfer*, *12*, 265–375.

Naylor, D., & Machin, A. D. (2001). The accuracy of beam averaged interferometric temperature measurement in a three-dimensional field. *Experimental Heat Transfer*, *14*, 217–228.

Naylor, D. (2003). Recent developments in the measurement of convective heat transfer rates by laser interferometry. *International Journal of Heat and Fluid Flow*, *24*, 345–355.

Newport, D., Sobhan, C. B., & Garvey, J. (2008). Digital interferometry: techniques and Trends for Fluid Measurement. *International Journal of Heat and Mass Transfer*, *44*, 535–546.

Nie, J. H., & Armaly, B. F. (2002). Three-dimensional convective flow adjacent to backward-facing step-effects of step height. *International Journal of Heat and Mass Transfer*, *45*(12), 2431–2438.

Nie, J. H., & Armaly, B. F. (2003). Reattachment of three-dimensional flow adjacent to backward-facing step. *Journal of Heat Transfer*, *125*(3), 422–428.

Rajabi, E., & Kavianpour, M. R. (2012). Intelligent prediction of turbulent flow over backward-facing step using direct numerical simulation data. *Engineering Applications of Computational Fluid Mechanics*, *6*(4), 490–503.

Rani, H. P., Sheu, T. W., & Tsai, E. S. (2007). Eddy structures in a transitional backward-facing step flow. *Journal of Fluid Mechanics*, *588*, 43–58.

Barbosa, J. G. S., Sarin, V., & Anand, N. K. (2005). Numerical simulation of mixed convective flow over a three-dimensional horizontal backward facing step. *Journal of Heat Transfer*, *127*(9), 1027–1036.

Shakher, C., & Nirala, A. K. (1999). A review on refractive index and temperature profile measurements using Laser-based interferometric techniques. *Optics and Lasers in Engineering*, *31*, 455–491.

Sehgal, S. S., Murugesan, K., & Mohapatra, S. K. (2012). Effect of channel and plenum aspect ratios on the performance of microchannel heat sink under different flow arrangements. *Journal of Mechanical Science and Technology*, *26*(9), 2985–2994.

Singh, A. P., Paul, A. R., & Ranjan, P. (2011). Investigation of reattachment length for a turbulent flow over a backward facing step for different step angle. *International Journal of Engineering, Science and Technology*, *3*(2), 84–88.

Tanda, G. (2004). Heat transfer in rectangular channels with transverse and V-shaped broken ribs. *International Journal of Heat and Mass Transfer*, *47*(2), 229–243.

Tihon, J., Pěnkavová, V., Havlica, J., & Šimčík, M. (2012). The transitional backward-facing step flow in a water channel with variable expansion geometry. *Experimental Thermal and Fluid Science*, *40*, 112–125.

Togun, H., Kazi, S. N., & Badarudin, A. (2011). A review of experimental study of turbulent heat transfer in separated flow. *Australian Journal of Basic and Applied Sciences*, *5*(10), 489–505.

Wang, Y., Zhong, K., Zhang, N., & Kang, Y. (2014) Numerical analysis of solar radiation effects on flow patterns in Street Canyons. *Engineering Applications of Computational Fluid Mechanics*, *8*(2), 252–262.

12

The simulation of a landslide-induced surge wave and its overtopping of a dam using a coupled ISPH model

Pengzhi Lin[a]*, Xin Liu[b] and Jianmin Zhang[a]

[a] *State Key Laboratory of Hydraulics and Mountain River Engineering, Sichuan University, Chengdu, China;* [b] *College of Harbour Coastal and Offshore Engineering, Hohai University, Nanjing, China*

In this paper, a landslide-induced surge wave and its overtopping of a dam is simulated using a coupled ISPH model. In the model, the mirror particle treatment was employed to simulate the moving structures and a particle symmetry criterion was enforced for free surface judgment. A telescopic boundary is presented for the landslide body motion treatment. The model is first used to simulate the two types of surge wave induced by submarine and aerial landslides and the simulation results are compared to the previously published experimental data. Then, in order to study landslide-induced wave propagation and especially the overtopping of a dam, laboratory experiments were conducted in which the landslide motion, the wave gauge results and the total overtopping volume were presented and compared to the numerical results. It was found that the present numerical model can accurately simulate the entire process of surge wave generation, propagation and overtopping of a dam.

Keywords: ISPH; landslide-induced waves; wave overtopping a dam; coupled body motion

1. Introduction

Landslide-induced tsunami have caused extensive loss of human life and devastating damage around the world. Besides the direct destruction to nearby towns, landslide masses collapse into reservoirs, costal areas, river channels or other shallow-water regions, which usually results in a huge surge wave – a landslide-induced surge wave. On 9 October 1963, a 300 million m^3 landslide body slipped into the Vaiont reservoir in Italy, with an entry velocity of about 30 m/s. The maximum surge wave was about 260 m high from the reservoir water level, and 100 m high from the dam crest. The surge wave propagated 1.4 km within 7 minutes, killing 3000 people in five towns. Comparatively speaking, the landslide-induced surge wave has been more frequently observed in coastal area than on those in proximity to reservoirs. On 17 July 1998, a large tsunami was generated in Papua New Guinea after a M7.0 earthquake, destroying seven villages and taking 2200 lives. After the disaster, researchers found that both the arrival time and the large wave height could not be directly generated by an earthquake, and that underwater landslide might have been the main cause (Tappin, Watts, McMurtry, Lafoyd, & Matsumoto, 2001). This conclusion was supported by other studies (Okal, 2003). The largest tsunami recorded in history was the Lituya tsunami, which occurred on 9 July 1953 in Alaska, with a maximum run-up height of 524 m. The main cause was also considered to be an earthquake-induced landslide (Fritz, 2001).

Since the complex generation process and propagation properties of the surge wave, many experimental studies had been conducted on the landslide-induced wave. Wiegel (1955) presented the results of his experiment in a 2D flume, in which he showed that the landslide surge wave might be more serious than that directly caused by the earthquake, and he suggest that people to pay more attention to this issue. Heinrich (1992) conducted a 2D experiment on the landslide surge wave, and a video camera was used to record the surge wave surface. In the experiment, two types of landslide body were studied, which resulted in two different surge wave profiles. In another 2D experiment presented by Walder, Watts, Sorensen, and Janssen (2003), the landslide body consisted of a triangular front and a rectangular tail, which made the body volume adjustable. 3D experiments have also been carried out (Grilli & Watts, 2005; Panizzo, De-Girolamo, & Petaccia, 2005), and researchers have found that the resistance from the liquid to landslide body is of paramount importance to surge wave generation (Grilli & Watts, 2005). Alongside experimental studies, numerical investigation has become increasingly popular and various numerical models have been presented. Lynett and Liu (2002) developed a Boussinesq model for this problem, in which they

*Corresponding author. Email: cvelinpz@126.com

used a predetermined bed deformation equation to represent the landslide motion. Watts, Grilli, Kirby, Fryer, and Tappin (2003) employed a hyperbolic cosine function to simulate the landslide motion, and their wave model was also based on the Boussinesq equation. Heinrich (1992) conducted a numerical study using his NASA-VOF2D model based on the Navier-Stokes (NS) equations. In this model, the motion of the landslide was set to be that recorded in the experiment, and a comparison between numerical and experimental results was presented. The numerical model for the landslide-induced wave requires skillful simulation of dynamic fluid-structure coupling and water wave propagation. In the above mesh-based models, the Lagrangian landslide body motion must be described in a Euler grid, and the movement of the body-fitted grid near the structure and the communication between the moving body and the Eulerian meshes obviously increase the algorithm complexity; as a result, a full Lagrangian model is necessary. Because of its Lagrangian and meshless features, smoothed particle hydrodynamics (SPH) is inherently well suited to the analysis of problems with a moving boundary and large free surface deformation (Monaghan, 1994). This method has been widely used in many hydrodynamics problems (Shao, 2009; Shao & Gotoh, 2004), but so far there has been little application of SPH to landslide surge wave simulation. Ataie-Ashtiani and Shobeyri (2008) developed an Incompressible SPH (ISPH) model for the simulation of a landslide-induced surge wave. In their simulation, they presented free surface deformation near landslide, wave run along a side wall, and the pressure field. Pastor, Haddad, Sorbino, Cuomo, and Drempetic (2009) used a depth-integrated SPH model for the 3D and large-scale problems.

The omissions in the previous studies lie in the following two aspects. Firstly, almost all of the SPH results in the literature are not focused on surge wave propagation (De-Girolamo et al., 2006; Panizzo, Cuomo, & Dalrymple, 2006; Panizzo & Dalrymple, 2004), which is the direct threat to hydraulic structures. The prediction of wave propagation and its *destructive power* are the primary goal of the numerical investigation, which relies on the accurate simulation of the fluid-landslide interaction. The second omission is that in almost all of the numerical studies, the landslide motion is treated as a known value or function. In these studies, the landslide motion is set as a measuring result (Sue, Nokes, & Davidson, 2011), a constant value or function (Enet, Grilli, & Watts, 2003), or calculated by a predesigned force-velocity relation (Guo & Wu, 2000). These treatments may obviously improve the results for the comparison between a simulation and its corresponding experiment. However, in real life, landslide motion is usually not known in advance. X. Liu, Xu, Shao, and Lin (2013) presented an ISPH model for the simulation of the motion of coupled structures in fluid; however, in their model, body contact between floating structure and fixed boundary was not allowed. Obviously, a landslide simulation will inevitably involve body contact between the landslide and the side slope. In addition, although there have been many numerical models for the simulation of wave overtopping using both the grid-based method (P. Liu, Lin, Chang, & Sakakiyama, 1999; Reeve, Soliman, & Lin, 2008) and the particle-based method (Shao et al., 2006), the overtopping of a landslide surge wave is still rarely studied.

In this paper, a coupled ISPH model will be developed for the simulation of a landslide-induced surge wave. In our study, the ISPH method following X. Liu, Lin, and Shao (2014) was employed for treating the large free surface deformation and the complex fluid-solid coupling. The model was applied to a landslide entry case with a coupled landslide motion in order to verify the model performance on surge wave propagation and landslide motion. Finally, an experiment was conducted together with the numerical simulation, which focused on the surge wave generation, propagation and overtopping of a dam. A comparison is then made of the body motion, free surface deformation, wave gauge results and the total overtopping volume.

2. The ISPH model

2.1. *Governing equations*

NS equations are solved in this model for the free surface flow in fluid-solid coupling problems. In the ISPH method, NS equations are written in Lagrangian form and the advection term is automatically calculated through the tracking of particle motion, hence, the numerical diffusion arising from the successive interpolation of the advection function in Eulerian grid-based methods is controlled without the need for a sophisticated scheme. The mass and momentum conservation equations are written as

$$\nabla \cdot \vec{u} = 0, \tag{1}$$

$$\frac{d\vec{u}}{dt} = -\frac{1}{\rho}\nabla p + \vec{g} + \upsilon_0 \nabla^2 \vec{u} + \frac{1}{\rho}\nabla \cdot \vec{\tau}, \tag{2}$$

where $\vec{u}$ is the particle velocity vector, t is the time, ρ is the fluid density, p is the particle pressure, $\vec{g}$ is the gravitational acceleration vector, υ_0 is the laminar kinematic viscosity and $\vec{\tau}$ is the sub-particle scale (SPS) turbulence stress (Lo & Shao, 2002).

The eddy viscosity assumption is often used to model the SPS turbulence stress $\vec{\tau}$ as

$$\tau_{ij} = \rho\left(2\upsilon_t S_{ij} - \frac{2}{3}k\delta_{ij}\right), \tag{3}$$

where υ_t is the turbulence eddy viscosity, $S_{ij} = (\partial u_i/\partial x_j + \partial u_j/\partial x_i)/2$, k is the turbulence kinetic energy and δ_{ij} is the Kronecker delta function. Here the turbulence eddy viscosity υ_t is calculated using a modified Smagorinsky model

(Li, Troch, & Rouck, 2004),

$$\upsilon_t = (C_S d_0)^2 |S|, \tag{4}$$

where C_S is the Smagorinsky constant ($= 0.1$), d_0 is the particle spacing representing the characteristic length scale of the small eddies and $|S| = \sqrt{|2S_{ij}S_{ij}|}$, which is the local strain rate.

The two-step projection method is employed here to solve Equations (1) and (2) (Chorin, 1968). The computation is composed of two steps. The first step is an explicit integration of velocity in time without considering the pressure and gravity terms,

$$\vec{u}_* = \vec{u}_t + \left(\upsilon_0 \nabla^2 \vec{u} + \frac{1}{\rho}\nabla \cdot \vec{\tau}\right)\Delta t, \tag{5}$$

$$\vec{r}_* = \vec{r}_t + \vec{u}_* \Delta t, \tag{6}$$

where $\vec{u}_t$ and $\vec{r}_t$ are the particle velocity and position at time t, $\vec{u}_*$ and $\vec{r}_*$ are the intermediate particle velocity and position and Δt is the time increment.

Assuming that $\Delta\vec{u}_{**}$ is the changed particle velocity contributed by the remaining pressure and gravity terms,

$$\vec{u}_{t+1} = \vec{u}_* + \Delta\vec{u}_{**} = \vec{u}_* + \left(\vec{g} - \frac{1}{\rho}\nabla p_{t+1}\right)\Delta t. \tag{7}$$

Through combination with the mass conservation Equation (1), the Pressure Poisson Equation (PPE) is obtained as follows (X. Liu et al., 2013):

$$\nabla \cdot \left(\vec{g} - \frac{1}{\rho}\nabla p_{t+1}\right) = -\frac{\nabla \cdot \vec{u}_*}{\Delta t}. \tag{8}$$

After obtaining the pressure field, the particle velocity is updated by Equation (7) and the positions of the particles are centered in time as

$$\vec{r}_{t+1} = \vec{r}_t + \left(\frac{\vec{u}_t + \vec{u}_{t+1}}{2}\right)\Delta t. \tag{9}$$

2.2. *ISPH formula*

The viscous and turbulent stress terms in Equation (2) (Lo & Shao, 2002) are given as follows:

$$(\upsilon_0 \nabla^2 \vec{u})_i = \sum_j \left(\frac{4m_j(\mu_i + \mu_j)\vec{r}_{ij} \cdot \nabla_i W_{ij}}{(\rho_i + \rho_j)^2(|\vec{r}_{ij}|^2 + \eta^2)}\right)\vec{u}_{ij}, \tag{10}$$

$$\left(\frac{1}{\rho}\nabla \cdot \vec{\tau}\right)_i = \sum_j m_j \left(\frac{\vec{\tau}_i}{\rho_i^2} + \frac{\vec{\tau}_j}{\rho_j^2}\right) \cdot \nabla_i W_{ij}, \tag{11}$$

where i and j are a reference particle and its neighbor, m is the particle mass, W is the kernel function, μ is the dynamic viscosity and equal to $\rho\upsilon_0$. $\vec{r}_{ij} = \vec{r}_i - \vec{r}_j$, $\vec{u}_{ij} = \vec{u}_i - \vec{u}_j$, η is $0.1h$ to keep the denominator nonzero with h as the smoothing length, and $\nabla_i W_{ij}$ is the kernel gradient.

The pressure term (Khayyer & Gotoh, 2008, 2009) is expressed as

$$\left(\frac{1}{\rho}\nabla p\right)_i = \sum_j m_j \left(\frac{p_i}{\rho_i^2} + \frac{p_j}{\rho_j^2}\right)\nabla_i W_{ij}. \tag{12}$$

The velocity divergence in the right-hand side of Equation (8) (Khayyer, Gotoh, & Shao, 2009) is discretized as

$$-(\nabla \cdot \vec{u}_*)_i = \sum_j \frac{m_j}{\rho_j}(\vec{u}_{*i} - \vec{u}_{*j}) \cdot \nabla_i W_{ij}, \tag{13}$$

and the left-hand side of Equation (8) is expressed as

$$\nabla \cdot \left(\frac{1}{\rho}\nabla p\right)_i = \sum_j \frac{8m_j}{(\rho_i + \rho_j)^2} \frac{p_{ij}\vec{r}_{ij}}{(|\vec{r}_{ij}|^2 + \eta^2)} \cdot \nabla_i W_{ij}, \tag{14}$$

where $p_{ij} = p_i - p_j$.

The choice of different kernel functions can greatly affect SPH performance, just like the different finite difference schemes in an Finite Different Method. In this proposed 2D ISPH model, the B-spline kernel function was employed (Monaghan & Pongracic, 1985):

$$W(q,h) = \begin{cases} \dfrac{15}{7\pi h^2}\left(\dfrac{2}{3} - q^2 + \dfrac{1}{3}q^3\right) & 0 < q < 1.0 \\ \dfrac{15}{7\pi h^2}\left(\dfrac{1}{6}(2 - q^3)^3\right) & 1.0 < q < 2.0, \\ 0 & q \geq 2.0 \end{cases} \tag{15}$$

where $q = \left|\vec{r}_i - \vec{r}_j\right| / h$.

2.3. *Boundary treatments*

2.3.1. *The mirror particle method*

Since its first presentation in Cummins and Rudman (1999) for solid boundary simulation, the mirror particle treatment has been employed by many SPH researchers (X. Liu et al., 2013; Lo & Shao, 2002). X. Liu et al. (2014) developed this method for coupling structure motion simulation. In their treatment, boundary particles (mirror particles) are automatically generated according to the fluid particle parameters, even when the solid boundary changes its position. Compared to the other solid boundary particle treatment (Vandamme, Zou, & Reeve, 2011), this method does not require adjusting the position of boundary particles and the boundary motion parameters are adaptively included in the mirroring rules, which makes the numerical algorithm more straightforward. In this study, the treatment in X. Liu et al. (2014) is employed for the moving boundary treatment.

2.3.2. The modified landslide motion model

In general, there are three forces acting on the landslide body and determining its motion: the fluid force F_f, the supported force F_s and the gravity G, as shown in Figure 1(a). The landslide will slide down the inclination under external forces. In this study, the incline surface friction is ignored for simplicity. Following the treatment in X. Liu et al. (2014), the fluid pressure around the moving body will be integrated using the pressure integration procedure to get the fluid force F_f on the rigid body. In this problem, the body sliding direction of the landslide is perpendicular to the F_s direction, which means that supported force makes no contribution to the sliding velocity calculation. Therefore, only the resultant force on the direction of movement (the sliding direction) is needed for the calculation of the sliding acceleration

$$F_{\text{res}} = G^{\text{incline}} - F_f^{\text{incline}}, \qquad (16)$$

where G^{incline} and F_f^{incline} are the G and F_f components, respectively. F_{res} is the resultant force on the direction of movement.

In most studies on fluid-solid coupling using the SPH method, the moving boundary is not allowed to contact the fixed boundary (Oger, Doring, Alessandrini, & Ferrant, 2006). For landslide problems, the moving body slides along the inclination, which makes the moving boundary truncate the fixed boundary, as shown in Figure 1(b). In this situation, the fixed boundaries (L_1 and L_2) need to be stretched or compressed during the coupling. In Figure 1(b), from time t to time t', the end point L_{1e} of the upper fixed boundary will move to L_{1e}' (this boundary is stretched), while the start point L_{2s} of the lower fixed boundary will move to L_{2s}' (this boundary is compressed). Due to self-adaption, the mirror particles will adapt the boundary deformation and generate or disappear in the deformed region. As shown in Figure 1(b), when the solid boundaries move from their old position (dashed) at time t to the new position (solid) at time t', the mirror particles (green hollow circles) near old boundaries disappear and new particles (green solid circles) generate near new boundaries to adapt fluid particles at time t'. With these procedures, the coupled motion of the landslide body can be well described.

2.3.3. The free surface treatment

On the free surface, the Dirichlet boundary condition 'p = constant' should be added, where the constant is usually the atmospheric pressure (zero in this paper). Following X. Liu et al. (2014), the following judgments are used for the free surface treatment:

$$(\rho_{\text{number}})_i = \sum_j W(|\vec{r}_i - \vec{r}_j|, h)\mathrm{d}V, \qquad (17)$$

$$(\rho_{\text{sym}})_i = \sum_j \frac{m_j}{\rho_j}\frac{x_{ij} + z_{ij}}{\mathrm{d}_0} W(|\vec{r}_i - \vec{r}_j|, h), \qquad (18)$$

where ρ_{number} is the particle number density that is close to 1.0 in the inner fluid region, $\mathrm{d}V$ is the volume of the particle and $\mathrm{d}V = m_j/\rho_j$. When $\rho_{\text{number}} < 0.9$, the particle is classified as a surface particle. $x_{ij} = x_i - x_j$, $z_{ij} = z_i - z_j$. ρ_{sym} denotes the quantification of the particle distribution asymmetry, and the critical value is chosen as 0.1 for the inner fluid particle in the present model.

3. Model application

In this section, the present model will be applied to landslide-induced surge wave simulation. First, a similar

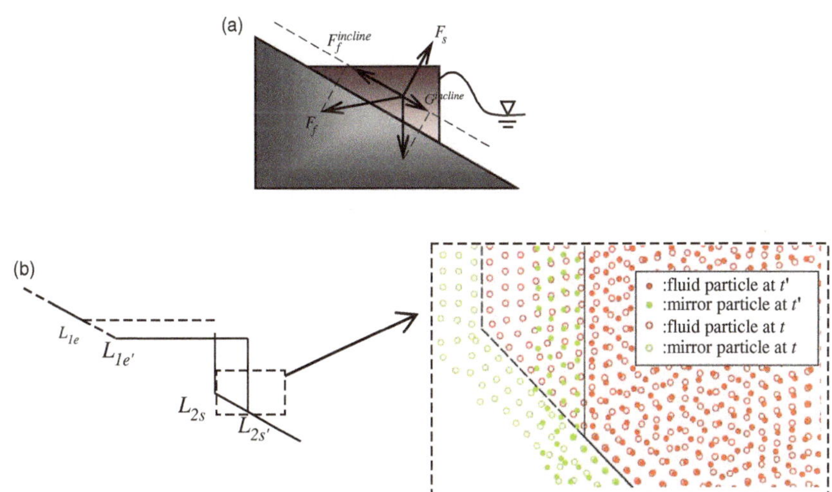

Figure 1. Landslide motion treatment: (a) fluid force calculation; (b) telescopic boundary and update of the mirror particles.

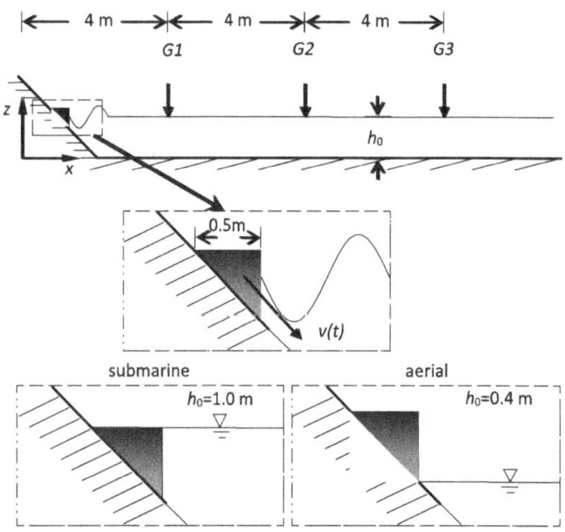

Figure 2. Set-up of landslide simulation in Heinrich (1992).

simulation as that in Heinrich (1992) is presented as the verification of the model accuracy, except that the landslide motion will be simulated in couple. After this, the experimental and numerical study are presented and the comparisons focus on surge wave propagation and overtopping.

3.1. *The simulation of surge wave propagation*

In this section, the water entry of a landslide and its induced waves are simulated by the present ISPH coupling model for verification. This problem was studied experimentally and numerically by Heinrich (1992) and also by Ataie-Ashtiani and Shobeyri (2008) using a standard ISPH model. The box that modeled the landslide was triangular in cross section (0.5 m × 0.5 m) and has the same width as the flume. The physical experiment generated the water waves by allowing a rigid box to slide down an inclined plane with a 45° slope. Here two different experimental situations are considered: submarine and aerial landslides. For the submarine landslide, the surge wave was generated with the submerged geometrical top of the box parallel to the still water surface, while for the aerial landslide, the bottom of the box was initially placed above the still free-surface, generating an aerial landslide-induced wave. The coordinate origin is located at the bottom of the flume and the left end of the horizontal section of the flume was at (2.0, 0.0). The box motion was calculated in couple instead of that given by a predefined velocity. Three wave gauges (G1, G2 and G3) were placed at 4 m, 8 m, and 12 m from the coordinate origin, as shown in Figure 2, where h_0 is the still water depth.

In the submarine landslide simulation, the water depth $h_0 = 1.0$ m and the top of the box was initially 0.01 m below the free surface. While in the aerial landslide simulation the still water depth is fixed at $h_0 = 0.4$ m and the bottom of the box is initially placed above the still water surface. The initial particle distance of two cases are both $d_0 = 0.02$ m and a total of 45,900 and 18,190 particles are used in the submarine and aerial cases, respectively. A variable time step is employed to accelerate computation time. In Heinrich's (1992) study, landslide motion simulation was attempted using the coupled model, but it was found that the landslide velocity would become unstable after 0.4 s due to the fluid incompressibility. Accordingly, they used a prescribed landslide motion obtained from the experiment to model the coupled structure motion. In the present model, the landslide motion will be calculated using the truly coupled method in which the body motion is computed by the fluid dynamics.

The computed motion of the landslide body is shown in Figure 3 and compared with the experimental data in Heinrich (1992). In both sets of results the body movement by ISPH is slightly faster than that in the experiment, which may be due to neglecting the friction effect. In the aerial landslide result, the discrepancy is found to be larger, which may be due to the smaller water depth and the lower number of SPH particles. Therefore, a high resolution case with $d_0 = 0.005$ m is presented, in which a smaller discrepancy is observed. In the simulation, the effect on surge wave propagation is negligible. In general, the comparison between the numerical and experimental results is accepted.

The computational surge wave profiles of the present ISPH model are shown in Figure 4(a)–4(f) at three different time intervals and compared with the experimental data presented by Heinrich (1992). In the numerical simulations, the moving landslide suddenly stopped when the leading edge of the box arrives at the flume bottom, similar to the effect in the experiment. From the figure, it is shown that the computed profiles of the surge wave compare well with the experimental data, except that some tiny differences are found at the water interface on the left slope wall and the cavity region over the top of solid boundary. Figure 5(a)–5(c) shows the ISPH computed free surface variations on wave gauge G1, G2 and G3 (solid line) for both submarine and aerial cases, compared with the experimental data (circle) and numerical results of Heinrich's study (triangle) using the mesh-based method. In the figure, the agreements on all three wave gauges are shown to be quite good. Moreover, for the wave phase comparison at wave gauge G2, the ISPH simulation achieved a much better accuracy than Heinrich's mesh-based method. In the case of the submarine landslide, the maximum surge wave appeared in the second phase, which was caused by the wave reflection from the left side of the flume, while in the case of the aerial landslide, the entry of the moving body directly caused the maximum surge wave that appeared in the first phase. By use of the proposed fluid-structure coupling model, accurate agreement between the ISPH and experimental results on surge wave propagations were obtained.

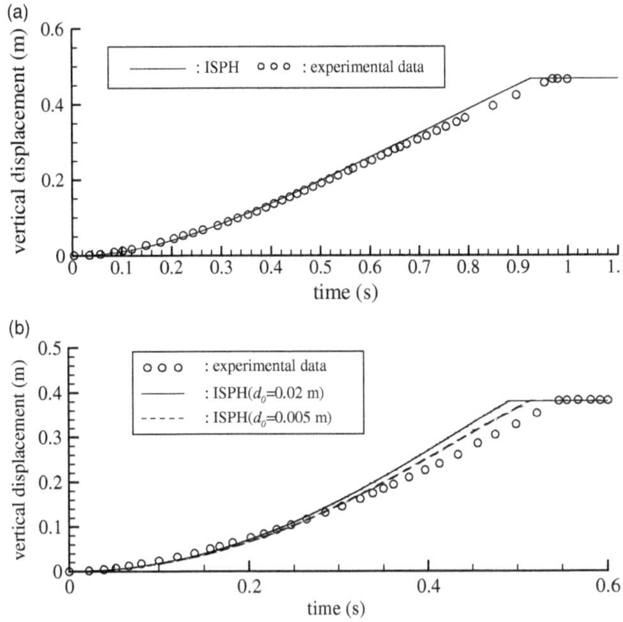

Figure 3. Computed landslide motion of the ISPH model compared with the experimental data of Heinrich (1992) for (a) submarine landslide motion and (b) aerial landslide motion.

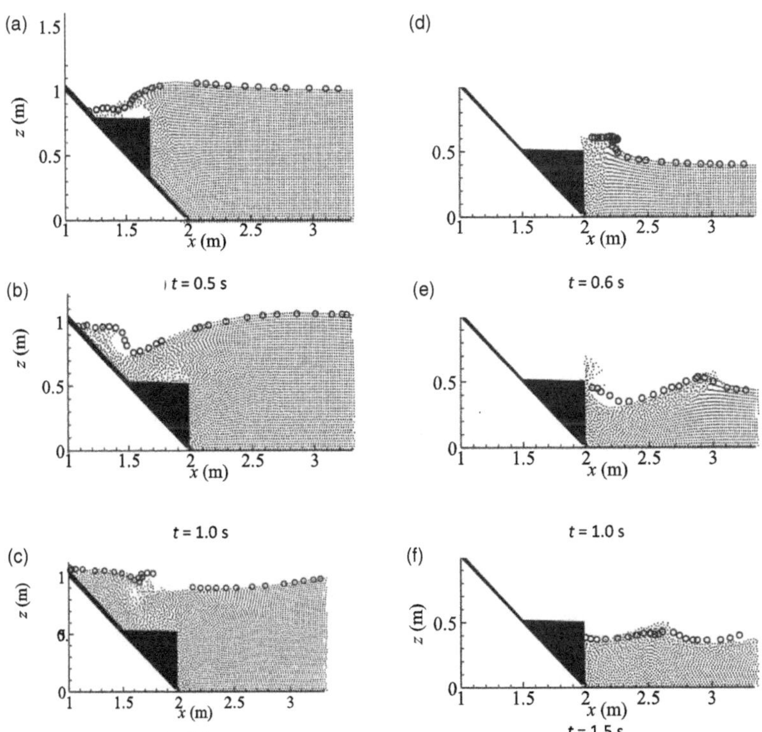

Figure 4. ISPH computed surge wave profiles at 0.5 s (0.6 s), 1.0 s and 1.5 s, compared with the experimental data (circles) of Heinrich (1992) (a–c: submarine; d–f: aerial).

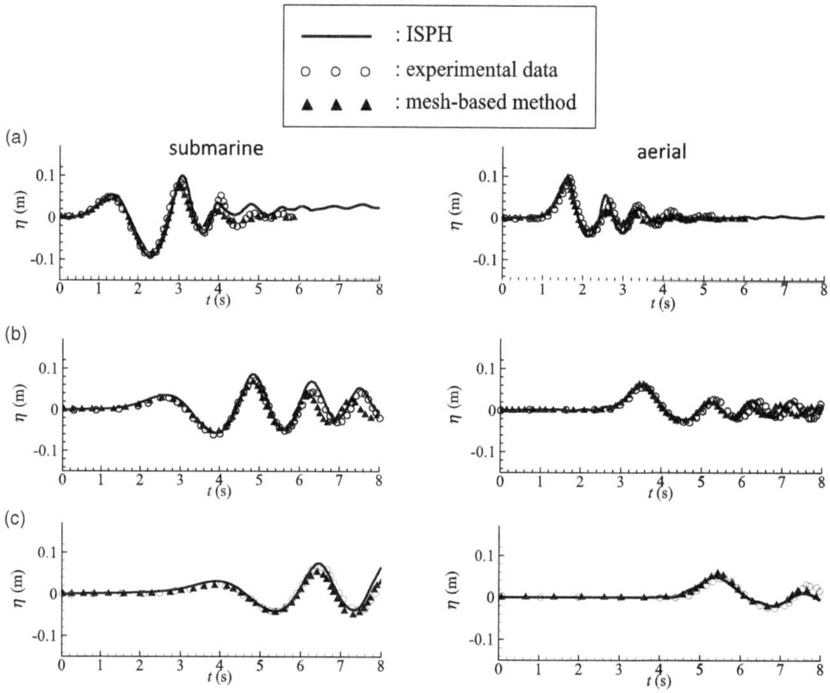

Figure 5. ISPH computed free surface elevations of submarine landslide at (a) $x = 4$ m, (b) $x = 8$ m and (c) $x = 12$ m, compared with the experimental data and mesh-based results (triangles) of Heinrich (1992).

3.2. Experimental and numerical study of coupling landslide-induced surge wave

In this section, the results of a laboratorial study on surge wave generation, propagation and its overtopping are presented for the verification of the present model. The coupling motion of the landslide body was simulated and compared with the experimental data, which is relatively rare in other similar studies, especially for the SPH method.

The experiment was conducted in the State Key Laboratory of Hydraulics and Mountain River Engineering, Sichuan University. Similar to Heinrich's (1992) set-up, the incline platform used to model the slope was fixed at the left side of the flume, as shown in Figure 6. The platform was made of board with an angle of 46° with respect to the horizon. The landslide body was made of PVC plat of a triangular geometric shape section. The horizontal and vertical edges were 251 mm and 260 mm, respectively. The two aluminum alloy guideways were embedded in the platform to connect the incline and the landslide body, which allowed the landslide to freely slide along the plat. The total mass of the sliding device was 55.9 kg. The area used for the experiment was 1.195 m wide, 8.0 m long and 1.0 m high (the total length of the flume was about 18.0 m). The still water depth $h_0 = 0.38$ m. The Plexiglas dam was fixed at the far end of the area and the detailed geometry is shown in Figure 6. A video camera was placed at the landslide section to record the body velocity, and three wave gauges were fixed at 1.6 m, 3.55 m and 6.55 m in the flume for the water surface elevation measurements. The bottom of the triangle section landslide was initially placed above the still water surface. At the beginning of the experiment, the landslide body slid down into the water, resulting in the large surface deformation. A buffer device was placed at the end of the guide 3 cm above the flume to prevent the landslide hitting the flume bottom. The surge wave propagated along the flume and finally impacted the dam, which led to the wave overtopping the dam. A water collecting device was placed behind the dam to measure the total volume of overtopped water.

In most studies on landslide-induced wave simulation (Ataie-Ashtiani & Shobeyri, 2008; Cremonesi, Frangi, & Perego, 2011), the landslide body motion was described by a prescribed time-velocity relation. In reality, however, the structure motion is usually unknown and is very hard to predict without the use of a dynamic-coupled model. In this section, the coupled motion of the landslide body will be modeled together with the surge wave propagation and overtopping. The calculation set-up is shown in Figure 6, as described in above paragraph. The initial particle distance was 0.005 m and a total of 106,680 water particles were employed. Two right-angled sides of the landslide body were designated as the moving boundary

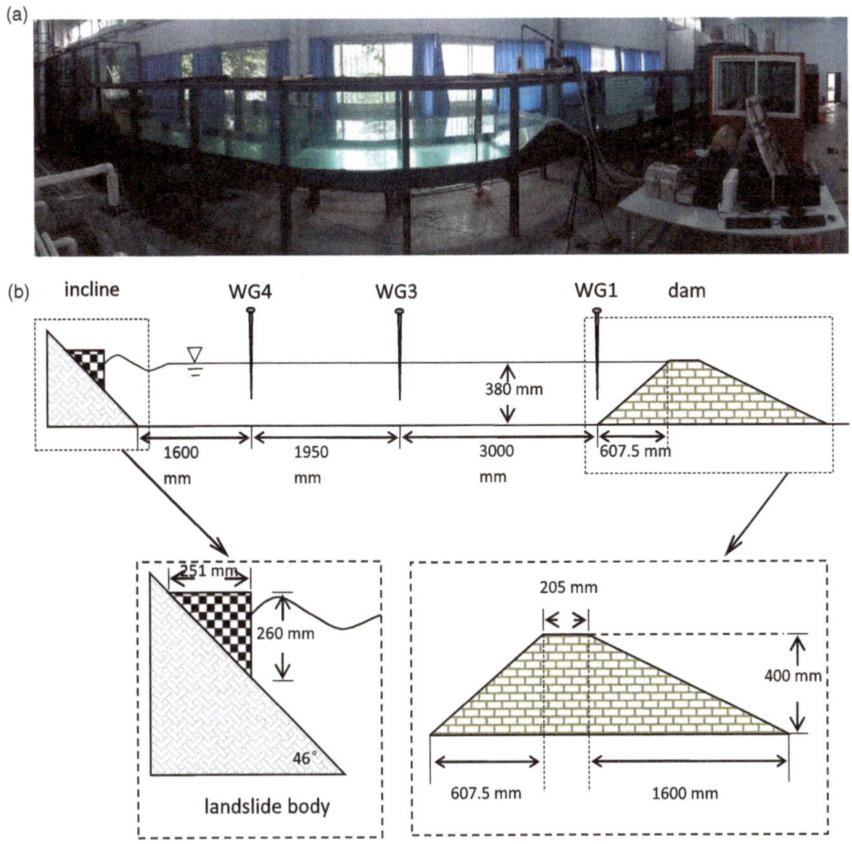

Figure 6. Set-up of coupled landslide motion: (a) photograph of the experiment flume; (b) dimensions of the model.

and connected to the incline using the telescopic boundary mentioned in section 2.3.2.

The simulated surface profile and velocity field near the moving landslide are shown in Figure 7. At about $t = 0.5$ s, the landslide stopped due to hitting the buffer device on the flume bottom and a large surge wave was generated in front of the landslide. The water passed over the top side of the landslide and impacted the left incline, after which the surge wave reversed its direction with a wave breaking on the top, generating the second and third phases of the surge wave, etc. With the use of the surface symmetry criterion, Equation (18), a clear free surface profile will be observed and surface particle judgment will be improved. In the simulation, almost all of the surface particle errors near the solid boundary which may be mistakenly classified as surface particles by Equation (17) are corrected. The surface profile in the present model was clear and smooth. After the landslide entry into the water, the maximum fluid velocity occurred near to the solid body and free surface. The fluid velocity near the leading edge of the landslide caused the first surge wave propagated from the landslide to the dam, while the fluid velocity near the surface flowed over the landslide top and impacted the side slope. The reflection wave broke on top of the landslide. It was observed that the maximum surge wave appeared in the first phase, which is the wave that will cause the largest damage.

In the simulation, the landslide motion was modeled interactively, in which the landslide velocity was updated by the fluid force dynamically. The time series of the landslide motion is shown in Figure 8 and compared with the experimental data recorded by video camera. It is seen from the figure that the difference between the numerical and experimental results is very slight. In terms of the accuracy prediction of the landslide motion, the landslide-induced surge wave was simulated well.

The free surface elevation at the three gauges is shown in Figure 9(a)–9(c) and compared with the experimental data. In the experiment, it was observed that only the first and second wave profiles were directly generated by the landslide body entry. Since the landslide motion was accurately simulated by the coupling model, the discrepancy between the numerical and experimental data for the first two waves is very small. A smaller amplitude was found for the third, fourth and fifth wave phases in the numerical results in comparison to the experimental data, while for the phase that occurred after 8 s the difference in wave

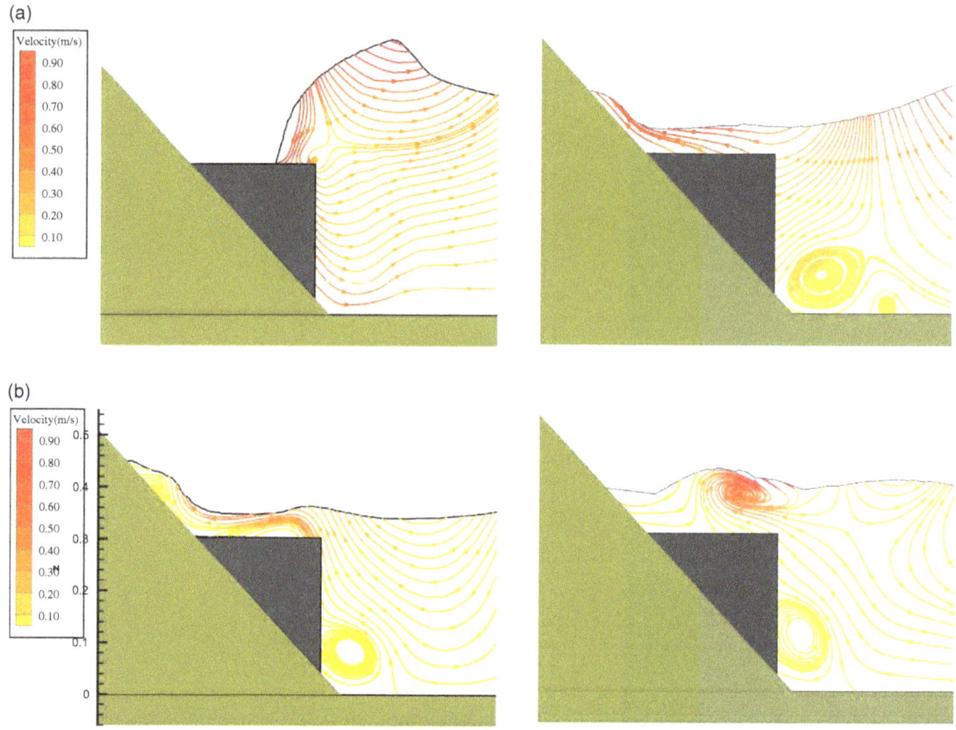

Figure 7. Surface profile and velocity field at time (a) 0.5 s, (b) 0.74 s, (c) 1.00 s and (d) 1.26 s.

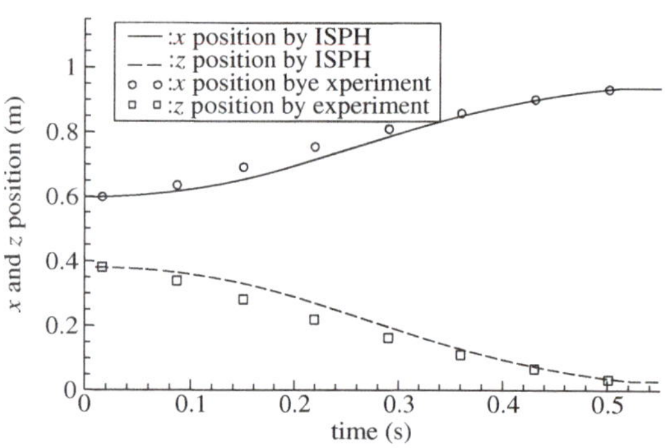

Figure 8. Landslide motion over time.

height is larger. This is probably due to the complex interactions between the incident wave, the reflected wave and the dam.

For the earth-rock dam, the overtopping volume of the surge wave is usually of paramount importance in dam risk assessment. The overtopping phenomena was simulated in the present model and is illustrated in Figure 10(a)–10(e). It was observed that the first surge wave approached the dam at about 3.0 s in Figure 10(b), together with a large overtopping volume. From 4.0 s to 4.6 s (Figure 10(c) and 10(d)), the wave crest passed over the dam and the water level in front of the dam reduced. It was found that the largest overtopping velocity occurred at both sides of the dam surface, which may destroy the earth-rock dam. At about 6.1 s, the second surge wave approached the dam and similar overtopping was observed (Figure 10(e)). During the second surge wave overtopping, the wave run up and overtopping volume were significantly less than the first wave, but a large overtopping velocity was still observed. The results were similar for the rest of the waves and the dam surface was continually scoured.

Inthe experiment, the overtopping water was collected by a water-collecting device and the total volume was measured after the experiment. In the present simulation, the Lagrangian particles flowing over the left side of the dam crest were summed into the overtopping volume and this volume was recorded over time. Figure 11 shows the overtopping results in the simulation (the solid line C-1). From the figure, it is observed that the first surge wave approached the dam at about 3.0 s and flowed over, and this phenomenon can also be observed in Figure 10. The total overtopping volume was caused by about six waves, of which the first wave contributes the primary flow rate. As shown in the line C-1 in the figure, more than 80% of the overtopping volume was collected before 6.5 s, which shows that the largest damage is usually caused during the first and second wave phases. The wave overtopping finished after six or seven wave phases, and a total of 13.789 L water was recorded as having overtopped the dam. The

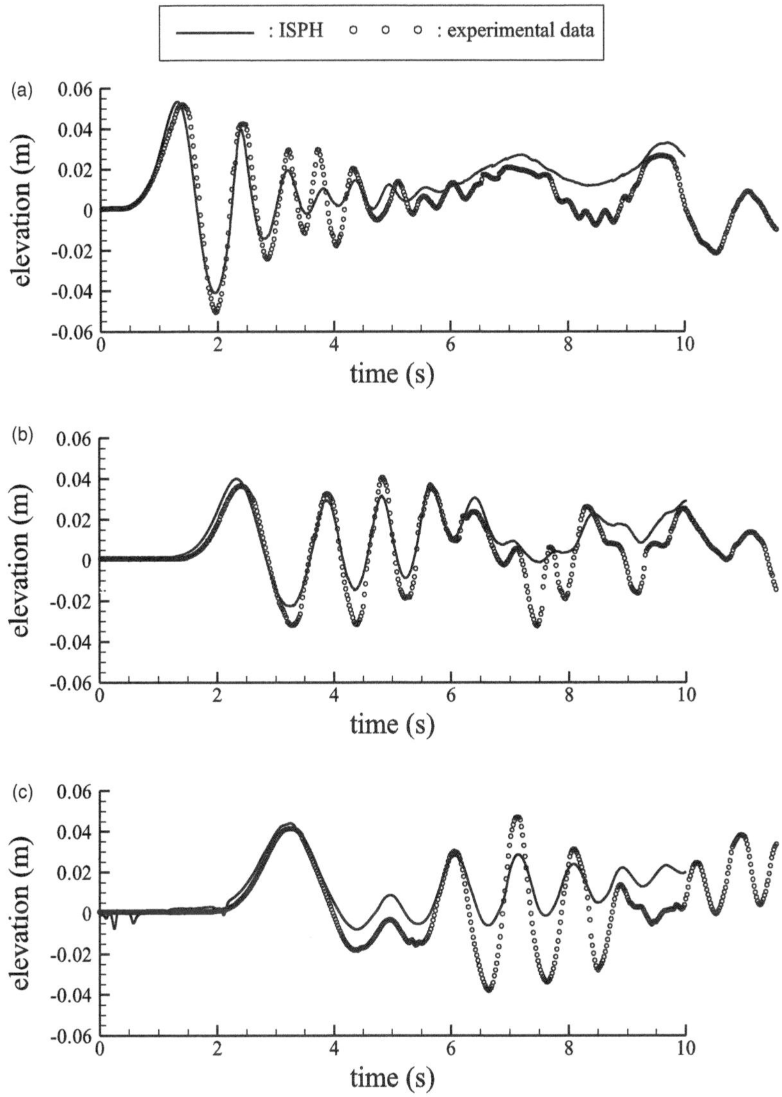

Figure 9. Free surface elevation comparison between ISPH and the experimental data at (a) wave gauge 4, (b) wave gauge and (c) wave gauge 1.

difference between the numerical and experimental results is acceptable, although a lower total volume was obtained in the simulation. This discrepancy may due to calculating errors in the later stages of the surge wave. Fortunately, this effect is very small because that these late stage waves make a very small contribution to the overtopping volume.

After verifying the accuracy of the numerical model, additional cases were simulated with different landslide positions and still water depths in order to investigate their effect on overtopping volume. These parameters are presented in Table 1 in four additional cases (C-2 to C-5), while the overtopping volume results are illustrated in Figure 10. From the simulation results, it is found that for a higher position landslide (C-2 and C-3), the total overtopping volume increases, but the difference is relatively small. This is because more violent surge wave breaking happens for higher landslide impactions, which may dissipate more energy. However, for the deeper water depth (C-4 and C-5), the total volume obviously increases. Similar to the result in C-1, most of the overtopping volume is caused by the first two or three wave phases, especially for C-1 to C-3. From the above results, it is known that most of the surge wave damage is caused by the first few surge waves, and for the same landslide body, there is a greater risk with a deeper reservoir.

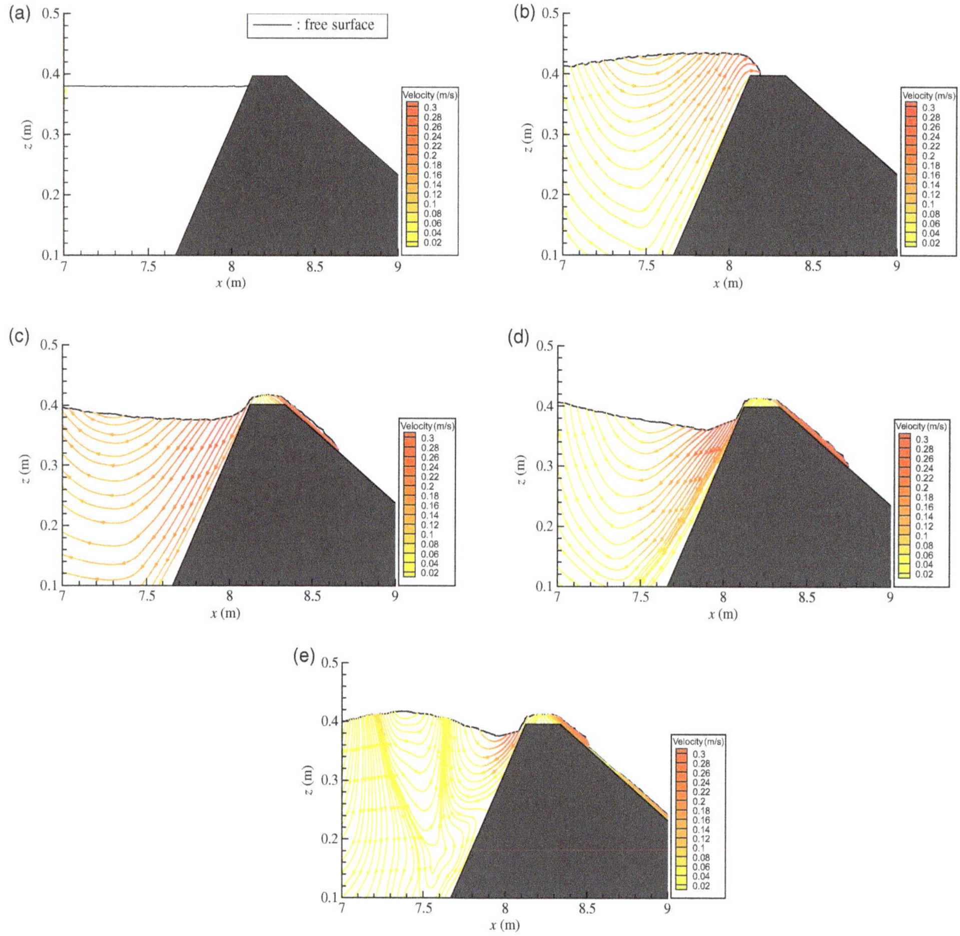

Figure 10. Surge wave overtopping of the dam.

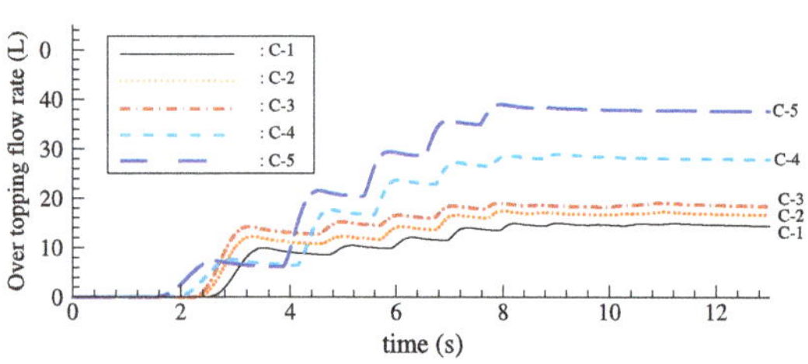

Figure 11. Overtopping flow rate by present ISPH model for five cases C-1 to C-5 (the total overtopping volume for C-1 recorded in the experiment is 13.789 L).

Table 1. Landslide position and still water depth in overtopping simulation cases.

Case	Landslide vertical offset from C-1 (m) h_l	Still water depth (m) h_0
C-1	0.000	0.38
C-2	0.095	0.38
C-3	0.190	0.38
C-4	0.000	0.57
C-5	0.000	0.76

4. Conclusions

In this paper, a landslide-induced surge wave and its overtopping of a dam were investigated using a coupled ISPH model in which all of the landslide body motions were simulated using the mirror particle treatment and a telescopic boundary was presented for the body sliding simulation.

In landslide-induced surge wave investigation, coupled motion was simulated using the present model to verify model accuracy, in which submarine and aerial landslide bodies ware modeled. Good results were obtained in the comparison with the experimental data on landslide body motion and wave propagation. Finally, both experimental and numerical investigations were carried out to investigate the surge wave generation, together with the wave overtopping of a dam. Good comparisons were achieved for the landslide body motion and acceptable results were obtained for the wave propagation and wave overtopping, which proves the present model's performance in the simulation of landslide-induced surge waves.

Acknowledgements

Part of the experiment in section 3.2 was made by Mr Guanqing Zhang. Part of the present model references the ISPH model by Dr Songdong Shao.

Disclosure statement

No potential conflict of interest was reported by the authors.

Funding

This work was supported by the National Key Basic Research Program of China [grant number 2013CB036401]; the China Postdoctoral Science Foundation [grant number 2014M552358]; the National Natural Science Foundation of China [grant number NSFC 51479126], [grant number NSFC 51279120], [grant number NSFC 51279118].

References

Ataie-Ashtiani, B., & Shobeyri, G. (2008). Numerical simulation of landslide impulsive waves by incompressible smoothed particle hydrodynamics. *International Journal of Numerical Methods in Fluids*, *56*, 209–232.

Chorin, A. J. (1968). Numerical solution of the Navier-Stokes equations. *Mathematics of Computation*, *22*, 745–762.

Cremonesi, M., Frangi, A., & Perego, U. (2011). A Lagrangian finite element approach for the simulation of water-waves induced by landslides. *Computers and Structures*, *89*, 1086–1093.

Cummins, S. J., & Rudman, M. (1999). An SPH projection method. *Journal of Computational Physics*, *152*, 584–607.

De-Girolamo, P., Wu, T. R., Liu, P. L. F., Panizzo, A., Bellotti, G., & Di-Risio, M. (2006). *Numerical simulation of three dimensional Tsunamis water waves generated by landslides: Comparison between physical model results, VOF, SPH and depth-integrated models*. ICCE Conference, San Diego.

Enet, F., Grilli, S. T., & Watts, P. (2003). *Laboratory experiments for tsunamis generated by underwater landslides: Comparison with numerical modeling*: Proceedings of the Thirteenth (2003) International Offshore and Polar Engineering Conference. Vol 3. Cupertino: International Society Offshore&Polar Engineerings, 372–379.

Fritz, H. M. (2001). Lituya bay case rockslide impact and wave run-up. *Science of Tsunami Hazards*, *19*(1), 3–22.

Grilli, S. T., & Watts, P. (2005). Tsunami generation by submarine mass failure. I: Modeling experimental validation, and sensitivity analyses. *Journal of Waterway, Port, Coastal and Ocean Engineering*, *131*(6), 283–297.

Guo, H. W., & Wu, C. C. (2000). Reservoir landslide generated wave model and its application. *Journal of North China Institute of Water Conservancy and Hydroelectric Power*, *21*(1), 24–27. (in chinese).

Heinrich, P. (1992). Nonlinear water waves generated by submarine and aerial landslides. *Journal of Waterway, Port, Coastal and Ocean Engineering*, *118*, 249–266.

Khayyer, A., & Gotoh, H. (2008). Development of CMPS method for accurate water-surface tracking in breaking waves. *Coastal Engineering Journal*, *50*(2), 179–207.

Khayyer, A., & Gotoh, H. (2009). Modified moving particle semi-implicit methods for the prediction of 2D wave impact pressure. *Coastal Engineering*, *56*(4), 419–440.

Khayyer, A., Gotoh, H., & Shao, S. D. (2009). Enhanced predictions of wave impact pressure by improved incompressible SPH methods. *Applied Ocean Research*, *31*, 111–131.

Li, T. Q., Troch, P., & Rouck, J. D. (2004). Wave overtopping over a sea dike. *Journal of Computational Physics*, *198*, 686–726.

Liu, P. L.-F., Lin, P., Chang, K.-A., & Sakakiyama, T. (1999). Numerical modeling of wave interaction with porous structures. *Journal of Waterway, Port, Coastal, and Ocean Engineering*, *125*(6), 322–330.

Liu, X., Lin, P. Z., & Shao, S. D. (2014). An ISPH simulation of coupled structure interaction with free surface flows. *Journal of Fluids and Structures*, *48*, 46–61.

Liu, X., Xu, H. H., Shao, S. D., & Lin, P. Z. (2013). An improved incompressible SPH model for simulation of wave-structure interaction. *Computers and Fluids*, *71*, 113–123.

Lo, E. Y. M., & Shao, S. D. (2002). Simulation of near-shore solitary wave mechanics by an incompressible SPH method. *Applied Ocean Research*, *24*, 275–286.

Lynett, P., & Liu, P. (2002). A numerical study of submarine-landslide-generated waves and run-up. *Proceedings of the Royal Society of London Series a-Mathematical Physical and Engineering Sciences*, *458*(2028), 2885–2910.

Monaghan, J. J. (1994). Simulating free surface flow with SPH. *Journal of Computational Physics*, *110*, 399–406.

Monaghan, J. J., & Pongracic, H. (1985). Artificial viscosity for particle methods. *Applied Numerical Mathematics*, *1*, 187–194.

Oger, G., Doring, M., Alessandrini, B., & Ferrant, P. (2006). Two-dimensional SPH simulations of wedge water entries. *Journal of Computational Physics*, *213*, 803–822.

Okal, E. A. (2003). T waves from the 1998 Papua New Guinea earthquake and its aftershocks: Timing the tsunamigenic slump. *Pure and Applied Geophysics*, *160*(10–11), 1843–1863.

Panizzo, A., Cuomo, G., & Dalrymple, R. A. (2006). *3D-SPH simulation of landslide generated waves*. ICCE 2006 Conference, San Diego.

Panizzo, A., & Dalrymple, R. A. (2004). *SPH modelling of underwater landslide generated waves*. ICCE 2004 Conference, Lisbon.

Panizzo, A., De-Girolamo, P., & Petaccia, A. (2005). Forecasting impulse waves generated by subaerial landslides. *Journal of Geophysical Research: Oceans*, *110*(C12).

Pastor, M., Haddad, B., Sorbino, G., Cuomo, S., & Drempetic, V. (2009). A depth-integrated, coupled SPH model for flow-like landslides and related phenomena. *International Journal for Numerical and Analytical methods in Geomechanics*, *33*, 143–172.

Reeve, D. E., Soliman, A., & Lin, P. (2008). Numerical study of combined overflow and wave overtopping over a smooth impermeable seawall. *Coastal Engineering*, *55*, 155–166.

Shao, S. D. (2009). Incompressible SPH simulation of water entry of a free-falling object. *International Journal of Numerical Methods in Fluids*, *59*, 91–115.

Shao, S. D., & Gotoh, H. (2004). Simulating coupled motion of progressive wave and floating curtain wall by SPH-LES model. *Coastal Engineering Journal*, *46*(2), 171–202.

Shao, S. D., Ji, C., Graham, D. I., Reeve, D. E., James, P. W., & Chadwick, A. J. (2006). Simulation of wave overtopping by an incompressible SPH model. *Coastal Engineering*, *53*(9), 723–735.

Sue, L. P., Nokes, R. I., & Davidson, M. J. (2011). Tsunami generation by submarine landslides: Comparison of physical and numerical models. *Environmental Fluid Mechanics*, *11*(2), 133–165.

Tappin, D. R., Watts, P., McMurtry, G. M., Lafoyd, Y., & Matsumoto, T. (2001). The Sissano, Papua New Guinea tsunami of July 1998 – offshore evidence on the source mechanism. *Marine Geology*, *175*(1), 1–23.

Vandamme, J., Zou, Q. P., & Reeve, D. E. (2011). Modeling floating object entry and exit using smoothed particle hydrodynamics. *Journal of Waterway, Port, Coastal and Ocean Engineering*, *137*(5), 213–224.
Walder, J. S., Watts, P., Sorensen, O. E., & Janssen, K. (2003). Tsunamis generated by subaerial mass flows. *Journal of Geophysical Research*, *108*(B5).
Watts, P., Grilli, S. T., Kirby, J. T., Fryer, G. J., & Tappin, D. R. (2003). Landslide tsunami case studies using a boussinesq model and a fully nonlinear tsunami generation model. *Natural Hazards and Earth System Science*, *3*, 391–402.
Wiegel, R. L. (1955). Laboratory studies of gravity waves generated by the movement of a submerged body. *American Geophysical Union-Transaction*, *36*(5), 759–774.

13

Improvement of a solution of inviscid compressible flows using a mixed wall boundary condition

A. Madadi*, M.J. Kermani and H. Khazaei

Department of Mechanical Engineering, Amirkabir University of Technology (Tehran Polytechnic), Tehran, Iran, P. Code 15875-4413

Boundary conditions have a vital role in the numerical solution of the partial differential equations governing fluid flows, and they have a great influence over the numerical stability and accuracy of the final solutions. One of the most important physical boundary conditions in flow field analysis, especially for inviscid flows, is the wall boundary condition imposed on body surfaces. To solve a three-dimensional compressible Euler equation (with five PDE's), a total of five boundary conditions on the body surface should be prescribed. The wall's velocity magnitude is one of the parameters to be determined, and the way this velocity magnitude is calculated affects the accuracy and stability of the numerical approach. In this paper, four different methods for calculation of the wall velocity magnitude are introduced, tested and compared against several test cases of subsonic and supersonic flows. Since there are many problems where both subsonic and supersonic flows coexist, a mixed boundary condition takes into account the flow Mach number is proposed. The mixed boundary condition is applied to several test cases and the stability of the method is examined.

Keywords: wall boundary condion; inviscid; mixed boundary condition

1. Introduction

Design and analysis of systems which deal with fluid flows are of interest to many researchers. With the increased calculation speed of new computers, CFD has become an important tool for fluid engineers. This increased speed has facilitated the research of flow patterns in a number of studies. Yu, Shademan, Barron, and Balachandar (2012) used CFD to study the effect of nozzle geometry on incompressible flow through four nozzles with different turbulence models and compared the results to previous experimental data. Other researchers applied CFD to design and optimization of geometries used in industrial systems. Idahosa, Golubev, and Balabanov (2008) developed and automated design tools for turbo machinery parts of propulsion systems. They used CFD coupled with optimization tools to improve the efficiency of a low speed fan. Liu, Dang, and Xi (2008) used CFD to improve the performance of a centrifugal fan. They optimized the profile of linkage between the inlet duct and impeller.

The accuracy and stability of CFD models are very important especially in optimization systems. Several researchers have focused on development of new numerical methods and CFD models to improve the accuracy of simulations. For example, Qamar, Hasan, and Sanghi (2010) presented a new spatial descritization method for convective terms in the hyperbolic equations of fluid dynamics.

It is obvious that boundary conditions have a significant role in the stability and accuracy of numerical methods. An improper boundary condition may lead to a non-physical solution and also the divergence of the numerical method. The body surface is one of the most important boundary conditions which should be implemented carefully. Kim, Jeon, Kim, Kwon, and Lee (2001) showed that the solution accuracy and stability of Roe's FDS scheme is affected by the wall boundary condition.

The effect of different wall boundary conditions in two-dimensional coordinates with three types of upwind schemes; Stager-Warming's flux vector splitting, Roe's flux difference splitting, and Liou's advection upstream splitting method (AUSM) (Liou & Steffen, 1993), was investigated in Kim and Kim (2000).

Wang and Yuzhi (2002) developed a curvature-based wall boundary condition for inviscid flows. They used the projected adjacent node velocity vector of the wall, and using the surface curvature and computed velocity magnitude on the wall, obtained the wall pressure.

Marshall and Ruffin (2004) presented a new wall boundary condition for subsonic and supersonic flows using an embedded boundary Cartesian solver. For subsonic flows, they also used the projected velocity from the adjacent node, and for computation of pressure and temperature on the wall the gas dynamic relation between static and total pressures and temperatures was applied. In the supersonic cases, they used the oblique shock and expansion fan gas dynamic relationship regarding the angle between the adjacent velocity vector and the surface direction.

*Corresponding author. Email: ali.madadi@gmail.com

Khazaei, Madadi, and Kermani (2011) applied four types of wall boundary conditions on subsonic and supersonic test cases. They used four methods to modify velocity magnitude on the walls and found that the method of correction of the velocity magnitude greatly affected the accuracy and stability of the solution. They proposed two different methods for subsonic and supersonic flows.

The objective of this paper is to suggest an accurate, stable, and physical wall boundary condition for three-dimensional, compressible, and inviscid subsonic and supersonic flow analysis. On the wall, five boundary conditions should be imposed. The momentum equation in the normal direction is used to compute the pressure on the wall (Hoffmann & Chiang, 2000). For the no-heat source and sink conditions, the total temperature through the domain remains constant. Hence, as a second condition, the total temperature on the wall is set equal to the inlet T_0.

To compute the velocity components, the no penetration condition through the wall is imposed as the third equation. To complete the equations, the velocity vector from an adjacent node, is projected onto the wall and the velocity direction on the wall is obtained. Four different methods for correcting the wall velocity magnitude introduced by Khazaei et al. (2011) are investigated here. These four methods are applied to six benchmarks which consist of the following subsonic and supersonic test cases. Two consecutive 90^0 elbows, NACA65-410 airfoil, and subsonic flow over a bump are used to study subsonic flows; and a converging diverging nozzle with supersonic outflow, transonic, and supersonic flow over a bump are used as supersonic test cases.

The results show that the third method is more convenient for subsonic flows while for supersonic problems the second type gives better accuracy and stability. Since both subsonic and supersonic flows can be presented in the same simulation, it is important to provide the correct boundary conditions on the wall. Therefore, in this work, we derive a new wall boundary condition that results from a combination of methods two and three (explained before), taking into account the Mach number on the adjacent node. This mixed boundary condition was applied to different test cases and produced results with acceptable accuracy and stability.

NOMENCLATURE

C_p	Specific heat in constant pressure [J/kg.K]
E	x-Direction flux vector
e_t	Total energy [J/kg]
F	y-Direction flux vector
G	z-Direction flux vector
h_t	Total enthalpy [J/kg]
n	Unit vector
NC	Not Corrected
P	Pressure [Pa]
Q	Conservative vector
R	Gas constant [J/kg.K]
T	Temperature [K]
t	Time [s]
u,v,w	Velocity components [m/s]
x,y,z	Physical Cartesian coordinates [m]
$\vec{V}$	Velocity vector
V_t	Velocity magnitude [m/s]
α	The correction factor of the wall velocity magnitude
ρ	Density [kg/m^3]
ξ,η,ζ	Generalized coordinates
Subscripts	
Adj	Adjacent to the wall
Ber	Bernoulli
c	Contravariant components
Corr	Corrected
In	Inlet conditions
NC	Not Corrected
u	Upstream
W	Wall
x,y,z	Derivatives in physical domain
0	Total condition
ξ,η,ζ	Derivatives in generalized coordinate system

2. Governing equations

The Euler equations in three-dimensional, unsteady, inviscid and conservative form for compressible flow are given as follows:

$$\frac{\partial Q}{\partial t}+\frac{\partial E}{\partial x}+\frac{\partial F}{\partial y}+\frac{\partial G}{\partial z}=0 \tag{1}$$

where,

$$Q=\begin{bmatrix}\rho\\ \rho u\\ \rho v\\ \rho w\\ \rho e_t\end{bmatrix},\ E=\begin{bmatrix}\rho u\\ \rho u^2+P\\ \rho uv\\ \rho uw\\ \rho uh_t\end{bmatrix},$$
$$F=\begin{bmatrix}\rho v\\ \rho uv\\ \rho v^2+P\\ \rho vw\\ \rho vh_t\end{bmatrix},\quad G=\begin{bmatrix}\rho w\\ \rho uw\\ \rho vw\\ \rho w^2+P\\ \rho wh_t\end{bmatrix} \tag{2}$$

Q represents the conservative vector; and E, F, and G represent the flux vectors in the x, y, and z-directions, respectively. The equations are completed with equation of state,

$$P=\rho RT \tag{3}$$

3. Solver description

To solve the inviscid flow field, a recently developed in-house code based on the flux difference splitting (FDS) scheme of Roe (1981) is used. The governing equations are discretized in the computational domain using formulations presented by Kermani and Plett (2001a). In this paper, the primitive variables are extrapolated to the cell faces by the MUSCL idea using the third order upwind biased scheme. Time integration is done using a fourth order explicit Runge – Kutta scheme with the local time stepping method.

The van Albada, van Leer, and Roberts (1982) flux limiter is used to prevent spurious numerical oscillations. Metrics of transformation are used to convert the equations from the physical domain to the computational domain. The Roe scheme gives non-physical expansion shocks in the regions where the eigenvalues of the Jacobian matrix vanish, e.g., the sonic regions and the stagnation points. To obtain a physical solution under the abovementioned conditions, an entropy correction formula from Kermani and Plett (2001b) is used.

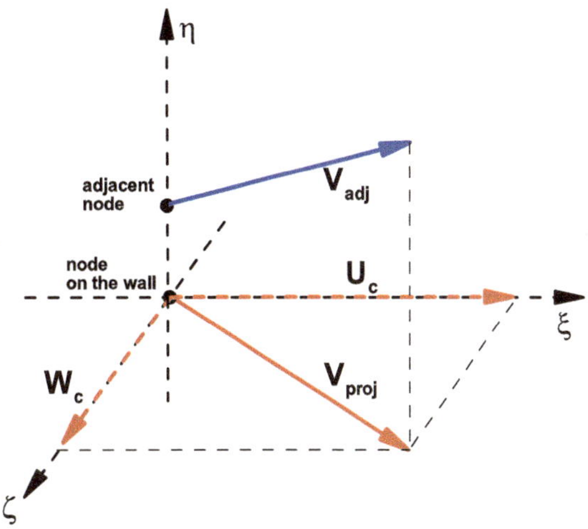

Figure 1. Extrapolation of velocity direction from the adjacent node to the wall.

4. Grid generation

In the present work an in-house combined algebraic-elliptic algorithm is developed and used for grid generation (Hoffmann & Chiang, 2000). To impose the grid orthogonality on the body surfaces and clusterings near the wall, the corresponding control functions are considered in the elliptic algorithm for complicated geometries. Because the governing equations are discretized in the computational domain, metrics of transformation are needed to transform equations from the physical domain to the computational domain (Kermani & Plett, 2001a). These metrics of transformation are computed using a second-order finite difference formula.

5. Inviscid wall boundary conditions

To solve three-dimensionalEuler equations five boundary conditions on each surface, such as a wall, are needed. For an inviscid flow velocity components on the wall can be calculated using the slip condition. Therefore, the contravariant velocity components, except normal to the surface, can be extrapolated to the wall from the interior. This is shown in Figure 1. The equations for contravariant velocity components on the wall are given here:

$$\text{on the Wall}\begin{cases} u_{c,w} = (\xi_x u + \xi_y v + \xi_z w)_w \\ v_{c,w} = (\eta_x u + \eta_y v + \eta_z w)_w \\ w_{c,w} = (\zeta_x u + \zeta_y v + \zeta_z w)_w \end{cases} \quad (4)$$

where ξ_x, ξ_y, ξ_z, η_x, η_y, η_z, ζ_x, ζ_y, and $\zeta_z$are the metrics of transformation. Because no penetration through the solid wall is permitted, the normal component of velocity vector is set to zero, i.e:

$$\vec{V}.\hat{n}_W = 0 \quad (5)$$

where$\hat{n}_W$ is the unit vector normal to the wall.Hence, one of the following cases will occur:

(i) the ξ-constantsurface is a wall $\rightarrow$ $u_{c,w} = 0$
(ii) the η-constant surface is a wall $\rightarrow$ $v_{c,w} = 0$
(iii) the ζ-constant surface is a wall $\rightarrow$ $w_{c,w} = 0$

For example, consider a η-constant wall, the velocity equations on the wall are rewritten as:

η – constant wall

$$\times\begin{cases} u_{c,w} = (\xi_x u + \xi_y v + \xi_z w)_w = (\xi_x u + \xi_y v + \xi_z w)_{adj} \\ v_{c,w} = (\eta_x u + \eta_y v + \eta_z w)_w = 0 \\ w_{c,w} = (\zeta_x u + \zeta_y v + \zeta_z w)_w = (\zeta_x u + \zeta_y v + \zeta_z w)_{adj} \end{cases} \quad (6)$$

The Crammer formula can be applied to the system of linear equations to compute the velocity components on the wall surface. Theses equations force the flow vector to be tangent to the wall surface.

Since there are no heat sources or sinks in the test cases studied here and the flow is steady, the total temperature on the wall is equal to the incoming flow. So the wall static temperature can be computed from the stagnation temperature as:

$$T_W = T_{0,\,In} - \frac{u_w^2 + v_w^2 + w_w^2}{2C_p} \quad (7)$$

Because of the wall surface curvature the zero pressure gradient normal to the wall could not be used here. In Hoffmann and Chiang (2000), using the momentum equation in the direction normal to the solid wall the relation for the pressure in two-dimensional coordinates is given. Similarly, the computations for three-dimensional cases are performed and applied here.

With the slip condition, no-penetration, constant total temperature, and pressure relation described above, the boundary conditions on the wall can be determined. The wall velocity magnitude (Equation (8)) is an important parameter which affects the flowfield results.

$$V_t = \sqrt{u_w^2 + v_w^2 + w_w^2} \tag{8}$$

Four different methods for correction of the velocity magnitude on the wall are investigated in this paper. To consider the velocity magnitude on the wall, a correction factor is defined as:

$$\alpha_{corr} = \frac{V_{t,corr}}{V_{t,NC}} \tag{9}$$

where, $V_{t,NC}$ refers to the uncorrected velocity magnitude computed from Equations (6) and (8), and $V_{t,corr}$ is the corrected velocity magnitude. In the first method, the projected velocity vector with no correction is used. In this method one of the contravariant velocity components is set to zero; therefore the computed velocity magnitude on the wall is less than the adjacent node which may not be compatible with the flow physics in some flow fields.

In the second method, the same velocity magnitude of the adjacent node is applied to the wall. The velocity magnitude on the curved wall would not be remained constant; hence, this method has non-physical results in test cases with curved walls.

The constant mass flux condition for correcting velocity magnitude is used in the third method. In this method the following constraint is applied:

$$\rho_{Adj} V_{t,Adj} = \rho_w V_{t,w} \tag{10}$$

or

$$V_{t,w} = \frac{\rho_{Adj} V_{t,Adj}}{\rho_w} \tag{11}$$

The correction factor for this type can be computed from the following equation:

$$\alpha_{III} = \frac{V_{t,w}}{V_{t,NC}} = \frac{\rho_{Adj} V_{t,Adj}}{\rho_w V_{t,NC}} \tag{12}$$

Applying constant total pressure is the fourth method considered here. If the grid size on the wall is fine, the Bernoulli equation for steady frictionless ($P_0 = cte$) flow along a streamline can be written as:

$$dV_t + \frac{dP}{\rho V_t} = 0 \tag{13}$$

so, the increment of velocity magnitude of the two neighbor nodes on the wall is,

$$V_t = \sqrt{u_w^2 + v_w^2 + w_w^2} \tag{14}$$

In this case, according to the velocity direction (upwinding scheme) along each grid line on the wall, for example ξ-Line and ζ-Line on a η-Constant wall, the velocity magnitude is computed.

$$\begin{cases} V_{t\xi} = \left(V_{tAdj} - \frac{P_W - P_{Adj}}{\rho V_{tu}}\right)_\xi \\ V_{t\zeta} = \left(V_{tAdj} - \frac{P_W - P_{Adj}}{\rho V_{tAdj}}\right)_\zeta \end{cases} \tag{15}$$

and,

$$V_{tBer} = \frac{|u_c| V_{t\xi} + |w_c| V_{t\zeta}}{|u_c| + |w_c|} \tag{16}$$

And α_{IV} is computed as,

$$\alpha_{IV} = \frac{V_{tBer}}{V_{t,NC}} \tag{17}$$

Since the correction factor is computed, the velocity vector can be modified as below,

$$\vec{V}_{W_{Cor}} = \alpha \vec{V}_{W_{NC}} \tag{18}$$

It should be noted that the corrected velocity should be used to compute the wall static temperature and pressure. These four methods of velocity magnitude correction are summarized in Table 1.

6. Results and discussion

Here, the effects of the introduced wall boundary conditions on the accuracy and convergence rate of the flow field solutions are investigated. To do so, four subsonic and supersonic test cases are studied:

- Subsonic flow through two consecutive 90^0 elbows.
- Isentropic flow through a converging-diverging nozzle with an outlet-throat area ratio of 16.

Table 1. The correction factors for each calculation method of the wall velocity magnitude.

No.	The Correction Method of the Wall Velocity Magnitude	Correction Factor
I	Not Corrected (NC)	$\alpha = 1$
II	Constant Velocity Magnitude ($V_t = cte$)	$\alpha = \frac{V_{tAdj}}{V_{t,NC}}$
III	Constant Mass Flux ($\rho V_t = cte$)	$\alpha = \frac{(\rho V_t)_{Adj}}{\rho_w V_{t,NC}}$
IV	Constant Total Pressure ($P_0 = cte$)	$\alpha = \frac{V_{tBer}}{V_{t,NC}}$

- Subsonic, transonic and supersonic flow over a bump.
- Subsonic flow through a NACA65-410compressor cascade.

The flow field of the test cases is solved using all introduced types of wall boundary conditions and the performance of the approach is studied. Here, the results for each test case are presented and compared.

Each test case has been studied using different initial conditions. If the solution converges for all initial conditions the method is called "stable" in the following sections. On the other hand, if the solution diverges for all initial conditions it is called "unstable," and if the method converges for some initial conditions and diverges for other ones it is called "sensible to initial conditions."

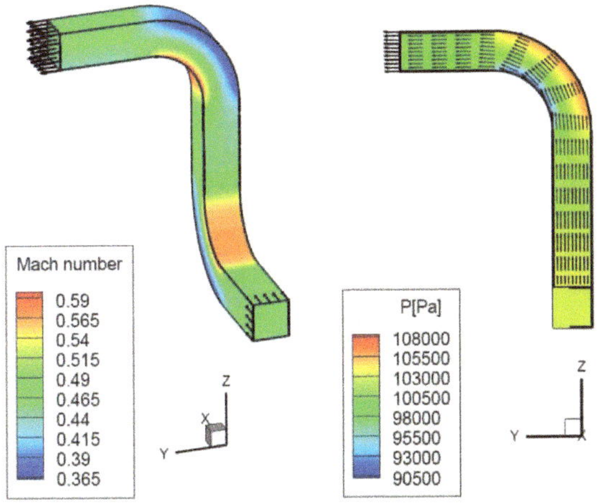

Figure 2. Contours of static pressure and Mach number for the 90^0 elbows.

6.1. *Subsonic flow through two consecutive 90^0 elbows*

In this section, the subsonic flow field through two consecutive 90^0 elbows is studied. For the elbow test case, all types of wall boundary conditions are imposed and the solution is achieved. A sample of the results of the elbows in Figure 2 is presented by contours of static pressure and Mach number containing velocity vectors. As shown, the velocity on the inner surface is greater than the outer surface which leads to higher static pressure on the outer wall.

In Figure 3, the velocity magnitude contours for four types of wall boundary conditions are presented. As it is seen in Figure 3(a) for the uncorrected method, the trend of velocity magnitude toward the wall is reversed. For the inner wall, the velocity magnitude on the wall is

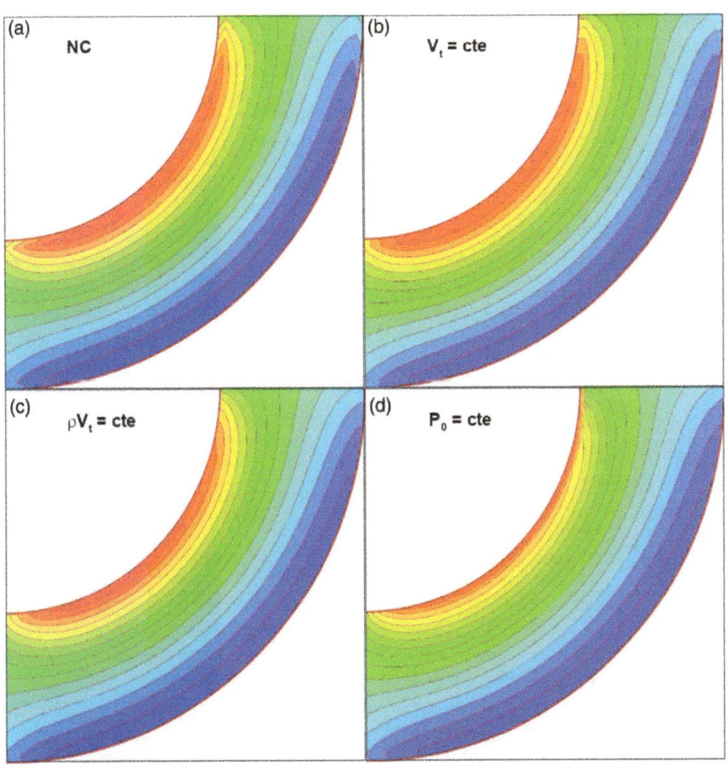

Figure 3. Velocity magnitude contours obtained by different types of wall boundary conditions for 90^0 elbows.

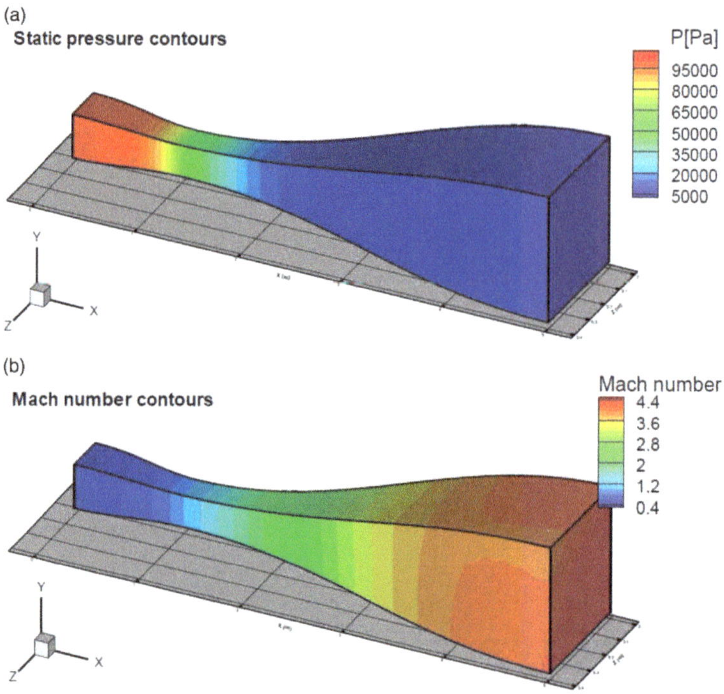

Figure 4. Static pressure and Mach number contours for the converging-diverging nozzle.

smaller than the adjacent interior node. Similarly, for the outer wall, the velocity magnitude on the wall is greater than the adjacent interior node which is incompatible with the expected physical trend. Here, there is a discontinuity in the constant velocity magnitude lines. In the second type of wall boundary conditions, the velocity magnitude is extrapolated from the adjacent interior node; hence in Figure 3(b) the constant velocity lines are normal to the wall. Similar to Figure 3(a), a nonphysical solution on the wall is achieved. In Figure 3(c) and (d), the physical trend is continued to the wall and the velocity contours are continuous. So, the third and fourth types of wall boundary conditions make a better prediction of the velocity magnitude on the wall. It should be noted that the solution with the fourth type of wall boundary condition was sensitive to the initial conditions, i.e., using some initial conditions the method diverges. In summary the third type of wall boundary condition, ($\rho V_t = cte$), gave the best accuracy and stability for the elbow test case.

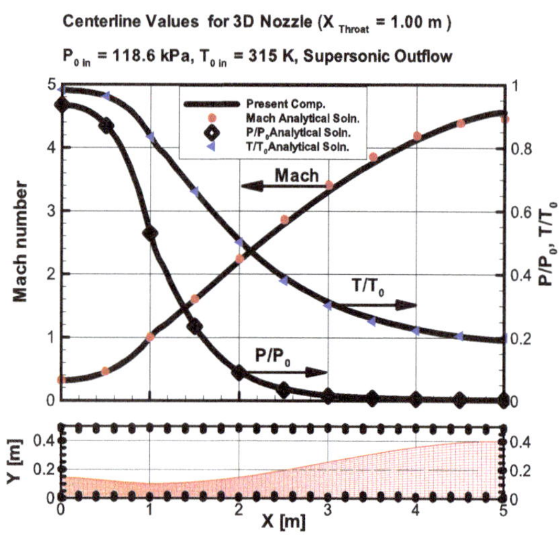

Figure 5. Comparison between numerical and analytical results for the nozzle test case.

6.2. *Isentropic flow through a converging-diverging nozzle*

In this section, flow through a converging-diverging nozzle is investigated. For this test case, the third method of wall boundary conditions ($\rho V_t = cte$) diverges and the results for the other three methods are presented here.

In Figure 4, samples of the results are presented by static pressure and Mach number contours. In Figure 5, the numerical results on the nozzle center line are compared with the analytical solution of isentropic flow. The results show good agreement. In the diverging section of the nozzle, the flow is supersonic and at the outlet the Mach number is about 4.5.

In Figure 6, contours of velocity magnitude for the nozzle are shown. As it is seen, the first and second type of wall boundary conditions make smoother contours, the

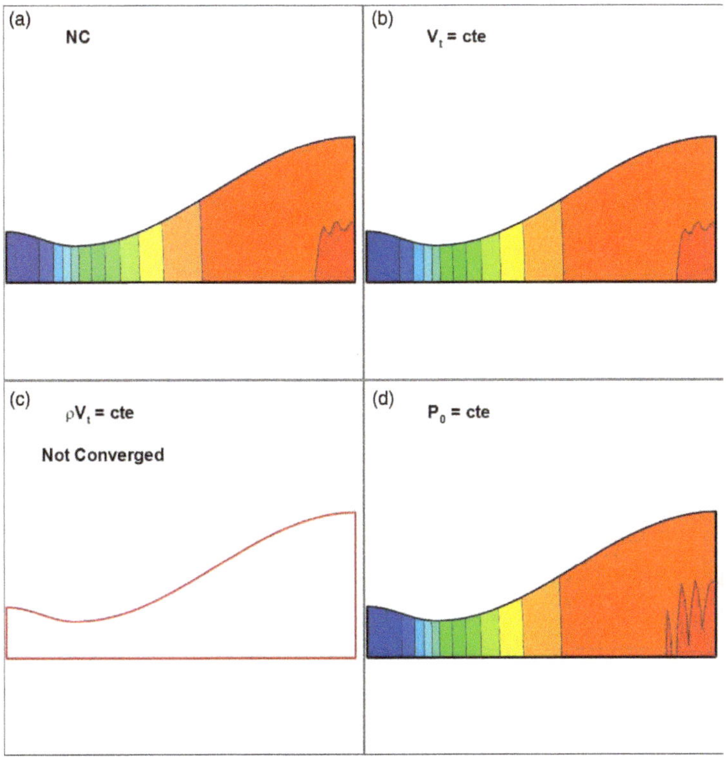

Figure 6. Contours of velocity magnitude for the converging-diverging nozzle.

third type results in divergence of the solution, and for the fourth type there are some oscillations in the results.

As a conclusion, it can be noted that for the converging-diverging nozzle, the first and second types of wall boundary conditions gives better stability and continuity of the results.

6.3. *Subsonic channel flow over a bump*

The subsonic channel flow over a bump is a bench mark for validation of numerical methods and boundary conditions for inviscid flow regimes. A 10% thick bump at the bottom of the channel is studied for subsonic and transonic flow. If the inlet Mach number is less than a certain (or critical) value, the flow is subsonic in the channel. If a decrease in back pressure results in the increase of the inlet Mach number to the critical value, the flow is transonic and a sonic line appears over the bump. So increasing the inlet Mach number causes a supersonic pocket to grow on the top of the bump. For the channel flow over the bump, the results are compared with numerical solutions presented by Ni (1982).

Here, the back pressure to the inlet total pressure ratio is selected in such a way that the inlet Mach number is equal to 0.5. The flow through the duct is subsonic. The solution with all types of wall boundary conditions converged. In Figure 7, the Mach number contours for the four types of boundary conditions are shown. As can be seen in Figure 7(c), the third type (Figure 7(c)) gives smoother contours, while a discontinuity occurred near the wall for the others. The discontinuity is more obvious in Figure 7(d) where the disturbances are expanded through the domain.

In Figure 8, the Mach number distribution on the upper and lower walls for all four types of boundary conditions are shown and compared with the results of Ni (1982). As seen, there is a major difference between the results of the fourth type and the others that is consistent with Figure 7. The oscillation in the results for the fourth type of wall boundary condition is more obvious here.

Finally, it can be concluded that for a subsonic flow over the bump, the third type of wall boundary conditions shows better accuracy, while its stability is as good as the first and second types.

6.4. *Transonic channel flow over a bump*

The transonic flow over the bump is studied in this section. As mentioned before, by decreasing the back pressure, the inlet Mach number will increases and reach a critical Mach number. In this case, a normal shock appears on the bump. The ratio of back pressure to the inlet total pressure is selected to have an inlet Mach number of 0.675.

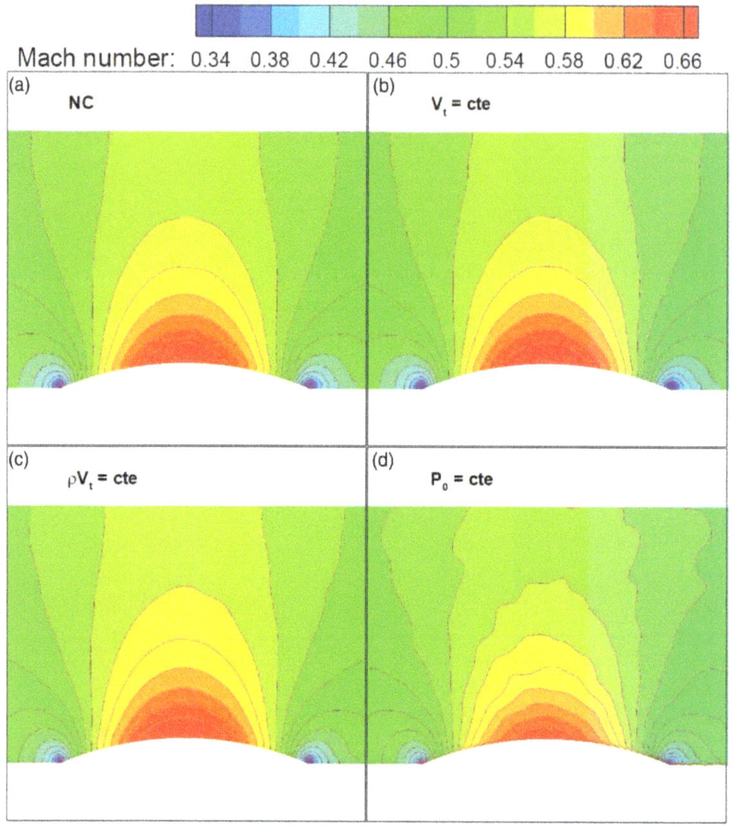

Figure 7. Subsonic flow field over the bump shown by Mach number contours for all types of boundary conditions.

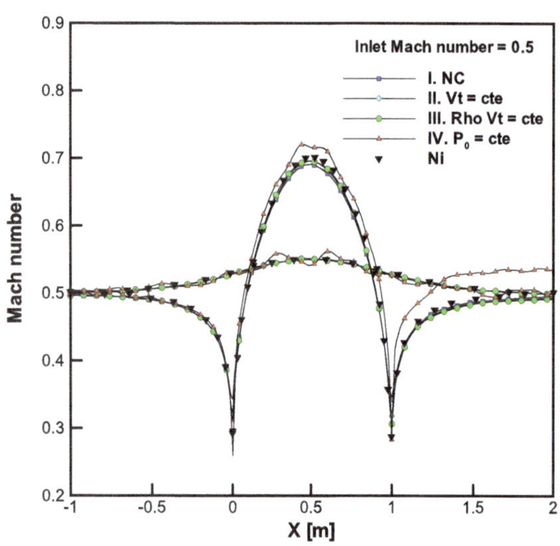

Figure 8. Comparison between Mach number over the lower and upper walls for all types of boundary conditions for subsonic bump problem.

In Figure 9, the Mach number contours of transonic flow over the bump for all types of boundary conditions are presented. As you can see, a normal shock occurs over the bump and the flow become subsonic. In Figure 9(a), the trend of the Mach number near the wall is reversed. Physically, the velocity magnitude near the wall must be increased and a constant total temperature leads to a lower static temperature; so, a higher Mach number near the bump surface is anticipated. As mentioned above, in the first type of boundary condition the velocity magnitude on the wall is less than the adjacent node and in this case it is not consistent with the physical trend.

In Figure 9(b), as expected, the iso-Mach number line is normal to the wall. Although the trend seen in Figure 9(b) does not match with the physics, with respect to the first type of boundary conditions it gives a better solution. In Figure 9(c), the best compatibility with the physical trend is observed and the iso-Mach number lines are continuous through the wall. In Figure 9(d), the Mach number on the wall is over estimated and again a discontinuity occurs in the Mach number contours.

In Figure 10, comparison between Mach number distribution over the upper and lower walls for the four types of boundary conditions are presented and compared with the results of Ni (1982). As expected, the first type gives the lowest Mach number over the walls. If the wall is straight or its curvature radius is very large, the difference between velocities obtained using the four types

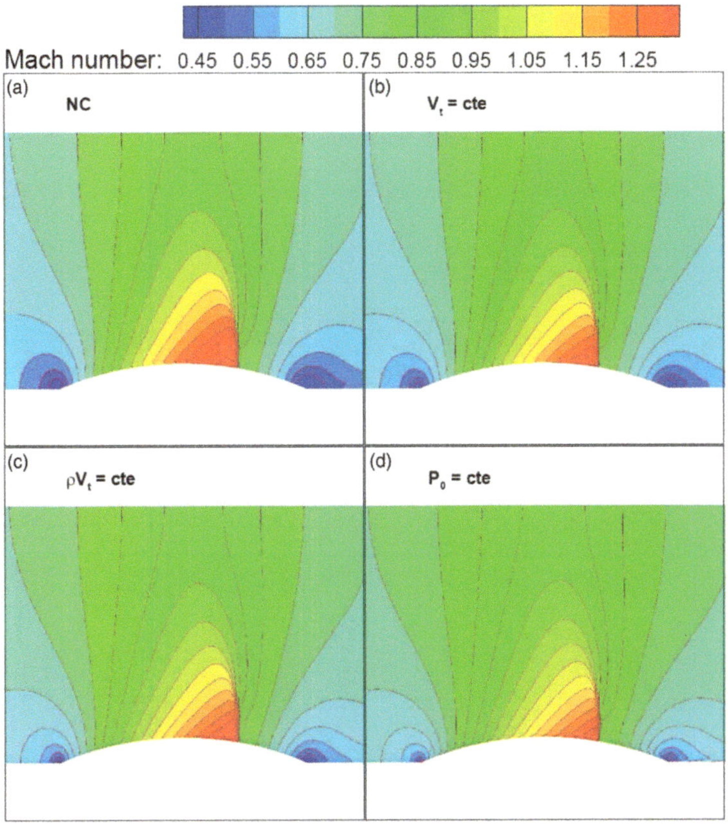

Figure 9. Transonic flow field over the bump shown by Mach number contours for all types of boundary conditions.

of boundary conditions studied in this paper is not sensible. Because the upper wall is straight and the lower wall has a curved surface, the difference between Mach number distributions is more visible over the lower wall on the bump section.

In conclusion, for this test case, the third type has the best prediction of velocity on the wall surfaces.

6.5. *Supersonic channel flow over a bump*

In this section, the supersonic flow over the bump is studied. For the supersonic flow, the thickness of the bump is selected as 4% of the channel height. In this case the oblique shock waves and the reflections from the walls are investigated. The inlet Mach number is set at 1.4 and the outlet is set as supersonic outflow boundary conditions.

For the supersonic flow, the solution with the third type of boundary conditions does not converge; hence, its results are not presented here.

In Figure 11, the Mach number contours are presented. For all types of wall boundary conditions, the Mach number contours near the wall are discontinuous. The discontinuity for the first and second types is in the reversed direction and the Mach number on the wall is underestimated, while for the fourth type the predicted Mach number on the wall is greater than the physically expected value. To summarize, it can be said that the

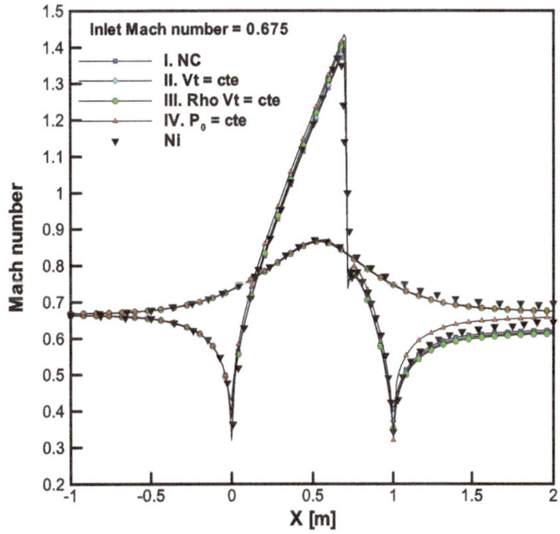

Figure 10. Comparison between Mach number over the lower and upper walls for all types of boundary conditions for transonic bump problem.

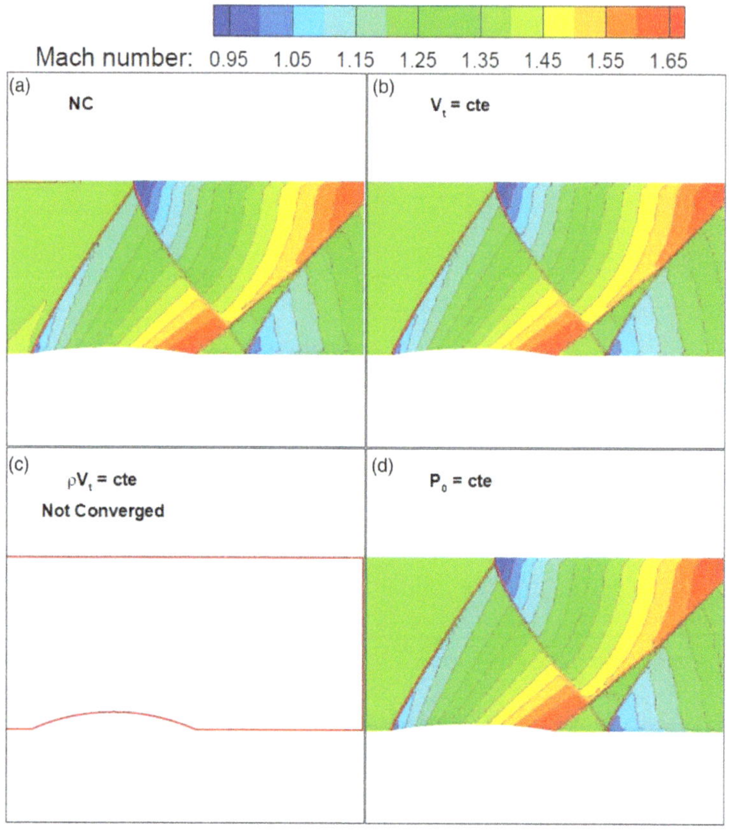

Figure 11. Supersonic flow field over the bump shown by Mach number contours for all types of boundary conditions.

second type of boundary conditions with the least deviation gives the best trend of Mach number through the wall.

In Figure 12, the comparison between Mach number distributions on the upper and lower walls for three types of boundary conditions is presented and compared with the numerical results of Ni (1982). As seen, the fourth type gives a non-smooth distribution especially near the outlet.

We concluded that for the supersonic flow over the bump, the second type of wall boundary conditions has the best stability and accuracy.

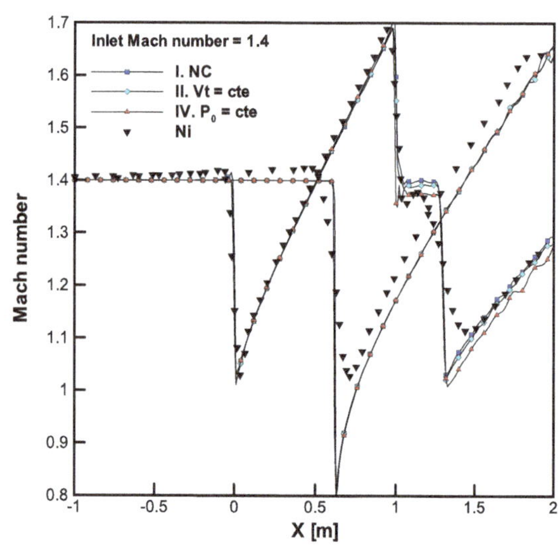

Figure 12. Comparison between Mach number over the lower and upper walls for all types of boundary conditions for supersonic bump problem.

6.6. *Flow through a NACA65-410 compressor cascade*

To assess the performance of the wall boundary conditions for an applicable test case, inviscid flow through a NACA65-410 compressor cascade, is studied here. In Figure 13, a sample of the results for a NACA65-410 cascade is shown by Mach number contours. As can be seen in the figure, the flow is subsonic and the Mach number decreases from inlet to outlet of the cascade as expected for a compressor airfoil. For the NACA65410 test case, the solution with the fourth type of boundary conditions does not converge. For previous test cases this method was sensible for the initial conditions.

In Figure 14, the velocity magnitude contours around the leading edge of the blade are shown. As shown, the third type ($\rho V_t = cte$) makes a smoother distribution of

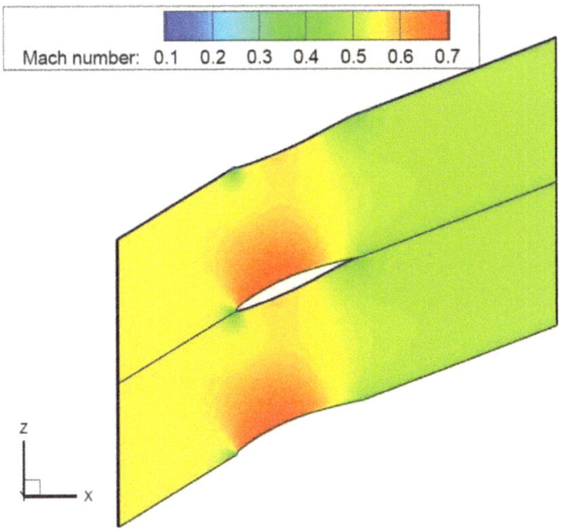

Figure 13. Mach number contours for NACA65-410 cascade.

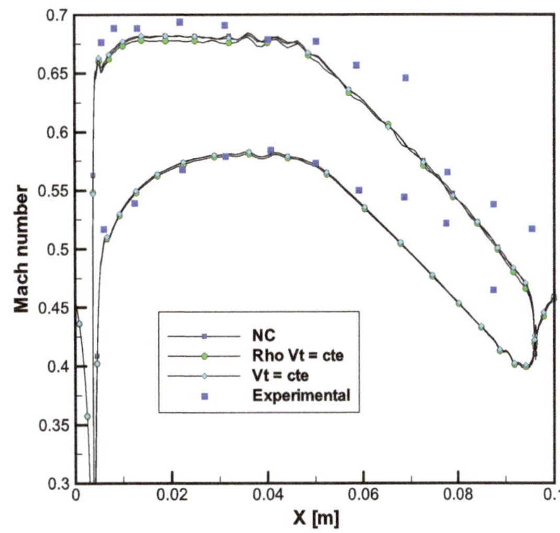

Figure 15. Mach number distribution on the blade surfaces for all types of wall boundary conditions.

the velocity magnitude. In Figure 15, the Mach number distributions on the blade surfaces for three types of wall boundary condition are presented and compared with the isentropic Mach number calculated from experimental results of Emery, Herrig, Erwin, and Felix (1957). As seen in the previous test cases (channel flow over the bump), the results for the Mach number of the first, second, and third types of boundary conditions are close together.

From the continuity of velocity magnitude contours, it is concluded that for subsonic flow through the NACA 65-410 cascade, the third type of velocity correction gives better accuracy.

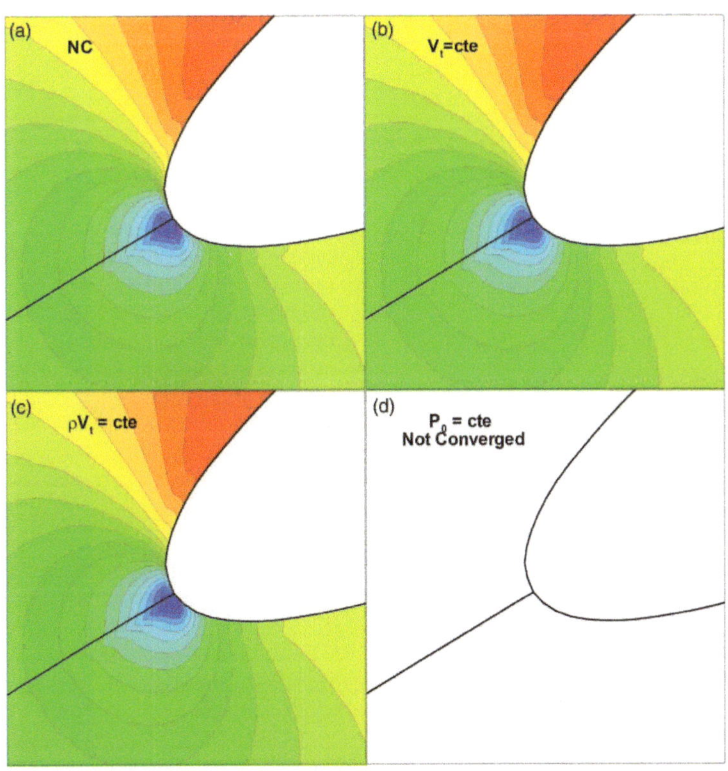

Figure 14. Contours of velocity magnitude obtained using different types of wall boundary conditions for NACA65-410 cascade.

7. Mixed boundary conditions

In previous sections, four types of wall boundary conditions are investigated and their stability and accuracy are compared. It has been concluded that for the subsonic test cases, the third type, which is the $\rho V_t = cte$ condition, has better accuracy and produces more compatible results with flow physics. On the other hand, for the supersonic test cases, the third method was unstable (diverged) and the second, ($V_t = cte$), shows better accuracy and stability.

Since there are many problems where some part of the domain is subsonic and some other parts are supersonic, one needs a boundary condition which supports both subsonic and supersonic flows. Hence, a mixed wall boundary condition is introduced and implemented here. To do so, for a node on the wall, the Mach number for the adjacent node is considered and in regard to the flow regime, the appropriate type of boundary condition is imposed. If the Mach number is less than unity, the third type ($\rho V_t = cte$) is applied and if the Mach number is greater than unity the second type ($V_t = cte$) isimposed.

To evaluate the mixed boundary condition, two transonic test cases are studied. In these cases both subsonic and supersonic flow regimes are observed. Based on the Mach number, in parts of the walls where the flow is subsonic the third method is applied and in the supersonic parts the second method is used. The first test case is the transonic flow over a bump, this case was has been previously studied for all the methods. In Figure 16, Mach number distribution over the walls for the second, third, and mixed types of boundary conditions are shown. As seen, the results of the mixed boundary conditions show good agreement with the second and third method.

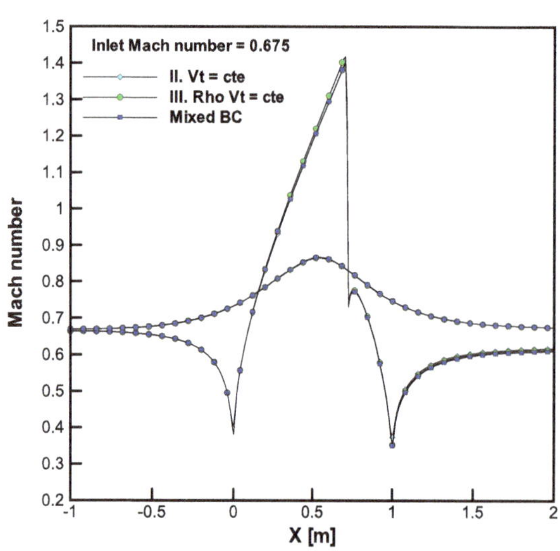

Figure 16. Comparison between Mach number over the walls obtained using 2nd, 3rd and mixed types of boundary conditions for transonic bump problem.

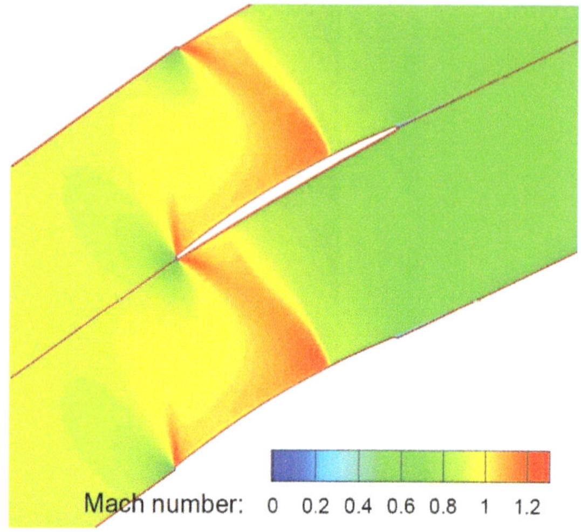

Figure 17. Contours of Mach number for DCA test case computed using mixed boundary conditions.

As a second test case, a double circular arc airfoil (DCA) is studied in a transonic flow regime. The DCA airfoils are usually used for transonic fans and compressors blade section profiles. In some part of the domain, the flow is supersonic and as the Mach number increases over the blade surface it passes from a normal shock to become subsonic. This airfoil is selected because of its application in compressors and also its transonic flow regime. In Figure 17, contours of the Mach number are presented for a DCA airfoil using mixed boundary conditions. No discontinuity or oscillation is observed in the domain.

8. Conclusions

Four types of wall boundary conditions are introduced. To assess the performance of each type, several subsonic and supersonic test cases are selected, the boundary conditions are implemented for each test case, and the flow fields are investigated using a recently developed inviscid flow solver.

In Table 2, the studied test cases are summarized for stability and accuracy. For subsonic and transonic flow the first, second and third conditions are robust while the fourth type of wall boundary conditions is more sensitive to the initial conditions. In comparison with the others, the third type or ($\rho V_t = cte$) has better accuracy.

For supersonic flows, the third method is unstable and results in divergence of the solution; the fourth method has less stability and is more sensitive to initial conditions. The second method ($V_t = cte$) shows better accuracy for supersonic test cases.

To have a convenient boundary condition, a combination of the second and third type of flow regimes called the mixed boundary condition is introduced and tested in two

Table 2. Summary of test cases.

Test Case	Flow Regime	Stability with regard to initial conditions				Best accuracy
		I	II	III	IV	
		Not Corrected (NC)	V_t = cte	ρV_t = cte	P_0 = cte	
90° Elbows	Subsonic	Robust	Robust	Robust	Sensible	3rd type
Nozzle	Supersonic	Robust	Robust	Diverged	Sensible	1st and 2nd types
Bump	Subsonic	Robust	Robust	Robust	Sensible	3rd type
	Transonic	Robust	Robust	Robust	Sensible	3rd type
	Supersonic	Robust	Robust	Diverged	Sensible	2nd type
NACA 65-410	Subsonic	Robust	Robust	Robust	Diverged	3rd type

test cases. The results show that the mixed boundary condition is applicable for both subsonic and supersonic flow regimes.

References

Emery, J. C., Herrig, L. J., Erwin, J. R., & Felix, A. R. (1957). *Systematic two dimensional cascade tests of NACA 65-series compressor blades at low speeds*. NACA Reports 1368.

Hoffmann, K. A., & Chiang, S. T. (2000). *Computational fluid dynamics* (Vol. I, 4th ed.). Wichita, KS: Engineering Education System.

Idahosa, U., Golubev, V. V., & Balabanov, V. O. (2008). An automated optimization of fan blade using an integrated CFD/MDO computer environment. *Journal of Engineering Applications of Computational Fluid Mechanics*, *2*(2), 141–154.

Kermani, M. J., & Plett, E. G. (2001a, January 8–11). *Roe scheme in generalized coordinates; part I- formulations*. 39th AIAA Science Meeting and Exhibition, Reno, NV, 2001–0086.

Kermani, M. J., & Plett, E. G. (2001b, January 8–11). *Modified entropy correction formula for the Roe scheme*. 39th AIAA Science Meeting and Exhibition, Reno, NV, 2001–0083.

Khazaei, H., Madadi, A., & Kermani, M. J. (2011, July 24–29). *Assessment of various inviscid-wall boundary conditions: applications to NACA65 compressor blade*. ASME-JSME-KSME Joint Fluids Engineering Conference, AJK2011-FED, Hamamatsu, Shizuoka, Japan.

Kim, M. S., Jeon, B. W., Kim, Y. N., Kwon, H. B., & Lee, D. H. (2001). An implementation of the robust inviscid wall boundary condition in high-speed flow calculations. *KSME International Journal*, *15*(5), 671–680.

Kim, M. S., & Kim, H. K. (2000). An investigation of characteristics of Roe's FDS by using the compressible Euler equation. *Journal of the Korean Society for Aeronautical and Space Science*, *28*(7), 1–9.

Liou, M. S., & Steffen, C. (1993). A new flux splitti ng scheme. *Journal of Computational Physics*, *107*(1), 23–39.

Liu, X., Dang, Q., & Xi, G. (2008). Performance improvement of centrifugal fan by using CFD. *Journal of Engineering Applications of Computational Fluid Mechanics*, *2*(2), 130–140.

Marshall, D. D., & Ruffin, S. M. (2004, January 5–8). *A new inviscid wall boundary condition treatment for embedded boundary Cartesian grid schemes*. 42nd AIAA Science Meeting and Exhibition, Reno, NV.

Ni, R. H. (1982). A multiple-grid scheme for solving the Euler equations. *AIAA Journal*, *20*(11), 1565–1571.

Qamar, A., Hasan, N., & Sanghi, S. (2010). A new spatial discretization strategy of the convective flux term for the hyperbolic conservation laws. *Journal of Engineering Applications of Computational Fluid Mechanics*, *4*(4), 593–611.

Roe, P. L. (1981). Approximate Riemann solvers, parameter vectors and difference schemes. *Journal of Computational Physics*, *43*, 357–372.

Van Albada, G. D., van Leer, B., & Roberts, W. W. (1982). A comparative study of computational methods in cosmic gas dynamics. *Astronomy and Astrophysics*, *108*(1), 76–74.

Wang, Z. J., & Yuzhi, S. (2002, January 14–17). *A curvature based wall boundary condition for the Euler equations on unstructured grids*. 40th AIAA Science Meeting and Exhibition, Reno, NV.

Yu, Y., Shademan, M., Barron, R. M., & Balachandar, R. (2012). CFD study of geometry variations on flow in a nozzle. *Journal of Engineering Applications of Computational Fluid Mechanics*, *6*(3), 412–425.

14

Efficient methods for prediction of velocity fields in open channel junctions based on the artifical neural network

Amir Hossein Zaji and Hossein Bonakdari*

Department of Civil Engineering, Razi University, Kermanshah, Iran

Measurement results, sediment transport and pollutant dispersion in confluence regions are highly affected by flow structures of open channel junction and it is necessary to find an efficient method that can fully describe the velocity fields in junctions. In this study the velocity field in an open channel junction was investigated by using Artificial Neural Network (ANN) and three dimensional modelling. First, a modified Genetic Algorithm (GA) was introduced, then, an ANN model was optimized and the flow velocity was predicted in a 90 degree channel junction. Also, for three dimensional simulation of the free surface flow in the considered junction ANSYS-CFX software was used. Comparison of the results obtained from ANN and CFX models with laboratory data with root mean square error of 0.094 and 0.182 and percent standard error of prediction of − 9.75 and − 25.71 respectively, showed that the introduced ANN perform better than CFX in modelling velocity field in open channel junction and it was able to predict accurate in different areas and flow-rates.

Keywords: open channel junction; artificial neural network; genetic algorithm; ANSYS-CFX; velocity fields

1. Introduction

Channel confluences are really significant in environmental and hydraulic engineering. Their flow structures highly affect measurement results, erosion, sediment transport and the dispersion of the pollutants in confluence regions. Open channel flow junctions, as considered in this paper, are those in which the flow from a lateral branch channel is entertained in a main channel.

In these flows, due to interaction between the main and branch flows, flow has a completely three dimensional behavior and calculating the characteristics of this complex flow requires precise and separate studies. The main flow features that characterize these configurations are zones of separation (or recirculation), the maximum and minimum velocity regions and a shear plane which is developed between the two joining flows downstream of the confluent channel. Numerous laboratory and theoretical analysis have been done to analyze and predict the characteristics of these flows (Taylor, 1944; Webber & Greated, 1966; Lin & Soong, 1979; Best & Reid, 1984; Ramamurthy, Carballada, & Tran, 1988; Biron, Best, & Roy, 1996; Wilcox, 1998; Chung-Chieh, Wen-Jung, & Cheng-Hsi, 1998). Conducting a laboratory study on the flow profile in a 90° channel junction, Weber, Schumate, and Mawer (2001) attempted to measure the velocity and depth in different flows and provide comprehensive information about the characteristics of flow in junctions. This extensive measurement provided an outstanding experimental data for validation.

For several decades experimental studies and theoretical analysis of flow junctions have been a great concern for engineers; however, three dimensional numerical modelling that employs computational fluid dynamic softwares is a recent phenomenon. Bradbrook, Biron, Lane, Richards, and Roy (1998), Bradbrook, Lane, and Richards (2000a), Bradbrook, Lane, Richards, Biron, and Ag (2000b) validated the application of three dimensional numerical model to simulate the flow structures in an open channel junction. Comparing the results obtained from their experiments Wang and Cheng (2000) investigated the performance of the three dimensional numerical model for the prediction of three dimensional side-discharge flows. Huang, Weber, and Lai (2002) used a three dimensional model and k-ω turbulence model to investigate the open channel junction flow and analyzed the effect of junction flow on flow characteristics. Shamloo and Pirzadeh (2008) used 2D numerical model to predict the velocity fields and dimension of separation zone in a T junction. Zhao, Zhu, and Rajaratnam (2008) established a computational fluid dynamics model to simulate fully surcharged flow at a 90 degree combining sewer junction. Shakibainia, Tabatabai, and Zarrati (2010) used a three dimensional numerical model, SSIIM2.0, and k-ω turbulence model to investigate the flow characteristics at separation zones, secondary currents, velocity profiles, and flow surface mapping indifferent confluence angles and discharge, ratios. Bonakdari, Lipeme-Kouyi, and Wang (2011) using a three dimensional

*Corresponding author. Email: bonakdari@yahoo.com

computational fluid dynamics model, analyzed the effect of discharge ratio on recirculation size and presented formulas to calculate the length and width of the separation zone in terms of discharge ratio for a 30° junction. Considering the distance from the junction, Dursun, Kaya, and Firat (2012) used 3D numerical model to analyze the junction's impact on the velocity fields to evaluate the error that is derived from calculating the flow rate close to the junction.

Due to the high capacity of soft computing method to analyze complex problems, this method has been used extensively in various hydraulic engineering problems such as discharge capacity of lateral weirs (Neary, & Sotiropoulos, 1996; Bilhan, Emiroglu, & Kisi, 2011), scour depth prediction (Muzzammil, 2010), flow characteristics in different open channels (Donmez, 2001; Grace & Priest, 1958; More, 1978), rainfall modeling and stream flow prediction (Asadi, Shahrabi, Abbaszadeh, & Tabanmehr, 2013; Chau, Wu, & Li, 2005; Chen & Chau, 2006; Cheng, Lin, Sun, & Chau, 2005; Firat & Gungor, 2008; Wu, Chau, & Li, 2009; Yurtseven, & Zengin, 2013), modelling coastal algal blooms (Muttil & Chau, 2006), evapotranspiration (Cobaner, 2011; Kisi, & Ozturk, 2007), combined open channel flow (Hager, 1987), sediment transport (Ebtehaj, & Bonakdari, 2013; Van Maanen, Coco, Bryan, & Ruessink, 2010), ground water level prediction (Taormina, Chau, & Sethi, 2012) and water demand forecasting (Pulido-Calvo & Gutierrez-Estrada, 2009; Tiwari, & Adamowski, 2013).

Artificial Neural Network (ANN) is one of the most applied methods for soft computing while it has comes in various types, with the simplest one being the Multi-Layer Perceptron (MLP). The present study was aimed to analyze the longitudinal velocity profile of flow in open channel junction and to measure its predictability by ANN and CFD models. To this end, the structure of a Levenberg-Marquardt ANN model was optimized by a new adaptive genetic algorithm. Also, a numerical model was made by ANSYS-CFX software for junctions' flow modeling. The accuracy of both ANN and CFX models were measured by laboratory results and accuracy of each model was evaluated in flow prediction. Finally, velocity fields in some cross sections where laboratory findings are not available are calculated by ANN and compared with the results obtained from three dimensional numerical modeling.

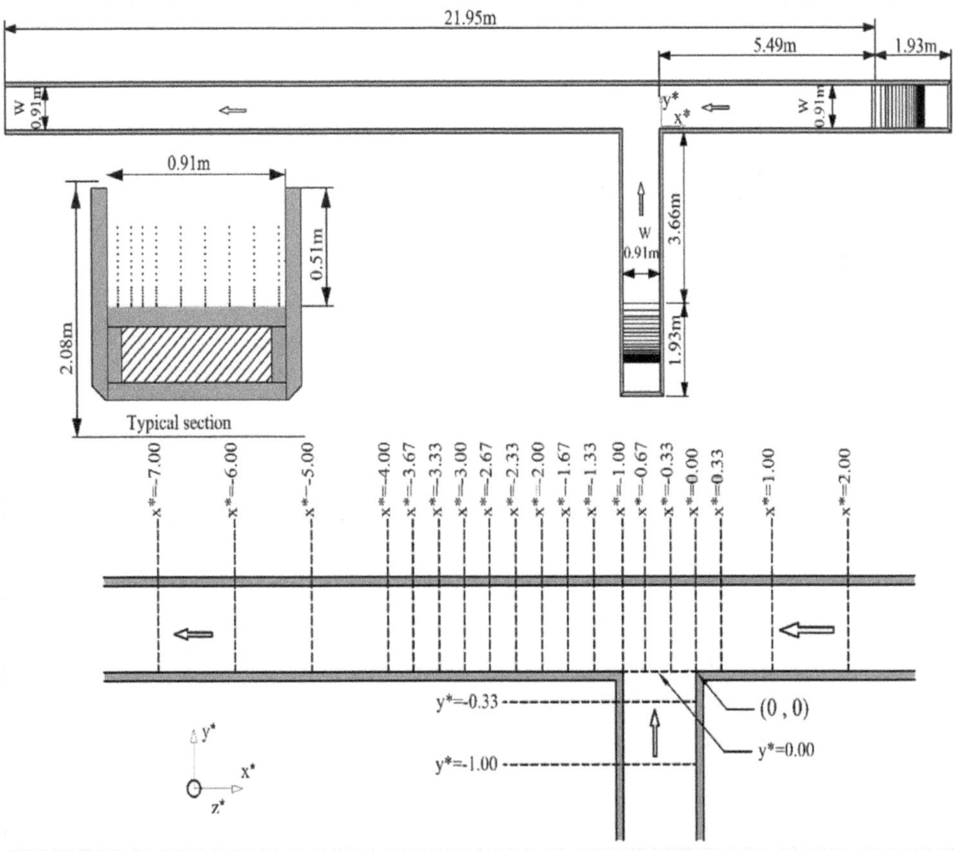

Figure 1. Experimental flume layout, Weber et al. (2001).

Table 1. Experimental flow conditions, Weber et al. (2001).

Q_m (m^3/s)	Q_b (m^3/s)	q*
0.014	0.156	0.083
0.042	0.127	0.250
0.071	0.099	0.417
0.099	0.071	0.583
0.127	0.042	0.750
0.156	0.014	0.917

2. Experimental model

In the present study, the laboratory results of Weber et al. (2001) and Shumate and Weber (1998) were used for validating ANN and numerical modeling results. The experiments were performed in a 90 degree combining flow flume as shown in Figure 1. While the junction has two branches, the upstream branches straight and the lateral branch has a junction angle of 90 degree. Two head tanks were placed in the entrance of main channel and branch channel that supplied the required flow into channels. Main and tributary channels had the same width (=0.91 m) and lengths of 21.95 m and 3.66 m, respectively. The slope of both channels was horizontal. The defined coordinate system had positive x-axis in the main channel's upstream direction. The positive y-direction points to the main channel wall which stands opposite to the positive z-axis upward in the vertical direction. All lengths to channel width (=0.91) were nondimensionalized. The nondimensionalized coordinates are shown with x*, y* and z* that are equivalent to x/W, y/W and z/W. Also, the measured velocities were nondimensionalized by downstream average velocity, (V_t=0.628 m/s), and u*, v* and w* are equivalent to u/V_t, v/V_t and w/V_t respectively. Upstream main channel flow and branch channel flow are indicated with Q_m and Q_b respectively. Flow ratio q* is the ratio of main channel's upstream flow to total outflow of main and branch channels from the end of the channel ($q^* = Q_m/(Q_m + Q_b)$). The flow velocity is measured in flow conditions indicated in Table 1 in the marked cross sections (see Figure 1). More details about experiments are found in Weber et al. (2001). It should be noted that in all stages of the study, CFX and ANN numerical modeling, defined nondimensional coordinate system (x*, y*, z*) and nondimensional velocities (u*, v*, w*) were applied.

3. Methods

Initially, through ANSYS-CFX software, a three dimensional model was made and was compared with laboratory results. Then, through a genetic algorithm, an ANN was optimized in each layer for number of the hidden layer neurons. Using a part of laboratory data, ANN model was trained and tested with the other part. After evaluating ANN and CFX modeling error accuracy, under conditions in which no laboratory data are available, the results of ANN model were compared with those of CFX numerical simulation.

3.1. *Model performance*

The performance of an ANN is determined by comparing the predicted and observed outputs. There are several methods to quantify the difference between predicted and observed data, one of the most common of which is Mean Square Error (MSE) that defined as average of squares of the errors. In this study, model performance in ANN training process and GA objective function is calculated with MSE (Equation 1).

$$MSE = \frac{1}{N}\sum_{i=1}^{N}(Y_{i_{observed}} - Y_{i_{estimated}})^2 \quad (1)$$

where N is the number of output and Y_i the velocity along x*. To analyze the accuracy of ANN in train and test data, MSE was computed separately.

To evaluate the accuracy of ANN model, Root Mean Square Error (RMSE), Mean Absolute Error (MAE), coefficient of determination (R^2), coefficient of efficiency (E_j), percent Standard Error Prediction (%SEP), Average Relative Variance (ARV), and average absolute deviation (%δ) were used. The advantage of RMSE and MAE models in evaluating the model error is that both use the same unit of variables. R^2 provides a measure of how well the observed outcomes are replicated by the model. The coefficient of efficiency by using j=2 (E_2) has been extremely used to evaluate the model accuracy (Leavesley, Lichty, Troutman, & Saindon, 1983; Wilcox, Rawls, Brakensiek, & Wight, 1990). If E_2= 1, then ANN model has the perfect match to observed data, If it equals 0, then it indicates that the ability of ANN model in predicting longitudinal velocity is similar to average of observed data and if the value of E_2 is less than 1, the average of the observed data has higher accuracy than the ANN model (Nash & Sutcliffe, 1970). Due to squaring the difference terms in E_2, the high values of coefficient of efficiency were obtained (Willmott et al., 1985). In order to obtain the appropriate weighting of errors and decrease the effect of squared terms, the modified coefficient of efficiency by using j=1 (E_1) were used (Legates & McCabe, 1999). One of the significant advantages of %SEP and %δ is their nondimensionality which makes it possible to compare different models regardless of their dimension and size (Pulido-Calvo & Portela, 2007).

$$RMSE = \sqrt{MSE} \quad (2)$$

$$MAE = \frac{1}{N}\sum_{i=1}^{N}|Y_{i_{estimate}} - Y_{i_{observed}}| \quad (3)$$

$$R^2 = \left[\frac{\sum_{i=1}^{n} (Y_{i_{observed}} - \overline{Y}_{observed})(Y_{i_{estimated}} - \overline{Y}_{estimated})}{\sqrt{\sum_{i=1}^{n} (Y_{i_{observed}} - \overline{Y}_{observed})^2 \sum_{i=1}^{n} (Y_{i_{estimated}} - \overline{Y}_{estimated})^2}} \right]^2 \quad (4)$$

$$E_j = 1 - \frac{\sum_{i=1}^{N} |Y_{i_{observed}} - Y_{i_{estimated}}|^j}{\sum_{i=1}^{N} |Y_{i_{observed}} - \overline{Y}_{observed}|^j} \quad (5)$$

$$ARV = \frac{\sum_{i=1}^{N} |Y_{i_{observed}} - Y_{i_{estimated}}|^2}{\sum_{i=1}^{N} |Y_{i_{observed}} - \overline{Y}_{observed}|^2} = 1 - E_2 \quad (6)$$

$$SEP = \frac{100}{\overline{Y}_{observed}} \times RMSE \quad (7)$$

$$\delta = \frac{\sum_{i=1}^{N} |Y_{i_{estimate}} - Y_{i_{observed}}|}{\sum_{i=1}^{N} Y_{i_{estimate}}} \times 100 \quad (8)$$

In these equations, $\overline{Y}_{observed}$ and $\overline{Y}_{estimated}$ are the average of the observed and estimated velocity along x* direction respectively.

3.2. *Numerical model*

The three dimensional numerical modelling is performed by commercial Ansys-CFX software package, which solves the three-dimensional fundamental flow equations. The vital equations for fluid motion in the whole domain are: (1) the Eulerian approach's continuity equation for the incompressible fluid and (2) the three Reynolds-Averaged Navier-Stokes (RANS) momentum equations for the incompressible turbulent fluid. Solving these equations require a turbulence model to set the Reynolds stresses. In this study, k-ω turbulence model (Wilcox, 1998) was used in which two transport equations are solved, one for the turbulent kinetic energy, k, and the other one for the turbulent frequency, ω.

To assess the accuracy of different turbulence models, RNG k-ϵ, k-ϵ and k-ω models were compared with the experimental results in same meshing (Table 2). The results indicate that RNG k-ϵ and k-ω models give more agreeable results than k-ϵ model. Although the accuracy of RNG k-ϵ and k-ω models is much close to each other, in some standard error measures like RMSE, MAE and R^2, k-ω was more reliable than RNG k-ϵ. Also, this model is used by various authors, Huang et al. (2002) and Shakibainia, Tabatabai, and Zarrati (2010) for simulations of flow in open channel junctions.

At the inlet, uniform velocity and water depth h are specified, turbulent intensity sets 5% (Bonakdari, Larrarte, & Bardiaux, 2007) and sufficient length is provided upstream of the junction to get a fully developed velocity profile. A length (10 m) is provided upstream of each inlet branch in order to obtain a fully developed turbulent flow when reaching the junction.

At the outlet, pressure is specified at the cell face center, while Cartesian velocity components and turbulence quantities are extrapolated from the interior using a second-order extrapolation. For the pressure correction equation, the pressure increment is set to zero at the outlet, as the pressure should not change during the solution. The rigid walls were set to smooth with no slip condition because it was considered appropriate for a laboratory flume. In the near wall area, Launder and Spalding (1974) method of standard wall function was used; that is, the velocity components are equal to zero in the walls. The free surface location is determined by the Volume of Fluid (VOF) method. The VOF technique, as described in Hirt and Nichols (1981), has been incorporated into the solution of the Navier–Stokes equations in order to take into account the free surface effects on the Reynolds stresses and, thus, velocity distribution in the cross section.

To find the best gridding, three mesh states, coarser mesh, mediocre mesh and finer mesh were used with 340526, 994652, and 2876399 elements respectively. The results of comparisons this three CFX models with

Table 2. Performance of CFX model in various turbulence models.

Turbulence	RMSE	MAE	R^2	%SEP	E_1	E_2	ARV	%δ
k-ϵ	0.233	0.201	0.487	– 44.798	– 1.356	– 7.989	8.989	– 32.922
RNG k-ϵ	0.187	0.145	0.595	– 23.791	– 0.019	– 0.150	1.15	– 16.060
k-ω	0.182	0.138	0.579	– 25.717	– 0.153	– 0.481	1.481	– 17.409

Table 3. Performance of CFX model in various mesh size.

Mesh	RMSE	MAE	R^2	%SEP	E_1	E_2	ARV	%δ
Coarser	0.234	0.192	0.623	– 49.425	– 1.864	– 11.291	12.291	– 26.779
Mediocre	0.182	0.138	0.579	– 25.717	– 0.153	– 0.481	1.481	– 17.409
Finer	0.162	0.124	0.566	– 22.051	0.127	0.141	0.859	– 16.656

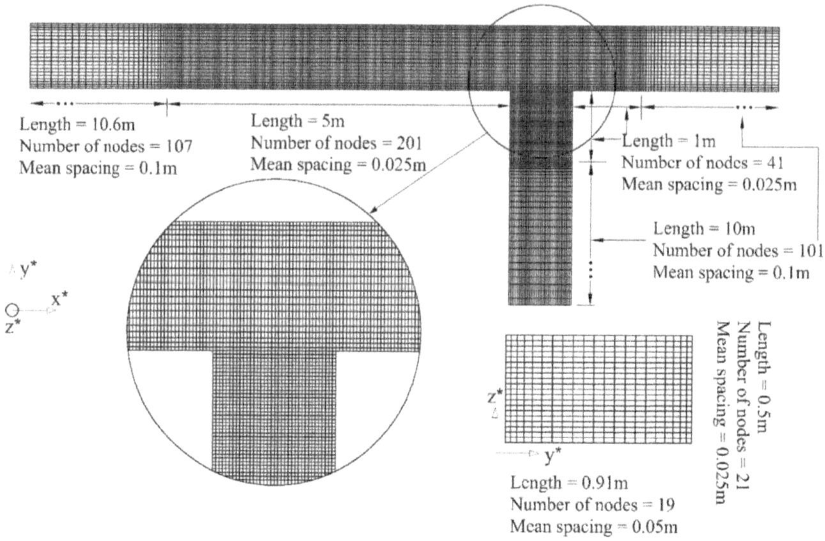

Figure 2. Details of computational mesh.

observed data in flow ratio of $q^* = 0.25$ shown in Table 3. Comparison of coarser mesh state with that of mediocre mesh state indicates that the overall accuracy of the model has increased because RMSE value has decreased from 0.23 to 0.18 and $\%\delta$ value has decreased from -27.66 to -17.40. On the contrary, the comparison of the mediocre mesh modeling results with those of finer mesh modeling indicates that RMSE error has decreased from 0.182 to 0.162 and $\%\delta$ error has reduced from -17.40 to -16.65 indicating that reducing the mesh size has no significant effect on improving the results. In this study mediocre mesh modelling with 9,94,652 elements and 9,31,307 nodes were used.

A smaller mesh size was used around the confluence and a larger mesh size was used for upper and lower parts as shown in Figure 2. The numerical solution was terminated so that normalized residual would reach 10^{-6} for governing equations.

3.3. *Prediction using ANN*

In the first section, ANN model has been studied and then to optimize neural network structure, a modified genetic algorithm was introduced in the second section.

3.3.1. *Artificial neural network*

The flexible structure of ANN provides the capacity to make complex and nonlinear models between input and output. The capacity to estimate the correct results using input is based on training and learning processes. Training the neural networks means acquiring network weights (w). Also, various ANN are classified based on the methods of acquiring weights and transfer functions. One of the most applicable neural networks is Multi-Layer Perceptron (MLP). A feed forward MLP consists of an input layer or one or more hidden output layers. Every layer comprises a number of neurons. Neurons number in input layers is the same as the number of inputs and outputs of the given issue, respectively. In this study, four pieces of data were considered as input and one piece of data was considered as output. The inputs are nondimensional coordinates (x^*, y^* and z^*) and flow ratio (q^*), and target is nondimensional longitudinal velocity in x direction (u^*). In the neural networks made for hidden-layer and output neurons, sigmoid and linear activation function are used, respectively. An f(x) function is of sigmoid type if it is bounded and is increased by increasing x value (Smith, 1993). However, diverse functions can be regarded as sigmoid function. In the present study, hyperbolic tangent was used in hidden layers as activation function. To train ANN, Levenberg-Marquardt method was applied (Guven & Kisi, 2011; More, 1978). In this method, back-propagation algorithm is used to find the weights and bias of neural network. It is one of the most useful algorithms in MLP neural network (Azamathulla & Ghani, 2011; Goh, 1995). This algorithm quantifies the difference between the outputs observed in laboratory studies and outputs of ANN model by determining weights and bias with high velocity.

To model the neural networks, the laboratory results obtained in flow ratio 0.083, 0.583 and 0.917, totally comprising 7348 pieces of data, are used. These data are divided into three parts: training, testing and validation. Data are randomly classified as 5142 (70%) into training data, 1103 (15%) into testing data and 1103 (15%) to validation data.

To investigate the potential of ANN model in predicting velocity profile in other flow ratios, in the next step, the

ANN model compare with the laboratory data in flow ratios q* = 0.25 and q* = 0.75 which virtually had no role in neural network train, validation and test was studied. Stop condition for ANN model is reaching 100 epochs.

So far, many parameters required were determined to define an MLP neural network like number and type of neurons in input and output layers, activation functions in hidden and output layers and training method. The number of hidden layers and the number of neurons in each layer are parameters important to define an MLP neural network. However, there is no definite principle to select these parameters and the number of layers and each layer's neurons should be tested to find the most optimized neural network (Bilhan et al., 2011; Melesse, Ahmad, McClain, Wang, & Lim, 2011). For example, if the required neural network can have 1–2 hidden layers and 1–20 neurons in each layer, we should test 420 states to find the optimized neural network, which is very difficult and time-consuming. In this study, the modified genetic algorithm was employed to identify the number of neurons in each of algorithm's layer.

3.3.2. Genetic algorithm

A common optimization algorithm in engineering sciences is genetic algorithm which was introduced by Holland (1975). The first step of the algorithm is creating a random initial population; then is creating a sequence of

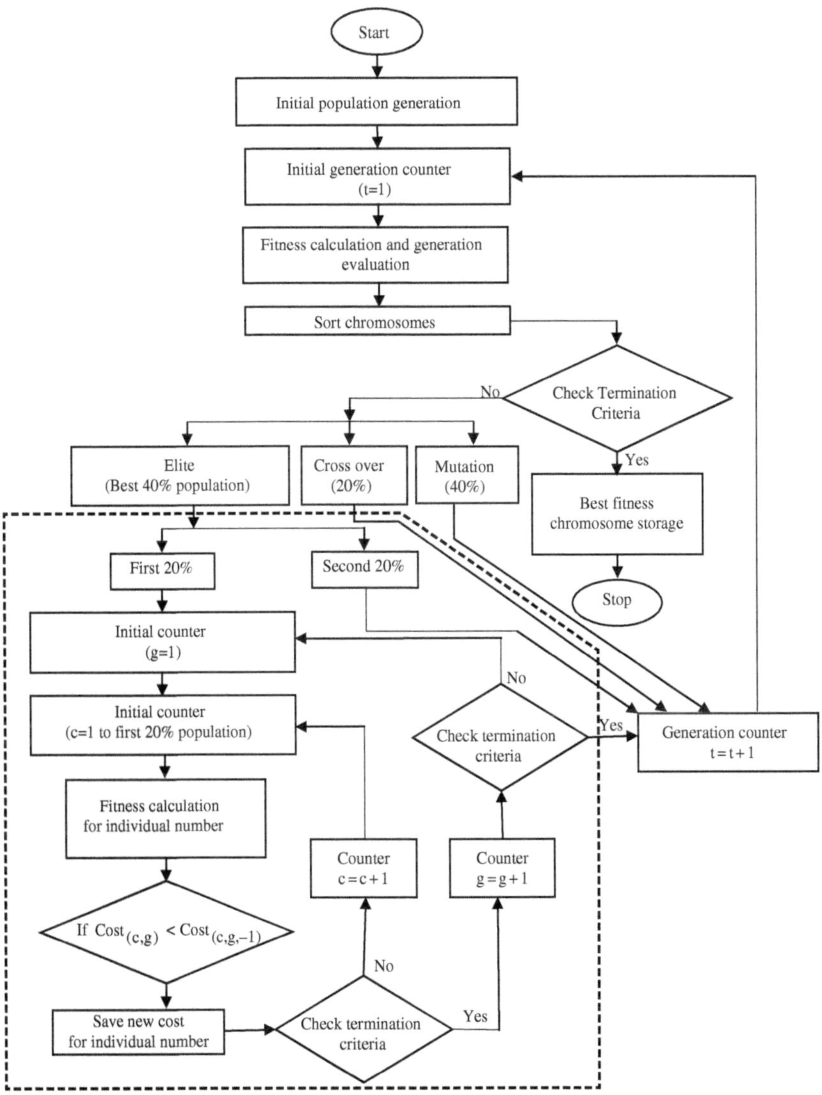

Figure 3. Flowchart of proposed GA.

Table 4. The MSE statistics of the two samples ANN models.

Runs	First ANN	Second ANN
1	0.0156	0.0098
2	0.0235	0.0129
3	0.1222	0.0234
4	0.0948	0.0345
5	0.0743	0.0143

new population, or generations. Each step consists of the algorithm's use of individuals in the current generation to produce the consequent generation. In creating a new generation, two steps are needed to be taken as shown in Figure 3, first, each member of the current population has to be scored by computing its fitness values; second, children should be reproduced from the current generation. To determine the individual population cost in genetic algorithm, the cost of every individual was calculated by Equation (9) in order to prevent the possible overtraining of ANN. Overtrain is a state in which there is a big difference between test and train data accuracy. A good ANN model is one in which test and train data accuracy are close to each other. To investigate the goodness of each chromosome in GA (ANN models with different structures), instead of using MSE_{test} or MSE_{train} separately, the sum of both coefficients is used. In GA model, a chromosome is considered to be superior in which the sum of its MSE_{test} and MSE_{train} is minimum (Equation 9). Also by giving different values to C_{test} and C_{train}, one can use MSE_{test} and MSE_{train} weighted sum. C_{test} and C_{train} parameters can be determined by trial and error method.

$$\cos t_i = C_{train} \times MSE_{i_{train}} + C_{test} \times MSE_{i_{test}} \quad (9)$$

C_{test} and C_{train} are coefficients that indicate the importance we give to the train and test data. They are determined experimentally in terms of neural network and data. In this study, they are given value 1. Children are reproduced via Elite, crossover and mutation. When the new population is created, it replaces the current population with the children so that creation of the next generation becomes possible. Meeting one of the stopping criteria leads in the stop of the algorithm.

Due to the random selection of neural network in selecting the test and train data from the present dataset, even when all parameters of an ANN like number of layers, neurons in each layer and activation function are retrained by ANN model, we do not obtain previous series of MSE; instead, we obtain a specific MSE in each training. For instance, in Table 4, two 4-layer neural networks (one input, two hidden and one output) are trained with the present data. In the first ANN in each hidden layer, 3 neurons are used and in the second ANN, 6 neurons are used in each hidden layer. Then, the produced neural networks are trained 5 times. Due to random selection of train and test data in each training, the MSEs obtained are not the same, Table 4 indicates that the second ANN provides better results than the first ANN. This feature of the neural network makes optimization algorithm used to optimize the neural networks is trapped by local minimums and produces incorrect responses. To solve this problem, it is necessary to make a little change in the elite section of creating the next generation, indicated in Figure 3 with dash line.

Elite children refer to individuals in the current generation that are identified as having the best fitness values. These individuals automatically pass to the subsequent generation. The changes made were intended to prevent elimination of good individuals in the transfer process to the next generation. By 40 times repetitions in the top 20% of the population, the best fitness value that each individual has gained in these repetitions is substituted for the fitness value of that individual. Thus, falling in a local minimum cannot make any changes in the top 20% of population and they only change when, due to mutation or crossover, an individual is produced with a fitness value better than that of the top 20% of the population after 40 repetitions. In the present model, having optimized the network with modified GA, the number of neurons in layers 2 and 3 was 9 in a 4-layer neural network. The performance of optimized ANN is shown in Table 5. In this table, closer the values of RMSE, MAE, %SEP, AVR and %δ get to 0 and closer the values of E_1, E_2, R^2 to 1, shown the model has higher accuracy. As seen from 5, as expected train data as compared to validation and test data has higher accuracy. The closeness of train, validation and test data indicates that ANN model does not trap in overtrain. Based on %SEP and %δ error in all data, one can conclude that on the whole optimized ANN model has %12.90 and %-9.45 error, respectively. This values indicating its agreeable accuracy in longitudinal velocity modelling (u*).

Table 5. Performance of optimized ANN model.

Data set	RMSE	MAE	R^2	%SEP	E_1	E_2	ARV	%δ
Train	0.086	0.064	0.963	− 12.652	0.833	0.963	0.036	− 9.349
Test	0.094	0.067	0.956	− 13.691	0.823	0.956	0.043	− 9.754
Validation	0.091	0.066	0.959	− 13.284	0.829	0.959	0.040	− 9.634
All data	0.088	0.064	0.962	− 12.908	0.831	0.962	0.038	− 9.453

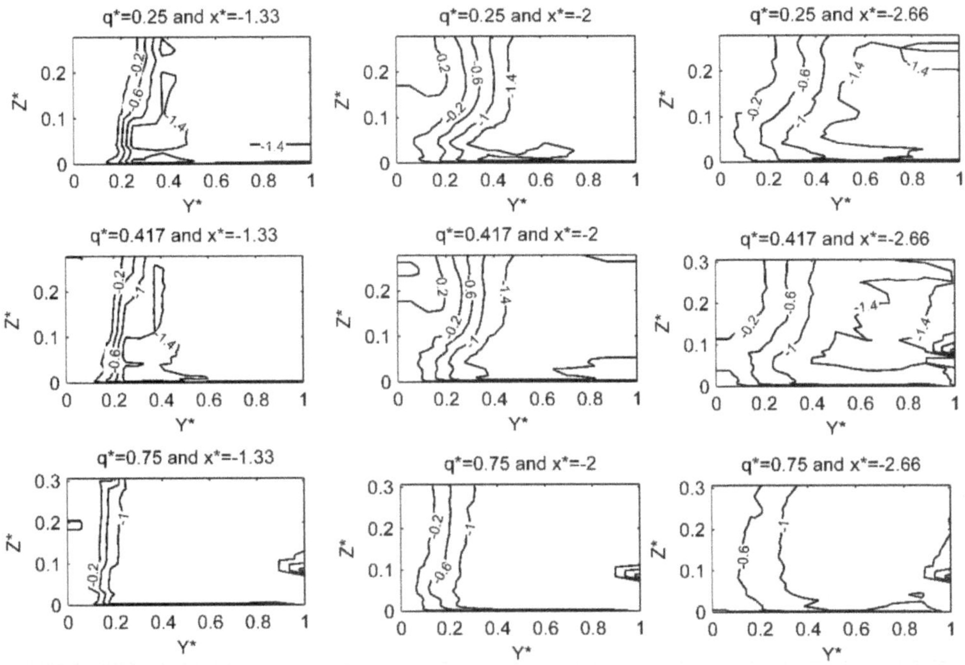

Figure 4. Contours of longitudinal velocity.

4. Results

4.1. *Investigation of variation of longitudinal velocity*

As mentioned, longitudinal velocity, u*, indicates nondimensionalized velocity along x. In Figure 4, the u*-velocity contours for $x^* = -1.33$, -2 and -2.66 are shown in $q^* = 0.250$, 0.417 and 0.750. Slightly after confluence in longitudinal walls of the channel, an area is created in whose boundaries velocity is decreased and in which recirculation occurs, indicated in Figure 4 with positive velocities, this region is called separation zone.

The comparison of Figure 4 in $q^* = 0.250$, 0. 417 and 0.750 indicated that by increasing the flow in the main channel and increasing q* in a constant x* (e.g. $x^* = -2$), the separation zone dimensions are decreased due to flow momentum decline in the branch channel owing to q* increase. On the other hand, when q* is low, due to high momentum of the branch flow, it is diverted to the downstream side of the main channel when it enters the main channel and two flows join each other. This increases the separation zone dimensions. Moreover, by moving from the channel floor to the surface of the water, the separation zone dimensions are increased and consequently recirculation and velocity are increased in the direction opposite to the flow. Another area that happens in the downstream junctions is constricted area that occurs in the right side of the sections (in y* between 0.5 and 1) (Figure 4). The more is the constriction of the flow, the higher is the flow velocity. The intensity of constriction is directly associated with separation zone width, and the constriction rate and flow velocity are increased by increasing the width of this area. Unlike open channels in which the highest velocity occurs near the free surface, in this area the maximum velocity occurs near the channel bed. However, by distancing the confluence, the velocity profile in the channel returns to normal state and the highest velocity happens in the flow surface.

4.2. *Results of CFX*

Longitudinal velocity distribution values obtained from the numerical model made by ANSYS-CFX software were compared with laboratory values in $x^* = -2$ and -3 and in $q^* = 0.250$ and 0.750 illustrated in Figure 5. In this figure, the horizontal axis indicates nondimensionalized values of the channel's width (y*) and the vertical axis shows nondimensionalized values along the water depth (z*). As it is indicated, the numerical and laboratory models were in good agreement and there were only differences between numerical and laboratory values in some points near the wall on $y^* = 0$. Figure 5 shows that the velocity value in the opposite direction of the flow even in $x^* = -2$, which is located in the middle of the separation zone, are very small compared to the longitudinal velocity and this value is decreased by moving from $x^* = -2$ to $x^* = -3$. In fact, by distancing the flow confluence, it tries to

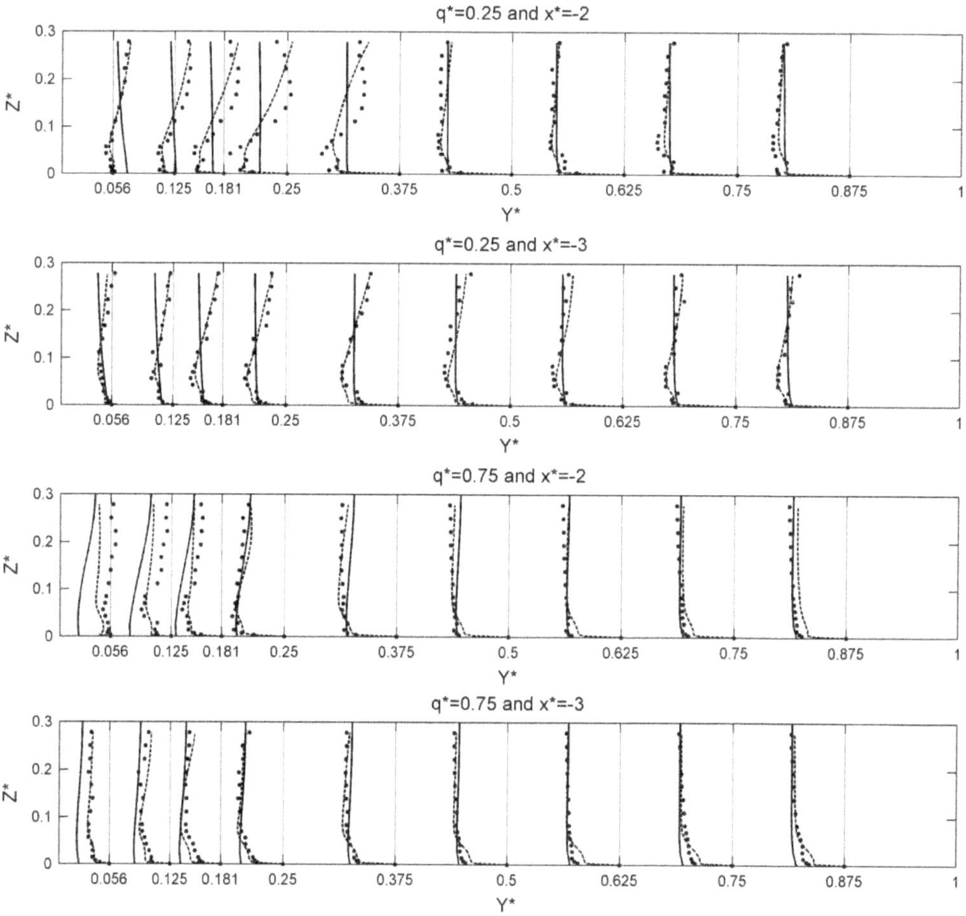

Figure 5. Longitudinal velocity profiles in measurements (. . .) and 3D CFX ( ) and ANN (—) simulations.

return to the typical state of flow in open channel, but this recovery is very slow and occurs in a length 3 times more than the confluence point to x* = − 2 point (Dursun et al., 2012; Huang et al., 2002). Contours a, b, c and d in Figure 6 indicate the error of the results obtained from CFX model compared to laboratory findings. Given these contours in q* = 0.250 and x* = − 2 and − 3, the typical error is about 7%, which is increased to 20% locally and 30% in very small areas. The higher accuracy of numerical modeling in q* = 0.750 than in q* = 0.250 is because of recirculation zone reduction in high q* owing to weakening of inflow from branch channel to main channel and decline of flow displacement in the main channel.

4.3. Results of ANN

As cited before, ANN model used in this study is a four-layer ANN that consists of an input layer, an output layer and two hidden layers; the input layer includes nondimensionalized coordinate of the given point (z*, y*, x*) and flow ratio in flow conditions (q*). The input layer has totally four inputs and output layer has one output, which is the longitudinal velocity of flow (u*). After optimization of ANN model by modified genetic algorithm, the number of neurons existing in the two hidden layers was 9. The characteristics of modified genetic algorithm are presented in Figure 3, laboratory data in q* = 0.083, 0.583 and 0.917 for ANN modeling were used and the input data were totally 7348. Since the input values of neural network (x*, y*, z* and q*) were nondimensionalized, the obtained neural network can be used for every open channel junction with desirable dimensions and flow ratios. Therefore, the Equation (10) was extracted in this study to determine the longitudinal velocities (u*) in every open channel junction for x*, y*, z* and q* inputs of trained ANN, where B_1, B_2 and B_3 stand as bias values for the first, second and third layers and W_1, W_2 and W_3 are weight values for first,

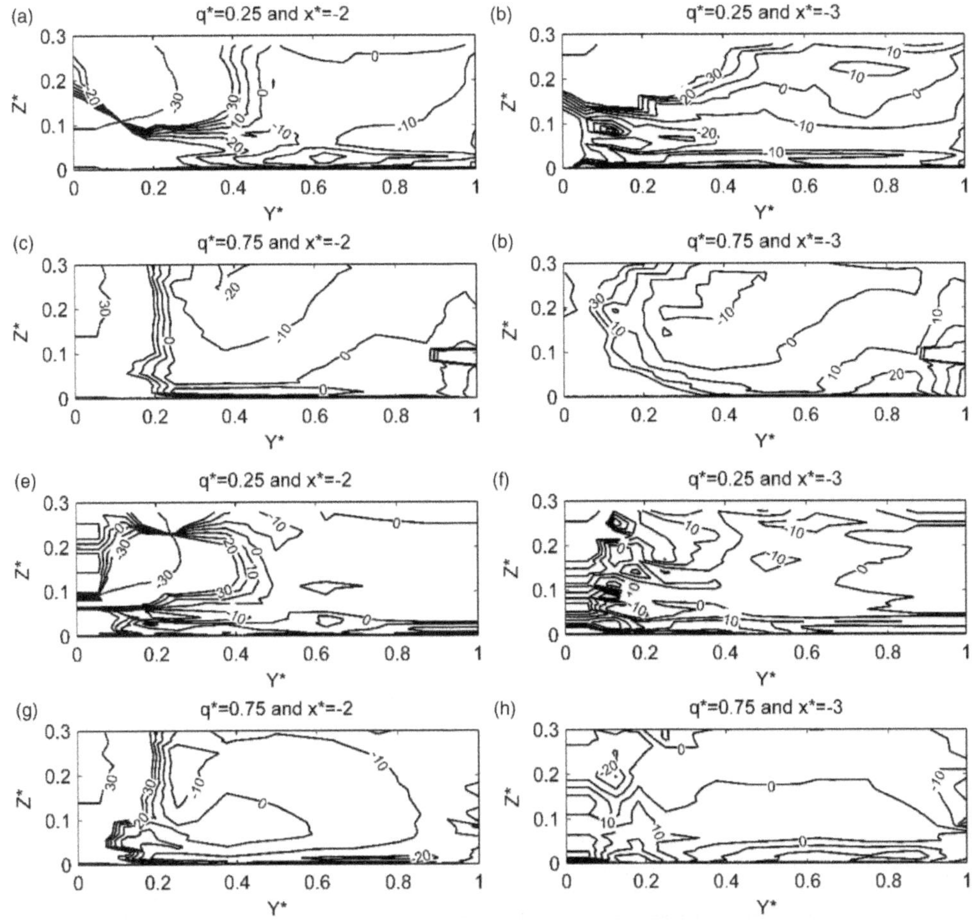

Figure 6. Error of the longitudinal velocity predicted (u^*_{calc}) by CFX model (a,b,c,d) and ANN model (e,f,g,h) compared to experimental data (u^*_{exp}) that given as $(u^*_{calc}-u^*_{exp})/u^*_{exp}$ in %.

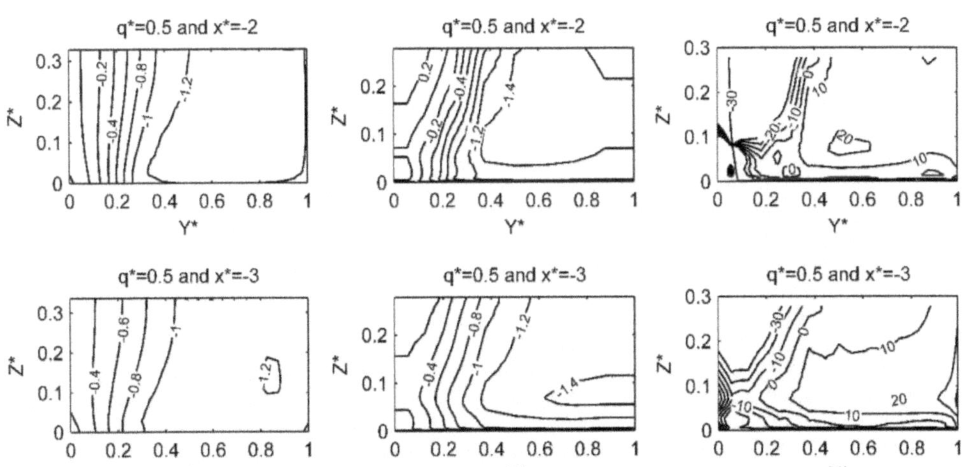

Figure 7. Longitudinal velocity profile where we don't have experimental data in CFX simulation (the left column) and ANN model (the middel column) and their deviation that given as (u*cfx-u*ann)/ u*cfx in % (the right column)

second and third layers, respectively, which are shown in Equations (11) to (16). Matrix I shows the inputs of the problem presented in Equation (17). Also, F_1, F_2 and F_3 are activation functions used in ANN model. As shown in Equations (18) and (19), F_1 and F_2 are sigmoid functions and the F_{out} function in the present model is a linear function. In Figure 5, the values of longitudinal velocities (u*) obtained from ANN are compared with those obtained from laboratory results. The findings of the error contours in this figure indicate that despite not using laboratory results for q*= 0.25 and 0.75 in the train and test of the neural networks, the results of ANN model in this q* are compatible with laboratory results. This shows that this model can be used to acquire the profiles of longitudinal velocities in every flow ratio (q*). The error contours of e, f, g and h in Figure 6, which indicate the error resulting from ANN model compared to laboratory results, show that the typical error is around 5% in both q*= 0.250 and q*= 0.750, and the local error in x*= −2, which has a remarkable recirculation, is 30% and in x*= −3 is 20% because it is slightly distanced from the separation zone. Based on the comparison of the results of CFX and ANN shown in Figure 5 and the error values of these models with those of laboratory model presented in Figure 6, it is obvious that ANN is more accurate than CFX in modelling the velocity field in open channel junction zone, which is more noticeable in recirculation zone that is the major deficiency of CFD models in flow modeling in confluence zone.

$$u^* = F_{out}(F_2(F_1(I_{(1\times4)} \times W1_{(4\times9)} + B1_{(1\times9)}) \times W2_{(9\times9)} + B2_{(1\times9)}) \times W3_{(9\times1)} + B3_{(1\times1)}) \tag{10}$$

$$B1_{(1\times9)} = [-2.3579 \;\; 2.9795 \;\; 26.9251 \;\; -5.3245 \;\; 1.3823 \;\; 0.1175 \;\; -2.0678 \;\; 0.8528 \;\; 5.3748] \tag{11}$$

$$B2_{(1\times9)} = [-11.3620 \;\; 3.9181 \;\; -1.9545 \;\; -1.8640 \;\; 9.8659 \;\; 1.9582 \;\; -0.8594 \;\; 2.4295 \;\; 0.6613] \tag{12}$$

$$B3_{(1\times1)} = [6.1646] \tag{13}$$

$$W1_{(4\times9)} = \begin{bmatrix} 7.6695 & -1.7634 & -0.0668 & 21.3580 & -0.6429 & -2.7991 & -0.6072 & 0.5022 & 0.4061 \\ 0.0291 & -0.0141 & -0.0136 & -0.1525 & -0.3292 & -0.3027 & -4.3405 & -0.5962 & 0.1870 \\ 0.2153 & -0.2092 & 25.8747 & 0.1791 & 0.7555 & -0.1116 & 0.8210 & -0.1163 & 6.9672 \\ 0.1834 & 2.9233 & -0.0213 & 0.2169 & 0.2895 & 0.0555 & -0.3815 & -1.8503 & -0.1503 \end{bmatrix} \tag{14}$$

$$W2_{(9\times9)} = \begin{bmatrix} -0.9104 & 1.2005 & 0.0835 & -0.1722 & -0.6964 & 0.0179 & -0.1343 & 1.0064 & 1.3996 \\ -0.1734 & 0.1189 & -0.0424 & 0.0362 & 0.0660 & -0.0033 & -0.0333 & 0.2678 & 0.0166 \\ -3.9389 & -1.9558 & 0.6751 & 1.1518 & -10.3368 & -0.9751 & 0.8998 & -0.9374 & 0.6126 \\ -0.0619 & 0.0640 & 0.2884 & -0.1684 & -0.0726 & 0.0373 & -0.6531 & 0.0336 & 0.5911 \\ -1.5376 & -0.4020 & 0.1708 & 0.3452 & -0.7708 & -0.2888 & 0.0275 & -0.2570 & 0.0704 \\ -0.1672 & 0.4823 & 0.4216 & -0.1457 & 1.4878 & -0.0677 & 0.6639 & -0.4353 & 0.4832 \\ -0.7921 & 0.5462 & 0.4421 & 0.1852 & 0.6746 & -0.2234 & 0.7911 & 0.6830 & -1.2219 \\ 0.1593 & -0.0391 & 0.1520 & 0.1050 & -0.5545 & -0.1418 & 0.0273 & -0.2191 & 0.3587 \\ -14.6429 & -0.0519 & -0.0186 & 0.0106 & -0.4693 & -0.0030 & -0.0376 & -0.0784 & 0.1619 \end{bmatrix} \tag{15}$$

$$W3_{(9\times1)} = [2.3321 \;\; -7.0601 \;\; -13.1806 \;\; -16.4147 \;\; 0.3578 \;\; -29.1941 \;\; 3.3368 \;\; 5.2391 \;\; 1.6925]^T \tag{16}$$

$$I_{(1\times4)} = [x* \;\; y* \;\; z* \;\; q*] \tag{17}$$

$$F_1(x) = F_2(x) = \tanh(x) \tag{18}$$

$$F_{out}(x) = x \tag{19}$$

4.4. *Comparison of results CFX and ANN*

The comparison of Figures 5 and 6 shows that the areas out of the recirculation zone of ANN and CFX models have rather the same accuracy, but the main difference is the accuracy of these models in obtaining longitudinal velocities (u*) in inner areas of recirculation in which ANN model is more successful than CFX model. The main purpose of CFX numerical modeling and ANN training and their validation with laboratory model is analyzing the reliability of them in obtaining longitudinal velocities in conditions where there is no laboratory studies. That is why in Figure 7 where no measurement has been done in laboratory studies for q*=0.500 (Shumate & Weber, 1998; Weber et al., 2001), the contours of longitudinal velocities (u*) are designed by numerical CFX model (left column) and trained by ANN model or by Equation (8) (middle column). If the deviation between the results of CFX and ANN is expressed by percentage (right column), it can be seen that both models have good agreement and the typical error in x*= -2 is approximately 15% and in x*= -3 around 10%, which is indicative of the reliability of both CFX and ANN models and their high accuracy in flow modeling in open channel junction.

5. Conclusion

In this study, GA-ANN, three dimensional numerical model and laboratory data were used to analyze and predict the longitudinal velocity in 90 degree open channel junction. First, the genetic algorithm was modified to be appropriate for optimization of the random structure of neural network and ANN model was optimized to predict the longitudinal velocity in different sections of an open channel junction in various flow ratios. To train the neural network, Levenberg-Marquardt method was applied. Then, ANN model was compared with laboratory results and its error was analyzed. The findings indicated the high accuracy of ANN model to predict the velocity profiles in junctions even in complex areas of flow like separation zone. Finally, the equation extracted from the produced neural network was presented for predicting the flow velocity that can be used for every 90 degree open channel junction with every flow ratio. A three dimensional two-phase flow (water + air) numerical model was made by ANSYS-CFX software and the results obtained to predict the longitudinal velocities were compared with laboratory findings. The findings indicated a good accuracy for CFX model. Moreover, the results of ANN and CFX models in $q^* = 0.500$ were compared and the results showed good agreement between the results obtained from both models. Finally, by comparing the results of CFX and ANN models, it can be argued that, except in recirculation zone areas where ANN model produces better results, the accuracy in other flow areas of both models are almost the same.

References

Asadi, S., Shahrabi, J., Abbaszadeh, P., & Tabanmehr, S. (2013). A new hybrid artificial neural networks for rainfall-runoff process modeling. *Neurocomputing*, *121*, 470–480.

Azamathulla, H. M., & Ghani, A. A. (2011). Genetic programming for predicting longitudinal dispersion coefficients in streams. *Water Resources Management*, *25*(6), 1537–1544.

Best, J. L., & Reid, I. (1984). Separation zone at open-channel junctions. *Journal of Hydraulic Engineering – ASCE*, *110*(11), 1588–1594.

Bilhan, O., Emiroglu, M. E., & Kisi, O. (2011). Use of artificial neural networks for prediction of discharge coefficient of triangular labyrinth side weir in curved channels. *Advances in Engineering Software*, *42*(4), 208–214.

Biron, P., Best, J. L., & Roy, A. G. (1996). Effects of bed discordance on flow dynamics at open channel confluences. *Journal of Hydraulic Engineering*, *122*(12), 676–682.

Bonakdari, H., Larrarte, F., & Bardiaux, J. (2007). Experimental and computational study of velocity fields in narrow or compound section sewers. *Water Practice & Technology*, *2*(2), 1–8.

Bonakdari, H., Lipeme-Kouyi, G., & Wang, X. (2011, May 22–26). *Experimental validation of CFD modeling of multiphase flow trough open channel confluence.* World Environmental and Water Resources Congress 2011: Bearing Knowledge for Sustainability, Palm Springs, CA, pp. 2176–2183.

Bradbrook, K. F., Biron, P. M., Lane, S. N., Richards, K. S., & Roy, A. G. (1998). Investigation of controls on secondary circulation in a simple confluence geometry using a three-dimensional numerical model. *Hydrological Processes*, *12*(8), 1371–1396.

Bradbrook, K. F., Lane, S. N., & Richards, K. S. (2000a). Numerical simulation of three-dimensional, time-averaged flow structure at river channel confluences. *Water Resources Research*, *36*(9), 2731–2746.

Bradbrook, K. F., Lane, S. N., Richards, K. S., Biron, P. M., & Ag, R. (2000b). Large Eddy Simulation of periodic flow characteristics at river channel confluences. *Journal of Hydraulic Research*, *38*(3), 207–215.

Chau, K. W., Wu, C. L., & Li, Y. S. (2005). Comparison of several flood forecasting models in Yangtze River. *Journal of Hydrologic Engineering*, *10*(6), 485–491.

Chen, W., & Chau, K. W. (2006). Intelligent manipulation and calibration of parameters for hydrological models. *International Journal of Environment and Pollution*, *28*(3–4), 432–447.

Cheng, C. T., Lin, J. Y., Sun, Y. G., & Chau, K. (2005, August 27–29). *Long-term prediction of discharges in Manwan Hydropower using adaptive-network-based fuzzy inference systems models.* First International Conference on Natural Computation, ICNC 2005, Changsha, pp. 1152–1161.

Chung-Chieh, H., Wen-Jung, L., & Cheng-Hsi, C. (1998). Subcritical open-channel junction flow. *Journal of Hydraulic Engineering*, *124*(8), 847–855.

Cobaner, M. (2011). Evapotranspiration estimation by two different neuro-fuzzy inference systems. *Journal of Hydrology*, *398*(3–4), 292–302.

Donmez, S. (2011). Using artificial neural networks for prediction of alternate depth shaped on rectangular channel in open channel flow. *Energy Education Science and Technology Part A: Energy Science and Research*, *28*(1), 339–348.

Dursun, O. F., Kaya, N., & Firat, M. (2012). Estimating discharge coefficient of semi-elliptical side weir using ANFIS. *Journal of Hydrology*, *426–427*, 55–62.

Ebtehaj, I., & Bonakdari, H. (2013). Evaluation of sediment transport in sewer using artificial neural network. *Engineering Applications of Computational Fluid Mechanics*, *7*(3), 382–392.

Firat, M., & Gungor, M. (2008). Hydrological time-series modelling using an adaptive neuro-fuzzy inference system. *Hydrological Processes*, *22*(13), 2122–2132.

Goh, A. T. C. (1995). Back-propagation neural networks for modeling complex systems. *Artificial Intelligence in Engineering*, *9*(3), 143–151.

Grace, J. L., & Priest, M. S. (1958). *Division of flow in open channel junctions.* Auburn, AL: Engineering Experiment Station, Alabama Polytechnic Institute.

Guven, A., & Kisi, O. (2011). Estimation of Suspended Sediment Yield in Natural Rivers Using Machine-coded Linear Genetic Programming. *Water Resources Management*, *25*(2), 691–704.

Hager, W. H. (1987). Discussion of "Separation Zone at Open-Channel Junctions" by James L. Best and Ian Reid (November, 1984). *Journal of Hydraulic Engineering*, *113*(4), 539–543.

Hirt, C. W., & Nichols, B. D. (1981). Volume of fluid (VOF) method for the dynamics of free boundaries. *Journal of Computational Physics*, *39*(1), 201–225.

Holland, J. (1975). Genetic Algorithms, computer programs that evolve in ways that even their creators do not fully understand. *Scientific American*, (267), 66–72.

Huang, J., Weber, L. J., & Lai, Y. G. (2002). Three-dimensional numerical study of flows in open-channel junctions. *Journal of Hydraulic Engineering*, *128*(3), 268–280.

Kisi, O., & Ozturk, O. (2007). Adaptive neurofuzzy computing technique for evapotranspiration estimation. *Journal of Irrigation and Drainage Engineering*, *133*(4), 368–379.

Launder, BE., & Spalding, D. B. (1974). The numerical computation of turbulent flows. *Computer Methods in Applied Mechanics and Engineering*, *3*(2), 269–289.

Leavesley, G. H., Lichty, R., Troutman, B., & Saindon, L. (1983). *Precipitation-runoff modeling system: User's manual.* Colorado, CO: US Geological Survey.

Legates, D. R., & McCabe, Jr G. J. (1999). Evaluating the use of 'goodness-of-fit' measures in hydrologic and hydroclimatic model validation. *Water Resources Research*, *35*(1), 233–241.

Lin, J. D., & Soong, H. K. (1979). Junction losses in open channel flows. *Water Resources Research*, *15*(2), 414–418.

Melesse, A. M., Ahmad, S., McClain, M. E., Wang, X., & Lim, Y. H. (2011). Suspended sediment load prediction of river

systems: an artificial neural network approach. *Agricultural Water Management*, *98*(5), 855–866.
More, J. J. (1978). The Levenberg-Marquardt algorithm: Implementation and theory, In G. A. Watson (Ed.), *Lecture Notes in Mathematics 630* (pp. 105–116). New York: Springer-Verlag.
Muttil, N., & Chau, K. W. (2006). Neural network and genetic programming for modelling coastal algal blooms. *International Journal of Environment and Pollution*, *28*(3–4), 223–238.
Muzzammil, M. (2010). ANFIS approach to the scour depth prediction at a bridge abutment. *Journal of Hydroinformatics*, *12*(4), 474–485.
Nash, J. E., & Sutcliffe, J. V. (1970). River flow forecasting through conceptual models part I - A discussion of principles. *Journal of Hydrology*, *10*(3), 282–290.
Neary, V. S., & Sotiropoulos, F. (1996). Numerical investigation of laminar flows through 90-degree diversions of rectangular cross-section. *Computers and Fluids*, *25*(2), 95–118.
Pulido-Calvo, I., & Gutierrez-Estrada, J. C. (2009). Improved irrigation water demand forecasting using a soft-computing hybrid model. *Biosystems Engineering*, *102*(2), 202–218.
Pulido-Calvo, I., & Portela, M. M. (2007). Application of neural approaches to one-step daily flow forecasting in Portuguese watersheds. *Journal of Hydrology*, *332*(1–2), 1–15.
Ramamurthy, A. S., Carballada, L. B., & Tran, D. M. (1988, December 12). Combining open channel flow at right angled junctions. *Journal of Hydraulic Engineering*, *114*, 1449–1460.
Shakibainia, A., Tabatabai, M. R. M., & Zarrati, A. R. (2010). Three-dimensional numerical study of flow structure in channel confluences. *Canadian Journal of Civil Engineering*, *37*(5), 772–781.
Shamloo, H., & Pirzadeh, B. (2008). Investigation of characteristics of separation zones in T-junctions. *WSEAS Transactions on Mathematics*, *7*(5), 303–312.
Shumate, ED., & Weber, L. J. (1998, August 3–7). *Experimental description of combining flows at an open channel junction*. Proceedings of the 1998 International Water Resources Engineering Conference. Part 2 (of 2). ASCE, Reston, VA, pp. 1679–1684.
Smith, M. (1993). *Neural networks for statistical modeling*. New York, NY: Thomson Learning.
Taormina, R., Chau, K. W., & Sethi, R. (2012). Artificial neural network simulation of hourly groundwater levels in a coastal aquifer system of the Venice lagoon. *Engineering Applications of Artificial Intelligence*, *25*(8), 1670–1676.
Taylor, EH. (1944). Flow characteristics at rectangular open-channel junctions. *Transactions of the American Society of Civil Engineers*, *109*(1), 893–902.
Tiwari, M. K., & Adamowski, J. (2013). Urban water demand forecasting and uncertainty assessment using ensemble wavelet-bootstrap-neural network models. *Water Resources Research*, *49*(10), 6486–6507.
Van Maanen, B., Coco, G., Bryan, K. R., & Ruessink, B. G. (2010). The use of artificial neural networks to analyze and predict alongshore sediment transport. *Nonlinear Processes in Geophysics*, *17*(5), 395–404.
Wang, X., & Cheng, L. (2000). Three-dimensional simulation of a side discharge into a cross channel flow. *The Computational Techniques and Applications Conference*, *29*(4), 415–433.
Webber, N. B., & Greated, C. (1966). An investigation of flow behaviour at the junction of rectangular channels. *ICE Proceedings*, *34*(3), 321–334.
Weber, L. J., Schumate, E. D., Mawer, N. (2001). Experiments on flow at a 90° open-channel junction. *Journal of Hydraulic Engineering*, *127*(5), 340–350.
Wilcox, B. P., Rawls, W. J., Brakensiek, D. L., & Wight, J. R. (1990). Predicting runoff from rangeland catchments: a comparison of two models. *Water Resources Research*, *26*(10), 2401–2410.
Wilcox, D. C. (1998). *Turbulence modeling for CFD* (2nd ed.). La Canada Flintridge, CA: D C W Industries Inc.
Willmott, C. J., Ackleson, S. G., Davis, R. E., Feddema, J. J., Klink, K. M., Legates, D. R., O'Donnell, J., & Rowe, C. M. (1985). Statistics for the evaluation and comparison of models. *Journal of Geophysical Research: Oceans (1978-2012)*, *90*(C5), 8995–9005.
Wu, C. L., Chau, K. W., & Li, Y. S. (2009). Predicting monthly streamflow using data-driven models coupled with data-preprocessing techniques. *Water Resources Research*, *45*(8), 1–23.
Yurtseven, I., & Zengin, M. (2013). Neural network modelling of rainfall interception in four different forest stands. *Annals of Forest Research*, *56*(2), 351–362.
Zhao, C. H., Zhu, D. Z., & Rajaratnam, N. (2008). Computational and experimental study of surcharged flow at a 90° combining sewer junction. *Journal of Hydraulic Engineering*, *134*(6), 688–700.

15

Rans, detached Eddy simulation and large Eddy simulation of internal Torque converters flows

Chunbao Liu, Changsuo Liu and Wenxing Ma*

School of Mechanical Science and Engineering, Jilin University, Changchun13002, China

Comparative analysis among the capabilities of the RANS, DES, and LES models to predict flow and turbulence distribution was conducted to come up with guidelines for hydraulic torque converter (TC) transient simulation. To ensure the accuracy of calculation, the complex geometry of hydrodynamic elements was accurately represented and the computational meshes of the structured hexahedron were appropriately distributed. Wall shear stress, pressure–streamline structure were analyzed. Compared with RANS, the transient vorticity features, including the birth, development, formation of a scroll; transportation along the blade surface; shedding and rupture at the trailing edge could be clearly captured by the LES and DES models. Rothalpy was used to quantitatively evaluate the hydraulic loss and a new computational formula was proposed to predict the efficiency of each element in TC. After the comparison of relative computing time, DES model was proved be a feasible method for efficiently and accurately simulating 3D unsteady turbulent flow of TC.

Keywords: hydraulic torque converter; transient simulation; DES model

1. Introduction

A hydraulic torque converter (TC) is a turbomachine that is widely used in automatic transmissions for automobiles. It consists of three major elements, namely, a pump, a turbine, and a stator. The functions of a TC include damping engine torque fluctuation, damping noise and vibration in the drive line, and automatically amplifying torque according to the difference in rotational speed between the input and output shafts without requiring external control (Kim, Ha, Lim, & Cha, 2008).

Various intrinsic flow motions, such as swirl, flow separation, and secondary flow, occur inside a TC. The accuracy of performance prediction heavily depends on the capability of the TC to capture and represent flow patterns and turbulence. The k-epsilon model, which is based on the average Reynolds number, is the most commonly used method to conduct numerical TC simulations. In recent years, however, more precise methods, including detached eddy simulation (DES) and large eddy simulation (LES), have been used to study the flow mechanism inside fluid machinery. Accordingly, these approaches have become the most effective tools for exploring the flow mechanism in a complex passages (Johnsen, Larsson, & Bhagatwala, 2010). LES captures cycle-to-cycle variation as opposed to the cycle-averaged mean computations of the Reynolds-averaged Navier–Stokes (RANS) equation, and therefore, is more accurate than the latter (Schmidt & Thiele, 2002). Wu, Wen, Yen, Weng, and Wang (2004) analyzed the transient separation process in the boundary layer at low Reynolds numbers. Su, Li, Li, and Wei (2012) made an assessment of the performance of the large eddy simulation (LES) method in simulating the complex vortical flow in a Francis hydraulic turbine. In the last decade, due to the intensive growth of computer capacity and parallelized numerical modeling solutions, the development and application of large eddy simulation (LES) became a convenient tool for the calculation of flow patterns around single and multiple cylinders, which result in very complex unsteady, three-dimensional flows that are dominated by large-scale coherent vortices (Palau-Salvador, Stoesser, Froehlich, Kappler, & Rodi, 2010; Stoesser, Kim, & Diplas, 2010). Matsuura and Kato (2006, 2007) performed a series of investigations on compressible transition flow in the low-pressure turbo cascade, T106. They summarized the changes in transient characteristic, including the propagation of pressure waves and flow separation, after considering and neglecting the effects of turbulent intensity.

Since its inception in 1997, the DES model, whose main idea is to solve the RANS equations in near-wall regions while applying the LES method in other areas, has been gradually applied in the field of aviation and fluid machinery (Squires, Krishnan, & Forsythe, 2008). This model is suitable for simulating separated flow under any Reynolds number. Caruelle and Durcos (2003) clarified the application of the DES model at the boundary layer. Xu, Chen, and Lu (2007) simulated flow separation with

*Corresponding author. Email: mawx@jlu.edu.cn

the large-eddy and detached-eddy models at the boundary layer of a cylinder and captured the transient characteristic of flow separation. So far, several numerical attempts have been made to investigate the flow field inside the torque converter passages. Most previous studies, however, adopted a simplified geometry such as the one-passage model and ignored temporal variation of the flow field such as the unsteady blade interaction, which characterizes the actual flow pattern inside the passage of a torque converter. (Flack, 2005) simulated unsteady flow with a moving grid and a complete TC model. They observed that unsteady effects are most important in the region wherein the interaction of the pump and the turbine occurs. This interaction is a flow phenomenon that cannot be predicted with steady calculation. Jung, Kang, and Hur (2011) performed a comparative study to investigate the effects of different methods for handling the relative motion of parts. The accuracy of performance prediction was improved. However, the effect of the turbulence model was not considered in the study of Jung et al., who also failed to investigate the accuracy of the turbulence model applied in performance prediction and turbulence distribution in the boundary layer.

In the present study, the complete geometry of a TC passage was meshed into a hexahedron via the high grid resolution in near-wall regions. Because the blades were distorted seriously, much effort and time were used to improve the quality of grid in order to ensure the accuracy of calculation. Through the comparison with experimental results, influence of various turbulence models on predicting TC performance was evaluated. Wall shear stress, pressure-streamline structure, vorticity and velocity distribution were also analyzed. Compared with RANS, the LES and DES models can capture transient vorticity features, including the birth, development, and formation of a scroll; transportation along the blade surface; and shedding and rupture at the trailing edge. Moreover, DES model was proven to be a feasible method for accurately and efficiently simulating 3D unsteady turbulent flow in the complex passage of a TC.

2. Numerical simulation

2.1. *Three kinds of turbulence models*

The Reynolds-averaged continuity and momentum equations for incompressible flow are as follows:

$$\frac{\partial \rho}{\partial t} + \frac{\partial (\rho u_i)}{\partial x_i} = 0 \qquad (1)$$

$$\frac{\partial (\rho u_i)}{\partial t} + \frac{\partial (\rho u_i u_j)}{\partial x_j} = -\frac{\partial p}{\partial x_i} + \frac{\partial}{\partial x_j}\left(\mu \frac{\partial u_i}{\partial x_j}\right) + \frac{\partial \tau_{ij}}{\partial x_j} \qquad (2)$$

where the Reynolds stress τ_{ij} is:

$$\tau_{ij} = -\rho \overline{u_i' u_j'} \qquad (3)$$

LES is a popular technique for simulating turbulent flows. An implication theory of self-similarity states that large flow eddies depend on geometry, whereas smaller ones are more universal. This feature allows explicitly solving for large eddies during calculation and implicitly accounting for small eddies by using a subgrid-scale model (Yang, Gupta, Kuo, & Gopalakrishnan, 2014), the subgrid-scale Reynolds stress is

$$\tau_{ij} = -2\upsilon_T \overline{S_{ij}} + \frac{1}{3}\tau_{kk}\delta_{ij} \qquad (4)$$

where the deformation rate tensor $\overline{S_{ij}}$ is:

$$\overline{S_{ij}} = \frac{1}{2} \cdot \left[\left(\frac{\partial \overline{u_i}}{\partial x_j}\right) + \left(\frac{\partial \overline{u_j}}{\partial x_i}\right)\right] \qquad (5)$$

and eddy viscosity coefficient υ_T is

$$\upsilon_T = (C_s \Delta)^2 |\overline{S}| \qquad (6)$$

where the magnitude of deformation rate tensor. $|\overline{S}| = (2\overline{S_{ij}}\,\overline{S_{ij}})^{1/2}$, Δ is the filtering scale, C_s is Smagorinsky coefficient.

Difficulties associated with using the standard LES model, particularly in near-wall regions, led to the development of the DES model, which attempts to combine the best aspects of RANS and LES methodologies in a single solution strategy. In this study, the Realizable k-epsilon model was used in near-wall regions, whereas the Smagorinsky–Lilly subgrid-scale model in LES (Breuer, Jovičić, & Mazaev, 2003) was used in separated flow regions.

The length size of eddy in turbulence is:

$$L_{DES} = k^{3/2}/\varepsilon \qquad (7)$$

It is used for replacing the item ε in equation of turbulent kinetic energy in Realizable k-epsilon model, then, equation of turbulent kinetic energy in DES model is:

$$\frac{\partial (\rho k)}{\partial t} + \frac{\partial (\rho k u_i)}{\partial x_i} = \frac{\partial}{\partial x_j}\left[\left(\mu + \frac{\mu_1}{\sigma_k}\right)\frac{\partial k}{\partial x_j}\right] + G_k - \rho \frac{k^{3/2}}{L_{DES}} \qquad (8)$$

The renormalization group (RNG) model was developed by using RNG methods to renormalize the N–S equations to account for the effects of small scales of motion. In the standard k-epsilon model, eddy viscosity is determined from a single turbulence length scale. Hence, the calculated turbulent diffusion refers to diffusion that occurs only at a specified scale. By contrast, all scales of motion contribute to turbulent diffusion in reality. Similar to k-epsilon, the RNG approach is a mathematical technique that can be used to derive a turbulence model. This approach results in a modified form of the epsilon equation, which attempts to account for different scales of motion through changes in the production term (Galvan, Reggio, & Guibault, 2011). Equation of (RNG) k-ε is as followed:

Equation of turbulent kinetic energy:

$$\rho\frac{\partial k}{\partial \tau}+\rho\overline{u_j}\frac{\partial k}{\partial x_j}=\frac{\partial}{\partial x_j}\left[\left(\mu+\frac{\mu_i}{\sigma_k}\right)\frac{\partial k}{\partial x_i}\right]+\mu_i\frac{\partial \overline{u_i}}{\partial x_j}\left(\frac{\partial \overline{u_i}}{\partial x_j}+\frac{\partial \overline{u_i}}{\partial x_i}\right)-\rho\varepsilon \quad (9)$$

Equation of turbulent dissipation rate:

$$\rho\frac{\partial \varepsilon}{\partial \tau}+\rho\overline{u_j}\frac{\partial \varepsilon}{\partial x_j}=\frac{\partial}{\partial x_j}\left[\left(\mu+\frac{\mu_t}{\sigma_\varepsilon}\right)\frac{\partial \varepsilon}{\partial x_j}\right]+\frac{c_1\varepsilon}{k}\mu_i\frac{\partial \overline{u_i}}{\partial x_k}\left(\frac{\partial \overline{u_i}}{\partial x_k}+\frac{\partial \overline{u_k}}{\partial x_i}\right)-c_2\rho\frac{\varepsilon^2}{k} \quad (10)$$

where the turbulent viscosity coefficient μ_t is:

$$\mu_t=\rho C_\mu\frac{k^2}{\varepsilon} \quad (11)$$

the fluctuation kinetic energy dissipation rate of per unit in turbulence ε is :

$$\varepsilon=\nu\overline{\left(\frac{\partial \mu_i'}{\partial x_k}\right)\left(\frac{\partial \mu_t'}{\partial x_k}\right)} \quad (12)$$

The following values are used for RNG k-ε model:

$C_\mu=0.09, C_1=1.44, C_2=1.91, C_K=1.00, C_\varepsilon=1.30$

The enhanced wall function method in the RANS model that employs the double zone model to divide turbulent flow into the viscous effect region and the fully turbulent flow region was adopted to ensure that the viscosity effect in the boundary layer can be reflected. Herein, the high Reynolds number RNG k-epsilon and Wolfstein (Chen & Patel, 1988) models were used to solve equations. However, the enhanced wall function method increases the requirement for mesh density in near-wall regions for grid encryption and smooth growth. In this study, the boundary layer had an initial size of 0.05 mm, a growth ratio of 1.1, and a maximum size of 0.60 mm.

2.2. *Hexahedral mesh generation and mesh test*

The complex geometry of hydrodynamic elements was accurately represented and the computational meshes were appropriately distributed with ANSYS ICEM computational fluid dynamics (CFD) software to generate the structured hexahedral grid of the entire TC. Mesh-point clustering was performed near the boundary layer, and cell deformity was checked with the software. Figure 1 describes the generation process of the structured grid of entire passage. A mesh sensitivity test in the RNG k-epsilon model was performed to minimize the effect of number error. The number of grids is 720,866, 1,530,270, 2,378,562 and 3,032,834 respectively. Figure 2 shows the grid distribution of 2.3 million.

The performance of a TC is defined by several operating parameters that are described as follows:

$$SR\text{ (speed ratio)}:\quad SR=\frac{N_{\text{turbine}}}{N_{\text{pump}}}, \quad (13)$$

TR (torque ratio between the pump and turbine):

$$TR=\frac{T_{turbine}}{T_{pump}}, \quad (14)$$

$$CF\text{ (capacity factor)}:\quad CF=\frac{T_{\text{pump}}\times 10^6}{N_{\text{pump}}^2}, \quad (15)$$

$$\eta\text{(efficiency)}:\quad \eta=\frac{N_{\text{turbine}}\times T_{\text{turbine}}}{N_{\text{pump}}\times T_{pump}}=SR\times TR, \quad (16)$$

where N is the rotating speed (rpm) and T is the torque of each component. To compare prediction accuracy under different grid densities, the absolute error of prediction

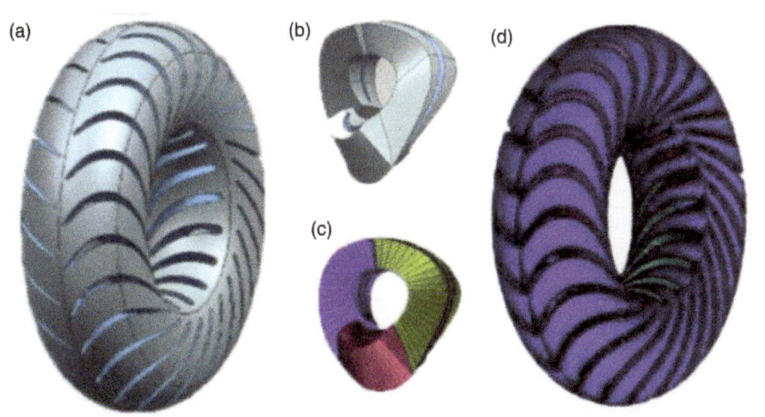

Figure 1. Generation process of the structured grid: (a) entire passage, (b) single passage, (c) single structured grid model and (d) entire passage of the structured hexahedral.

Total cell member	Mesh cut plane	Stator surface mesh	Stator leading edge
2,378,562			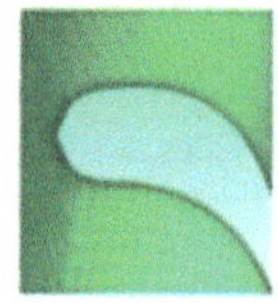

Figure 2. Grid distribution of 2.3 million.

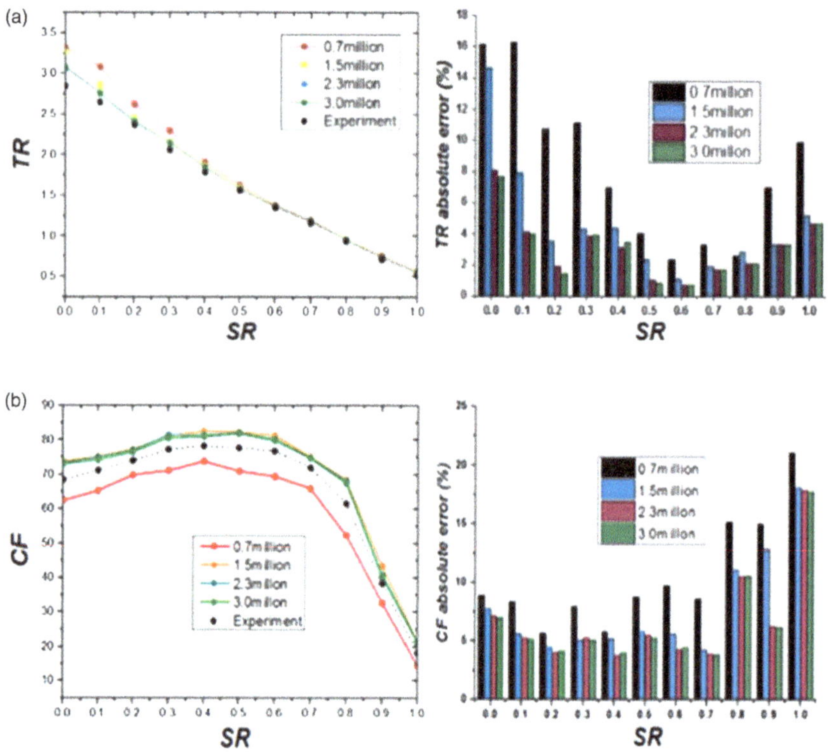

Figure 3. Results of CFD analysis of different mesh densities: (a) CF and CF absolute error, and (b) TR and TR absolute error.

among different turbulence models is defined as follows:

$$Et = \frac{|I_{TM} - I_{Exp}|}{I_{Exp}} \times 100\%, \tag{17}$$

where I represents the capacity coefficient, torque ratio, efficiency, and other physical quantities; and the subscripts TM and Exp denote the turbulence model and test data, respectively.

The YJ315X manufactured in Hangzhou Advance Gearbox Group Co., Ltd is used in this paper, which had been tested in the Supervise and Inspect Center of Hydraulic Parts in China National Machinery Industry Corporation in August 2011. Figure 3 illustrates that the simulated result moved closer to the experimental data with increasing element numbers. Only a subtle change was observed when element numbers increased from 2.3 million to 3.0 million. After an evaluation of computation amount and accuracy, 2.3 million was used in subsequent CFD simulations of three turbulent models.

2.3. *Computational setup*

ANSYS FLUENT software was employed to solve the incompressible Navier–Stokes equations without heat transfer. During calculation, the pump and the turbine rotate along the gird interface according to a certain time step size (0.0005s). The solution was assumed to have converged when the rate of change in the mass flow rate in two consecutive circulatory iterations and all normalized residuals in mass and momentum conservation equations were less than 10^{-4} for the differences of results in these criteria is not obvious (Wood, Yan, & Wei, 2013), the total time

Table 1. CFD model description.

Analysis Type	Transient State
Turbulence Mode	LES, DES, RNG k-epsilon Model
Pressure–Velocity Coupling	SIMPLE
Transient Formulation	Second-Order Implicit
Pump Status	Fixed at 2000 rpm
Turbine Status	Varied from 0 rpm to 2000 rpm
Stator Status	Stationary

step is 200. The rotational speed of pump is the same with experiment. Mesh interfaces were arranged on the intersecting surfaces of different impellers. The properties of CFD model are summarized in Table 1.

3. Results and analysis

3.1. *Verification of the wall function model*

The value of y-plus should be discussed through the post-process to investigate whether the grid satisfies the requirements of the enhanced wall function method or the mesh in near-wall regions. Figure 4 shows that all values of y-plus on the blades are less than 2. This result indicates that the mesh density in near-wall regions is acceptable (Barony, Olsen, Stoesser, & Sturm, 2012).

3.2. *Wall shear stress on blade surfaces*

The pressure surface of the pump blade was regarded as the positive face of the blade group (a single pump blade, turbine blade, and stator blade). Figure 5 indicates the wall shear stress distribution in the blade group.

Viscous effects facilitate the emergence of wall shear stress between the surface of the blades and the fluid. This stress attaches the fluid in near-wall regions to the surface of the blades, and thus, energy is consumed. The higher wall shear stress is located in the entrance region of the stator and the pump, whereas the lower one is located on the pressure surface of the turbine. The distributions of the RNG k-epsilon [Figure 5(c)] and DES models [Figure 5(b)] are similar, but significantly different from that of the LES model [Figure 5(a)]. This finding indicates that energy consumption is relatively larger in LES model than in other two models. The LES approach is always used to calculate the flow region at high Reynolds numbers. An extremely fine boundary layer mesh and a huge computational expense are required when the LES approach is adopted in near-wall regions at low Reynolds numbers. Therefore, the RANS model with enhanced wall function method or the DES model is recommended to simulate the near-wall region in most cases.

3.3. *Pressure–streamline structure*

Under stall condition (SR = 0), fluid flow from the turbine affects the leading edge of stator wherein the region with velocity stagnation and high vorticity is located. A part of fluid moves along the suction surface of the stator. Under the viscous force and inverse interaction of the pressure gradient, the velocity of the fluid particle within the boundary layer gradually decreases, whereas the fluid particle

Figure 4. Y-plus on blades surface.

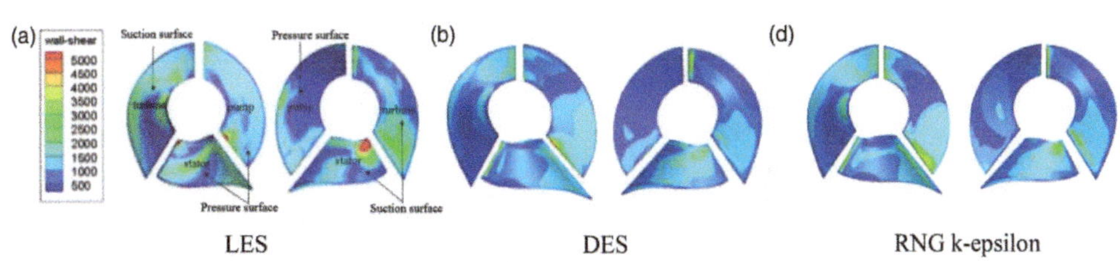

Figure 5. Wall shear stress in the blade group.

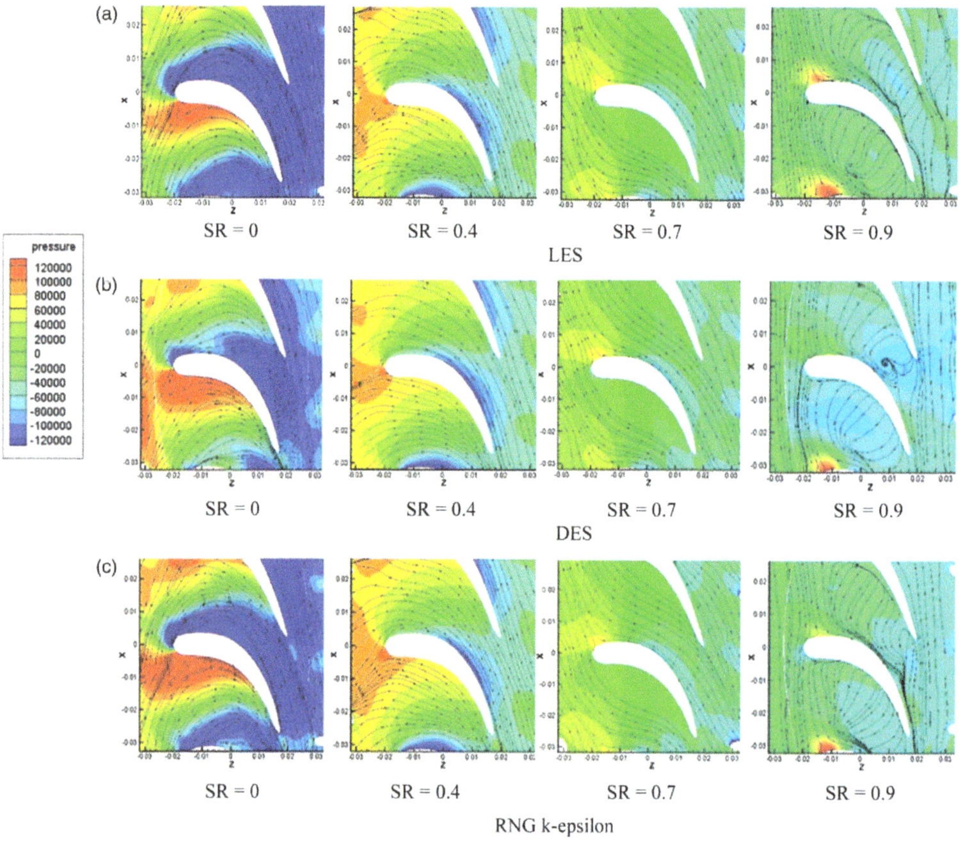

Figure 6. Pressure–streamline structure of the suspended plane in the stator (Y = 0.09 m).

remains stagnant. Subsequently, the separation point is formed. The downstream fluid particle at the boundary layer is reversed to form the vortex flow, which pushes the upper fluid out of the surface and simultaneously causes flow separation, that is, the two-flow phenomenon in a wide range of downstream area. Figure 6 describes the distributions of pressure and streamline of fluid flow under four kinds of operating conditions, which are stuck in the transient time of 0.1 s. the direction in which the liquid flow from turbine blades impacts the stator leading edge deflects clockwise with the increase of SR. Meanwhile, the pressure gradient is evidently becoming gentle. A slight difference in the pressure gradient was observed among RANS, LES, and DES models only when SR = 0 (the stall condition). Overall flow, pressure trends, and flow separation region described in the three turbulent models are basically the same, which suggests that all simulations perform efficiently in capturing the main flow structures.

3.4. *Vorticity distribution*

The vorticity (1/s) is a description of the physical quantity of vortex motion, which is defined as the curl of a velocity field, and used to measure the magnitude and direction of vortex. In an enclosed turbo-machinery, the cyclical phenomenon like flow separation and reattachment always occurs in boundary layer of the suction surface due to the effect of upstream impeller. Figures 7 and 8 depict the transient evolutionary process of vorticity (1/s) on the suspended cylindrical surface in the stator and the interaction between the pump and the turbine under the stall condition (SR = 0) at different transient time respectively. In particular, Figure 8 shows that the outflow with high velocity in circular motion from the pump affects the leading edge of the turbine blades, which results in the flow separation phenomenon on the suction surface of the turbine blades.

In Figures 7 and 8, the LES and DES models clearly display transient vortex features, including the birth development, and formation of the scroll; transportation along the suction surface; and shedding and rupture at the trailing edge. LES and DES full geometry capture the local peaks and troughs in contours. Whereas RANS contours miss out on that and give out the relatively smooth pictures, the features exhibited during the evolution of vortices in the transient simulation in the RANS model may or may not be entirely accurate. Hence, RANS fails to capture this important trends.

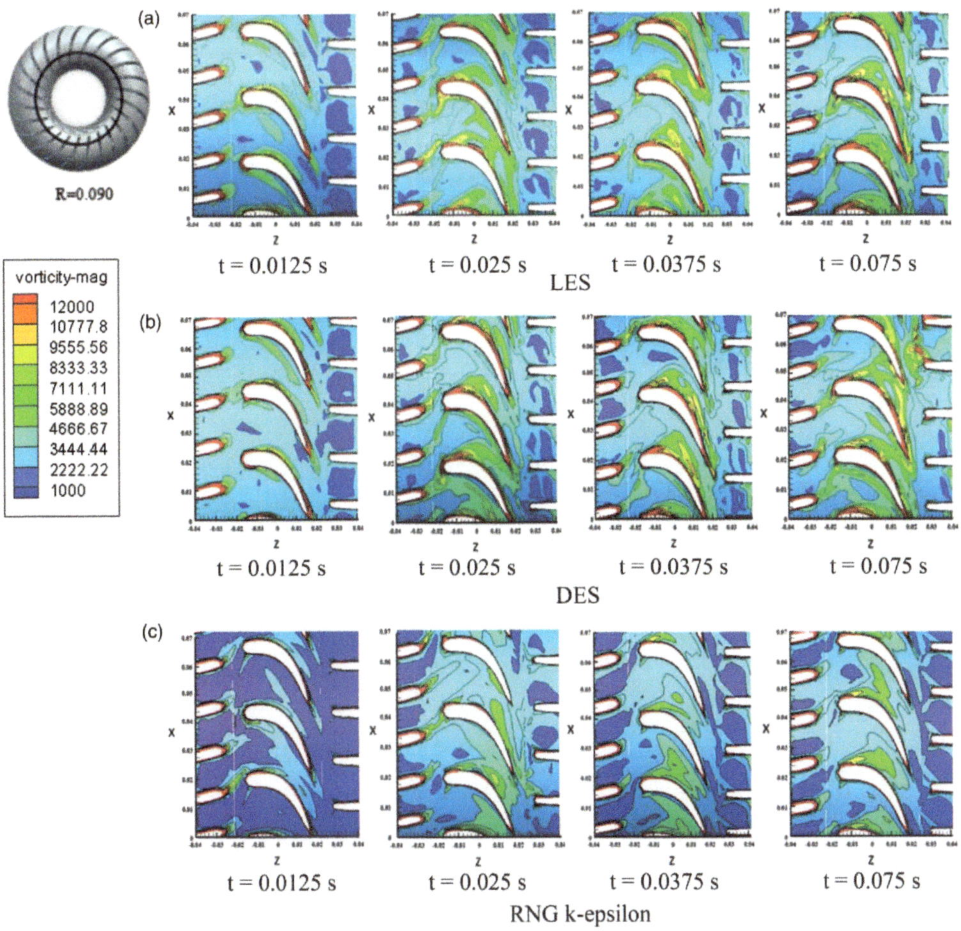

Figure 7. Vorticity distribution on the suspended cylindrical surface in the stator.

3.5. *Velocity distribution*

The observations in Figure 9 suggest that all simulations perform well in predicting the magnitude and levels of velocity, however, only LES and DES models capture the low-velocity zones in the downstream region of suction surface which is reasonable owing to the shedding and rupture of vortex features at the trailing edge expressed in Figure 7. In Figure 10, three kinds of turbulent models differ relatively small in velocity distribution, however, after the scrutiny of low-velocity structures, It indicates the impetus to separate from the suction surface in LES and DES models while the stability to stay still in RANS models respectively. These results further reveal that LES and DES models are better able to predict the transient flow structures as compared to the RANS model.

Rothalpy was introduced to quantitatively evaluate loss and conduct the direct analysis on flow. For viscous flow, the difference of the parameter between two points on a streamline represents hydraulic loss, Rothalpy is defined as follows:

$$R = \frac{P}{\rho} + \frac{1}{2}\omega^2 - \frac{1}{2}\mu^2, \tag{18}$$

where P is the pressure, ρ is the density, ω is the relative velocity, and μ is the circumferential velocity.

Figure 11 displays the change in Rothalpy from the inlet to the outlet on the suction surface of the stator under stall condition (SR $= 0$). "0" and "1" represent the location of inlet and the outlet of stator respectively. Considering the retarded effect of fluid along the surface of the blades, Rothalpy value was gradually reduced. However, Rothalpy fluctuated because of the continuous evolution of the vortex on the suction surface under instantaneous time. The region between 0 and 0.7 was considered as the flow separation region wherein the kinetic energy of the fluid is low. Meanwhile, the position 0.7 was the reattachment point of the flow. Where after the Rothalpy increases significantly because of fluid flow along the surface. Losing the obvious characteristics of the vortex in the region of the outlet,

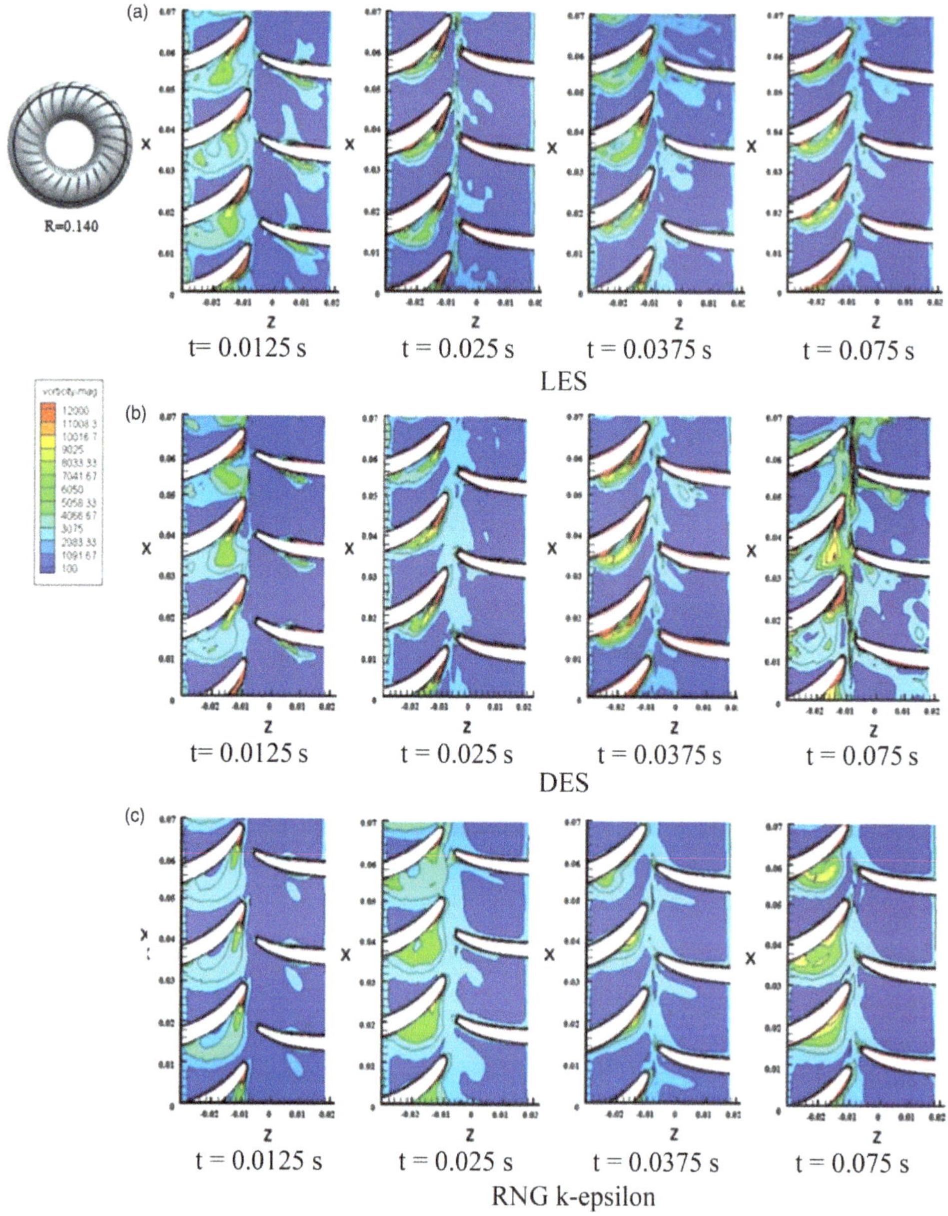

Figure 8. Vorticity distribution during the interaction between the pump and the turbine.

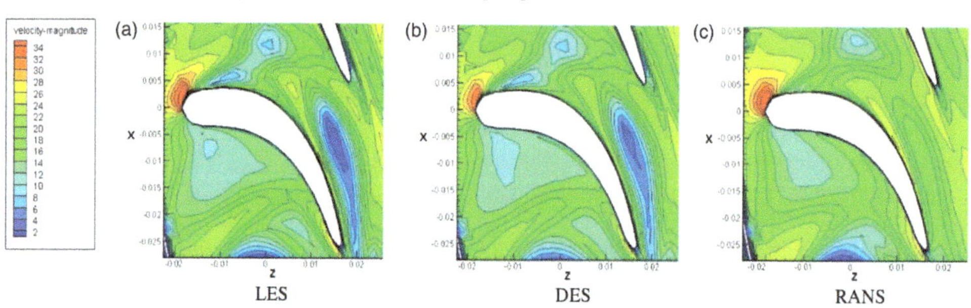

Figure 9. 2D contours of velocity magnitude at Y = 0.09 m plane (SR = 0, T = 0.075s)

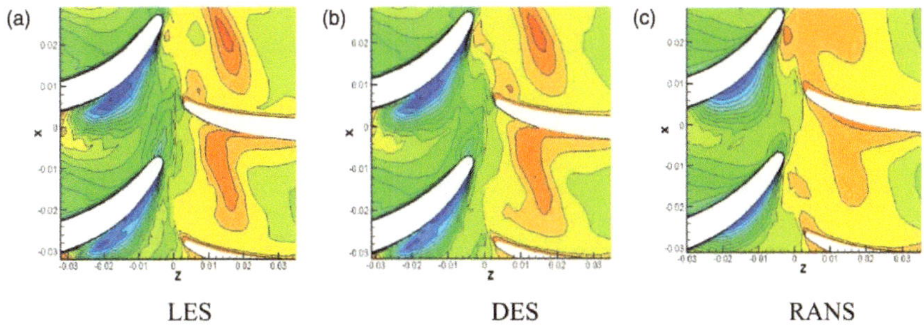

Figure 10. 2D contours of velocity magnitude at Y = 0.140 m plane (SR = 0, T = 0.075s).

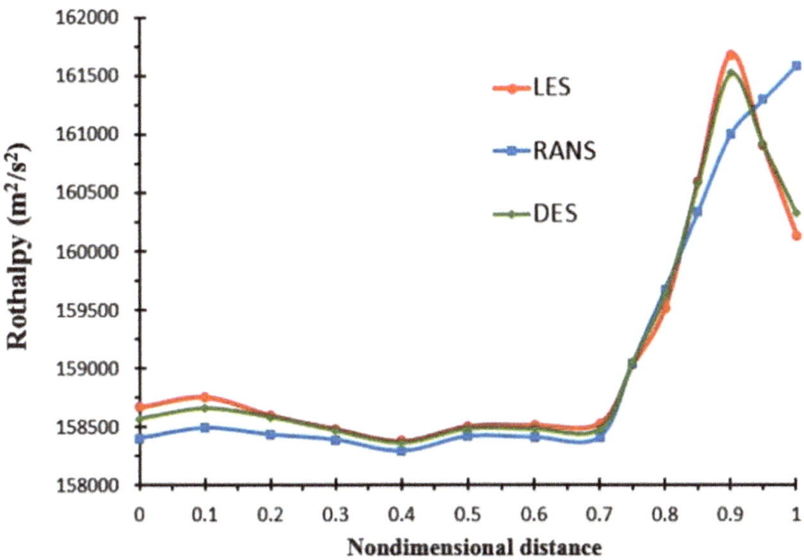

Figure 11. The change of Rothalpy from the inlet to the outlet on the suction surface of the stator

as exhibited in Figure 7(c), resulted in a significant difference between the value of the RANS model and those of the DES and LES models.

3.6. Efficiency of elements

When operating, the performance of a TC can be predicted through experiments. However, the efficiency of each element can only be evaluated by performing a numerical calculation. The DES model was used as the example to predict the efficiency of each element.

The individual efficiencies of the pump, turbine, and stator (η_P, η_T, and η_S, respectively) and the overall efficiency η_{TC} are defined by the following expressions:

$$\eta_p = (p_{p2} - p_{p1})/\rho w_1(r_{p2}V_{p2} - r_{p1}V_{p1}), \quad (19)$$

$$\eta_T = \rho w_2(r_{t2}V_{t2} - r_{t1}V_{t1})/(P_{t1} - P_{t2}), \quad (20)$$

$$\eta_S = 1 - (P_{t1} - P_{t2})/(P_{b2} - P_{b1}), \quad (21)$$

$$\eta_{TC} = \eta_P\eta_T\eta_S, \quad (22)$$

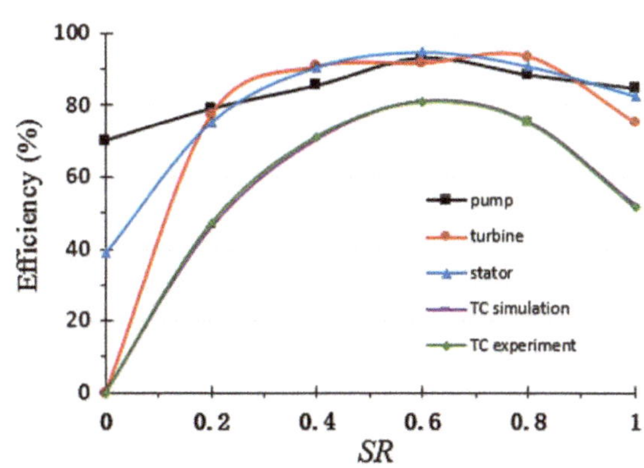

Figure 12. Efficiency of elements under different speed ratios.

where subscripts 1 and 2 refer to the inlet and outlet of each element, respectively; P is the total pressure; r is the radius; and V is the velocity. Figure 12 displays the efficiency of various elements under different speed

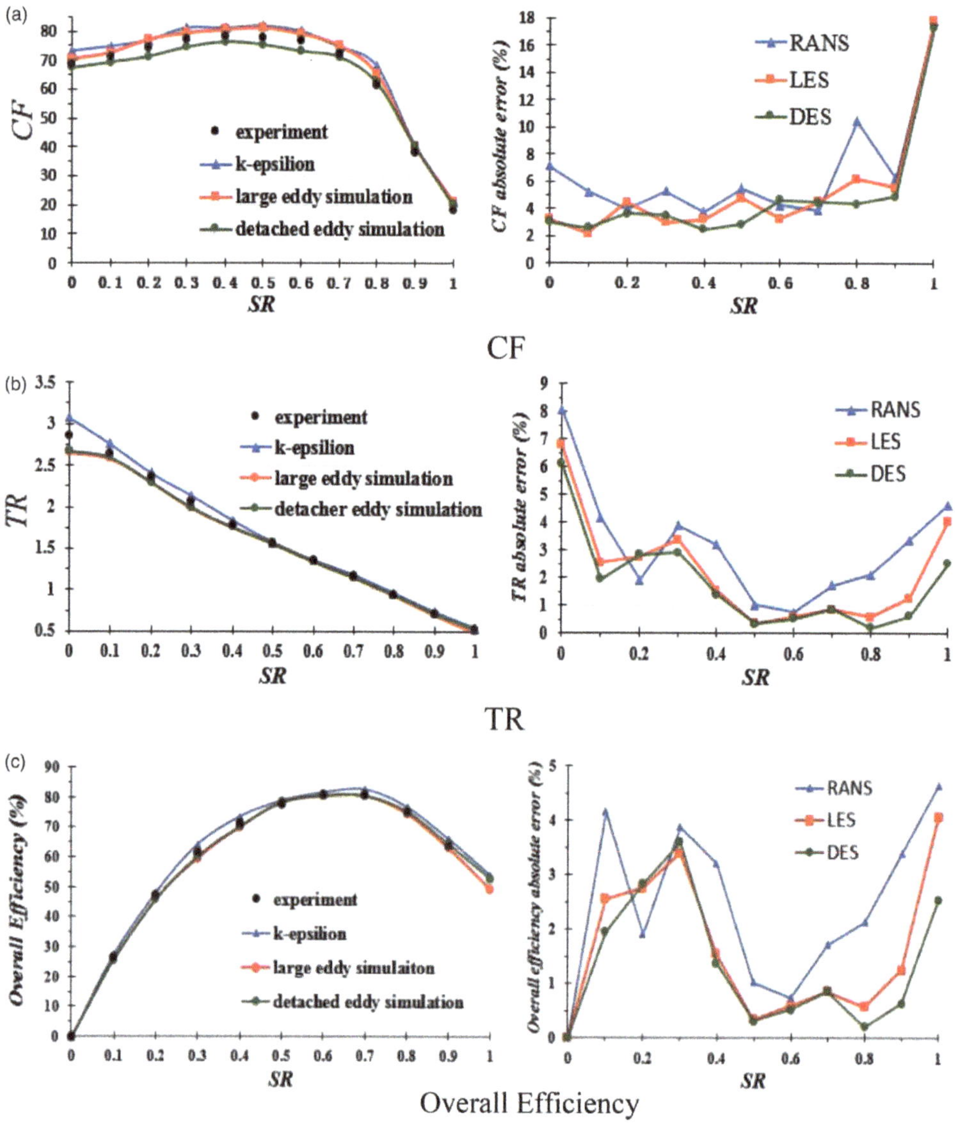

Figure 13. Prediction performance and absolute error of three turbulent models.

ratios. The highest efficiency of a single element is approximately 93%, whereas TC efficiency is 81.2%. The overall efficiency η_{TC} undoubtedly coincides well with the experimental data. Thus, this method can be used to assess the efficiency of individual elements in a TC, which can clearly indicates the reasons for inefficiency.

3.7. *Prediction performance of the three turbulence models*

As shown in Figure 13(a), the capacity factor of the LES and RNG k-epsilon models are relatively large, whereas that of the DES model is relatively small. However, both DES and LES models are more accurate in predicting CF than the RANS model, whose average absolute error is 2%. In predicting torque ratio, the absolute errors of the three models are high when SR is between 0 and 0.4, particularly under stall condition. The absolute errors of the RANS, LES, and DES models are 7.96%, 6.78%, and 6.12%, respectively. When SR is between 0.4 and 0.9, the absolute errors of the LES and DES models are less than 2%. By contrast, the data of the RANS model remain large although less than 5% [Figure 13(b)]. Only a slight difference exists between the highest efficiency data of the three

Table 2. Comparison of the computation time among different models (3.33 GHz, 4 CPU).

Model	CPU time (h)
LES	22.2
DES	14.7
RANS	8.6

models and the experiment data. [Figure 13(c)] Overall, the original characteristics of three turbulence models are basically the same, which can satisfy engineering requirements. Compared with the absolute error of RNG k-epsilon model, those of DES and LES models are closer to the experimental date.

The computation time was compared in Table 2, the RANS simulation is about the 2.5 times faster than the LES simulation. In industries, most of the CFD simulations are carried out through RANS approach which is computationally less expensive but gives lower accuracy levels than LES. LES models are more accurate, however, large computational time makes it impractical to use it for routine engineering applications. In addition, by combing the advantage of LES and RANS, DES model appears to be a good option in CFD simulations of the complex passage for the less computation time, but the more computational accuracy.

4. Conclusion

This paper performs an unsteady flow simulation with an entire passage model of the hexahedral mesh, a comparative analysis among the capabilities of the RANS, DES, and LES models in general to predict flow and turbulence distribution is conducted to come up with guidelines to use RANS/DES/LES for TC transient simulation. Scholars may resort to using the RANS model if they are interested in obtaining reasonably accurate estimates of velocity magnitudes and flow structures. However, to acquire the more accurate turbulent distributions, LES or DES approach of complete geometries should be recommended. A torque converter is a closed geometry. Typically, the temperature of Auto Transmission Fluid rises up to 80 ~ 120°C by the shear heating effect, which would result in a decrease of the torque by a small viscosity. In this study, the accuracy of performance prediction will be improved if the change of temperature could be considered. By combining the advantages of the RANS and LES methods, a complex vortex flow can be accurately simulated by DES model, which exhibits the characteristic of simulating large-scale vortices within the boundary layer of the LES method. The DES model has been proved be a feasible method for accurately and efficiently simulating 3D unsteady turbulent flow in complex passages. Yet, theoretical researches on this aspect are rather little, which would be a direction of future work in CFD numerical simulation of hydraulic torque converter.

Acknowledgements

This work was supported by the National High-tech R&D Program of China (2014AA041502) and Basic Research Programs of Jilin Province (120102JC01032095).

Nomenclature

CF	Capacity factor
Cs	Smagorinsly coefficient
k	Turbulent kinetic energy [$m2/s2$]
N	Rotating speed [rpm]
P	Fluid pressure [pa]
r	Radius [m]
SR	Speed ratio
T	Torque [n·m]
TR	Torque ratio between pump and turbine
u_i	Time-averaged velocity [m/s]
X_i	Coordinate [m]

Greek letters

η	Efficiency
ε	Turbulent dissipation rate [m^2/s^3]
μ	Circumferential velocity [m/s]
μ_t	Turbulence viscosity coefficient
ρ	Density [kg/m]
υ_T	Eddy viscosity coefficient
ν	Turbulent viscosity [m^2/s]
ω	Relative velocity [m/s]

Subscripts

P	Pump
S	Stator
T	Turbine
TC	Torque converter

References

Barony, S., Olsen, N. R. B., Stoesser, T., & Sturm, T. (2012). Three-dimensional RANS modeling of flow around circular piers using nested grids. *Engineering Applications of Computational Fluid Mechanics*, *6*(4), 648–662.

Breuer, M., Jovičić, N., & Mazaev, K. (2003). Comparison of DES, RANS and LES for the separated flow around a flat plate at high incidence. *International Journal for Numerical Methods in Fluids*, *41*(4), 357–388.

Caruelle, B., & Ducros, F. (2003). Detached-eddy simulations of attached and detached boundary layers. *International Journal of Computational Fluid Dynamics*, *17*(6), 433–451.

Chen, H. C., & Patel, V. C. (1988). Near-wall turbulence models for complex flows including separation. *AIAA Journal*, *26*(6), 641–648.

Flack, R. (2005). Experimental flow fields in an automotive torque converter-an invited summary and review paper. *International Journal of Vehicle Design*, *38*(2), 240–258.

Galvan, S., Reggio, M., & Guibault, F. (2011). Assessment study of k-ϵ turbulence models and near-wall modeling for steady state swirling flow analysis in draft tube using FLUENT. *Engineering Applications of Computational Fluid Mechanics*, *5*(4), 459–478.

Johnsen, E., Larsson, J., & Bhagatwala, AV. (2010). Assessment of high-resolution methods for numerical simulations of compressible turbulence with shock waves. *Journal of Computational Physics*, *229*(4), 1213–1237.

Jung, J. H., Kang, S., & Hur, N. (2011). A numerical study of a torque converter with various methods for the accuracy improvement of performance prediction. *Progress in Computational Fluid Dynamics, an International Journal*, *11*(3), 261–268.

Kim, B. S., Ha, S. B., Lim, W. S., & Cha, S. W. (2008). Performance estimation model of a torque converter, Part I: Correlation between the internal flow field and energy loss coefficient, *International Journal of Automotive Technology*, *9*(2), 141–148.

Matsuura, K., & Kato, C. (2006). Large-eddy simulation of compressible transitional cascade flows with and without incoming free-stream turbulence. *JSME International Journal Series B*, *49*(3), 660–669.

Matsuura, K., & Kato, C. (2007). Large-eddy simulation of compressible transitional flows in a low-pressure turbine cascade. *AIAA Journal*, *45*(2), 442–457.

Palau-Salvador, G., Stoesser, T., Froehlich, J., Kappler, M., & Rodi, W. (2010). Large eddy simulations and experiments of flow around finite-height cylinders. *Flow Turbulence Combustion*, *84*(2), 239–275.

Schmidt, S., & Thiele, F. (2002). Comparison of numerical methods applied to the flow over wall-mounted cubes. *International Journal of Heat and Fluid Flow*, *23*(3), 330–339.

Squires, K. D., Krishnan, V., & Forsythe, J. R. (2008). Prediction of the flow over a circular cylinder at high Reynolds number using detached-eddy simulation. *Journal of Wind Engineering and Industrial Aerodynamics*, *96*(10), 1528–1536.

Stoesser, T., Kim, S., & Diplas, P. (2010). Turbulent flow through idealized emergent vegetation. *Journal of Hydraulic Engineering*, *136*(12), 1003–1017.

Su, W. T., Li, F. C., Li, X. B., & Wei, X. Z. (2012). Assessment of LES performance in simulating complex 3D flows in turbomachines. *Engineering Applications of Computational Fluid Mechanics*, *6*(3), 356–365.

Wood, H. G., Yan, Q., & Wei, W. (2013). Parametric analysis and optimization of inlet deflection angle in Torque converters. *Journal of Fluid Engineering*, *137*, 031101.

Wu, M. H., Wen, C. Y., Yen, R. H., Weng, M. C., & Wang, A. B. (2004). Experimental and numerical study of the separation angle for flow around a circular cylinder at low Reynolds number. *Journal of Fluid Mechanics*, *515*, 233–260.

Xu, C., Chen, L., & Lu, X. (2007). Large-eddy and detached-eddy simulations of the separated flow around a circular cylinder. *Journal of Hydrodynamics, Ser. B.*, *19*(5), 559–563.

Yang, X., Gupta, S., Kuo, T. W., & Gopalakrishnan, V. (2014). RANS and large Eddy simulation of internal combustion engine flows-a comparative study. *Journal of Engineering for Gas Turbines and Power*, *136*(5), 051507.

16

Estimation of scour depth below free overfall spillways using multivariate adaptive regression splines and artificial neural networks

Mehrshad Samadi[a,b], Ebrahim Jabbari[b]*, H.M. Azamathulla[c] and Mohammad Mojallal[d]

[a]Department of Civil Engineering, AHAB Company, Tehran, Iran; [b]School of Civil Engineering, Iran University of Science and Technology, Tehran, Iran; [c]University of Tabuk, Tabuk 50060, Saudi Arabia; [d]School of Civil Engineering, University of Tehran, Tehran, IRAN

Erosion and scouring caused by the outflow jets of hydraulic structures is one of the most important topics in hydraulic engineering. In free overfall spillways, a water jet impacts the erodible downstream bed almost vertically and creates a scour hole. The scour hole can affect the safety and stability of the dam. In the present paper, the multivariate adaptive regression splines (MARS) approach has been adopted as a new soft computing tool for estimating the equilibrium scour depth below free overfall spillways. Using experimental data and dimensionless parameters, the MARS model has been developed to predict scour depth. Results obtained from the MARS approach were compared with those from the artificial neural network (ANN) and decision-tree algorithm. Statistical indicators demonstrated that the MARS approach had a good performance and presented competitive results, slightly better than ANN, for the prediction of this phenomenon.

Keywords: artificial neural network; free overfall spillway; multivariate adaptive regression splines; scour depth; decision tree

1. Introduction

In the structure of the dams, spillways are constructed to discharge the water excess to the reservoir capacity. In free overfall spillways, waterfalls over the crown of the spillway almost vertically and impacts the downstream bed of the dams. Due to the high velocity and energy of the flow which impacts the erodible downstream bed, it may cause scouring close to the foundation of the dam and consequently threaten the stability of the dam. Thus, an accurate estimation of the scour depth is one of the important topics in hydraulic engineering. In order to estimate the scour depth in downstream spillways, empirical equations based on regression methods have been proposed by many researchers (Mason & Arumugam, 1985). Figure 1 displays a schematic view of the local scour downstream of a free overfall spillway.

At present, the development of soft computing models and the use of these methods in modeling complex and non-linear issues has increased considerably. A review of previous studies shows the application of soft computing techniques in many fields of civil engineering, such as water resources engineering, and in the simulation of engineering problems (Chau, Wu, & Li, 2005; Chen & Chau, 2006; Cheng, Chau, Sun, & Lin, 2005; Taormina, Chau, & Rajandrea, 2012; Wu, Chau, & Li, 2009), geotechnical engineering (Alavi & Gandomi, 2011a, 2011b), earthquake engineering (Gandomi & Alavi, 2012a, 2012b), transportation engineering (Haleem, Abdel-Aty, & Santos, 2010; Yang, Gandomi, Talatahari, & Alavi, 2012), structures and infrastructures (Gandomi, Yang, Talatahari, & Alavi, 2013), coastal and ocean engineering (Chau, 2010; Etemad-Shahidi & Bali, 2012; Jabbari & Talebi, 2011; Kamranzad, Jabbari, & Samadi, 2013), construction engineering (Afshar & Amiri, 2010; Eshtehardian, Afshar, & Abbasnia, 2009), and environmental engineering (Ebtehaj & Bonakdari, 2013; Muttil & Chau, 2006; Saadatpour, Afshar, & Afshar, 2011).

In the last two decades, a new modeling technique called multivariate adaptive regression splines (MARS) has been introduced (Friedman, 1991). This technique has been applied successfully in various scientific fields. Among the notable applications of this method are the prediction of maximum shear modulus and minimum dam ratio (Samui & Kothari, 2012), the prediction of the maximum magnitude of a reservoir-induced earthquake (Pijush & Dookie, 2012), rainfall-runoff modeling (Sharda, Prasher, Patel, Ojasvi, & Chandra, 2008), the prediction of the friction capacity of driven piles in clay (Samui, 2011), the determination of the uplift capacity of suction caisson in clay (Samui, Das, & Kim, 2011), stream flow forecasting (Coulibaly & Baldwin, 2005), the prediction of sub-daily rainfall (Beuchat, Schaefli, Soutter, & Mermoud, 2011),

*Corresponding author. Email: jabbari@iust.ac.ir

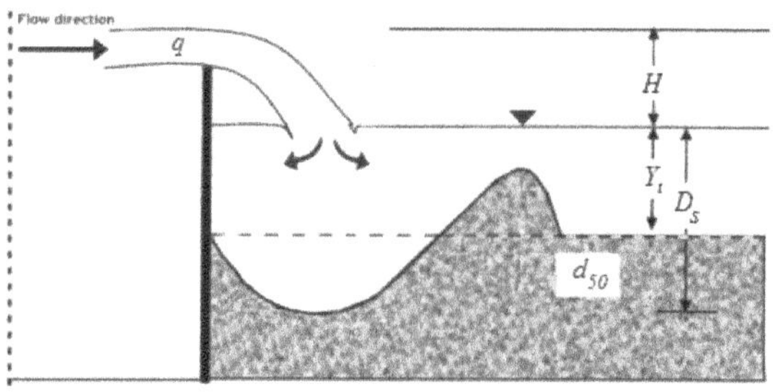

Figure 1. Schematic illustration of scouring below free overfall spillways (Samadi, Jabbari, & Azamathulla, 2014). (With permissions)

downscaling daily precipitation (Nasseri, Tavakol-Davani, & Zahraie, 2013), and pavement performance modeling (Attoh-Okine, Cooger, & Mensah, 2009).

In order to model scouring, a variety of soft computing schemes have been used in recent years, such as an artificial neural network (ANN) to predict scour below a ski-jump bucket spillway (Azmathullah, Deo, & Deolalikar, 2005) and culvert outlets (Azamathulla & Haque, 2012), classification and regression trees to estimate wave-induced scour around a circular pile (Ayoubloo, Etemad-Shahidi, & Mahjoobi, 2010), an M5′ model tree to estimate scour downstream of a ski-jump bucket spillway (Goyal & Ojha, 2011), an adaptive network based on fuzzy systems to determine scour around an arch-shaped bed sill (Keshavarzi, Gazni, & Homayoon, 2012), linear genetic programming to predict scour around a circular pile (Guven, Azamathulla, & Zakaria, 2009), and the group method of data handling to predict scour downstream of a ski-jump bucket spillway (Najafzadeh, Barani, & Hessami Kermani, 2014).

The above-mentioned studies reflect the successful application of soft computing techniques in scour modeling.

Recently, Samadi et al. (2014) employed two decision-tree algorithms, namely classification and regression trees (CART) and the M5′ model tree, for the prediction of equilibrium scour depth below free overfall spillways. These two algorithms were employed in order to discover and extract knowledge as a set of rules from the database used in order to provide mathematical expressions for the prediction of scour depth below free overfall spillways.

Equations derived from M5′ and CART can be employed in scour calculations as easily as empirical and traditional methods. In addition, a previous study (Samadi et al., 2014) showed that soft computing approaches are more accurate than empirical and traditional methods for the prediction of scour depth below free overfall spillways.

The CART and M5′ models expand in the form of a hierarchical inverted binary tree to predict scour depth. The main difference between these two methods, however, is that the M5′ model tree approach provides practical equations for the prediction of scour, while the regression trees give values for output parameters. Hence, it can be said that the M5′ model tree is more interpretable compared with CART. Furthermore, it seems that the equations resulting from M5′ can give better engineering judgment about the relation between the parameters and their importance (Samadi et al., 2014).

ANNs are one of the most widely used soft computing models applied to solve engineering problems. One of the weaknesses that usually has to be considered for the ANN model is that it acts like a black box in that finding a relationship between independent variables and output parameters is difficult and it therefore does not have the ability to interpret and understand the results, although there have been attempts in some studies to convert the ANN from a black-box model into an explicit formulation (Alavi & Gandomi, 2011a; Alavi, Gandomi, Mollahasani, Heshmati, & Rashed, 2010). However, it seems that the method of extracting knowledge and obtaining equations from the M5′ model tree and CART is easier to understand, more convenient to use and provides a clearer relationship between input variables and output parameters compared with ANN.

The MARS approach is a novel technique that uses a large number of piece-wise regression equations for the prediction of output parameters based on input parameters. It is notable that some studies have shown that the MARS approach can provide competitive or even better results compared to those of ANN (Abraham, Steinberg, & Philip, 2001; Samui et al., 2011; Yang, Prasher, Lacroix, & Kim, 2004).

The primary aims of the present study are to develop practical equations for the prediction of scour depth below free overfall spillways using the MARS method and to evaluate the performance of this method. Moreover, the accuracy of the MARS and ANN models developed in the present study are compared with the results obtained from two decision-tree algorithm methods, namely M5′ and CART, which have already been performed by Samadi et al. (2014).

2. Multivariate adaptive regression splines (MARS)

Developed in the early 1990s by statistician Jerry Friedman, MARS is a fast, flexible and precise technique for solving regression and classification problems (Friedman, 1991). MARS is a new soft computing technique which combines several simple linear regression schemes. This technique divides the solution domain into multiple ranges of predictive variables (input). Moreover, it can automatically search for the non-linear input/output relationship hidden in a large database and create an explicit model. It can also determine the importance of the weights of the inputs with regard to the corresponding output (Yang, Prasher, Lacroix, & Kim, 2003). It makes basis functions for each input variable and then generates flexible regression models for the prediction of the output parameters by combining the obtained basis functions.

The method is applicable for large or even small datasets with a free distribution type. MARS divides the space of the predictors into multiple knots and then fits a spline function between these knots. The elements used to fit a MARS model are called basis functions, each of which can be a main effect or an interaction of variables (Haleem, Gan, & Lu, 2013).

The basis function represents the information about the independent variable (input) and its relation with the output parameter and is defined by the following equations (Friedman, 1991):

$$h_m(x) = \max(0, c - x), \tag{1}$$

or

$$h_m(x) = \max(0, x - c), \tag{2}$$

where x is the input variable, c is the threshold value for input variable of x, and $h_m(x)$is the basis function.

The general form of the MARS model is introduced as:

$$Y = f(x) = \beta_0 + \sum_{m=1}^{M} \beta_m h_m(x), \tag{3}$$

where$Y$is the output parameter, $\beta_0$is the constant value, $M$is the number of basis functions, $h_m(x)$is the mth basis function and $\beta_m$is the corresponding coefficient of $h_m(x)$(Friedman, 1991).

Two important steps are taken in the development of MARS in order to produce the optimal structure for the prediction of phenomena. In the first step, all the possible basis functions are introduced to define Eq. (3). As a result, certain overfitted models with many basis functions are developed (forward step). In the second step, to prevent overfitting, the basis functions of less importance are discarded via the generalized cross-validation (GCV) measure and omitted from Eq. (3).

GCV was introduced by Craven and Wahba (1979) and extended by Friedman (1991) for MARS.

The expression of the GCV measure is described as follows:

$$GCV = \frac{(1/n)\sum_{i=1}^{n} [y_i - f(x_i)]^2}{[1 - (C(B)/n)]^2}. \tag{4}$$

In the above equation, n denotes the number of observations and $C(B)$ denotes a complexity penalty that increases with the number of basis functions in the model. It is defined as:

$$C(B) = (B + 1) + dB, \tag{5}$$

where d denotes a penalty for each basis function included in the model and B denotes the number of basis functions.

More information about d and the details of MARS has been provided by Friedman (1991).

3. Artificial neural networks (ANNs)

The ANN scheme is one of the most popular soft computing tools and has been used in a broad spectrum of engineering problems. The ANN is a nonlinear mathematical model with the ability to simulate the biological neurons of the human brain (Kamranzad, Etemad-Shahidi, & Kazeminezhad, 2011). Each neuron input in hidden layers or output layer is multiplied by a corresponding interconnection weight, and then a bias (threshold) is added to the sum of the products:

$$I_j = (\sum_i W_{ij} x_i + \theta_j). \tag{6}$$

Next, the result is passed through a transfer function which is usually a sigmoidal function (F)to determine an output,O_j, defined respectively by:

$$F(x) = \frac{1}{1 + \exp(-x)}, \tag{7}$$

$$O_j = F(I_j), \tag{8}$$

where O_j is the output from the jth neuron, $W_{ij}$is the weight of the connection from neuron i in the previous layer to neuron$j$, x_i is the output of the neuron from the previous layer, and θ_j is the bias (threshold) of the jth neuron (see Figure 2).

The most common neural network is a multilayer perceptron (MLP). An MLP consists of an input layer, one or more hidden layers and an output layer. Each layer contains neurons (nodes) and the number of nodes in the input and output layers represents the input variables (predictor variables) and dependent variables (target variables), respectively. The number of neurons in the hidden layers corresponds to the complexity of a given problem and is determined in the training stage via a trial-and-error approach.

In a feed-forward network, input flows are unidirectional from the input to the output layer and each node is connected solely to one node in the next layer; there are no links among the nodes in the same layer.

Once the configuration of the network is specified and the network is trained, the training process searches for the smallest error function (mean sum squared) as defined by:

$$E = \frac{1}{2}\sum (O_n - t_n)^2, \tag{9}$$

where O_n and t_n are the network output and target output of the nth output node, respectively.

At the commencement of training, the weights are randomly initialized, then fixed by the following process:

$$\Delta W(k + 1) = \alpha \Delta w(k) - \eta \partial E / \partial W, \tag{10}$$

where$\Delta w(k)$ indicates the change in the weight at the kth iteration and $\Delta w(k + 1)$is the change in the weight at the $(k+1)$th iteration; α indicates the momentum factor and η indicates the learning rate.

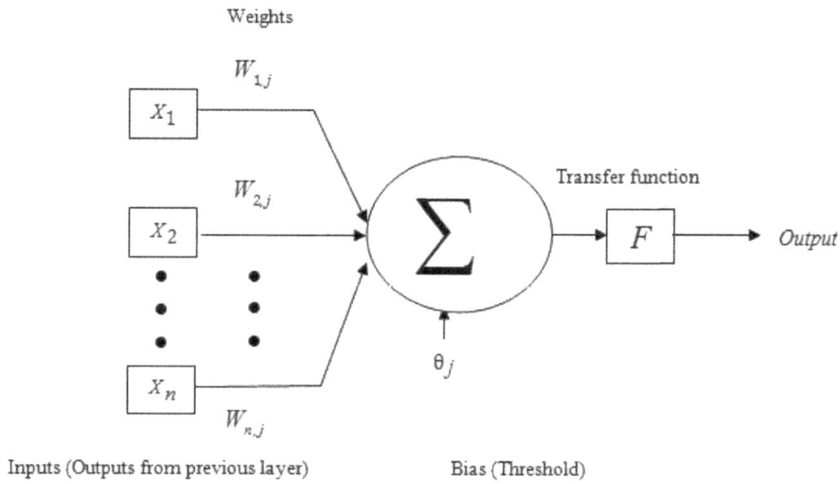

Figure 2. A typical neuron.

4. Relevant parameters and dimensional analysis

The equilibrium scour depth below a free overfall spillway can be a function (f) of certain parameters expressed by the following relations (Azar, 1998; Ghodsian & Azar, 2002):

$$D_s = f\,(q, H, Y_t, d_{50}, g, \rho_w, \rho_s), \tag{11}$$

where D_s is the depth of scour measured from the surface of tail water, q is the unit discharge, $H$is the total head or the difference between the upstream and downstream elevations, $Y_t$is the depth of the tail water, $d_{50}$is the mean diameter of the sediment grains in the downstream bed, $g$is the gravitational acceleration and ρ_w and ρ_s are the mass density of the water and sediment, respectively.

Applying dimensional analysis, the governing dimensionless equation is written as:

$$\frac{D_s}{H} = f\,(\frac{q}{\sqrt{gH^3}}, \frac{d_{50}}{H}, \frac{H}{Y_t}, \frac{\rho_w}{\rho_s}). \tag{12}$$

Furthermore, by discarding the constant parameter $\frac{\rho_w}{\rho_s}$ and combining $\frac{q}{\sqrt{gH^3}}$ and $\frac{d_{50}}{H}$, it is possible to rewrite Eq. (12) as (Azar, 1998; Ghodsian & Azar, 2002):

$$\frac{D_s}{H} = f\,(F_1, \frac{H}{Y_t}), \tag{13}$$

where $F_1 = \frac{q}{d_{50}\sqrt{gH}}$ is a densimetric Froude number (Ghodsian & Azar, 2002).

5. Data collection

In the present study, in order to estimate the depth of scour below a free overfall spillway, a database comprised of 104 data entries from laboratory experiments was taken from the work of Samadi et al. (2014). A brief description of the data used for development of the MARS and ANN models is given below.

5.1. Mahboobi experiments

Mahboobi (1997) conducted experiments to measure the maximum scour depth downstream of a free overfall spillway. He applied two different discharges per width of the spillway, 0.01 and 0.015 $m^3/s.m$, three drop heights of 45 cm, 60 cm, and 75 cm, and a constant tailwater depth of 10 cm. Uniform sediment with a mean size of 1 mm, 2.57 mm, 7.15 mm, and 11 mm was used below the spillways bed material. Mahboobi obtained 24 sets of data from his experiments.

5.2. Azar experiments

Azar (1998) measured scour downstream of a free overfall spillway. The total discharge of the spillway he applied ranged between 5 and 20 L/s. Drop heights ranged between 26 cm and 51 cm and tailwater depth varied from 5 to 25 cm. Azar employed uniform sediment with a mean size of 2.9 mm, 5 mm, 8 mm, and 15.6 mm downstream of the spillway. He obtained 80 sets of data from his experiments.

6. Development of the models

All of the 104 laboratory data entries that were collected were randomly divided into two groups – the training set and the testing set. Two thirds of them (69 data entries) were selected for training and the rest (35 data entries) were used to test the models. Table 1 shows the statistical parameters of the input and output variables of the datasets used in this study.

6.1. MARS model development

In order to develop the MARS model, with regard to Eq. (13), two parameters of F_1 and $\frac{H}{Y_t}$were used as input parameters for the estimation of the relative scour depth, $\frac{D_s}{H}$.

Table 1. Statistical parameters of input–output parameters used for the prediction of $\frac{D_s}{H}$ (Samadi et al., 2014).

Statistical parameters	Training set F_1	Training set $\frac{H}{Y_t}$	Training set $\frac{D_s}{H}$	Testing set F_1	Testing set $\frac{H}{Y_t}$	Testing set $\frac{D_s}{H}$
Minimum	0.320	1.055	0.203	0.375	1.055	0.234
Maximum	8.470	10.120	1.891	7.853	10.120	1.683
Standard deviation	1.977	2.946	0.427	1.956	2.995	0.426
Average	2.424	4.337	0.782	2.427	4.482	0.769
Median	1.852	4.138	0.708	1.720	4.320	0.687

Table 2. Basis functions $\{h_m(x)\}$and their corresponding coefficients $\{\beta_m\}$.

Basis functions$\{h_m(x)\}$	Equation	Coefficient$\{\beta_m\}$
$h_1(x)$	$\max(0, \frac{H}{Y_t} - 2.547)$	$\beta_1 = -0.590$
$h_2(x)$	$\max(0, 2.547 - \frac{H}{Y_t})$	$\beta_2 = 0.794$
$h_3(x)$	$\max(0, F_1 - 1.434)$	$\beta_3 = 0.182$
$h_4(x)$	$\max(0, 1.434 - F_1)$	$\beta_4 = -0.210$
$h_5(x)$	$\max(0, \frac{H}{Y_t} - 1.66)$	$\beta_5 = 0.458$
$h_6(x)$	$\max(0, \frac{H}{Y_t} - 4.32)$	$\beta_6 = 0.252$
$h_7(x)$	$\max(0, F_1 - 0.758)$	$\beta_7 = -0.108$
$h_8(x)$	$\max(0, \frac{H}{Y_t} - 9.276)$	$\beta_8 = 0.058$
$h_9(x)$	$\max(0, \frac{H}{Y_t} - 4.5)$	$\beta_9 = -0.155$

To develop the MARS model, in the first step (forward step) thirteen basis functions were obtained with this method. In the second step (backward step), four basis functions was removed. In the end, the optimal MARS model was obtained with nine basic functions, as described below:

$$\frac{D_s}{H} = 0.383 + \sum_{M=1}^{9} \beta_m h_m(x). \quad (14)$$

The expression for $h_m(x)$and values of β_m are presented in Table 2.

The general form of the MARS model that can be used to estimate scour depth downstream of a free overfall spillway is obtained from Eq. (14). As can be seen from Table 2, the $\frac{H}{Y_t}$ parameter appears in six of the nine basis functions obtained from the MARS model. The existence of the $\frac{H}{Y_t}$ parameter in h_1, h_2, h_5, h_6, h_8, and $h_9$in comparison with the existence of F_1in h_3, h_4, and h_7 shows the higher importance in scour estimation of the former parameter compared with the latter. Moreover, a previous study (Samadi et al., 2014) showed that the $\frac{H}{Y_t}$parameter is the most important splitting parameter in tree structures produced by the M5′ and CART models. Therefore, the result obtained from the MARS model is in agreement with the previous study (Samadi et al., 2014).

As can be seen, the MARS procedure divides the input domain into sub-domains and gives each region a basis function,$h_m(x)$. Finally, in combination with the basis functions, it provides a flexible equation for the estimation of scour depth downstream of a free overfall spillway in the form of Eq. (14).

The following explanation can be given of how to use the MARS model obtained in this study to estimate scour depth (Eq. (14)) according to the range of the input parameters (F_1 and$\frac{H}{Y_t}$) for estimation of$\frac{D_s}{H}$.

For example, if $\frac{H}{Y_t}$becomes smaller than 1.66, the values of basis functions h_1, h_5, h_6, h_8, and $h_9$will be zero. Moreover, if F_1 becomes smaller than 0.758, the values of basis functions h_3 and h_7 will be zero. Finally, using these values of the basis functions in Eq. (14), the following equation is obtained to estimate scour depth:

$$\frac{D_s}{H} = 0.383 + 0.794 \times (2.547 - \frac{H}{Y_t}) - 0.210 \times (1.434 - F_1). \quad (15)$$

As can be seen, the basis functions of MARS are determined by considering the range and values of the independent variables (F_1 and $\frac{H}{Y_t}$) and substituting in the general form of the scour depth equation, Eq. (14), which is transformed into the simple mathematical expression Eq. (15), which can easily be used for computation of the scour. As was illustrated, in order to estimate $\frac{D_s}{H}$with regard to the range of input variables and basis functions, the proposed MARS model could be transformed into simpler equations that can easily be employed to estimate scour depth.

6.2. *ANN model development*

The same training and testing data sets employed for the development of MARS were used for the development of ANN.

For the ANN technique in the present study, a simple MLP with three layers (input, hidden and output) with a back propagation (BP) learning rule was used to train the network. Also, the number of hidden-layer neurons was chosen based on the following criterion presented by Hecht-Nielsen (1987) and Rogers and Dowla (1994):

$$N^H \leq 2N^I + 1 \quad (16)$$

$$N^H \leq \frac{N^{TR}}{N^I + 1} \quad (17)$$

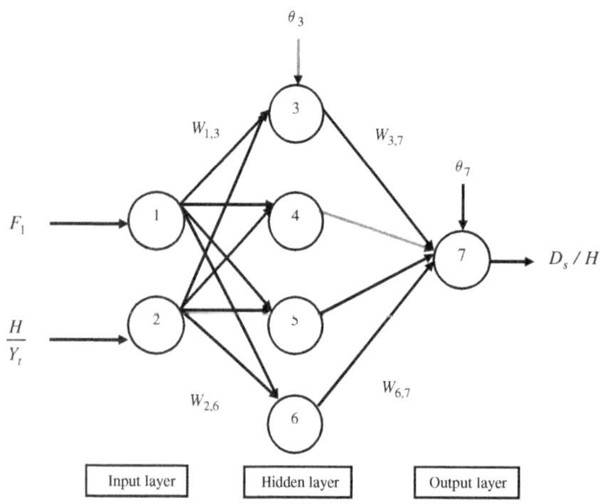

Figure 3. Architecture of the ANN model used for prediction of D_s/H.

where$N^H$is the number of nodes in the hidden layer, N^I is the number of inputs and N^{TR} is the number of training samples.

Recommendations have been made for the selection of the learning rate and the momentum factor (Basheer & Hajmeer, 2000). For example, Zupan and Gasteiger (1993) suggest that the summation of the selected learning rate and the momentum factor should be approximately equal to 1. Fu (1995) recommends a learning rate and momentum factor between 0 and 1. Furthermore, Swingler (1996) uses a learning rate of 0.25 and a momentum factor of 0.9 in order to solve all problems.

During the development of the ANN, different numbers of hidden layer neurons can be taken, using the Hecht-Nielsen (1987) method via a trial-and-error procedure, until the eventual optimum number of nodes in the hidden layer is found, which was four. The network training was terminated when 97% of the outputs were successfully matched with the actual values. In addition, we chose a learning rate and momentum coefficient of 0.25 and 0.9, respectively, as Swingler (1996) proposed. The illustration in Figure 3 shows the proposed ANN configuration used in this study.

The corresponding weights and biases extracted from the trained network, given in Table 3, could be beneficial and practical for the prediction of scour depth in the future.

Table 4. Statistical error measures of ANN approach trained by $\frac{H}{Y_t}$and F_1 individually for determination of $\frac{D_s}{H}$for testing data.

Input parameter	*CC*	*RMSE*
$\frac{H}{Y_t}$	0.9244	0.1610
F_1	0.4551	0.3849

Weights obtained from the optimal ANN in this study could be useful for other researchers seeking to reproduce the network on their own (Ayoubloo, Azamathulla, Jabbari, & Zanganeh, 2011).

Moreover, in order to find the importance of different input parameters based on output parameters, new models of ANN with$\frac{H}{Y_t}$ and F_1were constructed and evaluated individually to estimate$\frac{D_S}{H}$.

The results of these models are shown in Table 4. As can be seen, the model made only on the basis of parameter $\frac{H}{Y_t}$ has the highest accuracy compared to the model made only on the basis of parameter F_1. This fact shows the higher importance of $\frac{H}{Y_t}$ in the estimation of scour depth compared to F_1. It is noteworthy that the obtained results are consistent with those obtained from the MARS model and previous studies conducted by Samadi et al. (2014).

7. Analysis of the development models

In this study, in order to assess the performance and accuracy of the models quantitatively, statistical indicators such as the correlation coefficient (CC), root mean square error (RMSE) and mean absolute error (MAE) were used. These parameters are defined as below:

$$CC = \frac{\sum_{i=1}^{n}(x_i - \bar{x})(y_i - \bar{y})}{\sqrt{\sum_{i=1}^{n}(x_i - \bar{x})^2 \sum_{i=1}^{n}(y_i - \bar{y})^2}}, \tag{18}$$

$$RMSE = \sqrt{\frac{\sum_{i=1}^{n}(x_i - y_i)^2}{n}}, \tag{19}$$

$$MAE = \frac{1}{n}\sum_{i=1}^{n}|x_i - y_i|, \tag{20}$$

Table 3. Connection weights and biases of proposed ANN model (see Figure 3).

Input to hidden layer				Hidden to output layer	Bias
$W_{1,3} = 1.17$	$W_{1,4} = -7.86$	$W_{1,5} = 0.21$	$W_{1,6} = -1.57$	$W_{3,7} = 1.17$	$\theta_3 = -0.82$
$W_{2,3} = -0.05$	$W_{2,4} = 3.27$	$W_{2,5} = -11.45$	$W_{2,6} = 0.49$	$W_{4,7} = -6.19$	$\theta_4 = -3.05$
				$W_{5,7} = 9.40$	$\theta_5 = -1.12$
				$W_{6,7} = -1.86$	$\theta_6 = 1.04$
					$\theta_7 = -0.22$

where x_i and y_i are the measured and predicted parameters, respectively, $\bar{x}$ and $\bar{y}$ are the average of the measured and predicted values, respectively, and n is the total number of measurements. A model's performance is good when the amount of CC is close to one and the values of RMSE and MAE are close to zero.

8. Results and discussion

The accuracy of the MARS model developed in this study is verified by the testing data set and presented in Table 5. The results obtained from the MARS are compared with the results of the ANN obtained in this study and the M5′ model tree and CART reported by Samadi et al. (2014).

Considering the contents of Table 5, it is clear that compared with the other models for estimation of scour downstream of a free overfall spillway, MARS is superior and has a higher accuracy. Moreover, although the results obtained with the MARS and ANN models are very close and more or less similar, the MARS model results are slightly better in determining the value of $\frac{D_S}{H}$, as indicated by the high value of the CC and the low error indices (RMSE, MAE).

The MARS model provides a regression equation, employing many partial functions (basis functions) between the input variables and the output parameters, which results in a more transparent relation between the output and input parameters. This allows a straight estimation of the scour depth using only two predictor variables, i.e., $\frac{H}{Y_t}$ and F_1. The regression equation obtained from MARS can readily and easily be implemented and used for scour calculation below free overfall spillways, while ANN does not provide a clear relationship between input and output parameters for the prediction of scour depth and it seems it is difficult to extract understandable information about the relation between the input and output parameters.

In addition, in order to train the ANN it is necessary for the user to determine and select the optimum values of the parameters required by the network for proper performance. The number of hidden layers, the number of the nodes in the hidden layers, the rate of learning, the momentum coefficient, and the transfer function can be mentioned among the parameters that should be determined by the user in order to train the ANN.

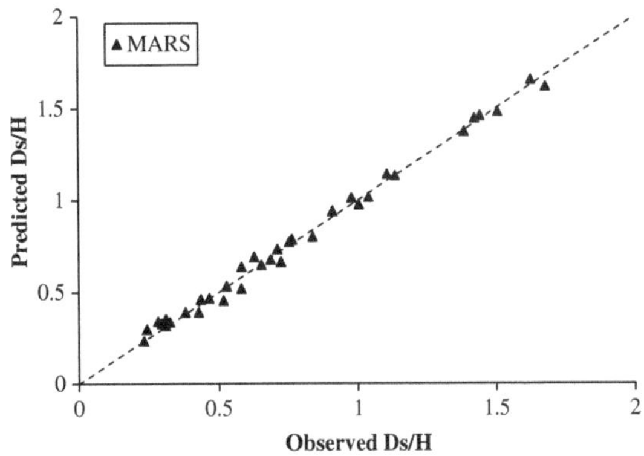

Figure 4. Comparison between the D_S/Hvalues estimated by MARS and the D_S/H values observed for the testing data.

Therefore, optimum selection of the values of these parameters usually requires the application of a trial-and-error procedure, which makes the construction of the ANN time consuming, while MARS can automatically search a large dataset and extract knowledge in the form of an explicit model that clearly shows the relation between input variables and output parameters (Yang et al., 2003). In addition, the obtained regression equations from MARS indicate the importance of input variables in the estimation of the output parameters, while in order to highlight the importance of each input variable ANN requires new models to be built, trained and tested individually for sensitivity analysis. Hence, determination of the importance of the input variables in the estimation of the output parameters in ANN compared to MARS is time consuming.

Additionally, comparison of MARS with the M5′ model shows that MARS results are significantly better than M5′ considering the error measures of RMSE and MAE, such that RMSE and MAE decreased by 53% and 50.9% respectively (see Table 5).

Figure 4 shows a comparison between the observed and predicted dimensionless scour depth by the MARS model for the testing data. Furthermore, a scatter plot of the ANN, illustrated in Figure 5, graphically displays the performance of the ANN in the prediction of the value of $\frac{D_S}{H}$. As is shown in Figures 4 and 5, MARS and ANN have good performance in the prediction of scour depth.

Table 5. Statistical error measures of the MARS, ANN, M5′ and CART approaches proposed in this study for the estimation of $\frac{D_s}{H}$.

	Training			Testing		
Approach	*CC*	*RMSE*	*MAE*	*CC*	*RMSE*	*MAE*
MARS	0.9950	0.0425	0.0319	0.9965	0.0360	0.0302
ANN	0.9916	0.0559	0.0417	0.9942	0.0459	0.0369
M5′ (Samadi et al., 2014)	0.9857	0.0716	0.0562	0.9833	0.0766	0.0615
CART (Samadi et al., 2014)	0.9260	0.1599	0.1321	0.9305	0.1546	0.1288

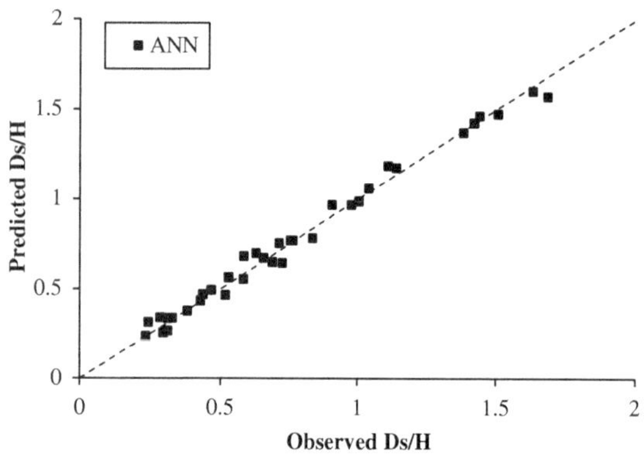

Figure 5. Comparison between the D_S/H values estimated by ANN and the D_S/H values observed for the testing data.

As can be seen in Figure 4, the MARS model predicts $\frac{D_S}{H}$ appropriately and predicts the trend of the observed data very well. This is confirmed by the values of the statistical parameters, such as CC, in Table 5.

9. Summary and conclusions

This work suggests the use of alternative soft computing tool approaches in order to improve the accuracy of the estimation of scour depth below free overfall spillways.

Two predictive models based on ANN and MARS were developed. For this purpose, the MARS technique was employed as a new soft computing tool capable of modeling complex non-linear phenomena.

Experimental data and dimensionless parameters were used to develop the MARS and ANN models. The performance of both models was comparable. Moreover, statistical indicators demonstrated that the results of MARS were better than other soft computing methods – namely ANN, M5′ and CART – when used for the prediction of scour depth below free overfall spillways.

In contrast to ANN, as a non-black-box model MARS is able to provide a clear and explicit relation between the input variables and output parameters for the prediction of scour depth.

Practical equations derived from MARS clearly showed the importance of input parameters in the phenomenon of scour depth. However, the same process in ANN needs sensitivity analysis and the building of new models for each input variable.

In the present study, it was demonstrated with reasonable accuracy that the MARS model can estimate scour depth using limited data and few inputs. Successful application of the MARS technique in the present study shows that MARS is a robust tool for the prediction of scour and suggests that the use of this technique is an appropriate practical method for researchers in other fields of hydraulic engineering.

Acknowledgements

The first author expresses special thanks to Ms Touran Amini because of her enormous effort. He would also like to thank Mr Abbas Amini, manager of AHAB company in the Shahre Aftab Tehran project, and Mr Amir Razzaghi.

Disclosure statement

No potential conflict of interest was reported by the authors.

Funding

The first author is grateful to the Iran Water Resource Management Company for their partial support of this work (Grant No. RIV4-89107).

References

Abraham, A., Steinberg, D., & Philip, N. S. (2001). Rainfall forecasting using soft computing models and multivariate adaptive regression splines. *IEEE SMC Transactions, Special Issue on Fusion of Soft Computing and Hard Computing in Industrial Applications 1*, 1–6.

Afshar, A., & Amiri, H. (2010). Risk-based approach to unbalanced bidding in construction projects. *Engineering Optimization*, *42*, 369–385.

Alavi, A. H., & Gandomi, A. H. (2011a). Prediction of principal ground-motion parameters using a hybrid method coupling artificial neural networks and simulated annealing. *Computers & Structures*, *89*, 2176–2194.

Alavi, A. H., & Gandomi, A. H. (2011b). A robust data mining approach for formulation of geotechnical engineering systems. *Engineering Computations*, *28*, 242–274.

Alavi, A. H., Gandomi, A. H., Mollahasani, A., Heshmati, A. A. R., & Rashed, A. (2010). Modeling of maximum dry density and optimum moisture content of stabilized soil using artificial neural networks. *Journal of Plant Nutrition and Soil Science*, *173*, 368–379.

Attoh-Okine, N. O., Cooger, K., & Mensah, S. (2009). Multivariate adaptive regression (MARS) and hinged hyperplanes (HHP) for doweled pavement performance modeling. *Construction and Building Materials*, *23*, 3020–3023.

Ayoubloo, M. K., Azamathulla, H. M., Jabbari, E., & Zanganeh, M. (2011). Predictive model-based for the critical submergence of horizontal intakes in open channel flows with different clearance bottoms using CART, ANN and linear regression approaches. *Expert Systems with Applications*, *38*, 10114–10123.

Ayoubloo, M. K., Etemad-Shahidi, A., & Mahjoobi, J. (2010). Evaluation of regular wave scour around a circular pile using data mining approaches. *Applied Ocean Research*, *32*, 34–39.

Azamathulla, H. M., & Haque A. A. M. (2012). Knowledge extraction from trained neural network scour model at culvert outlets. *Neural Computing and Applications*, *23*, 2137–2141.

Azmathullah, H. M., Deo, M. C., & Deolalikar, P. B. (2005). Neural networks for estimation of scour downstream of a ski-jump bucket. *Journal of Hydraulic Engineering*, *131*, 898–908.

Azar, F. A. (1998). *Effect of sediment size distribution on scour downstream of free overfall spillway* (Unpublished master's thesis). Tarbiat Modares University, Tehran, Iran.

Basheer, I. A., & Hajmeer, M. (2000). Artificial neural networks: Fundamentals, computing, design, and application. *Journal of Microbiological Methods*, *43*, 3–31.

Beuchat, X., Schaefli, B., Soutter, M., & Mermoud, A. (2011). Toward a robust method for subdaily rainfall downscaling from daily data. *Water Resources Research*, *47*, W09524.

Chau, K. W. (2010). *Modelling for coastal hydraulics and engineering*. Abingdon: Taylor & Francis.

Chau, K. W., Wu, C., & Li, Y. (2005). Comparison of several flood forecasting models in Yangtze River. *Journal of Hydrologic Engineering*, *10*, 485–491.

Cheng, C., Chau, K., Sun, Y., & Lin, J. (2005). Long-term prediction of discharges in Manwan Reservoir using artificial neural network models. In J. Wang, X.-F. Liao, & Z. Yi (Eds.), *Advances in neural networks – ISNN 2005: Second International Symposium on Neural Networks, Chongqing, China, May 30–June 1, 2005, proceedings, part III* (pp. 1040–1045). Berlin: Springer.

Chen, W., & Chau, K. W. (2006). Intelligent manipulation and calibration of parameters for hydrological models. *International Journal of Environment and Pollution*, *28*, 432–447.

Coulibaly, P., & Baldwin, C. K. (2005). Nonstationary hydrological time series forecasting using nonlinear dynamic methods. *Journal of Hydrology*, *307*, 164–174.

Craven, P., & Wahba, G. (1979). Smoothing noisy data with spline functions: Estimating the correct degree of smoothing by the method of generalized cross-validation. *Numerische Mathematik*, *31*, 317–403.

Ebtehaj, I., & Bonakdari, H. (2013). Evaluation of sediment transport in sewer using artificial neural network. *Engineering Applications of Computational Fluid Mechanics*, *7*, 382–392.

Eshtehardian, E., Afshar, A., & Abbasnia, R. (2009). Fuzzy-based MOGA approach to stochastic time–cost trade-off problem. *Automation in Construction*, *18*, 692–701.

Etemad-Shahidi, A., & Bali, M. (2012). Stability of rubble-mound breakwater using H50 wave height parameter. *Coastal Engineering*, *59*, 38–45.

Friedman, J. H. (1991). Multivariate adaptive regression splines. *The Annals of Statistics*, *19*, 1–141.

Fu, L. (1995). *Neural networks in computer intelligence*. New York, NY: McGraw-Hill.

Gandomi, A. H., & Alavi, A. H. (2012a). A new multi-gene genetic programming approach to nonlinear system modeling. *Part I: Materials and structural engineering problems. Neural Computing and Applications*, *21*, 171–187.

Gandomi, A. H., & Alavi, A. H. (2012b). A new multi-gene genetic programming approach to nonlinear system modeling. *Part II: Geotechnical and earthquake engineering problems. Neural Computing and Applications*, *21*, 189–201.

Gandomi, A. H., Yang, X. S., Talatahari, S., & Alavi, A. H. (2013). *Metaheuristic applications in structures and infrastructures*. Waltham, MA: Elsevier

Ghodsian, M., & Azar, F. A. (2002). Scour hole characteristics below free overfall spillway. *International Journal of Sediment Research*, *17*, 304–313.

Goyal, M., & Ojha, C. (2011). Estimation of scour downstream of a ski-jump bucket using support vector and M5′ model tree. *Water Resources Management*, *25*, 2177–2195.

Guven, A., Azamathulla, H. M., & Zakaria, N. A. (2009). Linear genetic programming for prediction of circular pile scour. *Ocean Engineering*, *36*, 985–991.

Haleem, K., Abdel-Aty, M., & Santos, J. (2010). *Multiple applications of the multivariate adaptive regression splines in predicting rear-end crashes at unsignalized intersections* (Transportation Research Records No. 2165, pp. 33–41). Washington, DC: Transportation Research Board of the National Academics.

Haleem, K., Gan, A., & Lu, J. (2013). Using multivariate adaptive regression splines (MARS) to develop crash modification factors for urban freeway interchange influence areas. *Accident Analysis & Prevention*, *55*, 12–21.

Hecht-Nielson, R. (1987). Kolmogorov's mapping neural network existence theorem. In *Proceedings of the international conference on Neural Networks* (Vol. 3, pp. 11–14). New York: IEEE Press.

Jabbari, E., & Talebi, O. (2011). Using artificial neural networks for estimation of scour at the head of vertical wall breakwater. *Journal of Coastal Research*, *SI64*, 521–526.

Kamranzad, B., Etemad-Shahidi, A., & Kazeminezhad, M. H. (2011). Wave height forecasting in Dayyer, the Persian Gulf. *Ocean Engineering*, *38*(1), 248–255.

Kamranzad, B., Jabbari, E., & Samadi, M. (2013). Assessment of soft computing models to estimate wave heights in Anzali port. *International Journal of Maritime Technology*, *9(17)*, 27–36.

Keshavarzi, A., Gazni, R., & Homayoon, S. R. (2012). Prediction of scouring around an arch-shaped bed sill using neuro-fuzzy model. *Applied Soft Computing*, *12*, 486–493.

Mahboobi, E. (1997). *The effect of sediment size on maximum scour depth in plunge pool* (Unpublished master's thesis). University of Science and Technology, Tehran, Iran.

Mason, P. J., & Arumugam, K. (1985). Free jet scout below dams and slip buckets. *Journal of Hydraulic Engineering*, *111*, 220–235.

Muttil, N., & Chau, K. W. (2006). Neural network and genetic programming for modelling coastal algal blooms. *International Journal of Environment and Pollution*, *28*, 223–238.

Najafzadeh, M., Barani, G. A., & Hessami Kermani, M. R. (2014). Group method of data handling to predict scour at downstream of a ski-jump bucket spillway. *Earth Science Informatics*, *7*, 231–248.

Nasseri, M., Tavakol-Davani, H., & Zahraie, B. (2013). Performance assessment of different data mining methods in statistical downscaling of daily precipitation. *Journal of Hydrology*, *492*, 1–14.

Pijush, S., & Dookie, K. (2012). Modeling of reservoir-induced earthquakes: A multivariate adaptive regression spline. *Journal of Geophysics and Engineering*, *9*(5), 494.

Rogers, L. L., & Dowla, F. U. (1994). Optimization of ground water remediation using artificial neural networks with parallel solute transport modeling. *Water Resources Research*, *30*, 457–481.

Saadatpour, M., Afshar, A., & Afshar, M. (2011). Fuzzy pattern recognition method for assessing soil erosion. *Environmental Monitoring and Assessment*, *180*, 385–397.

Samadi, M., Jabbari, E., & Azamathulla, H. M. (2014). Assessment of M5′ model tree and classification and regression trees for prediction of scour depth below free overfall spillways. *Neural Computing and Applications*, *24*(2), 357–366.

Samui, P. (2011). Multivariate adaptive regression spline applied to friction capacity of driven piles in clay. *Geomechanics and Engineering*, *3*, 285–290.

Samui, S., Das, S., & Kim, D. (2011). Uplift capacity of suction caisson in clay using multivariate adaptive regression spline. *Ocean Engineering*, *38*, 2123–2127.

Samui, P., & Kothari, D. P. (2012). A multivariate adaptive regression spline approach for prediction of maximum shear modulus (G max) and minimum damping ratio (ζ min). *Engineering Journal*, *16*(5), 69–77.

Sharda, V. N., Prasher, S. O., Patel, R. M., Ojasvi, P. R., & Chandra, P. (2008). Performance of multivariate adaptive regression splines (MARS) in predicting runoff in mid-Himalayan micro-watersheds with limited data. *Hydrological Sciences Journal*, *53*, 1165–1175.

Swingler, K. (1996). *Applying neural networks: A practical guide*. New York, NY: Academic Press.

Taormina, R., Chau, K., & Rajandrea, S. (2012). Artificial neural network simulation of hourly groundwater levels in a coastal aquifer system of the Venice lagoon. *Engineering Applications of Artificial Intelligence*, *25*, 1670–1676.

Wu, C. L., Chau, K., & Li, Y. S. (2009). Predicting monthly streamflow using data-driven models coupled with data-preprocessing techniques. *Water Resources Research*, *45*, W08432.

Yang, C.-C., Prasher, S. O., Lacroix, R., & Kim, S. H. (2003). A multivariate adaptive regression splines model for simulation of pesticide transport in soils. *Biosystems Engineering*, *86*, 9–15.

Yang, C.-C., Prasher, S. O., Lacroix, R., & Kim, S. H. (2004). Application of multivariate adaptive regression splines (MARS) to simulate soil temperature. *Transactions of the ASABE*, *47*, 881–887.

Yang, X. S., Gandomi, A. H., Talatahari, S., & Alavi, A. H. (2012). *Metaheuristics in water, geotechnical and transport engineering*. Waltham, MA: Elsevier.

Zupan, J., & Gasteiger, J. (1993). *Neural networks for chemists: An introduction*. New York, NY: Wiley.

Appendix

Nomenclature

ANN: artificial neural network.
MARS: multivariate adaptive regression splines.
CART: classification and regression trees.
BP: back propagation.
N: number of observations.
N^H: the number of the nodes in the hidden layer.
N^I: the number of the inputs.
N^{TR}: the number of the training samples.
MLP: multilayer perceptron.
CC: correlation coefficient.
RMSE: root mean square error.
L/s: liter per second
MAE: mean absolute error.
X: independent variable.
Y: output parameter.
W: weight of neuron.
D_s: the depth of scour measured from the surface of tail water.
q: unit discharge.
H: total head or the difference between the upstream and downstream elevations.
Y_t: depth of tail water.
d_{50}: mean diameter of the sediment grain.
g: gravitational acceleration.
F_1: densimetric Froude number.
c: the threshold value.
$h_m(x)$: basis function.
M: the number of the basis functions.
$h_m(x)$: the mth basis function.
GCV: generalized cross-validation.
W_{ij}: the weight of the connection from the neuron i in the previous layer to the neuron j
$\Delta w(k)$: the change in weight at the kth iteration.
$\Delta w(k+1)$: the change in weight at the $(k+1)$th iteration.
$m^3/s.m$: discharge per width of the spillway.

Greek letters

β_0: the constant value.
β_m: corresponding coefficient of $h_m(x)$.
θ_j: the bias (threshold) of the jth neuron.
α: momentum factor.
η: learning rate.
ρ_s: mass density of sediment.
ρ_w: mass density of water.

Permissions

All chapters in this book were first published in EACFM, by Taylor & Francis Online; hereby published with permission under the Creative Commons Attribution License or equivalent. Every chapter published in this book has been scrutinized by our experts. Their significance has been extensively debated. The topics covered herein carry significant findings which will fuel the growth of the discipline. They may even be implemented as practical applications or may be referred to as a beginning point for another development.

The contributors of this book come from diverse backgrounds, making this book a truly international effort. This book will bring forth new frontiers with its revolutionizing research information and detailed analysis of the nascent developments around the world.

We would like to thank all the contributing authors for lending their expertise to make the book truly unique. They have played a crucial role in the development of this book. Without their invaluable contributions this book wouldn't have been possible. They have made vital efforts to compile up to date information on the varied aspects of this subject to make this book a valuable addition to the collection of many professionals and students.

This book was conceptualized with the vision of imparting up-to-date information and advanced data in this field. To ensure the same, a matchless editorial board was set up. Every individual on the board went through rigorous rounds of assessment to prove their worth. After which they invested a large part of their time researching and compiling the most relevant data for our readers.

The editorial board has been involved in producing this book since its inception. They have spent rigorous hours researching and exploring the diverse topics which have resulted in the successful publishing of this book. They have passed on their knowledge of decades through this book. To expedite this challenging task, the publisher supported the team at every step. A small team of assistant editors was also appointed to further simplify the editing procedure and attain best results for the readers.

Apart from the editorial board, the designing team has also invested a significant amount of their time in understanding the subject and creating the most relevant covers. They scrutinized every image to scout for the most suitable representation of the subject and create an appropriate cover for the book.

The publishing team has been an ardent support to the editorial, designing and production team. Their endless efforts to recruit the best for this project, has resulted in the accomplishment of this book. They are a veteran in the field of academics and their pool of knowledge is as vast as their experience in printing. Their expertise and guidance has proved useful at every step. Their uncompromising quality standards have made this book an exceptional effort. Their encouragement from time to time has been an inspiration for everyone.

The publisher and the editorial board hope that this book will prove to be a valuable piece of knowledge for researchers, students, practitioners and scholars across the globe.

List of Contributors

Ungku Abdul Hafiz and Waqar Asrar
Department of Mechanical Engineering, Kulliyah of Engineering, International Islamic University, P. O. Box 10, 50728, Kuala Lumpur, Malaysia

Asif Hoda
Department of Mechanical Engineering, Jubail University College, Jubail, Kingdom of Saudi Arabia

A. Broatch, J. Galindo, R. Navarro and J. García-T.scar
CMT - Motores T.rmicos, Universitat Polit. cnica de Val.ncia, Camino de Vera S/N, 46022 Valencia, Spain

A. Daglish and R. K. Sharma
bJaguar Land Rover Limited, Abbey Road, Whitley, Coventry CV3 4LF, UK

Arun Kamath, Mayilvahanan Alagan Chella, Hans Bihs and Øivind A. Arntsen
Department of Civil and Transport Engineering, Norwegian University of Science and Technology, Trondheim, Norway

Danilo Sergi, Loris Grossi, Tiziano Leidi and Alberto Ortona
The iCIMSI Research Institute, The University of Applied Sciences and Arts of Southern Switzerland, Manno, Switzerland

F. Nietoa, D.M. Hargreavesb, J. S. Owenb and S. Hernándeza
School of Civil Engineering, University of A Coru.a, Spain; bFaculty of Engineering, University of Nottingham, UK

Farhad Hooshyaripor and Ahmad Tahershamsi
Faculty of Civil and Environment Engineering, Amirkabir University of Technology, Tehran, Iran

Azadeh Gholami, Hossein Bonakdari, Amir Hossein Zaji and Ali Akbar Akhtari
Department of Civil Engineering, Razi University, Kermanshah, Iran

Yan Chen and Yongsheng Lian
Department of Mechanical Engineering, University of Louisville, Louisville, KY, 40292, USA

B. G. Mulu
Vattenfall Research and Development AB, Laboratoriev.gen, SE-814 26 Alvkarleby, Sweden

M. J. Cervantes
Division of Fluid and Experimental Mechanics, Luleå University of Technology, SE-971 87 Luleå, Sweden; cDepartment of Energy and Process Engineering, Water Power Laboratory, Norwegian University of Science and Technology, Trondheim, Norway

F. Guibault and C. Devals
Computer and Software Engineering Department, Ecole Polytechnique de Montréal, Montréal, QC, Canada, H3C 3A7

T. C. Vu
Andritz Hydro Ltd., 6100 TransCanada Hwy, Pointe-Claire, QC, Canada

Weixing Yuan, Richard Lee and Luc Levasseur
Aerospace Portfolio, National Research Council (NRC) Canada, Ottawa, Ontario, K1A 0R6, Canada

G. L. Juste
Aerospace Propulsion and Fluid Mechanics, Universidad Politécnica de Madrid, Plaza Cardenal Cisneros 3, 28040 Madrid, Spain

P. Fajardo
Aerospace Engineering Group. Universidad Carlos III de Madrid, 28911 Leganés, Spain

Pengzhi Lin and Jianmin Zhang
State Key Laboratory of Hydraulics and Mountain River Engineering, Sichuan University, Chengdu, China

Xin Liu
College of Harbour Coastal and Offshore Engineering, Hohai University, Nanjing, China

A. Madadi, M. J. Kermani and H. Khazaei
Department of Mechanical Engineering, Amirkabir University of Technology (Tehran Polytechnic), Tehran, Iran, P. Code 15875-4413

Amir Hossein Zaji and Hossein Bonakdari
Department of Civil Engineering, Razi University, Kermanshah, Iran

Chunbao Liu , Changsuo Liu and Wenxing Ma
School of Mechanical Science and Engineering, Jilin University, Changchun13002, China

Mehrshad Samadi
Department of Civil Engineering, AHAB Company, Tehran, Iran
School of Civil Engineering, Iran University of Science and Technology, Tehran, Iran

Ebrahim Jabbari
School of Civil Engineering, Iran University of Science and Technology, Tehran, Iran

H. M. Azamathulla
University of Tabuk, Tabuk 50060, Saudi Arabia

Mohammad Mojallal
School of Civil Engineering, University of Tehran, Tehran, IRAN

Index